FLUID DYNAMICS AND HEAT TRANSFER

Fluid Dynamics and Heat Transfer

JAMES G. KNUDSEN

Professor of Chemical Engineering
Department of Chemical Engineering
Oregon State College

DONALD L. KATZ

Professor of Chemical Engineering
Chairman, Department of Chemical and Metallurgical Engineering
University of Michigan

ROBERT E. KRIEGER PUBLISHING COMPANY
HUNTINGTON, NEW YORK
1979

Original Edition 1958
Reprint Edition 1979

Printed and Published by
ROBERT E. KRIEGER PUBLISHING COMPANY, INC.
645 NEW YORK AVENUE
HUNTINGTON, NEW YORK 11743

Printed in the United States of America

Library of Congress Cataloging in Publication Data

Knudsen, James George, 1920-
Fluid dynamics and heat transfer.

Reprint of the edition published by McGraw-Hill, New York, in series: McGraw-Hill series in chemical engineering.
Includes index.
1. Fluid dynamics. 2. Heat—Transmission.
I. Katz, Donald La Verne, 1907- joint author.
II. Title.
[QA913.K75 1979] 536'.25 79-9748
ISBN 0-88275-917-5

PREFACE

A large portion of this text was first published in 1954 as Bulletin 37 of the Engineering Research Institute at the University of Michigan. The material in the original bulletin has been rearranged and expanded considerably to make it more suitable for class presentation. Some new material has been included.

Our purpose in preparing this text is essentially the same as that which prompted the writing of the original bulletin; that is, to present the fundamentals of fluid dynamics which are basic to an understanding of convection heat transfer. The material is designed for a one-semester graduate course in fluid flow and heat transfer, and this fact has necessitated the omission of many specialized topics ordinarily included in such a course. Such topics as settling, high-speed gas flow in nozzles and pipes, single- and two-phase flow through porous media, compressible flow, fluidization, mixing, conduction, radiation, and natural convection are either omitted or mentioned only briefly. The student continuing a study of these subjects will find the fundamental material presented here beneficial to him.

Our approach has been to present, wherever possible, the differential equations describing the particular fluid-flow or heat-transfer problem being discussed, along with the boundary conditions applicable to the problem. Detailed solutions of the differential equation are given in only a few instances. Whether detailed solutions are given or not, final relationships are given in a form that can be easily applied to specific fluid-flow or heat-transfer problems. We have attempted to compare theoretical equations with experimental results. Such comparison gives the student more faith in the theoretical approach and justifies many of the assumptions made in the analysis.

A number of illustrative examples have been included to demonstrate the application of relationships presented. The problems at the end of the book will also serve to acquaint the student with the application of the material.

We feel that the subjects are presented in logical order. The basic principles of various types of fluid flow are presented in the first eleven chapters, which serve as a foundation for the discussion of several aspects of forced-convection heat transfer in the last five chapters. Chapter 3 may be conveniently omitted from a course on fluid flow and heat transfer. Some instructors may prefer to present Chap. 10 after Chap. 2.

We are indebted to the Engineering Research Institute of the University of Michigan for permission to use the figures and much of the text which appeared in our original bulletin. Figures from other publications are also used, and appropriate credit is given in each case.

The literature of fluid dynamics and heat transfer is extensive. We have tried to include the major references, to which the reader is referred for more detailed discussion and bibliography on special topics.

We are indebted to the many workers, both past and present, without whose work the science of fluid dynamics and heat transfer would not be in its present advanced state.

James G. Knudsen
Donald L. Katz

CONTENTS

PART I

BASIC EQUATIONS AND FLOW OF NONVISCOUS FLUIDS

CHAPTER 1

FLUIDS AND FLUID PROPERTIES

1-1. Fluid Mechanics

The science of fluid mechanics is concerned with the motion of fluids and the conditions affecting that motion. Fluids at rest are a special case of fluid motion. Fluid kinematics is the subdivision of fluid mechanics which is restricted to a consideration of the geometry of motion and is not concerned with the forces involved. On the other hand, fluid dynamics is the subdivision of fluid mechanics concerned with the forces acting on the fluids. A further subdivision of fluid dynamics considers the state of fluid motion, i.e., fluids at rest or in uniform motion for which the forces are in equilibrium and fluids in unsteady motion for which the forces causing the flow are unbalanced.

The kinematics of fluid motion is concerned with such quantities as velocity, acceleration, and rate of discharge. To define these quantities, length and time scalars are necessary, and it is usual to express them in terms of some coordinate system, either rectangular, cylindrical, or spherical. The choice of coordinate system depends largely on the boundaries of the particular system being studied. Fluid dynamics, on the other hand, involves the application of Newton's second law of motion to the moving mass of fluid. This law states that the applied force is proportional to the rate of change of momentum. Such forces as pressure, shear, gravity, and inertia are involved. In the momentum term, the mass and velocity of the fluid must be known.

The various transfer processes which take place in fluids and between solids and fluids are momentum, mass, and heat transfer. During the flow of all real fluids in ducts or past immersed bodies, energy is dissipated through the action of viscosity, and this energy, which can be expressed in terms of rate of loss of momentum, represents the power required to pump the fluid. Heat transfer and mass transfer occur in the fluid under the influence of temperature and concentration differences respectively. The rate at which these transfer processes occur is determined by the mechanics of fluid flow, knowledge of which is basic in the fundamental study of the transfer processes and the relationship between them.

1-2. Fluid Properties

Certain physical properties of fluids are involved in any study of fluid mechanics and the related processes of momentum, mass, and heat transfer. These properties include density, viscosity, thermal conductivity, heat capacity, diffusivity, and surface tension.

The physical properties of a fluid or solid are a function of temperature and pressure. Numerous workers have measured physical properties, and attempts have been made to establish correlations between the various properties. Hougen and Watson [7] have presented a variety of empirical relationships for predicting the value of a particular physical property from other properties. Theoretical studies of physical properties, particularly those of Hirschfelder, Bird, and Spotz [6] and Lyderson and coworkers,[11,12] have produced relationships for predicting physical properties based on the molecular structure of compounds. These authors present relationships for obtaining critical properties, densities of gases and liquids, compressibility factors, viscosities, and thermal conductivities.

In the remainder of this chapter, the density, heat capacity, surface tension, viscosity, thermal conductivity, and diffusivity are defined, and their values are given for a few common substances. In most cases, the effect of pressure and temperature on the properties is indicated. Extensive tables of physical properties are given by Perry.[15] References containing lists of properties of materials other than those included here are also given.

1-3. States of Matter and Definition of a Fluid

All matter exists in either the solid or the fluid state. Physically, the solid state is characterized by relative immobility of the molecules in the solid. Each molecule has a fixed average position in space but vibrates and rotates about that average position. The fluid state is characterized by relative mobility of the molecules, which, in addition to rotation and vibration, also have translational motion; so they do not have fixed positions in the body of the fluid. Solids and fluids behave differently from each other when subjected to external forces. A solid has tensile strength, whereas a fluid has little or no tensile strength. A solid can resist compressive forces, up to a certain limit, while a fluid can resist compressive forces only if it is kept in a container and the compressive force is applied to all walls of the container. Solids can withstand shearing stress up to their elastic limit, but fluids, when subjected to a shearing force, deform immediately and continuously as long as the force is applied.

1-4. Gases and Liquids; Vapor Pressure

In the fluid state, matter exists either as a gas or a liquid. The molecules in gases are relatively far apart from each other and have high translational energy. Their translational energy is much greater than the energy of attraction between them. In liquids the molecules are mobile, and their translational energy is less than the energy of attraction between them, with the result that the average distance between them is small and very much less than the average distance between the molecules in a gas.

The distribution of translational energy among molecules in a liquid follows a Maxwell distribution curve. Some molecules, therefore, have sufficient translational energy to overcome the attractive forces of adjacent molecules, and they escape from the liquid and form a vapor. Liquids, consequently, exert a vapor pressure, which is the pressure attained by a vapor when left in contact with its liquid until equilibrium is attained, temperature being held constant.

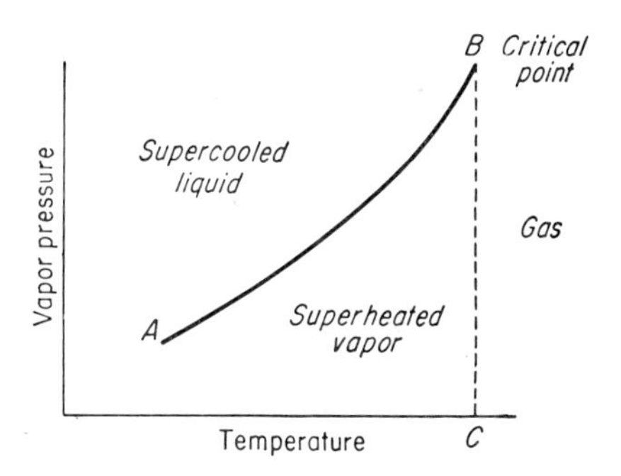

FIG. 1-1. Vapor-pressure curve for a liquid.

The vapor pressure is a function of temperature, as shown by the curve *AB* in Fig. 1-1. The area on the chart above *AB* is subcooled liquid. The curve terminates at *B*, which is the critical point. At this point, the translational energy of the molecules of the liquid becomes equal to the energy of attraction, and the liquid and vapor become identical with each other. The area *ABC* is the superheated-vapor region. A vapor is defined as a fluid that can be liquefied by compression alone. The area to the right of *BC* is the gaseous region. A gas is a fluid which cannot be liquefied no matter how much it is compressed.

The main difference between liquids and gases from the standpoint of fluid-flow studies is in their compressibility. Liquids at temperatures relatively far from the critical point can be compressed only slightly at very high pressures. For all practical purposes, they are incompressible. Gases and vapors, on the other hand, are very compressible, and in many flow problems this compressibility must be considered.

1-5. Density and Specific Gravity

The density of matter is expressed in terms of mass per unit volume (m/L^3). The specific gravity of a substance is the ratio of its density to the density of some reference substance. Pure water at 15.5°C is the reference substance for expressing the specific gravities of liquids and solids. Air is the reference substance for gases, the specific gravity of a gas being defined as the ratio of its density to the density of air, both at the same temperature and pressure. The gas gravity is also the molecular weight of the gas divided by the molecular weight of air.

1-6. Liquid Density

The effect of variations of pressure and temperature on the densities of liquids is generally considered insignificant in fluid flow; i.e., the flow is

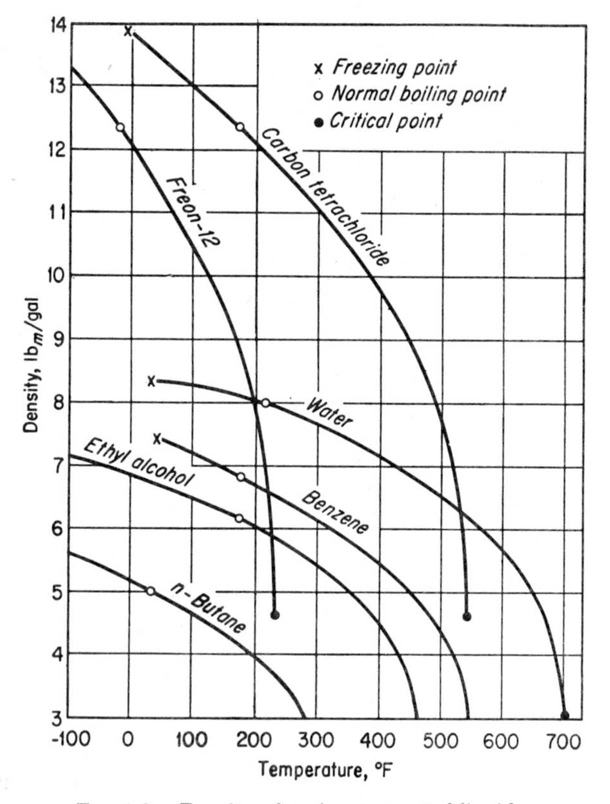

FIG. 1-2. Density of various saturated liquids.

incompressible. When the reduced temperature (ratio of temperature of the liquid to critical temperature of the substance) is 0.5 or below, the assumption of incompressibility involves no significant error. In the temperature range from 0 to 100°C, the coefficient of thermal expansion of liquids at constant pressure ranges from 2×10^{-4} to 16×10^{-4} (change in volume per unit volume per unit change in temperature). Likewise, the compressibility at constant temperature ranges from 2×10^{-6} to 16×10^{-6} (change in volume per unit volume per unit change in pressure) depending on the liquid, the temperature, and the pressure.

The density of a number of saturated liquids is shown as a function of temperature in Fig. 1-2. The freezing point, normal boiling point, and critical point are indicated on each curve. For many liquids, particularly the paraffin and olefin hydrocarbons, the specific gravity at the critical point is approximately 0.25. Othmer and Gilmont [13] present a nomograph for predicting the densities of a large number of liquids.

1-7. Gas Density

The common way of obtaining the density of a gas is through an equation of state relating the pressure, volume, and temperature. Perfect gases obey the equation of state

$$PV_m = RT \tag{1-1}$$

where P = pressure
V_m = volume per mole
T = absolute temperature
R = gas constant

all in appropriate units to make the equation dimensionally correct. Real gases also follow Eq. (1-1) with sufficient accuracy at reduced temperatures greater than 2 and reduced pressures less than 1.

Many equations of state have been developed for real gases, but they are quite complicated and difficult to use in engineering calculations. The simplest equation of state makes use of the compressibility factor Z.

$$PV_m = ZRT \tag{1-2}$$

Equation (1-2) is used to determine densities of gases under any condition of temperature and pressure. The compressibility factor Z may be obtained from a plot of the compressibility factor versus reduced pressure at constant values of reduced temperature. Such a plot is illustrated in Fig. 1-3. Densities obtained using Eq. (1-2) and Fig. 1-3 are accurate to about 5 to 10 per cent for all gases except hydrogen and helium.

For most organic compounds, the compressibility at the critical point is approximately 0.26.

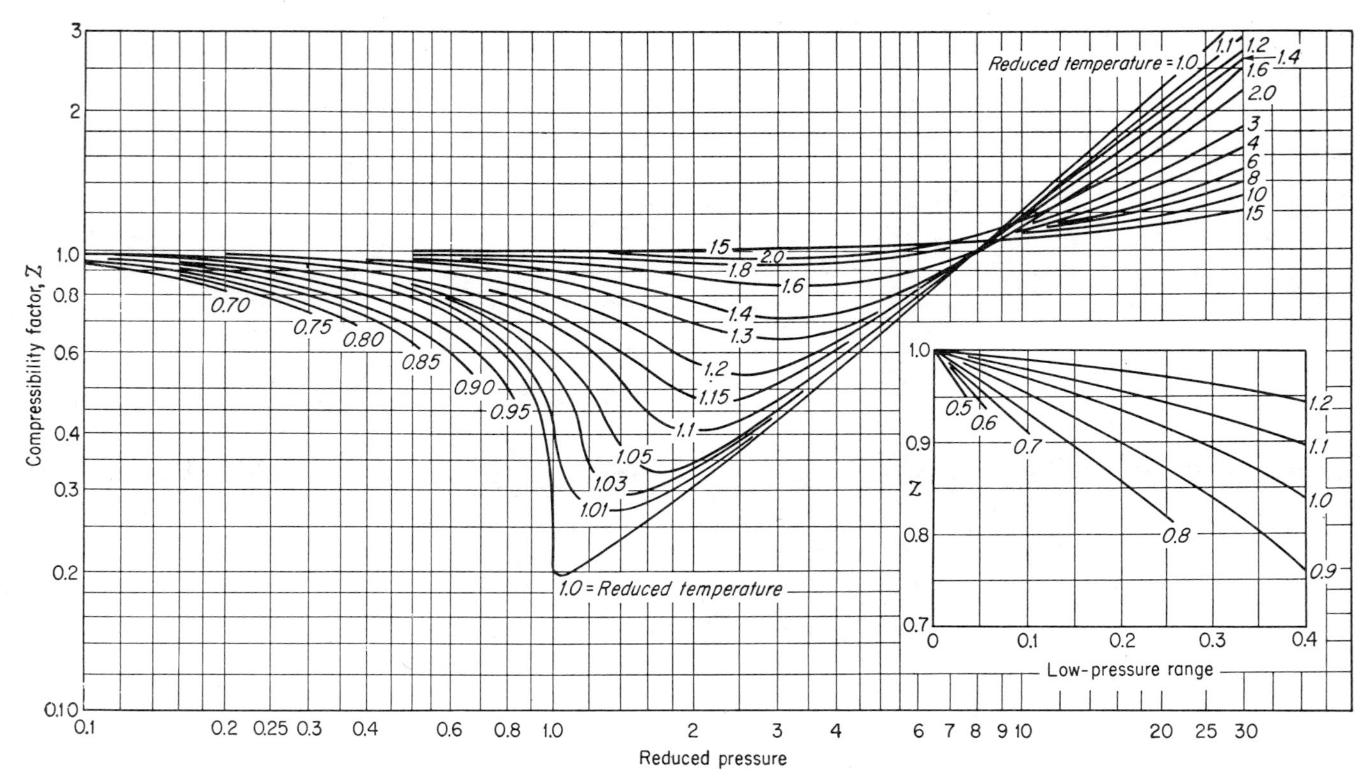

FIG. 1-3. Compressibility factors of gases and vapors. (*From O. A. Hougen and K. M. Watson, "Chemical Process Principles," John Wiley & Sons, Inc., New York.*)

1-8. Heat Capacity

Heat capacity is defined as the amount of heat required to increase the temperature of a material one degree. If the material is heated at constant pressure, the heat capacity becomes

$$\left(\frac{\partial H}{\partial T}\right)_p = C_p \tag{1-3}$$

and when the heating is carried out at constant volume,

$$\left(\frac{\partial E}{\partial T}\right)_v = C_v \tag{1-4}$$

where H = enthalpy
E = internal energy
C_p = heat capacity at constant pressure
C_v = heat capacity at constant volume

The dimensions of the heat capacity are energy per unit mass per unit temperature change (L^2/t^2T).

For perfect gases

$$C_p - C_v = R \tag{1-5}$$

while for liquids and solids C_p and C_v are very nearly equal. For most fluid-flow calculations C_p is used even though there may be pressure variation in the fluid. For a temperature change at constant pressure

$$H_2 - H_1 = \left(\int_{T_1}^{T_2} C_p\, dT\right)_P \tag{1-6}$$

When both pressure and temperature change, the enthalpy change is given by

$$dH = C_p\, dT + \left[-T\frac{\partial}{\partial T}\left(\frac{1}{\rho}\right) + \frac{1}{\rho}\right] dP \tag{1-7}$$

The last term of Eq. (1-7) gives the change in enthalpy due to pressure changes. Usually it is not significant except where large pressure changes are involved.

Heat capacities at constant pressure of some common liquids and gases are shown in Figs. 1-4 and 1-5. More extensive data are given by Perry.[15]

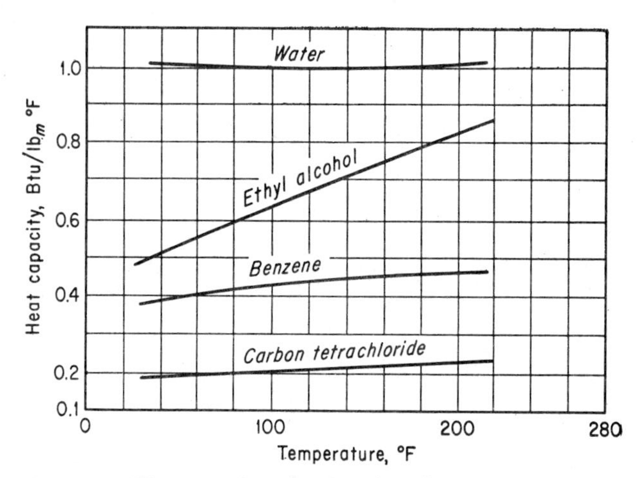

FIG. 1-4. Heat capacity of various liquids at 1 atm pressure.

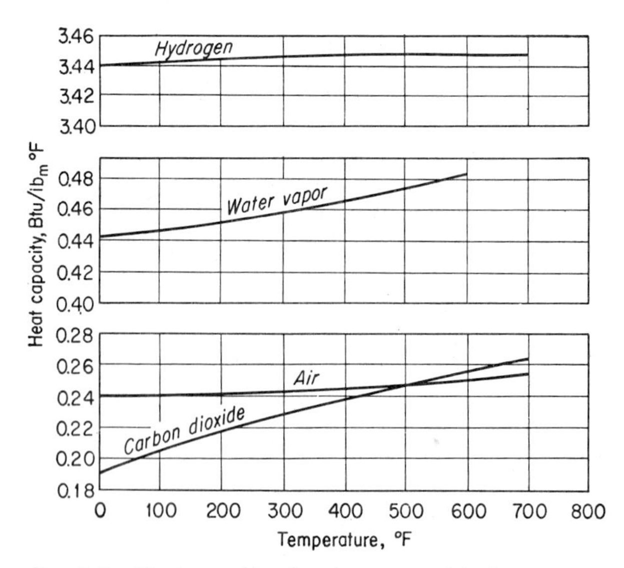

FIG. 1-5. Heat capacity of various gases at 1 atm pressure.

1-9. Surface Tension

Surface tension is defined as the amount of work required to increase the surface area of a liquid by one unit of area. The common unit of surface tension is dyne-centimeters per square centimeter, or dynes per centimeter. The nature of the fluid in contact with the liquid surface affects the surface tension, but this effect is slight for liquid surfaces in contact with gases. Othmer and Gilmont [13] present a nomograph for predicting surface tensions of a large number of liquids in contact with air. The interfacial tension of two immiscible liquids in contact is approximately the difference of their individual surface tensions when they are in contact with air.

1-10. Molecular-transport Properties of Fluids

The molecular-transport properties of fluids are those properties concerned with the rate of momentum, heat, and mass transfer by molecular motion. The rates of momentum, heat, and mass transfer in fluids may be expressed by analogous equations. In general, the rate is proportional to the potential gradient, the constant of proportionality being a physical property of the substance. The equations of molecular momentum, heat, and mass transfer are:

1. Momentum transfer:

$$\frac{F}{A} = \mu \frac{du}{dy} \tag{1-8}$$

$$\frac{\text{Momentum transfer}}{\text{(Unit area)(unit time)}} = \text{(viscosity)(velocity gradient)}$$

2. Heat transfer:

$$\frac{q}{A_q} = -k \frac{dT}{dy} \tag{1-9}$$

$$\frac{\text{Heat transfer}}{\text{(Unit area)(unit time)}} = \text{(thermal conductivity)(temperature gradient)}$$

3. Mass transfer:

$$\frac{N_m}{A_{N_m}} = -D \frac{dc_m}{dy} \tag{1-10}$$

$$\frac{\text{Mass transfer}}{\text{(Unit area)(unit time)}} = \text{(diffusion coefficient)(concentration gradient)}$$

The negative sign appears in Eqs. (1-9) and (1-10) since heat transfer and mass transfer occur only in the direction of decreasing temperature and concentration respectively.

1-11. Fluid Viscosity; Momentum Transfer

As a fluid is deformed because of flow and applied external forces, frictional effects are exhibited by the motion of molecules relative to each other. These frictional effects are encountered in all real fluids and are due to their *viscosity*. Consider a thin layer of fluid between two parallel planes placed a distance dy apart (see Fig. 1-6). One plane is fixed, and a shearing force F is applied parallel to the other plane. Since fluids deform continuously under shear, the movable plane moves steadily at a velocity du relative to the stationary plane. Under steady conditions, the external force F is balanced by an equal internal force due to the fluid viscosity. The shear force per unit area (F/L^2) is proportional to the velocity gradient in the fluid; i.e.,

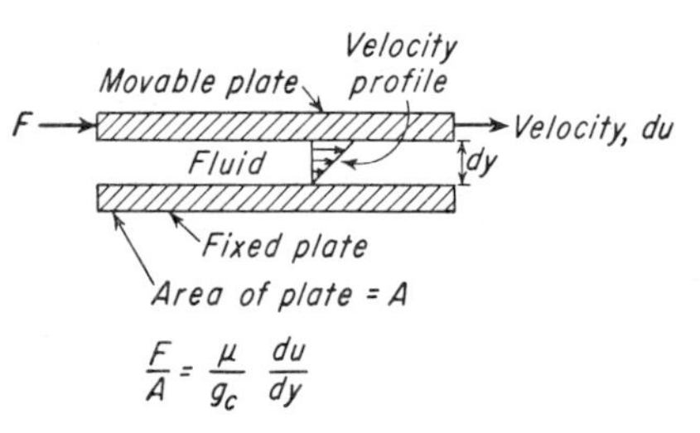

FIG. 1-6. Definition of viscosity.

$$\frac{F}{A} = \tau \propto \frac{du}{dy} \tag{1-11}$$

where τ is the shear force per unit area and du/dy is the velocity gradient (also called the *rate of shear*). The proportionality sign is removed by introducing the proportionality factor μ, which is the *Newtonian coefficient of viscosity*.

$$\tau = \frac{\mu}{g_c} \frac{du}{dy} \qquad \left[\frac{\text{lb}_f}{\text{ft}^2} = \frac{\text{lb}_m}{(\text{ft})(\text{sec})} \frac{\text{ft}}{\text{sec}} \frac{1}{\text{ft}} \frac{(\text{lb}_f)(\text{sec}^2)}{(\text{lb}_m)(\text{ft})} \right] \tag{1-12}$$

The use of the conversion factor g_c gives the viscosity dimensions of mass per unit length per unit time (m/Lt). In all flow where layers of fluid move relative to each other as indicated in Fig. 1-6, the shearing force exerted on the layers is defined by Eq. (1-12). The coefficient of viscosity μ is a characteristic physical property of fluids. Its value for a particular fluid is a function of temperature, pressure, and rate of shear. Physically, the viscosity is the tangential force per unit area exerted on layers of fluid a unit distance apart and having a unit velocity difference between them. The *kinematic viscosity* of fluids is the ratio of the viscosity to the density.

$$\nu = \frac{\mu}{\rho} \qquad \frac{\text{ft}^2}{\text{sec}} = \frac{\text{lb}_m}{(\text{ft})(\text{sec})} \frac{\text{ft}^3}{\text{lb}_m} \tag{1-13}$$

The viscous force between two layers of fluid may be also expressed as a rate of momentum transfer between the layers. By Newton's second law of motion, force is equal to the time rate of change of momentum.

Therefore, the shear stress defined in Eq. (1-12) is a force per unit area and is equivalent to a rate of change of momentum per unit area. Consider the adjacent layers of fluid shown in Fig. 1-7 moving at velocities u and $u + du$ respectively. Molecules of fluid move from layer 1 to layer 2 and from 2 to 1. The molecules moving from 1 to 2 have less momentum than those in 2 and exert a drag force on that layer.

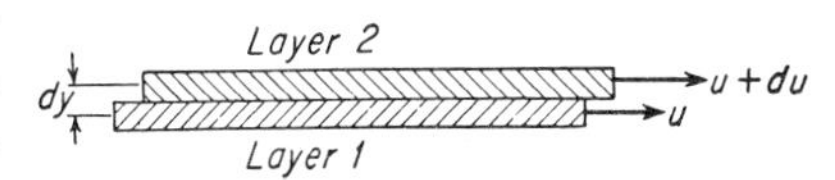

FIG. 1-7. Adjacent layers of fluid at different velocities to show momentum exchange.

1-12. Viscosity of Newtonian Fluids; Effect of Temperature and Pressure

The viscosity of all Newtonian liquids (see Sec. 1-15) decreases with an increase in temperature at constant pressure. Several empirical formulas have been proposed relating the viscosity of liquids to the temperature. The viscosity of gases increases as the temperature increases at constant pressure. This behavior is in accordance with the kinetic theory of gases, which predicts that the viscosity is proportional to the density, the average velocity of the molecules, and the mean free path of the molecules. As the temperature increases at constant pressure, the density decreases—but at a rate considerably less than the rate at which the velocity and mean free path increase, with the result that viscosity increases with temperature. The viscosities of some common liquids and gases are plotted as a function of temperature in Figs. 1-8 and 1-9. At temperatures above the normal boiling point, the pressure on the liquids is the saturation pressure.

For most liquids the viscosity increases with pressure at constant temperature; however, below the critical pressure, the effect of pressure on viscosity is small. The viscosity of gases also increases with pressure. According to the kinetic theory, the viscosity of gases should be independent of pressure. This is true for real gases at high reduced temperatures and low reduced pressures. Uyehara and Watson [16] have correlated viscosity data of fluids on a reduced basis as shown in Fig. 1-10, where μ/μ_c is plotted versus the reduced temperature T/T_C. Each curve is for a constant value of the reduced pressure P/P_c. The viscosity at the critical point is μ_c, and may be predicted from a relation proposed by Uyehara and Watson.[16]

$$\mu_c = \frac{61.6\sqrt{MT_C}}{V_C^{2/3}} \tag{1-14}$$

where M = molecular weight
T_C = critical temperature, °K
V_C = critical volume, cc/g mole
μ_c = critical viscosity, centipoises

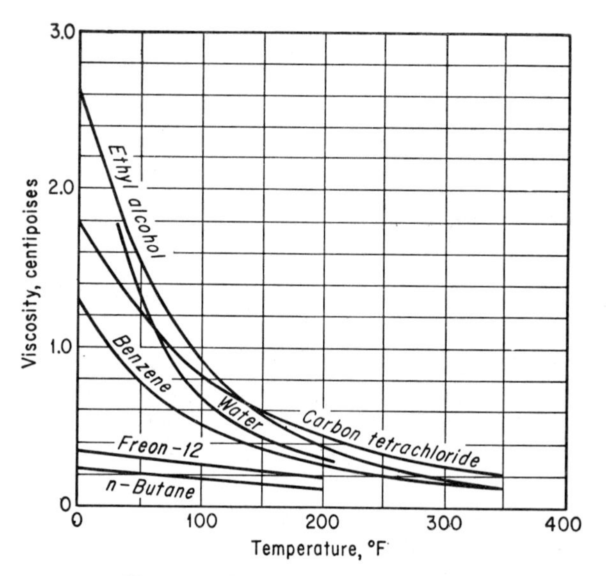

FIG. 1-8. Viscosity of various liquids.

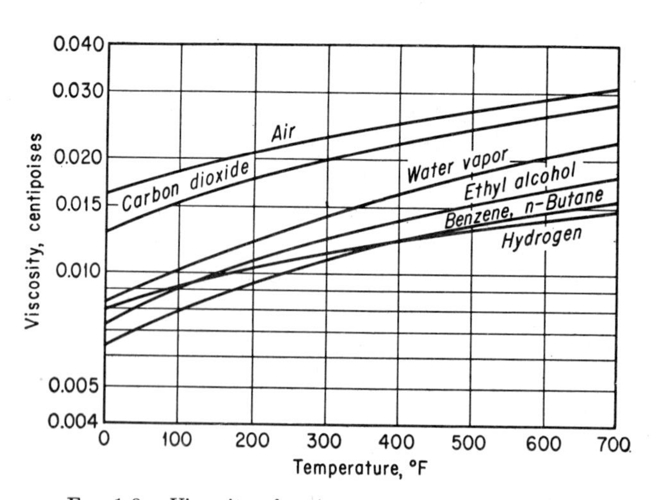

FIG. 1-9. Viscosity of various gases at 1 atm pressure.

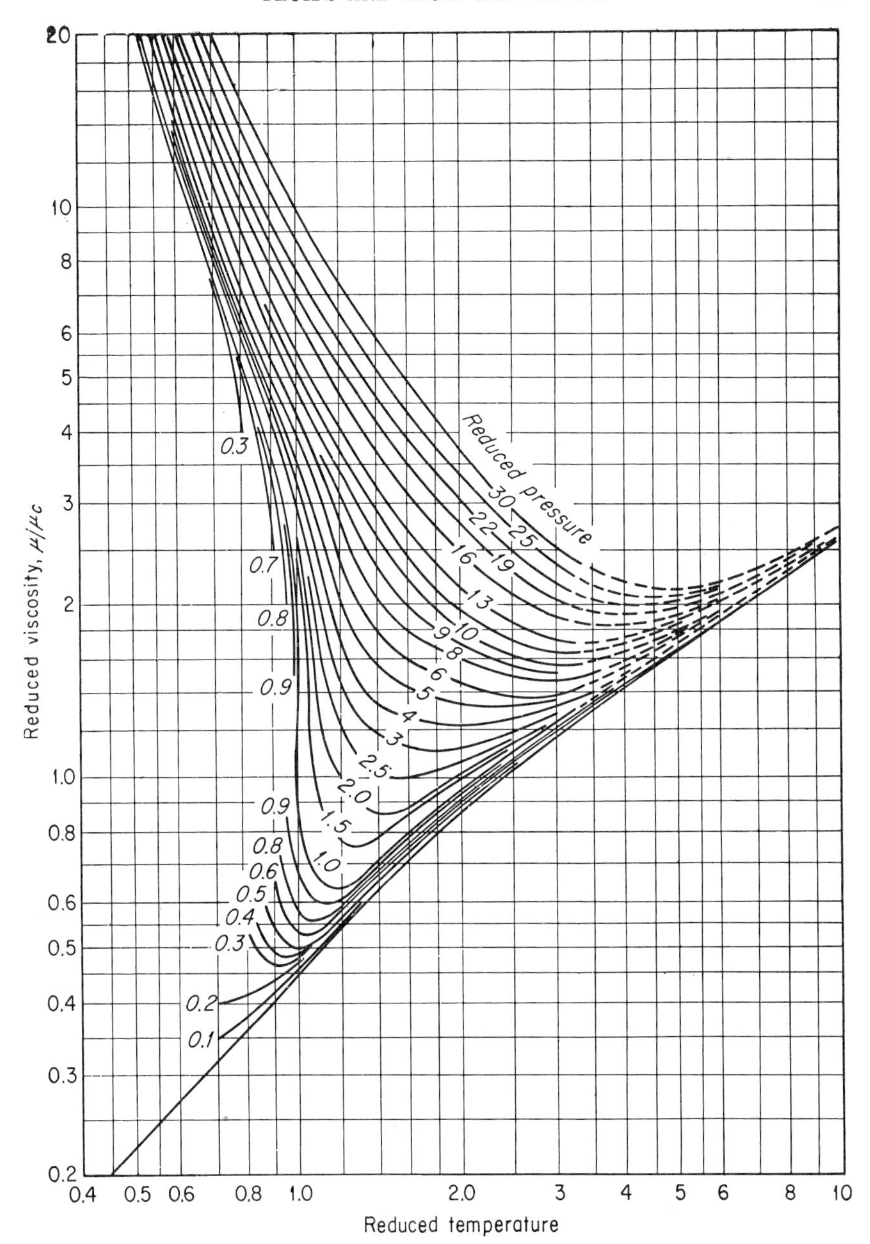

FIG. 1-10. Generalized reduced viscosities. (*From O. A. Uyehara and K. M. Watson, Natl. Petroleum News Tech. Sec.,* **36**:R764, *Oct.* 4, 1944.)

Figure 1-10 may be used to predict the viscosity at high pressure for both gases and liquids. Carr, Parent, and Peck [3] have presented a convenient chart to predict viscosities of gases at high pressures.

The viscosity of gaseous mixtures can be approximated by the relation

$$\mu_{\text{mix}} = \text{m.f.}_1\mu_1 + \text{m.f.}_2\mu_2 + \text{m.f.}_3\mu_3 + \cdots \tag{1-15}$$

and for liquids

$$\log \frac{1}{\mu_{\text{mix}}} = \text{m.f.}_1 \log \frac{1}{\mu_1} + \text{m.f.}_2 \log \frac{1}{\mu_2} + \cdots \tag{1-16}$$

where m.f. is the mole fraction of the component. Wilke [17] has proposed a more accurate relation than Eq. (1-15) for gases.

1-13. Thermal Conductivity

Thermal conductivity is a measure of the ability of a substance to transfer heat by molecular conduction. The differential equation for the one-dimensional molecular conduction of heat in a substance is

$$\frac{q}{A_q} = -k \frac{dT}{dy} \tag{1-9}$$

where q = rate of heat flow per unit time
A_q = area of flow
dT/dy = temperature gradient in material
k = thermal conductivity of substance

The sign is negative because heat is conducted from a higher temperature to a lower temperature. For Eq. (1-9) to be dimensionally correct, the units on the thermal conductivity are rate of energy transfer per unit cross-sectional area per unit temperature gradient (mL/t^3T). Figures 1-11 and 1-12 depict the thermal conductivities of some liquids and gases as a function of temperature. An approximate equation for predicting thermal conductivities of gases is suggested by Eucken.[4]

$$k = \mu \left(C_p + \frac{5R}{4M} \right) \tag{1-17}$$

where k is in Btu/(hr)(ft^2)(°F)/ft
μ is in lb$_m$/(ft)(hr)
C_p is in Btu/(lb$_m$)(°F)
M = molecular weight
R = 1.987 Btu/(lb mole)(°R)

For mixtures of gases, the thermal conductivity may be predicted by a relationship presented by Lindsay and Bromley.[10]

The effect of pressure on thermal conductivity of gases may be determined from a chart given by Lenoir, Junk, and Comings.[9] In general, the thermal conductivity is independent of pressure below reduced pressures of 0.2. Above this value it increases rapidly with pressure.

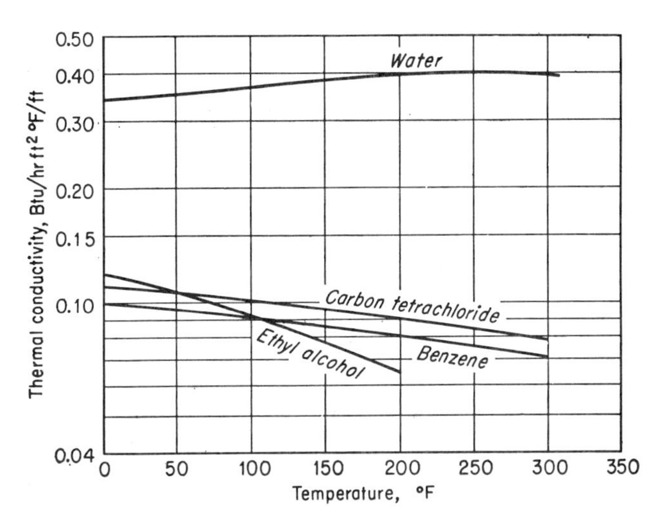

FIG. 1-11. Thermal conductivity of various liquids at 1 atm pressure.

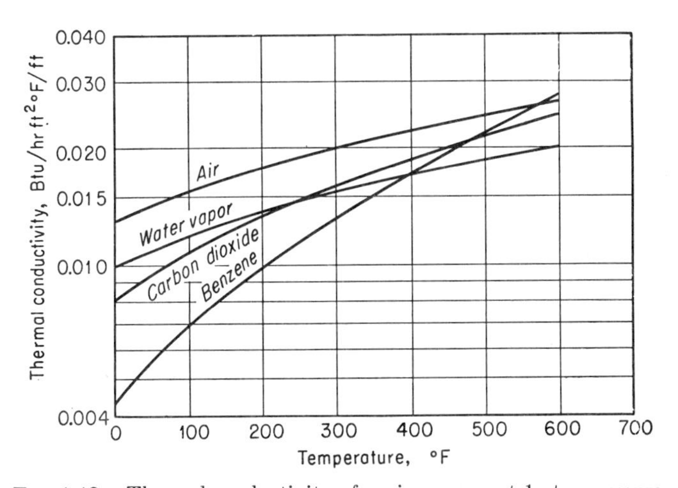

FIG. 1-12. Thermal conductivity of various gases at 1 atm pressure.

The thermal conductivity of liquids may be predicted by a relation given by Palmer.[14]

$$k = 22.9C_p \left(\frac{\rho}{M}\right)^{4/3} \frac{T_B}{\Delta H_v} \tag{1-18}$$

where C_p is in Btu/(lb$_m$)(°F)
ρ is in g/cc
M = molecular weight
ΔH_v = latent heat of vaporization at T_B, Btu/lb$_m$
T_B = normal boiling point, °F

Equation (1-18) is recommended when actual experimental data are lacking. For liquid mixtures Kern [8] recommends the approximate relation

$$k_{\text{mix}} = \text{w.f.}_1 k_1 + \text{w.f.}_2 k_2 + \cdots \tag{1-19}$$

where w.f. is the weight fraction.

The thermal conductivity of liquids is unaffected by pressure below reduced pressures of 0.1. Above these pressures it increases with increasing pressure. Bridgman [1] presents empirical data to show the effect of pressure on the thermal conductivity of liquids.

1-14. The Diffusion Coefficient

The diffusion coefficient in a system of two components is a measure of the rate of molecular diffusion (mass transfer) of either component under the influence of a concentration difference. Diffusion takes place in the direction of decreasing concentration. The differential equation for one-dimensional diffusion is

$$\frac{N_m}{A_{N_m}} = -D\frac{dc_m}{dy} \tag{1-10}$$

where N_m = molal rate of diffusion
A_{N_m} = area
$\dfrac{dc_m}{dy}$ = concentration gradient of diffusing substance
D = diffusion coefficient

The diffusion coefficient is dependent *on both* the components in the system. For gases, the empirical equation determined by Gilliland [5] may be used to predict diffusion coefficients.

$$D = 0.0043\frac{T^{3/2}}{P(V_{m_1}{}^{1/3} + V_{m_2}{}^{1/3})}\sqrt{\frac{1}{M_1} + \frac{1}{M_2}} \tag{1-20}$$

where V_{m_1}, V_{m_2} = respective molecular volumes of gases 1 and 2, cc
M_1, M_2 = molecular weights of gases
P = pressure, atm
T = temperature, °K
D = diffusivity, cm^2/sec

A more exact equation is presented by Hirschfelder, Bird, and Spotz.[6]

Diffusion coefficients for liquid systems are not plentiful. Wilke [17] has correlated the available data and obtained a relation for predicting diffusion coefficients in such systems.

1-15. Types of Fluids

An *ideal fluid* is one which is incompressible and has zero viscosity. With zero viscosity, the fluid offers no resistance to shearing forces, and hence during flow and deformation of the fluid all shear forces are zero. Many flow problems are simplified by assuming that the fluid is ideal. All *real fluids* have finite viscosity, and in most cases of flow in ducts and over immersed bodies it is necessary to consider the viscosity and the related shearing stresses associated with deformation of the fluid. Real fluids are also called *viscous fluids. Nonviscous fluids* are those having zero viscosity, but they may or may not be incompressible. Flow of an ideal fluid is called *nonviscous, incompressible flow,* while flow of a real fluid is called *viscous flow.*

Real fluids are further subdivided into two main classes. *Newtonian fluids* are those for which the viscosity coefficient is independent of the rate of shear (velocity gradient); i.e., the viscosity μ in Eq. (1-12) is a constant for each Newtonian fluid at a given temperature and pressure. A typical shear-stress–shear-strain diagram for such a fluid is shown in Fig. 1-13a. The shear stress τ is proportional to the shear strain du/dy,

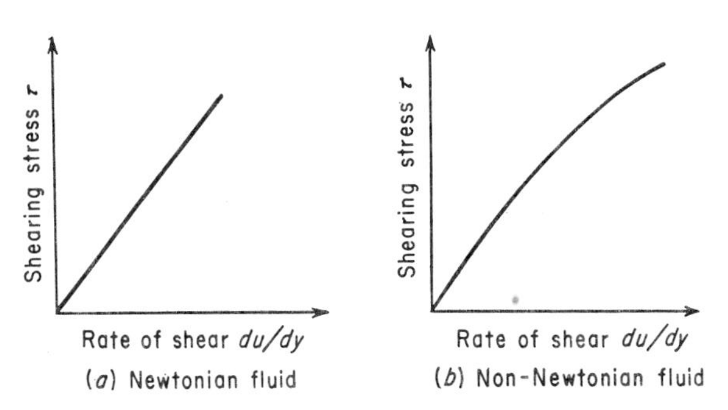

FIG. 1-13. Shear-stress rate of shear relationships for fluids.

the slope of the line being μ/g_c. *Non-Newtonian fluids* are those in which the viscosity at a given pressure and temperature is a function of the velocity gradient. Such fluids as colloidal suspensions, emulsions, and gels are included in this classification. The shear-stress–shear-strain diagram for a non-Newtonian fluid is shown in Fig. 1-13b. From the slope of the curve at any point the viscosity of the fluid may be determined. A plot of viscosity versus velocity gradient is shown in Fig. 1-14.

Non-Newtonian fluids may be further classified according to the manner in which the viscosity varies with the rate of shear. *Bingham plastics*, sometimes called *ideal plastics*, can withstand a certain amount of shearing stress. When the shearing stress has reached a certain yield value, the material deforms, giving the shear-stress–shear-strain diagram shown in Fig. 1-15 by curve B. The straight-line relationship indicates that, once the ideal plastic has been deformed, its viscosity is independent of the velocity gradient and is a function only of the temperature, pressure, and composition of the material. The relationship between shearing stress and shearing strain is

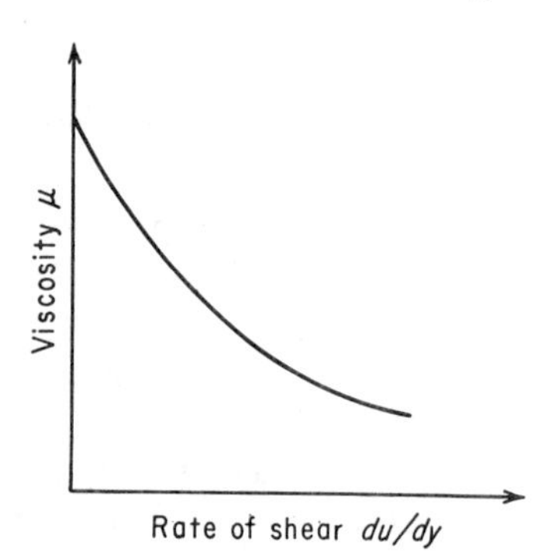

FIG. 1-14. Viscosity of a non-Newtonian fluid as a function of rate of shear.

$$\tau - \tau_0 = \frac{\mu}{g_c}\frac{du}{dy} \tag{1-21}$$

where τ_0 is the yield stress. Sewage sludge is a common example of a Bingham plastic. In most *real plastics* the viscosity does not become constant until fairly high rates of shear are attained, as is indicated by curve C in Fig. 1-15. Suspensions of clay in water behave like real plastics and are used extensively as drilling mud in the petroleum industry.

Pseudoplastic materials are those in which the viscosity decreases with rate of shear but the material deforms as soon as a shearing stress is applied. The viscosity becomes constant at high shear rates. Curve D in Fig. 1-15 shows the shear-stress–rate-of-shear relationship for pseudoplastic materials. The slope of the curve at the origin gives the viscosity of the material at zero rate of shear. Common pseudoplastic materials are gels, e.g., polystyrene in organic solvents and metallic soaps in gasoline.

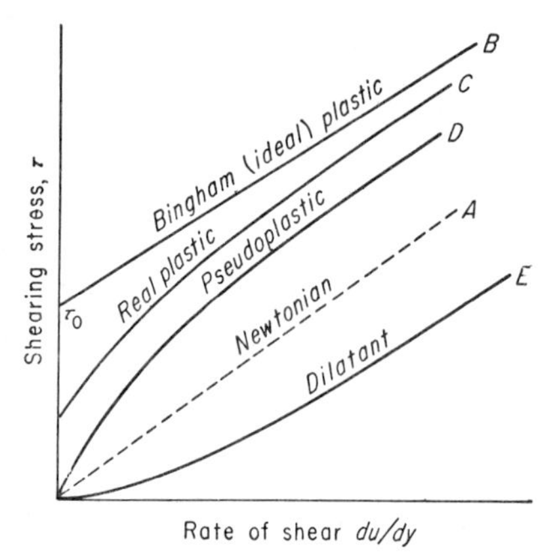

FIG. 1-15. Various non-Newtonian fluids.

Dilatant materials are those in which the viscosity increases with the rate of shear (see curve E in Fig. 1-15). Examples of dilatant materials are quicksand, butter, and starch suspensions.

Non-Newtonian fluids may be *thixotropic* or *nonthixotropic*. If the fluid possesses some sort of structure which is broken down when it is subjected to shear, then on removal of the shearing stress the viscosity, instead of being the same as at zero rate of shear, will change with time as the fluid builds up the structure it had prior to being deformed. If a thixotropic fluid is tested in an apparatus in which the rate of shear can be increased and then decreased, the relationship between the shear stress and the rate

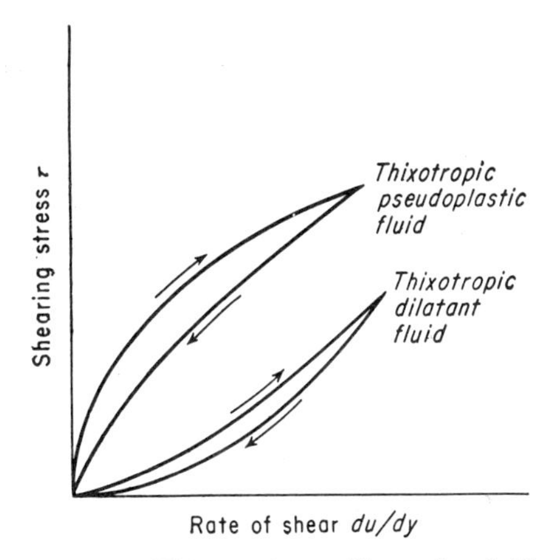

FIG. 1-16. Thixotropic non-Newtonian fluids.

of shear will be found to be different when the stress is increasing than when the stress is decreasing. Such curves for thixotropic pseudoplastic and dilatant materials are illustrated in Fig. 1-16.

1-16. Units and Dimensions

The system of dimensions used throughout this text is a combination of the two systems mass-length-time-temperature (m-L-t-T) and force-length-time-temperature (F-L-t-T). The use of pounds force (lb_f) and pounds mass (lb_m) is common in much engineering work. The conversion factor between the two systems of dimensions is the gravitational constant g_c, which has dimensions mL/Ft^2.

BIBLIOGRAPHY

1. Bridgman, P. W.: *Proc. Am. Acad. Arts. Sci.*, **50**:141 (1923).
2. Bromley, L. A., and C. R. Wilke: *Ind. Eng. Chem.*, **43**:1641 (1951).
3. Carr, N. L., J. D. Parent, and R. E. Peck: *Chem. Eng. Progr. Symposium Ser.*, [16] **51**:91 (1955).
4. Eucken, A.: *Physik. Z.*, **12**:1101 (1911).
5. Gilliland, E. R.: *Ind. Eng. Chem.*, **26**:681 (1934).
6. Hirschfelder, J. O., R. B. Bird, and E. L. Spotz: *Trans. ASME*, **71**:921 (1949).

7. Hougen, O. A., and K. M. Watson: "Chemical Process Principles," John Wiley & Sons, Inc., New York, 1947.
8. Kern, D. Q.: "Process Heat Transfer," McGraw-Hill Book Company, Inc., New York, 1950.
9. Lenoir, J. M., W. A. Junk, and E. W. Comings: *Chem. Eng. Progr.*, **49**:539 (1953).
10. Lindsay, A. L., and L. A. Bromley: *Ind. Eng. Chem.*, **42**:1508 (1950).
11. Lyderson, A. L., R. A. Greekhorn, and O. A. Hougen: *Univ. Wisconsin Eng. Expt. Sta. Rept.* 4, October, 1955.
12. Lyderson, A. L.: *Univ. Wisconsin Eng. Expt. Sta. Rept.* 3, April, 1955.
13. Othmer, D. F., and R. Gilmont: *Petroleum Refiner*, **31**(1):107 (1952).
14. Palmer, G.: *Ind. Eng. Chem.*, **40**:89 (1948).
15. Perry, J. H.: "Chemical Engineers' Handbook," 3d ed., McGraw-Hill Book Company, Inc., New York, 1950.
16. Uyehara, O. A., and K. M. Watson: *Natl. Petroleum News Tech. Sec.*, **36**:R764, Oct. 4, 1944.
17. Wilke, C. R.: *Chem. Eng. Progr.*, **45**:218 (1949).

CHAPTER 2

THE DIFFERENTIAL EQUATIONS OF FLUID FLOW

2-1 Introduction

Many physical problems that engineers must solve involve the evaluation of an unknown physical quantity. Frequently this physical quantity is a variable which is dependent on other physical quantities. The solution of the problem involves the determination of a functional relationship between the physical variables. In many cases, one is concerned with the rates of change of the function with respect to the variables. Equations involving an unknown function and its derivatives are differential equations.

In fluid flow there are several differential equations which result from the application of various physical laws. In these equations the independent variables are usually the space coordinates x, y, and z and time t. The dependent variables are velocity, temperature, pressure, and properties of the fluid. The important differential equations of fluid flow are:

1. The continuity equation, based on the law of conservation of mass
2. The momentum equation, based on Newton's second law of motion
3. The energy equation, based on the law of conservation of energy

2-2. The Continuity Equation for One-dimensional Flow

The continuity equation is the mathematical expression of the law of conservation of mass. Referring to Fig. 2-1, consider a fluid flowing parallel to the x axis. The mass flow of fluid through a cubical element of space having dimensions dx, dy, dz with its edges parallel to the x, y, z axes is to be determined. At the x face of the cube the fluid velocity and density are, respectively, u and ρ. At the $x + dx$ face the velocity and density are $u + (\partial u/\partial x)\,dx$ and $\rho + (\partial \rho/\partial x)\,dx$, where $\partial u/\partial x$ and $\partial \rho/\partial x$ are the rate of change of the velocity and density with respect to x. In this system u and ρ are the dependent variables, while x is an independent variable.

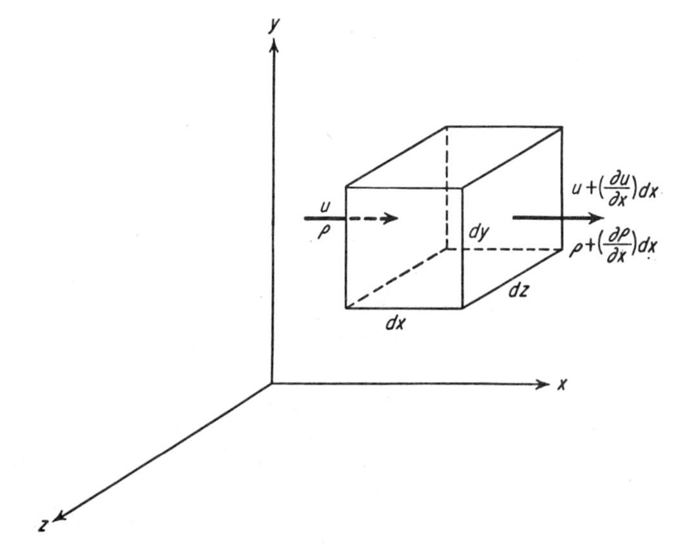

FIG. 2-1. One-dimensional flow through a differential element of space.

Since steady conditions do not necessarily exist, time is also an independent variable.

A mass balance on the element is made for a differential time dt. The mass *input* into the x face of the element in time dt is $\rho u\,dy\,dz\,dt$. The mass *output* from the $x + dx$ face in time dt is

$$\left(\rho + \frac{\partial \rho}{\partial x}\,dx\right)\left(u + \frac{\partial u}{\partial x}\,dx\right)dy\,dz\,dt$$

which becomes, on neglecting the term containing $(dx)^2$,

$$\left(\rho u + u\,\frac{\partial \rho}{\partial x}\,dx + \rho\,\frac{\partial u}{\partial x}\,dx\right)dy\,dz\,dt$$

The accumulation of mass in the cube in time dt is related to the time rate of change of density and is expressed as

$$\frac{\partial \rho}{\partial t}\,dx\,dy\,dz\,dt$$

Making a mass balance on the cube as follows,

$$\text{Input} = \text{output} + \text{accumulation} \tag{2-1}$$

$$\rho u\,dy\,dz\,dt = \left(\rho u + u\,\frac{\partial \rho}{\partial x}\,dx + \rho\,\frac{\partial u}{\partial x}\,dx\right)dy\,dz\,dt + \frac{\partial \rho}{\partial t}\,dx\,dy\,dz\,dt \tag{2-2}$$

Eq. (2-2) may be simplified to

$$-u\frac{\partial\rho}{\partial x} - \rho\frac{\partial u}{\partial x} = \frac{\partial\rho}{\partial t} \tag{2-3}$$

The left side of Eq. (2-3) is an exact derivative, and the continuity equation for one-dimensional flow becomes

$$-\frac{\partial(\rho u)}{\partial x} = \frac{\partial\rho}{\partial t} \tag{2-4}$$

Equation (2-4) is the differential equation expressing the law of conservation of mass for a fluid flowing parallel to the x-coordinate axis in which both the velocity and the density of the fluid are functions of x and t.

2-3. The Continuity Equation for Three-dimensional Flow

A similar analysis may be used to formulate the equation of continuity for flow in three dimensions. In this analysis, however, it is convenient to consider a point in a fluid where the velocity **V** may be represented by three component velocities u, v, and w, each parallel, respectively, to the x, y, and z axis of the rectangular coordinate system used, as illustrated in Fig. I-1 of Appendix I. The velocity **V** varies with time and position. This variation with position may be represented by the individual variation of the component velocities u, v, and w with respect to their individual directions. A mass balance is made on the differential element of space, but here, three directions of flow must be considered. The mass *input* to the element in time dt is

$$\rho u\,dy\,dz\,dt \text{ (for the } x \text{ direction)} + \rho v\,dx\,dz\,dt \text{ (for the } y \text{ direction)} + \rho w\,dx\,dy\,dt \text{ (for the } z \text{ direction)}$$

The mass *output* from the element in time dt is

$$\left(\rho + \frac{\partial\rho}{\partial x}dx\right)\left(u + \frac{\partial u}{\partial x}dx\right)dy\,dz\,dt \text{ (for the } x \text{ direction)}$$
$$+ \left(\rho + \frac{\partial\rho}{\partial y}dy\right)\left(v + \frac{\partial v}{\partial y}dy\right)dx\,dz\,dt \text{ (for the } y \text{ direction)}$$
$$+ \left(\rho + \frac{\partial\rho}{\partial z}dz\right)\left(w + \frac{\partial w}{\partial z}dz\right)dx\,dy\,dt \text{ (for the } z \text{ direction)}$$

The accumulation of mass in the cube in time dt is

$$\frac{\partial\rho}{\partial t}dx\,dy\,dz\,dt$$

Table 2-1. Forms of the Continuity Equation

	Unsteady state		Steady state	
	Compressible fluid	Incompressible fluid	Compressible fluid	Incompressible fluid
Three-dimensional flow	$\frac{\partial(\rho u)}{\partial x} + \frac{\partial(\rho v)}{\partial y} + \frac{\partial(\rho w)}{\partial z} = -\frac{\partial \rho}{\partial t}$	$\frac{\partial u}{\partial x} + \frac{\partial v}{\partial y} + \frac{\partial w}{\partial z} = 0$	$\frac{\partial(\rho u)}{\partial x} + \frac{\partial(\rho v)}{\partial y} + \frac{\partial(\rho w)}{\partial z} = 0$	$\frac{\partial u}{\partial x} + \frac{\partial v}{\partial y} + \frac{\partial w}{\partial z} = 0$
Two-dimensional flow in x and y direction	$\frac{\partial(\rho u)}{\partial x} + \frac{\partial(\rho v)}{\partial y} = -\frac{\partial \rho}{\partial t}$	$\frac{\partial u}{\partial x} + \frac{\partial v}{\partial y} = 0$	$\frac{\partial(\rho u)}{\partial x} + \frac{\partial(\rho v)}{\partial y} = 0$	$\frac{\partial u}{\partial x} + \frac{\partial v}{\partial y} = 0$
One-dimensional flow in x direction	$\frac{\partial(\rho u)}{\partial x} = -\frac{\partial \rho}{\partial t}$	$\frac{\partial u}{\partial x} = 0$	$\frac{\partial(\rho u)}{\partial x} = 0$	$\frac{\partial u}{\partial x} = 0$

A mass balance on the cube gives, after simplification,

$$-u\frac{\partial \rho}{\partial x} - v\frac{\partial \rho}{\partial y} - w\frac{\partial \rho}{\partial z} - \rho\left(\frac{\partial u}{\partial x} + \frac{\partial v}{\partial y} + \frac{\partial w}{\partial z}\right) = \frac{\partial \rho}{\partial t} \tag{2-5}$$

which may also be written as

$$\frac{\partial(\rho u)}{\partial x} + \frac{\partial(\rho v)}{\partial y} + \frac{\partial(\rho w)}{\partial z} = -\frac{\partial \rho}{\partial t} \tag{2-6}$$

Equation (2-6) is the general continuity equation for three-dimensional flow. It is a mathematical expression of the law of conservation of mass and involves no assumptions. Using vector notation,† Eq. (2-6) may be written

$$\text{div}\ (\rho \mathbf{V}) = -\frac{\partial \rho}{\partial t} \tag{2-7}$$

All problems in fluid flow require that the continuity equation be satisfied. If steady-state conditions prevail, all derivatives with respect to time are zero, and Eq. (2-6) becomes

$$\frac{\partial(\rho u)}{\partial x} + \frac{\partial(\rho v)}{\partial y} + \frac{\partial(\rho w)}{\partial z} = 0 \tag{2-8}$$

or

$$\text{div}\ (\rho \mathbf{V}) = 0 \tag{2-9}$$

If a fluid is compressible, the density will vary in space, so Eq. (2-9) applies for the steady-state flow of a compressible fluid. For the steady-state flow of an incompressible fluid the density is constant, and the continuity equation becomes

$$\frac{\partial u}{\partial x} + \frac{\partial v}{\partial y} + \frac{\partial w}{\partial z} = 0 \tag{2-10}$$

or

$$\text{div}\ \mathbf{V} = 0 \tag{2-11}$$

Table 2-1 gives the forms of the continuity equation which apply to various conditions of flow.

2-4. The Momentum Equations

Every particle of fluid at rest or in steady or accelerated motion obeys Newton's second law of motion, which states that the time rate of change

† A description of vector notation used in this and subsequent chapters may be found in Appendix I.

of momentum is equal to the external forces, i.e.,

$$\frac{d}{dt}(mu) = Fg_c \tag{2-12}$$

and since mass is constant,

$$m\frac{du}{dt} = Fg_c \tag{2-13}$$

or

$$(\text{Mass})(\text{acceleration}) = \text{external force} \tag{2-14}$$

The product of mass and acceleration is called inertial force. The momentum equations of fluid flow are a mathematical expression of Newton's second law applied to moving masses of fluid. The derivation of the equations involves determining the inertial force of the flowing fluid in each coordinate direction and equating it to the external forces acting on the fluid. The three main external forces which may act on the fluid are field forces (gravity forces), normal forces (pressure), and shear or tangential forces (caused by the resistance of the fluid to deformation).

Inertial Forces. The momentum equations will be derived for the x direction. Similar equations may be derived for the other two coordinate directions.† Consider a point in a moving fluid where the velocity is **V**. As pointed out in Sec. 2-3, this velocity may be represented by three component velocities u, v, and w. In the general case of three-dimensional unsteady flow these component velocities are functions of x, y, z, and t; i.e., for the x-coordinate direction

$$u = F_1(x, y, z, t) \tag{2-15}$$

Taking the differential of each side of Eq. (2-15),

$$du = \frac{\partial u}{\partial x}dx + \frac{\partial u}{\partial y}dy + \frac{\partial u}{\partial z}dz + \frac{\partial u}{\partial t}dt \tag{2-16}$$

and dividing by dt,

$$\frac{du}{dt} = \frac{\partial u}{\partial x}\frac{dx}{dt} + \frac{\partial u}{\partial y}\frac{dy}{dt} + \frac{\partial u}{\partial z}\frac{dz}{dt} + \frac{\partial u}{\partial t} \tag{2-17}$$

Since $u = dx/dt$, $v = dy/dt$, $w = dz/dt$,

$$\frac{du}{dt} = u\frac{\partial u}{\partial x} + v\frac{\partial u}{\partial y} + w\frac{\partial u}{\partial z} + \frac{\partial u}{\partial t} \tag{2-18}$$

du/dt is the acceleration in the x direction. The fluid in a cubical element of space of size dx, dy, dz has this acceleration in the x direction. The

† More detailed derivations of the momentum equation are found in refs. 1, 3, and 4.

inertial force in the x direction is the product of the mass and the acceleration; i.e., letting IF_x be the inertial force in the x direction,

$$IF_x = \frac{m}{g_c}\frac{du}{dt} \tag{2-19}$$

Putting $m = \rho\, dx\, dy\, dz$ and using Eq. (2-18),

$$IF_x = \frac{\rho}{g_c}\left(u\frac{\partial u}{\partial x} + v\frac{\partial u}{\partial y} + w\frac{\partial u}{\partial z} + \frac{\partial u}{\partial t}\right) dx\, dy\, dz \tag{2-20}$$

Similar equations may be obtained for the inertial forces in the y and z directions.

External Forces. 1. *Field forces.* If the fluid exists in a force field, such as a gravitational or electrostatic field or both, then each particle of fluid will have a potential energy which is a function of its position in the force field. The force potential of the field is Ω † and is defined as the energy stored in a unit mass of fluid in moving it from one point to the other in the force field. The force exerted on a unit mass is the rate of change of Ω with respect to distance. Therefore $\partial\Omega/\partial x$ is the force per unit mass exerted on the fluid in the x direction, and similar derivatives hold for the y and z directions. The field force exerted on the fluid in a spatial element $dx\, dy\, dz$ in the x direction is

$$FF_x = -\rho\frac{\partial\Omega}{\partial x}\, dx\, dy\, dz \tag{2-21}$$

2. *Normal and tangential forces.* The state of stress at a point in a fluid is completely defined by nine stress components,‡ as follows:

p_x = normal stress in direction of x axis
τ_{xy} = tangential stress parallel to x plane and in direction of y axis
τ_{yx} = tangential stress parallel to y plane and in direction of x axis
p_y = normal stress in direction of y axis
τ_{yz} = tangential stress parallel to y plane and in direction of z axis

† An example of a force field is the earth's gravitational field. If one considers a mass of fluid under the influence of gravity and selects some arbitrary plane where the potential energy of the fluid is zero, then clearly the potential energy of the fluid varies with the distance above the arbitrary plane. Letting Z be the distance above the arbitrary plane, the potential energy of the fluid per unit mass in terms of Z is gZ/g_c. This potential energy gZ/g_c is the same as the term Ω indicated above. Restricting changes to those in the vertical direction, the gravitational force exerted on a unit mass is $-\frac{d}{dZ}\left(\frac{gZ}{g_c}\right)$. It becomes $-(g/g_c)$, which has dimensions of force per unit mass. The force acts vertically downward, whereas the positive direction of Z is vertically upward; hence the negative sign.

‡ See, for example, ref. 4.

τ_{zy} = tangential stress parallel to z plane and in direction of y axis
p_z = normal stress in direction of z axis
τ_{zx} = tangential stress parallel to z plane and in direction of x axis
τ_{xz} = tangential stress parallel to x plane and in direction of z axis

The first subscript on the shear-stress component refers to the plane parallel to the stress, and the second subscript gives the direction in which the stress acts.

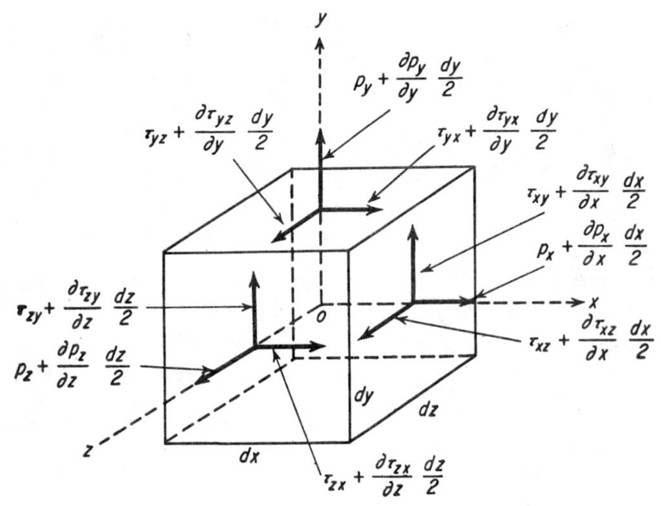

FIG. 2-2. External forces acting on the three positive faces of a small cubical element in space.

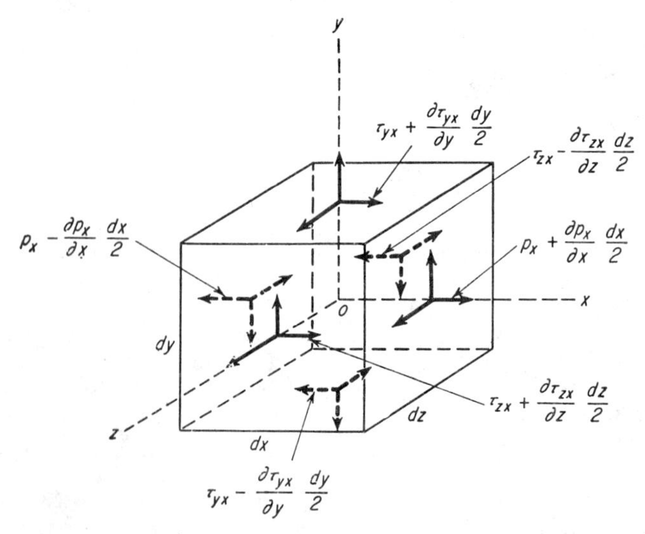

FIG. 2-3. External forces acting in the x direction on a small cubical element in space

Figure 2-2 shows all the stresses exerted on three positive faces of a cubical element in space. In Fig. 2-3 all the stresses exerted on the element are shown, but only those which act in the x direction are labeled.

Only three of the six shear stresses above are independent. This fact may be demonstrated by reference to Fig. 2-4, which shows the cross section of the fluid element of Fig. 2-3 at the plane $z = 0$.

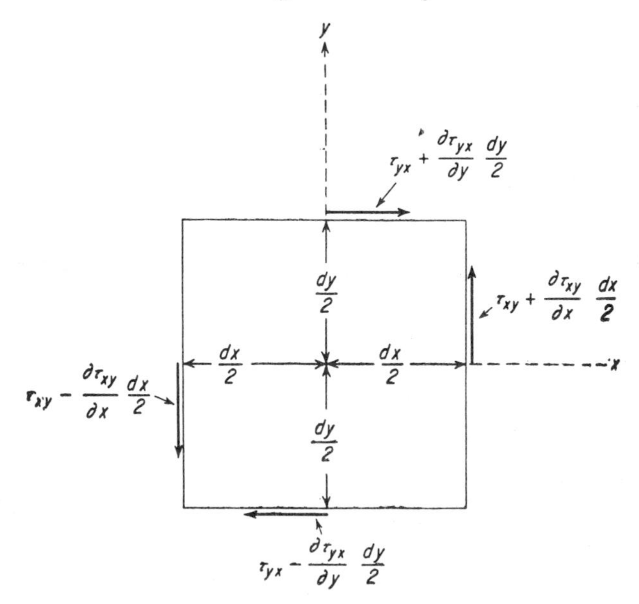

FIG. 2-4. External forces which have a moment about the z axis.

The algebraic sum of the moments of forces about the z axis equals the product of the mass, the square of the radius of gyration, and the angular acceleration. Since the normal stresses and gravity forces act through the center of the element, only the shear stresses have a moment about the z axis. Thus, considering the counterclockwise direction to be positive,

$$\left(\tau_{xy} + \frac{\partial \tau_{xy}}{\partial x}\frac{dx}{2} + \tau_{xy} - \frac{\partial \tau_{xy}}{\partial x}\frac{dx}{2}\right) dy\, dz\, \frac{dx}{2} - \left(\tau_{yx} + \frac{\partial \tau_{yx}}{\partial y}\frac{dy}{2} + \tau_{yx} - \frac{\partial \tau_{yx}}{\partial y}\frac{dy}{2}\right) dz\, dx\, \frac{dy}{2}$$
$$= \rho\, dx\, dy\, dz\, (\text{radius of gyration})^2 (\text{angular acceleration}) \qquad (2\text{-}22)$$

Therefore $\tau_{xy} - \tau_{yx} = \rho(\text{radius of gyration})^2(\text{angular acceleration})$ (2-23)

As the size of the element approaches zero, the right side of Eq. (2-23) becomes zero if the angular acceleration is finite. Thus

$$\tau_{xy} = \tau_{yx} \qquad (2\text{-}24)$$

Similarly, it may be shown by taking moments about the x and y axis respectively that

$$\tau_{yz} = \tau_{zy} \tag{2-25}$$

$$\tau_{zx} = \tau_{xz} \tag{2-26}$$

The summation of all the normal and tangential forces acting in the x direction gives

$$\begin{aligned} \mathrm{SF}_x = &\left(p_x + \frac{\partial p_x}{\partial x}\frac{dx}{2} - p_x + \frac{\partial p_x}{\partial x}\frac{dx}{2}\right) dy\, dz \text{ (on } x \text{ plane)} \\ &+ \left(\tau_{yx} + \frac{\partial \tau_{yx}}{\partial y}\frac{dy}{2} - \tau_{yx} + \frac{\partial \tau_{yx}}{\partial y}\frac{dy}{2}\right) dx\, dz \text{ (on } y \text{ plane)} \\ &+ \left(\tau_{zx} + \frac{\partial \tau_{zx}}{\partial z}\frac{dz}{2} - \tau_{zx} + \frac{\partial \tau_{zx}}{\partial z}\frac{dz}{2}\right) dy\, dx \text{ (on } z \text{ plane)} \end{aligned} \tag{2-27}$$

On addition Eq. (2-27) becomes

$$\mathrm{SF}_x = \left(\frac{\partial p_x}{\partial x} + \frac{\partial \tau_{yx}}{\partial y} + \frac{\partial \tau_{zx}}{\partial z}\right) dx\, dy\, dz \tag{2-28}$$

Application of Newton's Second Law. The application of Newton's second law requires that the inertial force of the element of fluid be equal to the external forces. Thus

$$\mathrm{IF}_x = \mathrm{FF}_x + \mathrm{SF}_x \tag{2-29}$$

Combining Eqs. (2-20), (2-21), and (2-28),

$$\begin{aligned} &\frac{\rho}{g_c}\left(u\frac{\partial u}{\partial x} + v\frac{\partial u}{\partial y} + w\frac{\partial u}{\partial z} + \frac{\partial u}{\partial t}\right) dx\, dy\, dz \\ &\qquad = -\rho\frac{\partial \Omega}{\partial x} dx\, dy\, dz + \left(\frac{\partial p_x}{\partial x} + \frac{\partial \tau_{yx}}{\partial y} + \frac{\partial \tau_{zx}}{\partial z}\right) dx\, dy\, dz \end{aligned} \tag{2-30}$$

Equation (2-30) is the mathematical expression of Newton's second law of motion for the forces exerted on fluid moving through a cubical element in space. Two similar equations may be derived for the y and z directions.

Relation between Shear Stress and Viscosity. As pointed out in Chap. 1, the viscosity of a fluid is that property which offers resistance to shear, and for Newtonian fluids the intensity of shear stress is a linear function of the time rate of angular deformation. This linear function is used in Eq. (2-30) to relate the shear stress to viscosity.

It is evident that the two-dimensional element shown in Fig. 2-5 is undergoing angular deformation. The velocities at three corners of the element are shown and are such that the element tends to assume the

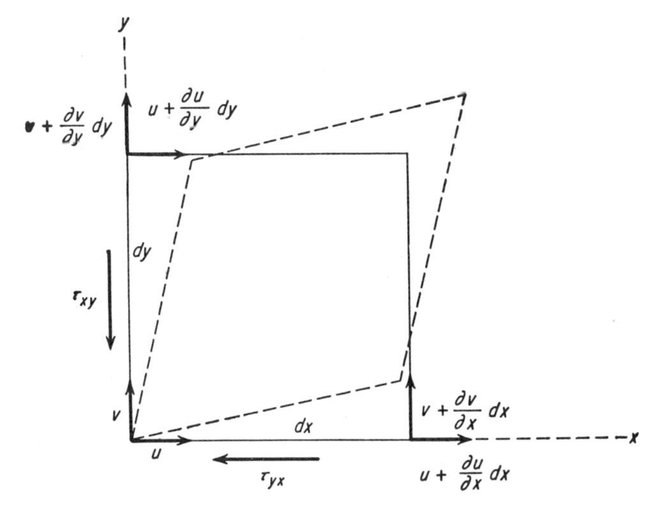

FIG. 2-5. Fluid element in two dimensions showing velocities causing angular deformation.

shape indicated by the broken lines. The angular velocity of the linear element dx is

$$\frac{v + (\partial v/\partial x)\,dx - v}{dx} = \frac{\partial v}{\partial x}$$

The angular velocity of the linear element dy is $-(\partial u/\partial y)$. The counterclockwise direction is considered positive. The net rate of angular deformation of the element is the difference of the angular velocities of the elements dx and dy.

$$\text{Rate of angular deformation} = \frac{\partial v}{\partial x} - \left(-\frac{\partial u}{\partial y}\right) = \frac{\partial v}{\partial x} + \frac{\partial u}{\partial y} \tag{2-31}$$

The relation between intensity of shear stress and viscosity is

$$\begin{aligned}
\tau_{xy} &= \tau_{yx} = \frac{\mu}{g_c}\left(\frac{\partial v}{\partial x} + \frac{\partial u}{\partial y}\right) \\
\tau_{yz} &= \tau_{zy} = \frac{\mu}{g_c}\left(\frac{\partial w}{\partial y} + \frac{\partial v}{\partial z}\right) \\
\tau_{zx} &= \tau_{xz} = \frac{\mu}{g_c}\left(\frac{\partial u}{\partial z} + \frac{\partial w}{\partial x}\right)
\end{aligned} \tag{2-32}$$

The normal stresses p_x, p_y, and p_z may be related to the viscosity in a similar manner. The normal stress is proportional both to the rate of linear deformation in the direction in which the normal stress acts and to the rate of volume deformation of the element. For any fluid at rest or in uniform motion the normal stresses are numerically equal to the static pressure $-P$. The sign is negative since the static pressure of the fluid is opposite to the direction of the normal stresses exerted on it. For arbitrary motion of the fluid the normal stresses differ from $-P$ by an amount dependent on the rate of linear and volume deformation of the fluid. Thus

$$\begin{aligned} p_x &= -P + \frac{2\mu}{g_c}\frac{\partial u}{\partial x} + \frac{\lambda}{g_c}\left(\frac{\partial u}{\partial x} + \frac{\partial v}{\partial y} + \frac{\partial w}{\partial z}\right) \\ p_y &= -P + \frac{2\mu}{g_c}\frac{\partial v}{\partial y} + \frac{\lambda}{g_c}\left(\frac{\partial u}{\partial x} + \frac{\partial v}{\partial y} + \frac{\partial w}{\partial z}\right) \\ p_z &= -P + \frac{2\mu}{g_c}\frac{\partial w}{\partial z} + \frac{\lambda}{g_c}\left(\frac{\partial u}{\partial x} + \frac{\partial v}{\partial y} + \frac{\partial w}{\partial z}\right) \end{aligned} \tag{2-33}$$

where $\partial u/\partial x$, $\partial v/\partial y$, and $\partial w/\partial z$ are the respective rates of linear deformation in the x, y, and z directions. It is convenient to include the factor 2 in the second terms on the right-hand side of the equations. The last term of Eq. (2-33) contains the rate of volume deformation $\partial u/\partial x + \partial v/\partial y + \partial w/\partial z$. The proportionality constant λ, which relates the normal stress to the rate of volume deformation, is of the nature of a bulk modulus. Adding Eq. (2-33),

$$p_x + p_y + p_z = -3P + \left(\frac{2\mu}{g_c} + \frac{3\lambda}{g_c}\right)\left(\frac{\partial u}{\partial x} + \frac{\partial v}{\partial y} + \frac{\partial w}{\partial z}\right) \tag{2-34}$$

From Eqs. (I-5)(Appendix I) and (2-7)

$$\frac{\partial u}{\partial x} + \frac{\partial v}{\partial y} + \frac{\partial w}{\partial z} = -\frac{1}{\rho}\frac{D\rho}{Dt}$$

Thus

$$p_x + p_y + p_z = -3P - \left(\frac{2\mu}{g_c} + \frac{3\lambda}{g_c}\right)\frac{1}{\rho}\frac{D\rho}{Dt} \tag{2-35}$$

Equation (2-35) states that the average normal stress is different from the static pressure by an amount proportional to the total derivative of the density. It is assumed [2] that the pressure is a function only of the density and not of the rate of change of the density. This leads to the requirement that the coefficient of the last term of Eq. (2-35) must be zero. Thus

$$\frac{2\mu}{g_c} + \frac{3\lambda}{g_c} = 0 \tag{2-36}$$

from which

$$\lambda = -\tfrac{2}{3}\mu \tag{2-37}$$

The shear stresses and normal stresses are now expressed in terms of the viscosity of the fluid. Combining Eqs. (2-30), (2-32), (2-33), and (2-37) gives

$$\frac{\rho}{g_c}\left(u\frac{\partial u}{\partial x} + v\frac{\partial u}{\partial y} + w\frac{\partial u}{\partial z} + \frac{\partial u}{\partial t}\right) dx\,dy\,dz = -\rho\frac{\partial \Omega}{\partial x}\,dx\,dy\,dz$$

$$+ \left\{\frac{\partial}{\partial x}\left[-P + \frac{2\mu}{g_c}\frac{\partial u}{\partial x} - \frac{2\mu}{3g_c}\left(\frac{\partial u}{\partial x} + \frac{\partial v}{\partial y} + \frac{\partial w}{\partial z}\right)\right]\right.$$

$$\left. + \frac{\partial}{\partial y}\left[\frac{\mu}{g_c}\left(\frac{\partial v}{\partial x} + \frac{\partial u}{\partial y}\right)\right] + \frac{\partial}{\partial z}\left[\frac{\mu}{g_c}\left(\frac{\partial u}{\partial z} + \frac{\partial w}{\partial x}\right)\right]\right\} dx\,dy\,dz \qquad (2\text{-}38)$$

Equation (2-38) may be reduced to

$$u\frac{\partial u}{\partial x} + v\frac{\partial u}{\partial y} + w\frac{\partial u}{\partial z} + \frac{\partial u}{\partial t} = -g_c\frac{\partial \Omega}{\partial x} - \frac{g_c}{\rho}\frac{\partial P}{\partial x}$$

$$-\frac{1}{\rho}\left\{\frac{2}{3}\frac{\partial}{\partial x}\left[\mu\left(\frac{\partial u}{\partial x} + \frac{\partial v}{\partial y} + \frac{\partial w}{\partial z}\right)\right]\right\} + \frac{1}{\rho}\left[\frac{\partial}{\partial x}\left(\mu\frac{\partial u}{\partial x}\right) + \frac{\partial}{\partial y}\left(\mu\frac{\partial u}{\partial y}\right)\right.$$

$$\left. + \frac{\partial}{\partial z}\left(\mu\frac{\partial u}{\partial z}\right) + \frac{\partial}{\partial x}\left(\mu\frac{\partial u}{\partial x}\right) + \frac{\partial}{\partial y}\left(\mu\frac{\partial v}{\partial x}\right) + \frac{\partial}{\partial z}\left(\mu\frac{\partial w}{\partial x}\right)\right] \qquad (2\text{-}39)$$

Written in vector notation, Eq. (2-39) becomes

$$u\frac{\partial u}{\partial x} + v\frac{\partial u}{\partial y} + w\frac{\partial u}{\partial z} + \frac{\partial u}{\partial t} = -g_c\frac{\partial \Omega}{\partial x} - \frac{g_c}{\rho}\frac{\partial P}{\partial x} - \frac{2}{3\rho}\frac{\partial}{\partial x}(\mu \operatorname{div} \mathbf{V})$$

$$+ \frac{1}{\rho}\left[\operatorname{div}(\mu \operatorname{grad} u) + \operatorname{div}\left(\mu\frac{\partial \mathbf{V}}{\partial x}\right)\right] \qquad (2\text{-}40)$$

Equation (2-40) is the momentum equation for viscous flow for the x-coordinate direction. Two similar equations may be derived for the y and z directions. The momentum equations represent the mathematical expression resulting from the application of Newton's second law of motion to the arbitrary flow of a fluid, and they involve the relation of the shear and normal stresses in the fluid to its viscosity.

Since Eq. (2-40) and the corresponding equations for the y and z directions are much too complicated to be solved analytically, they must be simplified. Any problem in fluid flow which involves the determination of fluid velocity as a function of space and time requires the solution of the momentum and continuity equations. In many problems it may be more convenient to use coordinate systems other than the rectangular system; e.g., in flow through circular tubes, cylindrical coordinates are most convenient, and for flow past spheres, spherical coordinates may be used

TABLE 2-2. MOMENTUM EQUATIONS FOR VARIOUS CONDITIONS OF FLUID FLOW

1. General equations (for Newtonian fluids): x direction	$\frac{Du}{Dt} = -g_c \frac{\partial \Omega}{\partial x} - \frac{g_c}{\rho}\frac{\partial P}{\partial x} - \frac{1}{\rho}\left[\frac{2}{3}\frac{\partial}{\partial x}(\mu \operatorname{div} \mathbf{V}) - \operatorname{div}(\mu \operatorname{grad} u) - \operatorname{div}\left(\mu \frac{\partial \mathbf{V}}{\partial x}\right)\right]$
y direction	$\frac{Dv}{Dt} = -g_c \frac{\partial \Omega}{\partial y} - \frac{g_c}{\rho}\frac{\partial P}{\partial y} - \frac{1}{\rho}\left[\frac{2}{3}\frac{\partial}{\partial y}(\mu \operatorname{div} \mathbf{V}) - \operatorname{div}(\mu \operatorname{grad} v) - \operatorname{div}\left(\mu \frac{\partial \mathbf{V}}{\partial y}\right)\right]$
z direction	$\frac{Dw}{Dt} = -g_c \frac{\partial \Omega}{\partial z} - \frac{g_c}{\rho}\frac{\partial P}{\partial z} - \frac{1}{\rho}\left[\frac{2}{3}\frac{\partial}{\partial z}(\mu \operatorname{div} \mathbf{V}) - \operatorname{div}(\mu \operatorname{grad} w) - \operatorname{div}\left(\mu \frac{\partial \mathbf{V}}{\partial z}\right)\right]$
2. Viscosity constant: x direction	$\frac{Du}{Dt} = -g_c \frac{\partial \Omega}{\partial x} - \frac{g_c}{\rho}\frac{\partial P}{\partial x} + \nu\left(\nabla^2 u + \frac{1}{3}\frac{\partial}{\partial x} \operatorname{div} \mathbf{V}\right)$
y direction	$\frac{Dv}{Dt} = -g_c \frac{\partial \Omega}{\partial y} - \frac{g_c}{\rho}\frac{\partial P}{\partial y} + \nu\left(\nabla^2 v + \frac{1}{3}\frac{\partial}{\partial y} \operatorname{div} \mathbf{V}\right)$
z direction	$\frac{Dw}{Dt} = -g_c \frac{\partial \Omega}{\partial z} - \frac{g_c}{\rho}\frac{\partial P}{\partial z} + \nu\left(\nabla^2 w + \frac{1}{3}\frac{\partial}{\partial z} \operatorname{div} \mathbf{V}\right)$
3. Viscosity constant, density constant (Navier-Stokes equations): x direction	$\frac{Du}{Dt} = -g_c \frac{\partial \Omega}{\partial x} - \frac{g_c}{\rho}\frac{\partial P}{\partial x} + \nu \nabla^2 u$
y direction	$\frac{Dv}{Dt} = -g_c \frac{\partial \Omega}{\partial y} - \frac{g_c}{\rho}\frac{\partial P}{\partial y} + \nu \nabla^2 v$
z direction	$\frac{Dw}{Dt} = -g_c \frac{\partial \Omega}{\partial z} - \frac{g_c}{\rho}\frac{\partial P}{\partial z} + \nu \nabla^2 w$

4. Zero viscosity (Euler equations): x direction	$\frac{Du}{Dt} = -g_c \frac{\partial \Omega}{\partial x} - \frac{g_c}{\rho} \frac{\partial P}{\partial x}$
y direction	$\frac{Dv}{Dt} = -g_c \frac{\partial \Omega}{\partial y} - \frac{g_c}{\rho} \frac{\partial P}{\partial y}$
z direction	$\frac{Dw}{Dt} = -g_c \frac{\partial \Omega}{\partial z} - \frac{g_c}{\rho} \frac{\partial P}{\partial z}$
5. Steady state	All derivatives with respect to time become zero
6. Two-dimensional flow in x and y directions	$w = 0$, and all derivatives with respect to z become zero
7. One-dimensional flow in x direction	$w = 0$, $v = 0$, and all derivatives with respect to y and z become zero
8. Steady laminar flow in a uniform horizontal circular tube with axis parallel to x axis	$v = 0$, $w = 0$ (cylindrical coordinates) $0 = -\frac{g_c}{\rho} \frac{\partial P}{\partial x} + \nu \left(\frac{1}{r} \frac{\partial u}{\partial r} + \frac{\partial^2 u}{\partial r^2} \right)$

to best advantage. In subsequent chapters various types of fluid flow are discussed, and in most cases the momentum and continuity equations are presented. In Chap. 3 flow of nonviscous fluids, wherein the viscosity may be assumed to be zero, is studied. In Chap. 4 and following chapters flow of viscous fluids in closed conduits and past immersed bodies is considered.

In many cases the nature of the flow is such that the equations may be considerably simplified. Table 2-2 gives the momentum equations for a variety of conditions of flow. When flow is one-dimensional in the x direction, various terms in Eq. (2-40) become zero, and the two momentum equations for the y and z directions are eliminated. For two-dimensional flow, two momentum equations are required, and those terms involving the coordinate direction in which no flow occurs become zero.

2-5. The Energy Equation for Three-dimensional Flow

When flow is nonisothermal, the temperature of the fluid is a dependent variable which is a function of x, y, z, and t. Just as the continuity equation is a mathematical expression of the law of conservation of mass and gives the velocity distribution in space, the energy equation is a mathematical expression for the law of conservation of energy and gives the temperature distribution in space. In making an energy balance on a spatial element of dimensions dx, dy, and dz, the following forms of energy must be considered:

1. Internal, or intrinsic, energy of the fluid E. This energy does not include energy of position or energy of motion.
2. Kinetic energy.
3. Pressure-volume energy.
4. Heat transferred by conduction.
5. Energy of position or potential energy.
6. Energy dissipated in the fluid by viscous action.
7. Energy generated by chemical reaction or electrical current.

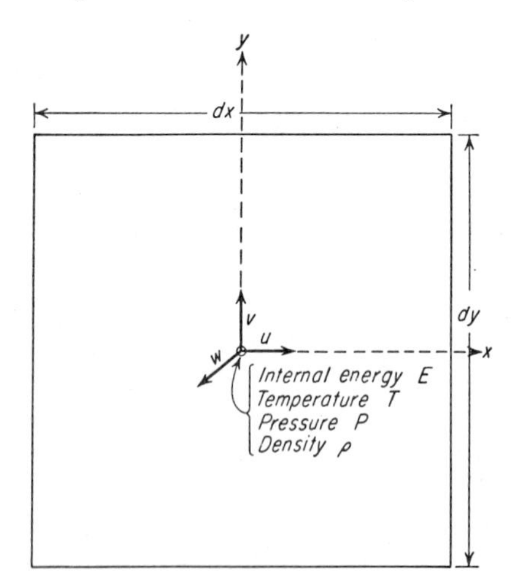

FIG. 2-6. Variables required in the derivation of the energy equation. (All quantities are considered to be a function of x, y, z, and t.)

In the following derivation of the energy equation external forces and potential energy are neglected. In Fig. 2-6 is shown the cross section (at $z = 0$) of a cubical element. At point (0,0) the various quantities required in making an energy balance are shown. The energy input to the element per unit time is

$$\Bigg(\overbrace{\left[u\rho E - \frac{\partial(u\rho E)}{\partial x}\frac{dx}{2}\right]}^{\text{internal energy}} + \overbrace{\frac{1}{2g_c}\left\{\rho u(u^2+v^2+w^2) - \frac{\partial}{\partial x}[\rho u(u^2+v^2+w^2)]\frac{dx}{2}\right\}}^{\text{kinetic energy}}$$

$$+ \overbrace{\left[Pu - \frac{\partial(Pu)}{\partial x}\frac{dx}{2}\right]}^{\text{pressure volume work}} - \overbrace{k\left[\frac{\partial T}{\partial x} - \frac{\partial^2 T}{\partial x^2}\frac{dx}{2}\right]}^{\text{heat conduction}}\Bigg) dy\, dz$$

$$+ \text{ similar terms for } y \text{ and } z \text{ directions} + q'\, dx\, dy\, dz + \Phi\, dx\, dy\, dz$$

where k = thermal conductivity of fluid
q' = time rate of energy generation (from chemical reaction) in fluid per unit volume
Φ = dissipation function

The dissipation function is defined as the time rate of energy dissipated per unit volume because of the action of viscosity. The energy output per unit time is

$$\Bigg(\left[u\rho E + \frac{\partial(u\rho E)}{\partial x}\frac{dx}{2}\right] + \frac{1}{2g_c}\left\{\rho u(u^2+v^2+w^2) + \frac{\partial}{\partial x}[\rho u(u^2+v^2+w^2)]\frac{dx}{2}\right\}$$

$$+ \left[Pu + \frac{\partial(Pu)}{\partial x}\frac{dx}{2}\right] - k\left[\frac{\partial T}{\partial x} + \frac{\partial^2 T}{\partial x^2}\frac{dx}{2}\right]\Bigg) dy\, dz$$

$$+ \text{ similar terms for the } y \text{ and } z \text{ directions}$$

The energy accumulated in the element consists of two parts, internal energy and kinetic energy, i.e.,

$$\text{Accumulation} = \left\{\frac{\partial(\rho E)}{\partial t} + \frac{1}{2g_c}\frac{\partial}{\partial t}[\rho(u^2+v^2+w^2)]\right\} dx\, dy\, dz$$

Making an energy balance according to Eq. (2-1),

$$k\left(\frac{\partial^2 T}{\partial x^2} + \frac{\partial^2 T}{\partial y^2} + \frac{\partial^2 T}{\partial z^2}\right) + q' + \Phi = \frac{\partial(Pu)}{\partial x} + \frac{\partial(Pv)}{\partial y} + \frac{\partial(Pw)}{\partial z} + \frac{\partial(\rho uE)}{\partial x}$$

$$+ \frac{\partial(\rho vE)}{\partial y} + \frac{\partial(\rho wE)}{\partial z} + \frac{1}{2g_c}\left\{\frac{\partial}{\partial x}[\rho u(u^2+v^2+w^2)] + \frac{\partial}{\partial y}[\rho v(u^2+v^2+w^2)]\right.$$

$$\left. + \frac{\partial}{\partial z}[\rho w(u^2+v^2+w^2)]\right\} + \frac{\partial(\rho E)}{\partial t} + \frac{1}{2g_c}\frac{\partial}{\partial t}[\rho(u^2+v^2+w^2)] \quad (2\text{-}41)$$

Differentiating the terms on the right side of Eq. (2-41) and making use of the continuity equation (2-6) results in †

$$k\left(\frac{\partial^2 T}{\partial x^2}+\frac{\partial^2 T}{\partial y^2}+\frac{\partial^2 T}{\partial z^2}\right)+q'+\Phi$$
$$=\frac{\partial(Pu)}{\partial x}+\frac{\partial(Pv)}{\partial y}+\frac{\partial(Pw)}{\partial z}+\frac{\rho}{2g_c}\frac{D}{Dt}(u^2+v^2+w^2)+\rho\frac{DE}{Dt} \qquad (2\text{-}42)$$

Equation (2-42) may be further simplified by considering the Euler equations shown in item 4 of Table 2-2. Multiplying the first equation by u, the second equation by v, and the third equation by w and adding the three equations gives (neglecting field forces)

$$\frac{1}{2}\frac{D}{Dt}(u^2+v^2+w^2)=-\frac{g_c}{\rho}\left(u\frac{\partial P}{\partial x}+v\frac{\partial P}{\partial y}+w\frac{\partial P}{\partial z}\right) \qquad (2\text{-}43)$$

but

$$\frac{\partial(Pu)}{\partial x}+\frac{\partial(Pv)}{\partial y}+\frac{\partial(Pw)}{\partial z}=P\left(\frac{\partial u}{\partial x}+\frac{\partial v}{\partial y}+\frac{\partial w}{\partial z}\right)+u\frac{\partial P}{\partial x}+v\frac{\partial P}{\partial y}+w\frac{\partial P}{\partial z} \qquad (2\text{-}44)$$

Substituting Eq. (2-43) into (2-44) and putting the result into Eq. (2-42) gives

$$\rho\frac{DE}{Dt}=k\left(\frac{\partial^2 T}{\partial x^2}+\frac{\partial^2 T}{\partial y^2}+\frac{\partial^2 T}{\partial z^2}\right)-P\left(\frac{\partial u}{\partial x}+\frac{\partial v}{\partial y}+\frac{\partial w}{\partial z}\right)+q'+\Phi \qquad (2\text{-}45)$$

From Eq. (1-4)

$$\frac{DE}{Dt}=C_v\frac{DT}{Dt} \qquad (2\text{-}46)$$

where C_v is the mean heat capacity at constant volume. Thus Eq. (2-45) becomes

$$\rho C_v\frac{DT}{Dt}=k\,\nabla^2 T-P\,\text{div}\,\mathbf{V}+q'+\Phi \qquad (2\text{-}47)$$

The solution of Eq. (2-47) gives the temperature distribution in the flowing fluid. Like the momentum equations, Eq. (2-47) is too complicated to solve analytically and consequently is greatly simplified in most flow problems.

† NOTE: The substantial derivative of u^2, $D(u^2)/Dt$ is

$$\frac{\partial(u^2)}{\partial t}+u\frac{\partial(u^2)}{\partial x}+v\frac{\partial(u^2)}{\partial y}+w\frac{\partial(u^2)}{\partial z}$$

For incompressible liquids Eq. (2-47) becomes, since div **V** = 0,

$$\rho C_v \frac{DT}{Dt} = k\,\nabla^2 T + q' + \Phi \tag{2-48}$$

The dissipation function Φ is of the following form:[1-3]

$$\Phi g_c = -\tfrac{2}{3}\mu\left(\frac{\partial u}{\partial x} + \frac{\partial v}{\partial y} + \frac{\partial w}{\partial z}\right)^2 + 2\mu\left[\left(\frac{\partial u}{\partial x}\right)^2 + \left(\frac{\partial v}{\partial y}\right)^2 + \left(\frac{\partial w}{\partial z}\right)^2\right]$$
$$+ \mu\left[\left(\frac{\partial w}{\partial y} + \frac{\partial v}{\partial z}\right)^2 + \left(\frac{\partial u}{\partial z} + \frac{\partial w}{\partial x}\right)^2 + \left(\frac{\partial v}{\partial x} + \frac{\partial u}{\partial y}\right)^2\right] \tag{2-49}$$

It is seen to be a function of the fluid viscosity and the linear deformation of the fluid. It has a value of zero for nonviscous fluids. For fluids of low viscosity and for velocities less than the sonic velocity, Φ has a value which is negligible compared to other terms in the equation. For velocities above that of sound, Φ is significant. It must also be considered in lubrication problems involving high-viscosity fluids.

Another form of Eq. (2-47) may be obtained by utilizing the relationship between C_v and C_p; i.e., for perfect gases

$$C_p - C_v = \frac{P}{\rho T} \tag{2-50}$$

Thus

$$C_v T = C_p T - \frac{P}{\rho} \tag{2-51}$$

and

$$C_v \frac{DT}{Dt} = C_p \frac{DT}{Dt} - \frac{D}{Dt}\left(\frac{P}{\rho}\right) \tag{2-52}$$

which becomes

$$C_v \frac{DT}{Dt} = C_p \frac{DT}{Dt} - \frac{1}{\rho}\frac{DP}{Dt} + \frac{P}{\rho^2}\frac{D\rho}{Dt} \tag{2-53}$$

From Eq. (2-5)

$$\frac{D\rho}{Dt} = -\rho\left(\frac{\partial u}{\partial x} + \frac{\partial v}{\partial y} + \frac{\partial w}{\partial z}\right) \tag{2-54}$$

Substituting Eqs. (2-53) and (2-54) into Eq. (2-47),

$$\rho C_p \frac{DT}{Dt} = k\,\nabla^2 T + \frac{DP}{Dt} + q' + \Phi \tag{2-55}$$

Equation (2-55) is the energy equation for perfect gases, and its solution will give the temperature distribution as a function of x, y, z, and t.

2-6. The Energy Equation for Steady Two-dimensional Flow

For flow past immersed bodies the energy equation for two-dimensional flow is used to determine the temperature distribution in the flowing fluid. The two-dimensional equation is approximately applicable for flow past bodies of revolution, such as spheres, where axial symmetry prevails in the direction of flow. For more precise results, however, the three-dimensional equation should be used and converted to a form employing spherical coordinates. For steady two-dimensional flow in the x and y directions Eq. (2-55) becomes

$$\rho C_p \left(u \frac{\partial T}{\partial x} + v \frac{\partial T}{\partial y} \right) = k \left(\frac{\partial^2 T}{\partial x^2} + \frac{\partial^2 T}{\partial y^2} \right) + u \frac{\partial P}{\partial x} + v \frac{\partial P}{\partial y} + q' + \Phi \quad (2\text{-}56)$$

Goldstein [1] points out that the term $v(\partial P/\partial y)$ is negligible and Φ can be approximated by the term $\mu(\partial u/\partial y)^2/g_c$. Thus, neglecting energy generation q', Eq. (2-56) becomes

$$\rho C_p \left(u \frac{\partial T}{\partial x} + v \frac{\partial T}{\partial y} \right) = k \left(\frac{\partial^2 T}{\partial x^2} + \frac{\partial^2 T}{\partial y^2} \right) + u \frac{\partial P}{\partial x} + \frac{\mu}{g_c} \left(\frac{\partial u}{\partial y} \right)^2 \quad (2\text{-}57)$$

Since the last two terms of Eq. (2-57) are usually negligible except above the sonic velocity, for low-speed flows (subsonic) Eq. (2-57) becomes

$$\rho C_p \left(u \frac{\partial T}{\partial x} + v \frac{\partial T}{\partial y} \right) = k \left(\frac{\partial^2 T}{\partial x^2} + \frac{\partial^2 T}{\partial y^2} \right) \quad (2\text{-}58)$$

Equation (2-58) is the common form of the two-dimensional energy equation and applies for the following conditions:

(1) Perfect gases (for liquids replace C_p by C_v)
(2) Steady flow
(3) $q' = 0$
(4) Φ negligible
(5) $u(\partial P/\partial x)$ negligible.

2-7. The Energy Equation for Steady Flow in Circular Tubes

For steady, laminar flow in circular tubes Eq. (2-55) becomes (neglecting the last three terms)

$$\rho C_p \left(u \frac{\partial T}{\partial x} \right) = k \left(\frac{\partial^2 T}{\partial x^2} + \frac{\partial^2 T}{\partial y^2} + \frac{\partial^2 T}{\partial z^2} \right) \quad (2\text{-}59)$$

This equation is most conveniently handled by employing cylindrical coordinates rather than rectangular coordinates (see Fig. 4-8).

$$\rho C_p \left(u \frac{\partial T}{\partial x} \right) = k \left[\frac{\partial^2 T}{\partial x^2} + \frac{1}{r} \frac{\partial}{\partial r} \left(r \frac{\partial T}{\partial r} \right) + \frac{1}{r^2} \frac{\partial^2 T}{\partial \theta^2} \right] \tag{2-60}$$

With symmetry about the axis of the tube the last term of Eq. (2-60) becomes zero.

BIBLIOGRAPHY

1. Goldstein, S. (ed.): "Modern Developments in Fluid Dynamics," Oxford University Press, London, 1938.
2. Kuethe, A. M., and J. D. Shetzer: "Foundations of Aerodynamics," John Wiley & Sons, Inc., New York, 1950.
3. Lamb, H.: "Hydrodynamics," 6th ed., Cambridge University Press, London, 1932.
4. Streeter, V. L.: "Fluid Dynamics," 2d ed., McGraw-Hill Book Company, Inc., New York, 1958.

CHAPTER 3

FLOW OF NONVISCOUS FLUIDS

3-1. Flow of Fluids with No Viscosity

This chapter deals with the flow of fluids having zero viscosity. To the engineer accustomed to calculating friction losses and familiar with velocity profiles and boundary layers, the value of relationships applying to nonviscous flow may be questionable, but in a large number of fluid-flow problems the effect of viscosity is insignificant compared to other quantities, such as pressure, inertia force, and field force, with the result that it may be neglected. This is particularly true of fluids having low viscosity, such as water and the common gases.

The solution of any fluid-flow problem involves the determination of the fluid velocity and fluid pressure as a function of time and space coordinates. This solution can be greatly simplified if the fluid viscosity is assumed to be zero. Typical problems where such an assumption is valid are in flow over weirs, through orifices, in large tanks, in channel and duct entrances, and in converging and diverging nozzles. In such problems the movement of the main mass of fluid is of particular interest, and conditions immediately adjacent to the solid boundary, where viscosity has a significant effect, are not of interest.

In the following sections the classical approach to nonviscous flow is presented. A number of examples are included to illustrate the application of the various relationships considered.

3-2. The Equations of Motion for Nonviscous Fluids (Euler Equations)

The *Euler equations* of motion for three-dimensional nonviscous flow are given in item 4 of Table 2-2. All the viscosity terms in the general equations of motion have disappeared, leaving only the inertia terms and the terms corresponding to field forces and normal forces.

$$\frac{\partial u}{\partial t} + u\frac{\partial u}{\partial x} + v\frac{\partial u}{\partial y} + w\frac{\partial u}{\partial z} = -g_c\frac{\partial \Omega}{\partial x} - \frac{g_c}{\rho}\frac{\partial P}{\partial x} \tag{3-1}$$

$$\frac{\partial v}{\partial t} + u\frac{\partial v}{\partial x} + v\frac{\partial v}{\partial y} + w\frac{\partial v}{\partial z} = -g_c\frac{\partial \Omega}{\partial y} - \frac{g_c}{\rho}\frac{\partial P}{\partial y} \tag{3-2}$$

$$\frac{\partial w}{\partial t} + u\frac{\partial w}{\partial x} + v\frac{\partial w}{\partial y} + w\frac{\partial w}{\partial z} = -g_c\frac{\partial \Omega}{\partial z} - \frac{g_c}{\rho}\frac{\partial P}{\partial z} \tag{3-3}$$

In addition the equation of continuity must be satisfied.

$$\frac{\partial(\rho u)}{\partial x} + \frac{\partial(\rho v)}{\partial y} + \frac{\partial(\rho w)}{\partial z} = -\frac{\partial \rho}{\partial t} \tag{2-6}$$

The number of unknowns in the above equations is five: u, v, w, P, ρ. Since it is usually possible to express the density ρ in terms of the pressure by means of an equation of state, the above four equations are sufficient to solve for all unknowns. However, in the solution of the equations, constants of integration appear which must be evaluated from the boundary conditions in any specific problem.

3-3. Velocity Potential and Irrotational Flow

In integrating Eqs. (3-1) to (3-3) it is convenient to introduce a new function ϕ, which is the *velocity potential*. The velocity potential in the velocity field is analogous to, but not the same as, the force potential Ω in the force field. Just as the derivative of Ω in any direction gives the force acting on the fluid in that direction, the derivative of ϕ in any direction gives the velocity in that direction. Therefore

$$\frac{\partial \phi}{\partial x} = u \tag{3-4}$$

$$\frac{\partial \phi}{\partial y} = v \tag{3-5}$$

$$\frac{\partial \phi}{\partial z} = w \tag{3-6}$$

Substituting Eqs. (3-4) to (3-6) into Eq. (2-6) gives

$$-\rho\left(\frac{\partial^2 \phi}{\partial x^2} + \frac{\partial^2 \phi}{\partial y^2} + \frac{\partial^2 \phi}{\partial z^2}\right) = u\frac{\partial \rho}{\partial x} + v\frac{\partial \rho}{\partial y} + w\frac{\partial \rho}{\partial z} + \frac{\partial \rho}{\partial t} = \frac{D\rho}{Dt} \tag{3-7}$$

Thus, for incompressible fluids Eq. (3-7) becomes

$$\nabla^2\phi = \frac{\partial^2 \phi}{\partial x^2} + \frac{\partial^2 \phi}{\partial y^2} + \frac{\partial^2 \phi}{\partial z^2} = 0 \tag{3-8}$$

Differentiating Eq. (3-4) with respect to y and Eq. (3-5) with respect to x gives

$$\frac{\partial^2 \phi}{\partial y\, \partial x} = \frac{\partial u}{\partial y} \tag{3-9}$$

$$\frac{\partial^2 \phi}{\partial x\, \partial y} = \frac{\partial v}{\partial x} \tag{3-10}$$

Subtracting Eq. (3-9) from (3-10),

$$\frac{\partial v}{\partial x} - \frac{\partial u}{\partial y} = 0 \tag{3-11}$$

Similarly it may be shown that

$$\frac{\partial u}{\partial z} - \frac{\partial w}{\partial x} = 0 \tag{3-12}$$

$$\frac{\partial w}{\partial y} - \frac{\partial v}{\partial z} = 0 \tag{3-13}$$

The terms on the left of Eqs. (3-11) to (3-13) are the *vorticity* components. It is shown below [Eq. (3-14)] that the vorticity components are twice the rate of rotation of the fluid element. Physically these relations mean

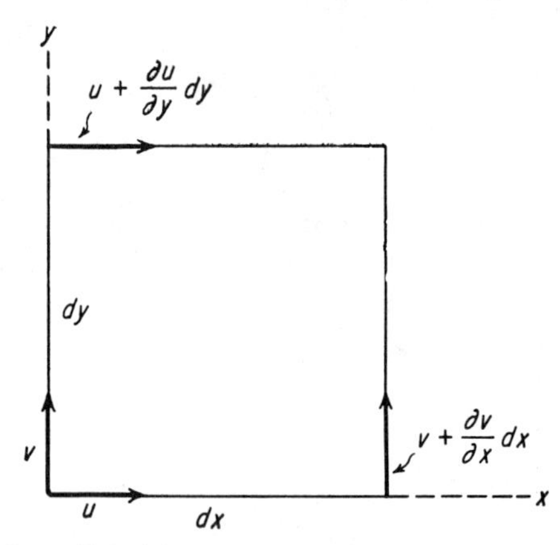

FIG. 3-1. Velocities causing rotation of a fluid element.

that the fluid element has no rotation in space. Therefore the assumption of a velocity potential as defined by Eqs. (3-4) to (3-6) requires that the flow be irrotational. Referring to Fig. 3-1, the rate of rotation of the dx element is

$$\frac{v + (\partial v/\partial x)\, dx - v}{dx} = \frac{\partial v}{\partial x} \qquad \text{radians/sec}$$

The rate of rotation of the dy element is

$$-\frac{u + (\partial u/\partial y)\,dy - u}{dy} = -\frac{\partial u}{\partial y} \qquad \text{radians/sec}$$

The counterclockwise direction is positive.

The net rate of rotation of the two-dimensional element is the average of the sum of the rate of rotation of the dx and dy elements; i.e.,

$$\frac{1}{2}\left(\frac{\partial v}{\partial x} - \frac{\partial u}{\partial y}\right) = \text{rate of rotation} \qquad \text{(3-14)}$$

Therefore Eqs. (3-11) to (3-13) mean that there is zero angular velocity of the fluid elements about their center. Since the fluid has zero viscosity, no tangential or shear stresses may be applied to the fluid elements. Pressure forces act through the center of the elements and can cause no rotation; therefore no torque may be applied to the fluid elements. If a fluid element is initially at rest, it cannot be set in rotation; if it is rotating, the rotation cannot be changed.

The fact that flow is irrotational does not preclude the possibility of deformation of the fluid element. In the flow of a nonviscous fluid between convergent boundaries, the elements of fluid deform as they flow through the channel, but there is no rotation about the axis of the element. The element in Fig. 3-2 is deformed in moving from B to C, but it has not been rotated in space.

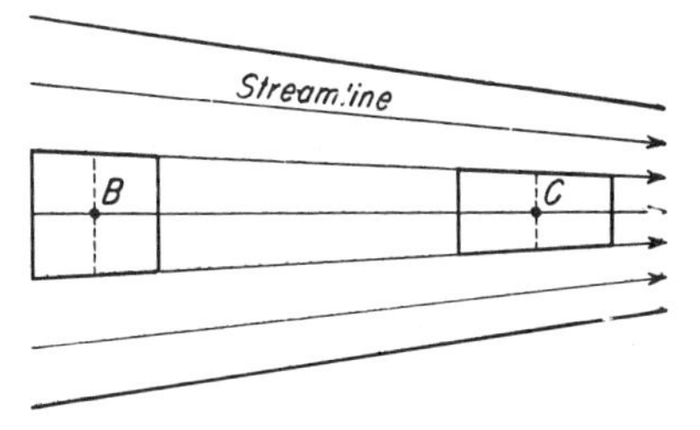

FIG. 3-2. Irrotational flow between convergent boundaries.

3-4. Streamlines

A satisfactory pictorial representation of fluid flow, particularly nonviscous flow, can be obtained by the use of streamlines. At every point in space the motion of the fluid can be indicated by means of a velocity vector showing the direction and magnitude of the velocity. In place of vectors, which are unwieldy, use is made of streamlines, which are lines drawn tangent to the velocity vector at every point in space. For all points on a streamline the velocity vectors meet the streamline tangentially, and therefore no fluid can cross the streamline. In Fig. 3-3 AB and CD are two arbitrary streamlines. Considering the sections ab and cd between them, the input of fluid through ab equals the output through cd plus the accumulation in the section *abcd*. If the fluid is incompressible, the input equals the output.

The components u, v, and w of the velocity $\mathbf{V}$ are the derivatives of distance with respect to time; i.e.,

$$u = \frac{dx}{dt} \qquad v = \frac{dy}{dt} \qquad w = \frac{dz}{dt}$$

Since the same element of time is considered in each case,

$$dt = \frac{dx}{u} = \frac{dy}{v} = \frac{dz}{w} \tag{3-15}$$

Equation (3-15) is the differential equation of the streamline in three-dimensional flow.

3-5. The Stream Function for Two-dimensional Incompressible Flow

For two-dimensional flow, the streamlines can all be represented on a two-dimensional plane. On the basis of the continuity equation and from the nature of the streamlines a *stream function* ψ may be defined which is related to the velocity of the fluid. In Fig. 3-3 let streamlines AB and CD represent the stream functions ψ_1 and ψ_2 respectively. Considering a unit

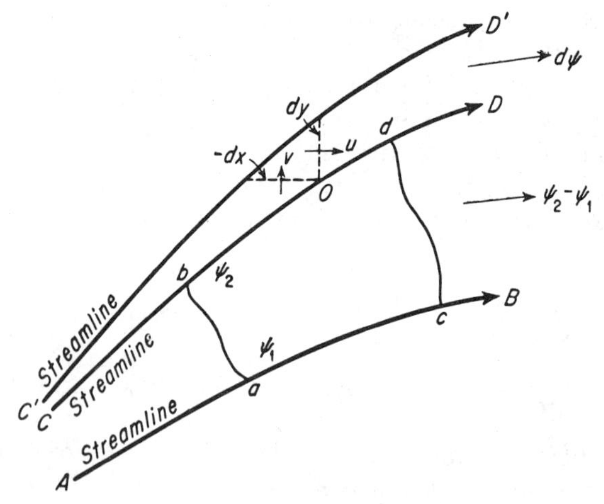

FIG. 3-3. Streamlines and the derivation of the stream function.

thickness of fluid, $\psi_2 - \psi_1$ is defined as the volume rate of fluid flow between the streamlines AB and CD. The streamline $C'D'$ is a differential distance away from CD, and the flow between streamlines CD and $C'D'$ is $d\psi$. At a point O on CD the distance elements $-dx$ and dy indicate the distance between CD and $C'D'$. The velocity of the fluid at O is u and v in the x and y directions respectively. Since no fluid crosses the streamlines, the volume rate of flow across the element dy is $u\,dy$, and the volume rate of flow across the element $-dx$ is $-v\,dx$. This volume rate of flow is $d\psi$ since flow is assumed incompressible. Thus

$$d\psi = u\,dy$$

$$= -v\,dx$$

Using partial derivatives, since ψ is a function of both x and y,

$$\frac{\partial \psi}{\partial y} = u \tag{3-16}$$

$$\frac{\partial \psi}{\partial x} = -v \tag{3-17}$$

Equations (3-16) and (3-17) define the stream function ψ for two-dimensional incompressible flow. Physically, the stream function is the volume rate of flow per unit distance normal to the plane of motion between a streamline in a fluid and an arbitrary base streamline.

The definition of the stream function does not require that flow be irrotational. However, if flow is irrotational, it may be shown that

$$\frac{\partial^2 \psi}{\partial x^2} + \frac{\partial^2 \psi}{\partial y^2} = 0 \tag{3-18}$$

3-6. Integration of Euler's Equations; Bernoulli's Equation

In order to integrate Eqs. (3-1) to (3-3) the velocity-potential equations (3-4) to (3-6) and the equations of irrotational flow (3-11) to (3-13) are used. Substituting these equations into Eq. (3-1) gives

$$\frac{\partial^2 \phi}{\partial t\,\partial x} + u\frac{\partial u}{\partial x} + v\frac{\partial v}{\partial x} + w\frac{\partial w}{\partial x} + g_c\frac{\partial \Omega}{\partial x} + \frac{g_c}{\rho}\frac{\partial P}{\partial x} = 0 \tag{3-19}$$

Equation (3-19) may be integrated with respect to x. The right side becomes a function of y, z, and t. It cannot be a function of x because the derivative with respect to x must be zero. Thus

$$\frac{\partial \phi}{\partial t} + \frac{u^2 + v^2 + w^2}{2} + g_c\Omega + \frac{g_c P}{\rho} = F_1(y,z,t) \tag{3-20}$$

Similarly Eqs. (3-2) and (3-3) become, on integration,

$$\frac{\partial \phi}{\partial t} + \frac{u^2 + v^2 + w^2}{2} + g_c\Omega + \frac{g_c P}{\rho} = F_2(x,z,t) \tag{3-21}$$

$$\frac{\partial \phi}{\partial t} + \frac{u^2 + v^2 + w^2}{2} + g_c\Omega + \frac{g_c P}{\rho} = F_3(x,y,t) \tag{3-22}$$

All the arbitrary functions of Eqs. (3-20) to (3-22) are equal; i.e.,

$$F_1(y,z,t) = F_2(x,z,t) = F_3(x,y,t) \tag{3-23}$$

Since F_1, F_2, and F_3 are each functions of only two distance variables but are all functions of t, the equality of Eq. (3-23) is satisfied if F_1, F_2, and F_3 are functions of t alone. The magnitude of the velocity vector $\mathbf{V}$ may be expressed in terms of its components:

$$|\mathbf{V}| = \sqrt{u^2 + v^2 + w^2} \tag{3-24}$$

Therefore

$$|\mathbf{V}|^2 = u^2 + v^2 + w^2 \tag{3-25}$$

The final integrated form of the Euler equations is

$$\frac{\partial \phi}{\partial t} + \tfrac{1}{2}\,|\mathbf{V}|^2 + g_c\Omega + \frac{g_c P}{\rho} = F_4(t) \tag{3-26}$$

Equation (3-26) is *Bernoulli's equation for unsteady flow.* It considers the rate of change of the velocity potential with respect to time, the magnitude of the velocity of the fluid, the potential energy of the fluid, and the static pressure. If flow is steady, Eq. (3-26) reduces to

$$\tfrac{1}{2}|\mathbf{V}|^2 + g_c\Omega + \frac{g_c P}{\rho} = \text{const} \tag{3-27}$$

Dividing by g_c gives

$$\frac{1}{2}\frac{|\mathbf{V}|^2}{g_c} + \Omega + \frac{P}{\rho} = \text{const} \tag{3-28}$$

If gravity is the only external force acting on the fluid, then the potential energy per unit volume becomes

$$\Omega = \frac{gZ}{g_c} \tag{3-29}$$

where Z is the distance above some base plane where the potential energy is zero. Substituting Eq. (3-29) into (3-28) gives the familiar form of *Bernoulli's equation for steady flow.*

$$\frac{1}{2}\frac{|\mathbf{V}|^2}{g_c} + \frac{gZ}{g_c} + \frac{P}{\rho} = \text{const} \tag{3-30}$$

The dimension of each term in Eq. (3-30) is energy per unit mass. Physically, the equation means that for steady, irrotational flow the sum of the kinetic energy $\frac{1}{2}(|\mathbf{V}|^2/g_c)$, the potential energy gZ/g_c, and the pressure energy P/ρ is the same at all points in space. For rotational flow, Eq. (3-30) can be applied to a particular streamline, the constant having different values for different streamlines [see also Eqs. (4-7) and (4-8)].

3-7. Two-dimensional, Irrotational, Incompressible Flow; the Flow Net

As pointed out in Sec. 3-3, irrotational flow implies the existence of a velocity potential ϕ, defined by Eqs. (3-4) to (3-6), which, when introduced into the continuity equation for two-dimensional incompressible flow, gives

$$\frac{\partial^2 \phi}{\partial x^2} + \frac{\partial^2 \phi}{\partial y^2} = 0 \tag{3-31}$$

In Sec. 3-5 it was shown that a stream function ψ, defined by Eqs. (3-16) and (3-17) for two-dimensional incompressible flow, satisfies Eq. (3-18) for irrotational flow; i.e.,

$$\frac{\partial^2 \psi}{\partial x^2} + \frac{\partial^2 \psi}{\partial y^2} = 0 \tag{3-18}$$

The determination of ϕ and ψ for any physical problem involves the solution of Eqs. (3-18) and (3-31). For two-dimensional motion, ϕ and ψ are both functions of x and y. The arbitrary constants resulting from the solution of Eqs. (3-18) and (3-31) are determined from the boundary conditions of the problem. Only one pattern of irrotational flow can exist for a given fixed boundary.

The particular solution of Eqs. (3-18) and (3-31) for any problem will yield ϕ and ψ as functions of x and y, which functions may be plotted on the xy plane at constant values of ϕ and ψ. The result is a *flow net* consisting of ϕ curves and ψ curves, which form an orthogonal system in that they intersect at right angles. In the limit, as the increments of ϕ and ψ between the curves approach zero, the flow net becomes a network of infinitesimal squares.

An example of a flow net is shown in Fig. 3-4, in which all lines intersect at right angles. The lines of constant ψ represent the direction of the velocity vector at any point. When ψ is obtained as a function of x and y, the velocity of the fluid at every point may be obtained as a function of x and y by Eqs. (3-16) and (3-17). Also

$$|\mathbf{V}| = \sqrt{u^2 + v^2} \tag{3-32}$$

Hence the velocity at any point may be obtained in terms of x and y.

Once the velocity is known at every point, the pressure may be determined by the application of Bernoulli's equation (3-30). The velocity relationships existing as two-dimensional flow occurs in the xy plane of Fig. 3-4 may be derived by considering the square meshes $abcd$ and $a'b'c'd'$ between the two streamlines ψ_3 and ψ_4, the respective velocities at the center of

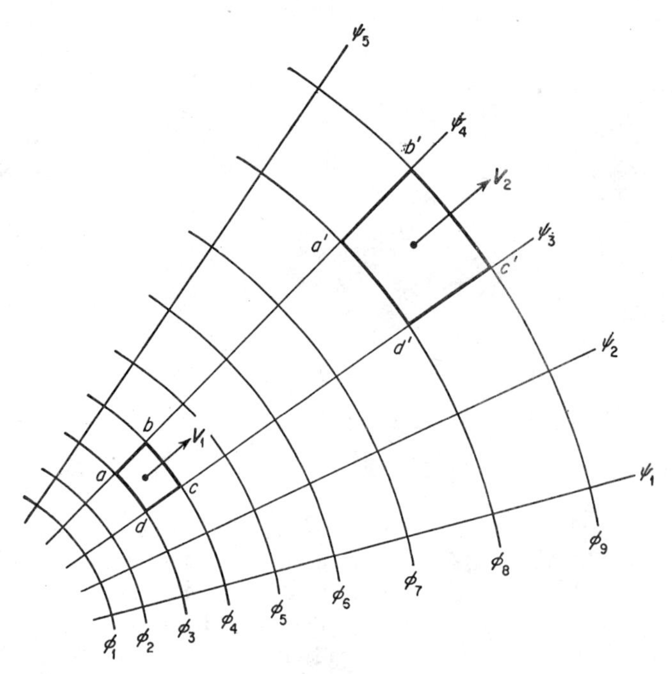

FIG. 3-4. The flow net.

each mesh being $\mathbf{V}_1$ and $\mathbf{V}_2$. Since no fluid crosses the streamlines ψ_3 and ψ_4, the amount of fluid flowing through $abcd$ is the same as through $a'b'c'd'$. The volume rate of flow per unit distance perpendicular to the xy plane is $|\mathbf{V}_1|ad$ for element $abcd$ and $|\mathbf{V}_2|a'd'$ for element $a'b'c'd'$. Thus

$$|\mathbf{V}_1|ad = |\mathbf{V}_2|a'd'$$

or

$$\frac{|\mathbf{V}_2|}{|\mathbf{V}_1|} = \frac{ad}{a'd'} = \frac{ab}{a'b'} \tag{3-33}$$

The last result of Eq. (3-33) comes from the fact that the meshes are square.

The lines representing constant values of ϕ indicate all points in space where the absolute velocity is the same.

3-8. Examples of Two-dimensional, Irrotational, Incompressible Flow

In a majority of fluid-flow problems encountered in practice it is not possible to assume that the viscosity of the fluid is zero, and hence the Euler equations do not adequately describe the flow. There are instances where irrotational flow is very nearly attained. Gas flow below the sonic velocity is very nearly incompressible, and in cases where the viscous effects are insignificant, one can assume irrotational flow and solve Eqs. (3-18) and (3-31) in order to obtain the velocity pattern and pressure pattern for the system. Instances where irrotational, incompressible flow can be assumed without too much error are the discharge of fluids from large reservoirs through small outlets, flow through small inlets into large reservoirs, flow through converging and diverging sections of large two-dimensional ducts, flow through sharp-edged orifices, flow over sharp-crested weirs, and flow past pitot tubes.

3-9. Analytical Solution of Some Problems in Two-dimensional, Incompressible, Irrotational Flow

In certain cases where the geometry of the system is quite simple it is possible to obtain analytical solutions for ψ and ϕ in terms of x and y. As shown in Appendix II, if a functional relationship exists between complex numbers

$$\phi + i\psi = F_1(x + iy) \tag{3-34}$$

then the Cauchy-Riemann equations

$$\frac{\partial \phi}{\partial x} = \frac{\partial \psi}{\partial y} \tag{3-35}$$

$$\frac{\partial \phi}{\partial y} = -\frac{\partial \psi}{\partial x} \tag{3-36}$$

result provided the derivative of the functional relationship is continuous. From the relations indicated in Eqs. (3-35) and (3-36) it may be shown that Eqs. (3-18) and (3-31) for two-dimensional, incompressible, irrotational flow are satisfied.

When the complex variable $\phi + i\psi$ is plotted on the $\phi\psi$ plane, the lines of constant ϕ and ψ will form a square-mesh flow net. A plot of $\phi(x,y)$ and $\psi(x,y)$ on the xy plane will produce a network having characteristics dependent on the functional relationship F_1 in Eq. (3-34). The meshes obtained on the xy plane will be square only if they are infinitesimal in size; however, all lines on the xy plane will intersect at right angles, and hence the requirements of a flow net will be satisfied. A few examples of functional relationships between $\phi + i\psi$ and $x + iy$ are given below, and the resulting flow nets are interpreted physically. A more detailed discussion of these examples is provided by Streeter.[1]

1. *One-dimensional Flow of an Infinite Fluid in the x Direction*

$$\phi + i\psi = U(x + iy) \qquad (3\text{-}37)$$

Since ϕ, ψ, x, and y are all real,

$$\phi = Ux$$

$$\psi = Uy$$

Plotting the flow net on the $\phi\psi$ plane and xy plane in Fig. 3-5 shows that in both cases the net is made up of square meshes.

The sides of the squares in the xy plane are $1/U$ times the sides of the squares in the $\phi\psi$ plane. The flow net shown on the xy plane depicts the flow of an infinite fluid in the x direction having a velocity U.

$$\frac{\partial \phi}{\partial x} = \frac{\partial \psi}{\partial y} = U = \text{velocity in } x \text{ direction}$$

$$\frac{\partial \phi}{\partial y} = -\frac{\partial \psi}{\partial x} = 0 = \text{velocity in } y \text{ direction}$$

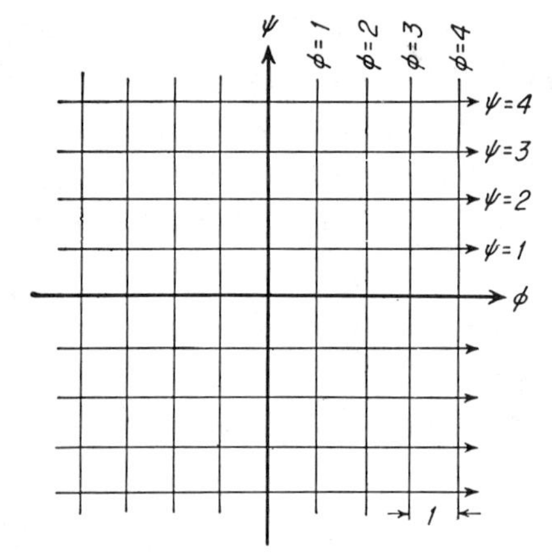

FIG. 3-5. Flow net of an infinite fluid flowing in the x direction.

In irrotational flow past solid boundaries, since the boundary is always a streamline, any streamline in the flow net may be replaced by a solid boundary, and the flow net beyond the boundary will remain unchanged. If two streamlines in Fig. 3-5 are replaced by solid boundaries, the flow net will represent flow in a two-dimensional channel with parallel walls.

2. *Flow along Two Inclined Planes.* The form of the functional relationship for flow along two inclined planes is

$$\phi + i\psi = a(x + iy)^{\pi/\alpha} \qquad 0 < \alpha < 2\pi \qquad (3\text{-}38)$$

where α is the enclosed angle between the two inclined planes. It is convenient to represent both complex variables of Eq. (3-38) in polar form so that the real and imaginary parts may be separated.

$$Re^{i\Theta} = ar^{\pi/\alpha}e^{i(\pi/\alpha)\theta_1} \tag{3-39}$$

where $R = \sqrt{\phi^2 + \psi^2}$

$r = \sqrt{x^2 + y^2}$

From Eq. (3-39)

$$R = ar^{\pi/\alpha}$$

$$\Theta = \frac{\pi\theta_1}{\alpha}$$

In the $\phi\psi$ plane, when Θ varies from 0 to π (corresponding to the upper half of the $\phi\psi$ plane), the angle θ_1 varies from 0 to α in the xy plane.

To obtain ϕ and ψ as functions of x and y Eq. (3-38) may be expressed as follows:

$$\phi + i\psi = ar^{\pi/\alpha}\left(\cos\frac{\pi\theta_1}{\alpha} + i\sin\frac{\pi\theta_1}{\alpha}\right) \tag{3-40}$$

from which

$$\phi = ar^{\pi/\alpha}\cos\frac{\pi\theta_1}{\alpha} \tag{3-41}$$

$$\psi = ar^{\pi/\alpha}\sin\frac{\pi\theta_1}{\alpha} \tag{3-42}$$

Example 3-1

An incompressible, nonviscous fluid is flowing along two planes inclined at an angle of 45° with each other. At a point 5 ft from the intersection of the planes the fluid in contact with the plane is flowing at a velocity of 5 ft/sec. Consider one plane to coincide with the x axis and the intersection of the planes to be at the point (0,0). Also assume that fluid flows in a positive direction along the x axis.

(*a*) Draw the flow net for this flow.

(*b*) Determine the velocity at the point (2, 0.5).

Solution

(*a*) From Eqs. (3-41) and (3-42), when $\alpha = \pi/4$,

$$\phi = ar^4\cos 4\theta_1$$

Similarly

$$\psi = ar^4\sin 4\theta_1$$

Since

$$\cos\theta_1 = \frac{x}{\sqrt{x^2 + y^2}} \quad \text{and} \quad \sin\theta_1 = \frac{y}{\sqrt{x^2 + y^2}}$$

$$\phi = a(x^4 - 6x^2y^2 + y^4)$$

$$\psi = 4a(x^3y - xy^3)$$

The flow net may be drawn from these relationships. From the statement of the problem the value of a may be determined.

$$u = \frac{\partial \phi}{\partial x} = a(4x^3 - 12xy^2)$$

$$v = \frac{\partial \phi}{\partial y} = a(4y^3 - 12x^2y)$$

At point (5,0)

$$u = 5 \text{ ft/sec}$$

$$v = 0$$

Thus

$$5 = a4(5)^3$$

From which

$$a = 5/500 = 0.01$$

The resulting flow net is indicated in Fig. 3-6.

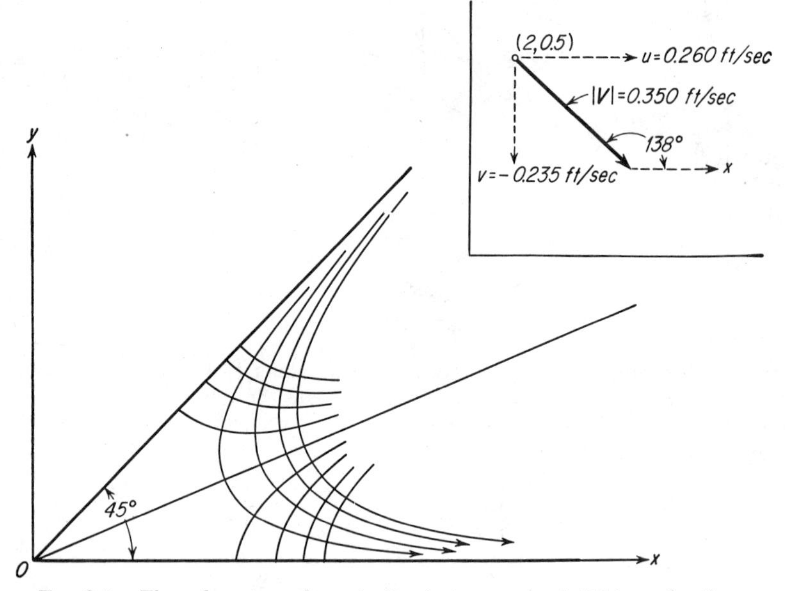

FIG. 3-6. Flow along two planes inclined at an angle of 45° to each other.

(b) The fluid velocity at the point (2, 0.5) may be calculated.

$$u = 0.01[(4)(2)^3 - (12)(2)(0.5)^2] = 0.260 \text{ ft/sec}$$

$$v = 0.01[(4)(0.5)^3 - (12)(2)^2(0.5)] = -0.235 \text{ ft/sec}$$

Thus

$$|\mathbf{V}| = \sqrt{(0.26)^2 + (0.235)^2} = 0.350 \text{ ft/sec}$$

The direction of **V** is toward the x axis and makes an angle of 138° with the x axis (Fig. 3-6).

3. *Point Source or Sink in an Infinite Fluid.* The function representing a point source in an infinite fluid is

$$\phi + i\psi = \frac{Q}{2\pi} \ln (x + iy) \tag{3-43}$$

where Q is the rate of fluid discharge from the point. Using polar coordinates,

$$\phi + i\psi = \frac{Q}{2\pi} \ln re^{i\theta_1}$$

$$= \frac{Q}{2\pi} \ln r + \frac{Q}{2\pi} i\theta_1$$

Since

$$r = \sqrt{x^2 + y^2}$$

$$\theta_1 = \tan^{-1} \frac{y}{x}$$

thus

$$\phi = \frac{Q}{2\pi} \ln \sqrt{x^2 + y^2} \tag{3-44}$$

$$\psi = \frac{Q}{2\pi} \tan^{-1} \frac{y}{x}$$

$$u = \frac{\partial \phi}{\partial x} = \frac{Q}{2\pi} \frac{x}{x^2 + y^2}$$

$$v = \frac{\partial \phi}{\partial y} = \frac{Q}{2\pi} \frac{y}{x^2 + y^2}$$

Therefore

$$|\mathbf{V}| = \sqrt{u^2 + v^2}$$

or

$$|\mathbf{V}| = \frac{Q}{2\pi} \frac{1}{\sqrt{x^2 + y^2}} = \frac{Q}{2\pi r} \tag{3-45}$$

The above results show that the magnitude of the velocity varies inversely as the radial distance from the point source. It is infinite at the source, and the direction of the velocity is always radially outward from the source. The flow net for a point source is shown in Fig. 3-7a, the velocity having a direction *away* from the origin. A point sink in an infinite fluid has the same-shaped flow net, as indicated in Fig. 3-7b, except that the direction of all velocity vectors is radially *toward* the sink. The functional relationship for a point sink is

$$\phi + i\psi = -\frac{Q}{2\pi}\ln(x + iy) \tag{3-46}$$

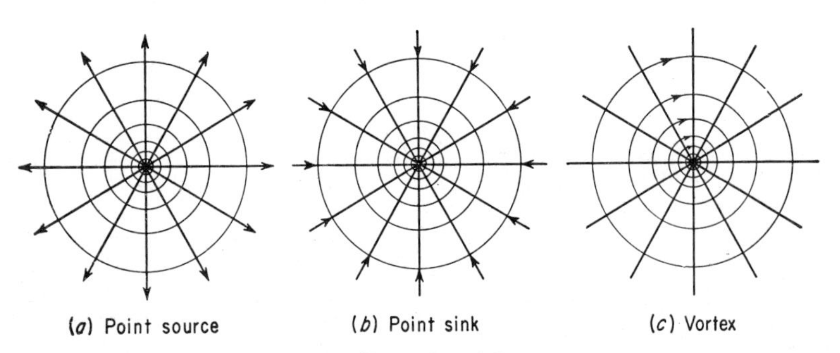

FIG. 3-7. Examples of flow nets.

4. *Combination of a Source and Sink.* Consider a source located at $x = -a$ and a sink located at $x = a$. Fluid flows at a rate of Q volume units per unit time per unit thickness from the source to the sink. The functional relation for this flow situation is

$$\phi + i\psi = \frac{Q}{2\pi}\ln\frac{x + a + iy}{x - a + iy} \tag{3-47}$$

Using polar coordinates,

$$\phi + i\psi = \frac{Q}{2\pi}\ln\frac{r_1 e^{i\theta_1}}{r_2 e^{i\theta_2}} \tag{3-48}$$

where $r_1 = \sqrt{(x + a)^2 + y^2}$

$$r_2 = \sqrt{(x - a)^2 + y^2}$$

$$\theta_1 = \tan^{-1}\frac{y}{x + a}$$

$$\theta_2 = \tan^{-1}\frac{y}{x - a}$$

From the above the absolute value of the velocity is

$$|\mathbf{V}| = \frac{Qa}{\pi r_1 r_2} \tag{3-49}$$

Stated physically, Eq. (3-49) means that the magnitude of the velocity vector in the region about a source and sink is proportional to the volumetric flow rate and one-half the distance between them and inversely proportional to the radial distances from the source and sink. A flow net of a source-and-sink combination is shown in Fig. 3-8.

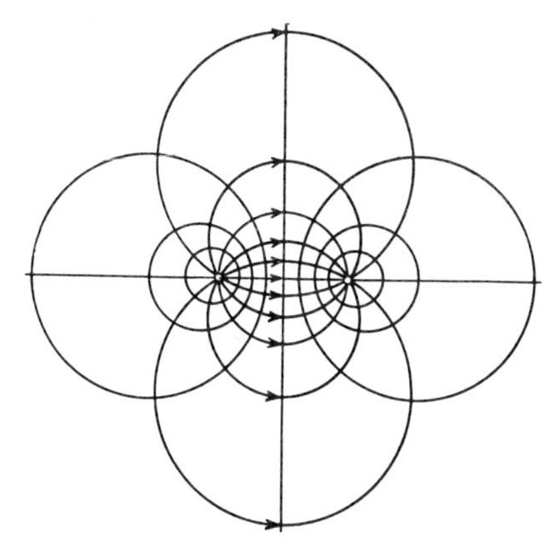

FIG. 3-8. Flow net for a source and sink of equal strength.

Example 3-2

Two perforated pipes are mounted vertically in an infinitely large tank. The pipes are situated so that the distance between them is 5 ft. Water discharges from one at the rate of 10 ft^3/(min)(ft of length) and enters the other at the same rate. Assuming two-dimensional irrotational flow, calculate the magnitude of the fluid velocity along a line joining the two pipes.

Solution

Equation (3-49) may be used to calculate the magnitude of the velocity vector at any point on a line joining the two pipes.

$$Q = 10 \text{ ft}^3/(\text{min})(\text{ft of length})$$

$$a = 2.5 \text{ ft}$$

At a point midway between the two pipes

$$r_1 = 2.5 \text{ ft}$$

$$r_2 = 2.5 \text{ ft}$$

Thus

$$|\mathbf{V}| = \frac{(10)(2.5)}{(\pi)(2.5)(2.5)} = 1.27 \text{ ft/min}$$

Table 3-1 gives the magnitude of the velocity of the fluid as a function of the distance from the source.

TABLE 3-1

Distance from source, ft	$\lvert\mathbf{V}\rvert$, *ft/min*
0	∞
0.5	3.53
1.0	1.99
1.5	1.51
2.0	1.33
2.5	1.27
3.0	1.33
3.5	1.51
4.0	1.99
4.5	3.53
5.0	∞

5. *Vortex.* If the function represented by Eq. (3-43) is modified as follows:

$$\phi + i\psi = \frac{\Gamma i}{2\pi} \ln (x + iy) \tag{3-50}$$

where Γ is the circulation, the resulting flow net (Fig. 3-7*c*) represents a vortex. Vortex flow is represented by circular streamlines, and the magnitude of the velocity varies inversely as the distance from the center of the streamlines. Geometrically this flow net is the same as for a source or sink except that for vortex flow the streamlines are circles, and the equipotential lines are radial lines emanating from the origin. Since no fluid can cross the streamlines, vortex flow means that fluid is circulating about a point.

An analysis similar to that for a point source shows that for a vortex

$$\lvert\mathbf{V}\rvert = \frac{\Gamma}{2\pi r} \tag{3-51}$$

indicating that the magnitude of the velocity is inversely proportional to the distance from the center of the streamlines.

6. *Flow around an Infinitely Long Cylinder.* For irrotational flow around an infinitely long cylinder of radius r_0, the function used to construct the flow net is

$$\phi + i\psi = U\left(x + iy + \frac{r_0^2}{x + iy}\right) \tag{3-52}$$

where U is the velocity of the undisturbed fluid flowing in the x direction. The axis of the cylinder coincides with the origin of the xy plane.

Separating the real and imaginary parts of Eq. (3-52) gives

$$\phi = Ux\left(1 + \frac{r_0^2}{x^2 + y^2}\right) \tag{3-53}$$

$$\psi = Uy\left(1 - \frac{r_0^2}{x^2 + y^2}\right) \tag{3-54}$$

The streamlines corresponding to $\psi = 0$ are the x axis and the boundaries of the cylinder. Figure 3-9 shows a flow net around a cylinder. From Eqs. (3-53) and (3-54)

$$u = \frac{\partial \phi}{\partial x} = U\left[1 + \frac{r_0^2}{x^2 + y^2} - \frac{2x^2 r_0^2}{(x^2 + y^2)^2}\right] \tag{3-55}$$

$$v = \frac{\partial \phi}{\partial y} = U\frac{-2r_0^2 xy}{(x^2 + y^2)^2} \tag{3-56}$$

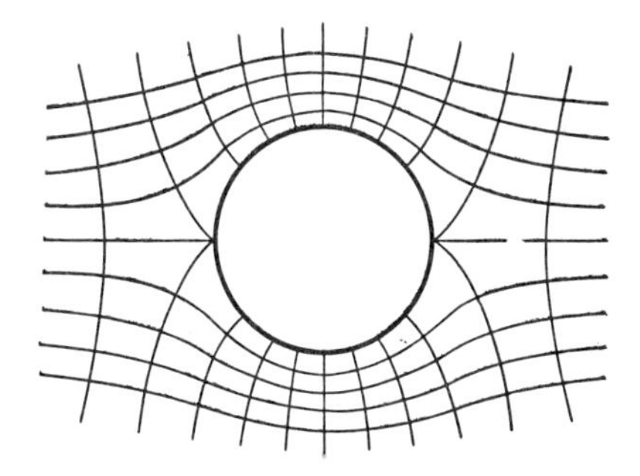

FIG. 3-9. Flow net for uniform flow past a circular cylinder.

It is of interest to show how the velocity varies at the surface of the cylinder. Since

$$r^2 = x^2 + y^2$$

$$x = r \cos \theta_1$$

and $\quad r = r_0$ at the surface of the cylinder

$$u = U(2 - 2\cos^2 \theta_1)$$

$$v = U(-2 \sin \theta_1 \cos \theta_1)$$

from which $\quad |\mathbf{V}| = 2U \sin \theta_1 = 2U \sin \theta \qquad \theta = \pi - \theta_1 \qquad$ (3-57)

Equation (3-57) gives the magnitude of the fluid velocity at the surface of the cylinder at any angle θ_1 measured in a positive direction from the positive x axis. Thus the forward part of the cylinder, which is encountered first by the fluid, corresponds to a value of $\theta_1 = \pi$.

Since $\sin \theta = \sin (\pi - \theta_1)$, Eq. (3-57) would be unchanged if the angle θ were measured in a clockwise direction from the negative x axis, i.e., measured from the forward part of the cylinder.

Example 3-3

An infinitely long cylinder 3 in. in diameter is placed in a wind tunnel in which air is flowing at the rate of 80 ft/sec. The air is at 60°F and 1 atm pressure. Assuming incompressible, irrotational flow, determine (*a*) the magnitude of the air velocity adjacent to the surface of the cylinder and (*b*) the pressure of the air adjacent to the cylinder.

Solution

(*a*) Measuring the angle θ from the forward point of the cylinder, Eq. (3-57) may be used to determine the velocity at the cylinder surface. A tabulation of the velocity as a function of θ is shown in Table 3-2.

TABLE 3-2

θ, deg, measured from forward point	$\mathbf{V}_\theta$, ft/sec, at surface of cylinder	$P_\theta - P$, lb_f/ft^2
0	0	7.62
30	80	0
60	139	−15.24
90	160	−22.86
120	139	−15.24
150	80	0
180	0	7.62

(*b*) The pressure distribution over the cylinder may be determined by Bernoulli's equation (3-30). Neglecting gravity forces, Eq. (3-30) becomes

$$\frac{1}{2}\frac{U^2}{g_c} + \frac{P}{\rho} = \frac{1}{2}\frac{|\mathbf{V}_\theta|^2}{g_c} + \frac{P_\theta}{\rho}$$

where $\mathbf{V}_\theta$ and P_θ are, respectively, the velocity and pressure at the angle θ measured from the forward point and U and P are the velocity and pressure of the undisturbed air. At 60°F and 1 atm the density of air is 0.0764 lb_m/ft^3. Values of $P_\theta - P$ are tabulated in Table 3-2.

The pressure distribution over the surface of the cylinder may be expressed in terms of θ by means of Eq. (3-30).

$$P_\theta - P = \frac{\rho}{2g_c}(U^2 - |\mathbf{V}_\theta|^2)$$

From Eq. (3-57)

$$P_\theta - P = \frac{\rho}{2g_c}(U^2 - 4U^2 \sin^2 \theta)$$

$$P_\theta - P = \frac{\rho U^2}{2g_c}(1 - 4 \sin^2 \theta) \tag{3-58}$$

At

$$\theta = 0, \ |\mathbf{V}_\theta| = 0$$

$$P_{\theta=0} - P = \frac{\rho U^2}{2g_c} \tag{3-59}$$

Equation (3-58) may be used to predict the pressure distribution over the surface of the cylinder for nonviscous flow. It is also applicable for the flow of viscous fluids up to a value of $\theta = 60°$. Equation (3-59) expresses the impact pressure at the leading edge of the cylinder. The impact pressure is the pressure resulting from the complete conversion of the kinetic energy of the fluid into pressure energy.

3-10. Approximate Construction of a Flow Net

In most cases the geometry of the system under consideration is such that an analytical solution of Eqs. (3-18) and (3-31) is extremely difficult, if not impossible. In these instances the flow net may be constructed by making use of the symmetry of the system and the fact that the fixed boundaries are also streamlines. The various examples of flow nets mentioned in Sec. 3-9 may be used to construct the flow net of a more complicated system. For example, in the case of a fluid discharging from a large tank through a small outlet one may consider that the outlet appears as a point sink to the fluid a large distance from it, and the flow net will be the same as for a point sink. Only in the vicinity of the outlet will the flow net differ from that for a point sink. The flow net is constructed by first drawing equally spaced streamlines, the number depending upon the accuracy required. The equipotential lines (constant ϕ) are then drawn at right angles to the streamlines so that approximately square meshes are obtained throughout the whole system. The equipotential lines are normal to the solid boundaries.

The accuracy of construction may be checked by also drawing the diagonals of the meshes, which should also form an orthogonal system of curves.

Example 3-4

Water is flowing out of an infinitely large reservoir through a rounded outlet having parallel sides 1 ft apart. The rounded portion of the outlet has a radius of

6 in. Velocity of flow in the outlet is 5 ft/sec. Assume flow is two-dimensional, incompressible, and irrotational.

(a) Draw the flow net.

(b) Show how the velocity changes along the center line as the water flows from the reservoir and through the outlet.

(c) Making use of Bernoulli's equation (3-30), show the pressure distribution along the center line of the outlet.

Solution

(a) The streamlines are first drawn. Referring to Fig. 3-10, the distance between the two boundaries of the outlet is divided into six equal spaces, and the streamlines A, B, C, D, E, F, and G are indicated as parallel lines in the outlet.

The corresponding streamlines in the infinite fluid on the left of the figure are A', B', C', D', E', F', G'. Streamlines $A'A$ and $G'G$ coincide with the fixed boundaries of the system. An infinite distance to the left of the outlet the fluid flows as if the outlet were a point sink at O. Consequently the flow net a large distance from the outlet is the same as shown in Fig. 3-7b for a point sink, with all streamlines pointing radially to the center of the outlet. These streamlines are shown as

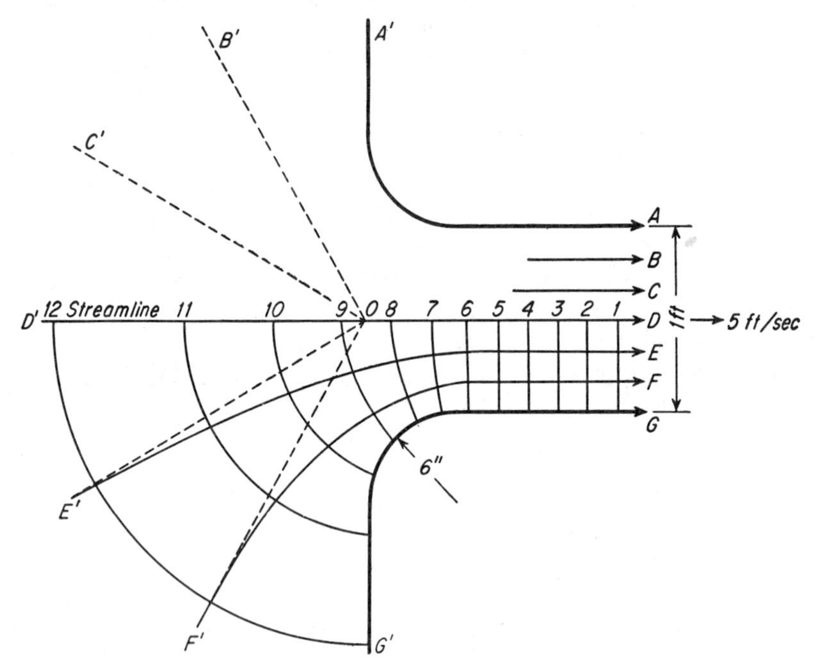

FIG. 3-10. Flow net for flow through a rounded outlet (Example 3-4).

broken lines in Fig. 3-10. To draw the actual streamlines it is necessary to construct smooth curves joining the radial streamlines A', B', C', D', E', F', G' to the parallel lines A, B, C, D, E, F, G. Since $A'A$ and $G'G$ coincide with the boundary and $D'D$ is the central streamline, only $B'B$, $C'C$, $E'E$, $F'F$ need be estimated. The equipotential lines are drawn so they intersect the streamlines at right angles. The completed flow net is shown in Fig. 3-10 for the lower half of the outlet.

(b) To obtain the velocity along the center $D'D$ consider the points 1 to 12 (Fig. 3-10). The spacing between the streamlines or equipotential lines may be used to give the velocity approximately midway between the lines. The velocity

TABLE 3-3

Points	Distance between equipotential lines on stream-line $D'D$, in.	Velocity on stream-line $D'D$ at mid-point between equi-potential lines, ft/sec	$P - P_{1\text{-}2}$ along streamline $D'D$, lb_f/ft^2
1-2	2.0	5.0	0
2-3	2.0	5.0	0
3-4	2.0	5.0	0
4-5	2.05	4.9	0.96
5-6	2.10	4.8	1.90
6-7	2.50	4.0	8.72
7-8	2.80	3.6	11.70
8-9	3.30	3.0	15.50
9-10	4.70	2.1	20.0
10-11	6.20	1.6	21.8
11-12	9.1	1.1	23.2

in the outlet is 5 ft/sec. In Table 3-3 the distances between equipotential lines are indicated, and the velocities midway between these lines are calculated from Eq. (3-33).

The distance between equipotential lines at points 1 and 2 (where the velocity is 5 ft/sec) is 2.0 in. At points 6 and 7 the distance between equipotential lines is 2.5 in. Thus the velocity $|\mathbf{V}_{6\text{-}7}|$ midway between points 6 and 7 is, by Eq. (3-33),

$$|\mathbf{V}_{6\text{-}7}| = \frac{(5)(2.0)}{2.5} = 4.0 \text{ ft/sec}$$

The direction of this velocity is the tangent streamline $D'D$. The velocity along streamline $D'D$ is plotted in Fig. 3-11.

(c) The pressure distribution along $D'D$ is obtained from the velocity distribution by means of Bernoulli's equation (3-30). Neglecting gravity forces, Eq. (3-30) becomes

$$\frac{1}{2}\frac{|\mathbf{V}_{1\text{-}2}|^2}{g_c} + \frac{P_{1\text{-}2}}{\rho} = \frac{1}{2}\frac{|\mathbf{V}|^2}{g_c} + \frac{P}{\rho}$$

where $\mathbf{V}_{1\text{-}2}$ and $P_{1\text{-}2}$ are the velocity and pressure, respectively, midway between points 1 and 2 and $\mathbf{V}$ and P are the velocity and pressure at any other point in the fluid. Therefore

$$P - P_{1\text{-}2} = \frac{\rho}{2g_c}(|\mathbf{V}_{1\text{-}2}|^2 - |\mathbf{V}|^2)$$

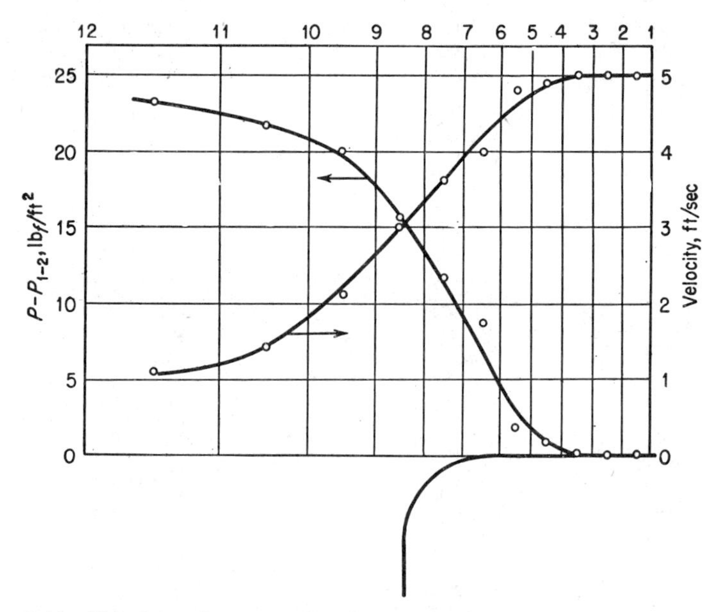

FIG. 3-11. Velocity and pressure distribution along streamline $D'D$ of Fig. 3-10.

This relation may be used to calculate the pressure along the streamline $D'D$. For example,

$$|\mathbf{V}_{6\text{-}7}| = 4 \text{ ft/sec}$$

$$P_{6\text{-}7} - P_{1\text{-}2} = \frac{62.4}{(2)(32.2)}[(5)^2 - (4)^2$$

$$= 8.7 \text{ lb}_f/\text{ft}^2$$

3-11. Separation

The solutions to the examples of Secs. 3-9 and 3-10 are based on irrotational flow. Flow is usually rotational, but nevertheless the flow net obtained for irrotational flow is very useful. If the fixed boundaries of the system are parallel or converge so that the streamlines converge, the actual flow will be very nearly irrotational except immediately adjacent to the solid boundary. However, if the streamlines of a flow net diverge rapidly, as they would for divergent boundaries, the actual flow will be quite different from that indicated by a flow net; i.e., the assumption that the limiting streamlines coincide with the fixed boundaries is not valid when the boundaries are divergent. The flowing fluid tends to leave the boundary whenever the streamlines are divergent. This is *separation*. The flow net obtained for irrotational flow is very useful in determining regions in a system where separation is likely to occur.

In the system shown in Fig. 3-10, as long as flow is from left to right, the streamlines are convergent, and there is no tendency for separation; however, if flow is from right to left, the streamlines diverge, and separation will occur. The region of separation for this type of inlet is shown in Fig. 3-12*a*. Separation also occurs during flow through abrupt contractions and enlargements in conduits (Fig. 3-12*b*), through abrupt bends (Fig. 3-12*c*), and past bodies of revolution (Fig. 3-12*d*).

The region between the separated fluid and the solid boundary is in a state of turbulent motion, a condition which causes a reduction of the efficiency of flow. It is therefore desirable to design boundaries which have the least tendency to cause flow separation. This is an important factor in the design of airfoils. As shown in Fig. 3-12*d*, there is considerable tendency for separation to occur during flow past a circular cylinder because of the rapidly diverging streamlines in the region past the center of the cylinder. If the cylinder were elliptical in shape, the streamlines would diverge more slowly, and the tendency for separation would be reduced. It is desirable to design the solid boundary so that the actual flow agrees closely with the flow net based on irrotational flow.

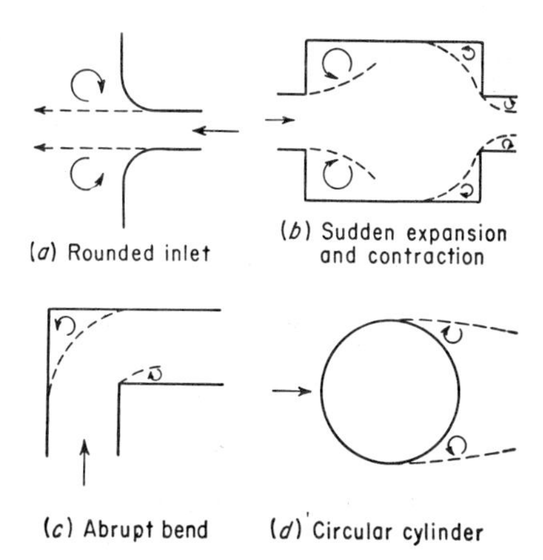

FIG. 3-12. Regions of separation (the broken lines indicate approximately the path of the separated fluid).

The fact that streamlines diverge in a flow net means that there is a *tendency* for separation to occur. Whether separation occurs or not depends on the velocity, density, and viscosity of the fluid. For a given fluid, separation will not occur until a certain velocity is reached.

3-12. Stagnation Points

A *stagnation point* is any point in a fluid where the velocity is zero. For two-dimensional, irrotational, incompressible flow, the conditions required for the velocity to be zero are

$$u = \frac{\partial \phi}{\partial x} = 0 \tag{3-60}$$

$$v = \frac{\partial \phi}{\partial y} = 0 \tag{3-61}$$

The *stagnation pressure* P_{st} is the pressure at the stagnation point and, from Eq. (3-30), may be expressed in terms of P and $\mathbf{V}$, the pressure and velocity at any other point in the field.

$$P_{st} = P + \frac{\rho|\mathbf{V}|^2}{2g_c} \tag{3-62}$$

An example of a stagnation point is the leading edge or forward point of a body immersed in a flowing fluid. Inspection of Eqs. (3-55) and (3-56) shows that at the point $x = -r_0$, $y = 0$ (the leading edge of the cylinder) both u and v are zero; hence this is a stagnation point. The stagnation pressure at this point is given by Eq. (3-59), which is equivalent to Eq. (3-62).

Example 3-5

An open tank car is 30 ft long, 8 ft wide, and 7 ft deep. It contains water to a depth of 5 ft and is in a train traveling at 60 miles/hr.

(*a*) Determine the slope of the surface of the liquid in the tank when the train goes around a curve having a radius of 1,000 ft.

(*b*) What is the minimum length of time at constant deceleration in which the train can stop without spilling the water?

Solution

In this problem the body of the fluid is being accelerated, so Eq. (3-1) reduces to

$$\frac{\partial u}{\partial t} = -g_c \frac{\partial \Omega}{\partial x} - \frac{g_c}{\rho}\frac{\partial P}{\partial x}$$

In the case of rotation about an axis, the acceleration $\partial u/\partial t$ is directed toward the center of rotation and is given by the relation

$$(\text{angular velocity})^2(\text{radius})$$

(*a*) The angular velocity of the tank car on the curve is

$$\frac{(60)(5{,}280)}{(3{,}600)(1{,}000)} = 0.088 \text{ radian/sec}$$

$$\frac{\partial u}{\partial t} = -(0.088)^2(1{,}000)$$

(Negative sign is used because acceleration is toward the center.) The slope of the liquid surface is required. Along the surface P is constant, so $\partial P/\partial x = 0$. Letting y be the height of the liquid surface above a datum plane of zero potential energy,

$$\Omega = \frac{yg}{g_c}$$

Thus,

$$\frac{\partial u}{\partial t} = -g\frac{\partial y}{\partial x}$$

(Partial derivatives may be changed to total derivatives since $\partial u/\partial t$ = constant.)

Therefore,

$$\frac{dy}{dx} = -\frac{1}{g}\frac{\partial u}{\partial t}$$

$$= \frac{(0.088)^2(1{,}000)}{32.2} = 0.240$$

which is the average slope of the liquid surface in the tank car.

(*b*) If the liquid is not to be spilled, the deceleration must be such that the surface of the liquid just reaches the top edge of the car as shown.

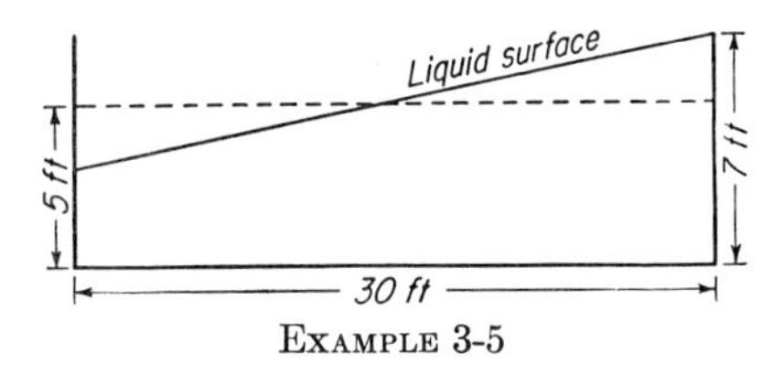

EXAMPLE 3-5

The slope of the surface dy/dx is 4/30. Along the surface (P = constant)

$$\frac{\partial u}{\partial t} = -g\frac{\partial y}{\partial x}$$

from which

$$\frac{\partial u}{\partial t} = \text{acceleration} = -\frac{(32.2)(4)}{30} = -4.29 \text{ ft/sec}^2$$

Thus the time to stop the train at 60 miles/hr (88 ft/sec) is

$$\frac{-88}{-4.29} = 20.5 \text{ sec}$$

Example 3-6

The pressure at the top of a natural gas well is 1,500 psia. The well contains essentially pure methane and is 10,000 ft deep. The average temperature in the well is 100°F. Compute the pressure at the bottom of the well by two methods:

(*a*) Using the average value of the compressibility factor and temperature.

(*b*) Using the average density of the gas in the well.

Solution

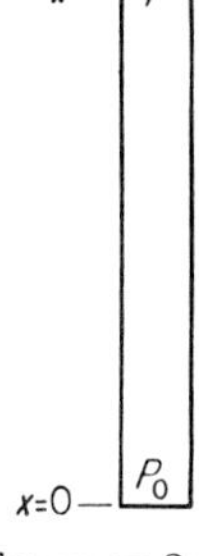

EXAMPLE 3-6

The pressure at the bottom of the well may be obtained from Eq. (3-1). (Take $x = 0$ at bottom of well and increasing as we move up well.) Since the fluid is stationary, Eq. (3-1) becomes

$$-g_c\frac{\partial \Omega}{\partial x} - \frac{g_c}{\rho}\frac{\partial P}{\partial x} = 0$$

In the earth's gravitational field

$$\Omega = \frac{xg}{g_c}$$

($x = 0$ is datum, or position of zero potential energy). Thus

$$-g\,dx - \frac{g_c\,dP}{\rho} = 0$$

Letting P_0 equal the pressure at $x = 0$ and P equal the pressure at $x = x$, integration gives

$$-\frac{gx}{g} - \int_{P_0}^{P} \frac{dP}{\rho} = 0$$

From Eq. (1-2)

$$\frac{P}{\rho} = ZnRT$$

Letting $$n = \frac{1}{29(Gr)}$$

where $Gr =$ the gas gravity (air = 1)

Thus $$\frac{1}{\rho} = \frac{RTZ}{29(Gr)}\frac{1}{P}$$

and $$-\frac{gx}{g_c} - \int_{P_0}^{P} \frac{ZRT}{29(Gr)}\frac{dP}{P} = 0$$

Integrating, assuming average values of Z and T,

$$-\frac{gx}{g_c} - \frac{ZRT}{29(Gr)} \ln \frac{P}{P_0} = 0$$

from which $$\ln \frac{P_0}{P} = \frac{29(Gr)}{ZRT}\frac{gx}{g_c}$$

or $$P_0 = P \exp\left[\frac{29(Gr)g}{ZRTg_c}\right] x$$

(*a*) For methane

$$Gr = 16/29 = 0.552$$

$$\text{Critical temperature} = 343°\text{R}$$

$$\text{Critical pressure} = 45.8 \text{ atm}$$

Assuming Z to be 0.90 (estimated from conditions at the top of the well),

$$\frac{29(Gr)gx}{ZRTg_c} = \frac{(29)(0.552)(32.2)(10{,}000)}{(0.90)(1.543)(560)(32.2)} = 0.206$$

Thus $$P_0 = 1{,}500e^{0.206}$$
$$= 1{,}845 \text{ psia}$$

To check assumption of compressibility factor:

$$\text{Average pressure} = \frac{1{,}845 + 1{,}500}{2} = 1.673 \text{ psia}$$

$$\text{Reduced pressure} = \frac{1{,}673}{(14.7)(45.8)} = 2.48$$

$$\text{Reduced temperature} = 560/343 = 1.63$$

From Fig. 1-3

$$Z = 0.90$$

This value checks with original assumption, so $P_0 = 1{,}845$ psia. The integration may be carried out using an average density. Thus

$$\frac{gx}{g_c} = \frac{P_0 - P}{\rho_{av}}$$

$$\frac{1}{\rho_{av}} = \frac{ZRT}{29(Gr)}\frac{1}{P_{av}}$$

(using average values of Z and T). Since

$$P_{av} = \frac{P_0 + P}{2}$$

it may be shown that

$$P_0 = P\,\frac{1 + [29(Gr)g/2ZRTg_c]x}{1 - [29(Gr)g/2ZRTg_c]x}$$

(*b*) Again assuming $Z = 0.90$,

$$\frac{29(Gr)gx}{2ZRTg_c} = 0.103$$

So

$$P_0 = \frac{(1{,}500)(1.103)}{0.897} = 1{,}845 \text{ psia}$$

Methods (*a*) and (*b*) check.

BIBLIOGRAPHY

1. Streeter, V. L.: "Fluid Dynamics," 2d ed., McGraw-Hill Book Company, Inc., New York, 1958.

PART II

THE FLOW OF VISCOUS FLUIDS

In the flow of all real, Newtonian fluids, the viscosity plays an important role in determining the flow pattern and the energy requirements to move the fluid. Indeed, a considerable portion of energy dissipated during fluid flow results from the fact that the fluid has a finite viscosity. As pointed out in Chap. 3, it is possible, in some instances, to neglect the viscous terms in the Navier-Stokes equations and from the resulting Euler equations to solve for the incompressible flow pattern and pressure distribution for a particular system. The solutions of the Euler equations give accurate flow patterns for certain types of boundaries, and the flow nets obtained are useful in predicting points of stagnation and regions of separation in the system.

As a nonviscous fluid flows past a solid boundary, the fluid immediately adjacent to the boundary has a finite velocity determined by the conditions of flow and the geometry of the boundaries. On the other hand, for the flow of viscous fluids it is generally assumed that there is no slip between the fluid and the solid boundary. In other words, as the fluid moves past a solid boundary, the fluid immediately adjacent to the boundary has the same velocity as the boundary. This assumption of no slip at the wall is valid except for the flow of rarefied gases and for fluids which do not wet the wall. The vicinity of the boundary is therefore a region where the fluid velocity changes from that of the boundary to that of the main stream of the fluid, and in this region velocity gradients exist in the fluid. As a consequence, the flow of the fluid is rotational, and shear stresses are present. This region of changing velocity is known as the *boundary layer*. The boundary layer extends from the solid boundary to the point in the fluid at which the velocity gradient becomes zero.

Heat and mass transfer take place to fluids from solid boundaries. In either case heat or mass must pass through the boundary layer. The manner of fluid flow in the boundary layer is important since it must be known in order to predict rates of heat and mass transfer to or from the fluid. In addition, the boundary layer is a region where energy is dissipated through viscous action, and the energy consumption in moving the fluid can be predicted from information about the flow in the boundary layer

The major portion of the remainder of this book is devoted to a consideration of the flow of viscous, Newtonian fluids, velocity profiles in boundary layers, friction losses during flow, and the relation of the rate of heat transfer to the velocity profiles and friction losses. For the most part, two main types of flow will be considered, *conduit flow*, or *internal flow* where the fluid completely fills a closed, stationary conduit, and *immersed flow*, or *external flow*, where the fluid flows past a stationary immersed solid.

CHAPTER 4

LAMINAR FLOW IN CLOSED CONDUITS

4-1. Introduction

Nearly every operation of fluid transport, heat transfer, and mass transfer involves the flow of viscous fluids in some form of a closed conduit. The frictional resistance of fluids as they flow through pipes and fittings and the rates of heat and mass transfer between fluids and pipe walls are important quantities in the design of pipelines and equipment. A complete knowledge of the mechanism of the flow of fluids in pipelines is basic to the understanding of the processes which occur during the flow. In this chapter laminar flow in several types of closed conduits is considered in detail. Theoretical expressions are derived for the velocity distribution in these conduits. Isothermal laminar-flow friction factors are derived, and theoretical values are compared with those obtained experimentally.

4-2. The Energy Equation for a Steady-flow Process

The majority of processes involving flow in conduits are steady-flow processes. The application of the law of conservation of energy to these processes gives the so-called flow equation. Figure 4-1 shows a system through which a fluid is flowing steadily between points 1 and 2. Between these points energy may enter or leave the fluid. Assuming no accumulation, the energy balance between points 1 and 2 is

$$E_2 + \frac{mU_2^2}{\alpha g_c} + \frac{mgZ_2}{g_c} + P_2V_2 = E_1 + \frac{mU_1^2}{\alpha g_c} + \frac{mgZ_1}{g_c} + P_1V_1 + q - w'' \tag{4-1}$$

where E = internal energy of fluid of mass m
m = mass of fluid
U = average velocity
Z = height above datum
P = static pressure

V = volume of mass m
q = energy in the form of heat added to fluid
w'' = energy in the form of work done by fluid

The subscripts 1 and 2 refer to points 1 and 2 respectively. Equation (4-1)

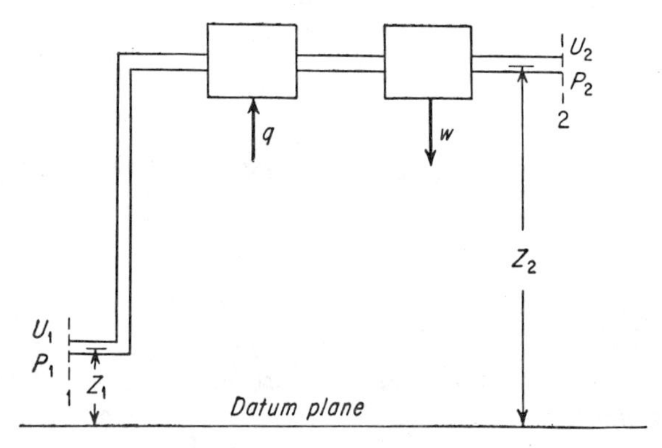

FIG. 4-1. Steady-flow system.

may be expressed in various forms.[5] Since $H = E + PV$, where H is the enthalpy of the fluid,

$$\Delta H + \Delta\left(\frac{mU^2}{\alpha g_c}\right) + \Delta\left(\frac{mgZ}{g_c}\right) = q - w'' \tag{4-2}$$

When limiting internal energy to heat and compression effects, Eq. (4-1) becomes

$$\int_1^2 T\,dS + \Delta\left(\frac{mU^2}{\alpha g_c}\right) + \Delta\left(\frac{mgZ}{g_c}\right) + \int_1^2 V\,dP = q - w'' \tag{4-3}$$

The flow of all viscous fluids is a process in which the irreversible dissipation of energy occurs. Thus $\int_1^2 T\,dS$, which is the increase in internal energy of the fluid due to heat effects, is equal to the heat absorbed from the surroundings plus all other energy dissipated into heat effects because of such irreversibilities as friction; i.e.,

$$\int_1^2 T\,dS = q + lw \tag{4-4}$$

where lw is the lost work, i.e., energy that could have done work but was dissipated in irreversibilities in the flowing material.

Combining Eqs. (4-3) and (4-4),

$$\int_1^2 V\,dP + \Delta\left(\frac{mU^2}{\alpha g_c}\right) + \Delta\left(\frac{mgZ}{g_c}\right) = -w'' - lw \tag{4-5}$$

Equation (4-5) is expressed in terms of a unit mass of material by dividing by m.

$$\int_1^2 \frac{dP}{\rho} + \frac{\Delta(U^2)}{\alpha g_c} + \frac{\Delta(gZ)}{g_c} = -\bar{w} - \overline{lw} \tag{4-6}$$

where $\rho = m/V$ and $\bar{w}$ and $\overline{lw}$ refer to a unit mass of fluid. Equations (4-1) to (4-6) assume steady flow and neglect chemical, electrical, and surface effects in defining the internal energy.

For the flow of an incompressible fluid and when $\bar{w}$ is zero, Eq. (4-6) becomes

$$\frac{\Delta P}{\rho} + \frac{\Delta(U^2)}{\alpha g_c} + \frac{g}{g_c}\Delta Z = -\overline{lw} \tag{4-7}$$

Bernoulli's equation (3-30) for the steady flow of an incompressible, nonviscous fluid may be written as

$$\frac{\Delta P}{\rho} + \frac{\Delta|\mathbf{V}|^2}{2g_c} + \frac{g}{g_c}\Delta Z = 0 \tag{4-8}$$

There are two main differences between Eqs. (4-7) and (4-8). The kinetic-energy term in Eq. (4-7) uses the average velocity of the fluid in the conduit. Since the velocity varies across the conduit, the kinetic energy of the whole stream is obtained by integration, and the value of α is obtained in this way. For laminar flow in circular tubes α has a value of 1 and for turbulent flow a value close to 2. In Eq. (4-8) the kinetic-energy term uses the absolute value of the point velocity. Equation (4-7) contains a dissipation term $\overline{lw}$ not appearing in Eq. (4-8). The dissipation term is due to the finite viscosity of the fluid and represents the energy dissipated by fluid friction. The dimensions on the terms in Eq. (4-7) are energy per unit mass (FL/m). It is usual to express $\overline{lw}$ in the form of a pressure loss due to friction, i.e.,

$$\int_1^2 \frac{-dP_f}{\rho} = \overline{lw} \tag{4-9}$$

where $-dP_f$ is the pressure decrease equivalent to the energy lost because of friction. From Eqs. (4-7) and (4-9), for a conduit of uniform cross section

$$\frac{\Delta P}{\rho} + \frac{g}{g_c}\Delta Z = \frac{\Delta P_f}{\rho} \tag{4-10}$$

4-3. Conduit Flow and the Formation of the Boundary Layer

When a viscous fluid flows in a closed conduit, the boundary layer extends from the wall to a point in the fluid where the velocity is a maximum. The velocity gradient at this point of maximum velocity is zero.

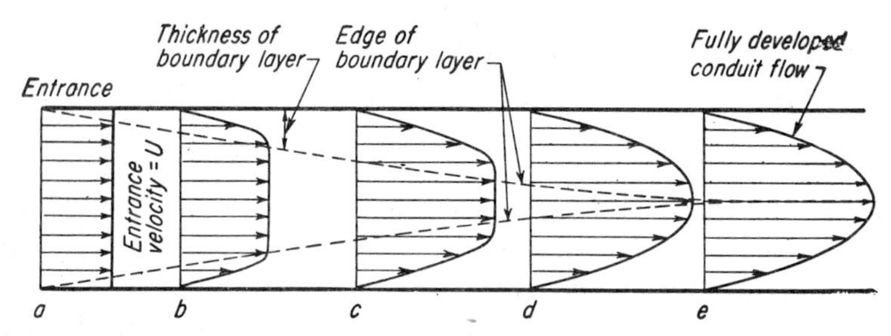

FIG. 4-2. Formation of the boundary layer and velocity profiles at the entrance of a circular tube.

The boundary layer does not form immediately at the entrance of the conduit but builds up in the entrance region. At the point where the fluid enters the conduit the boundary layer has zero thickness. The thickness increases along the length of the conduit and eventually becomes constant some distance from the entrance. Figure 4-2 illustrates the formation of the boundary layer in the entrance section of a conduit. The fluid enters the conduit at point a with a velocity of U. Velocity profiles are shown at points a, b, c, d, and e. At point a the velocity is uniform across the section of the conduit. At point e the boundary layer is completely formed, and from this point onward it has constant thickness. The dotted lines between a and e indicate the edge of the boundary layer. Beyond point e the influence of the entrance upon the pattern of flow has disappeared and *fully developed conduit flow* is said to exist.

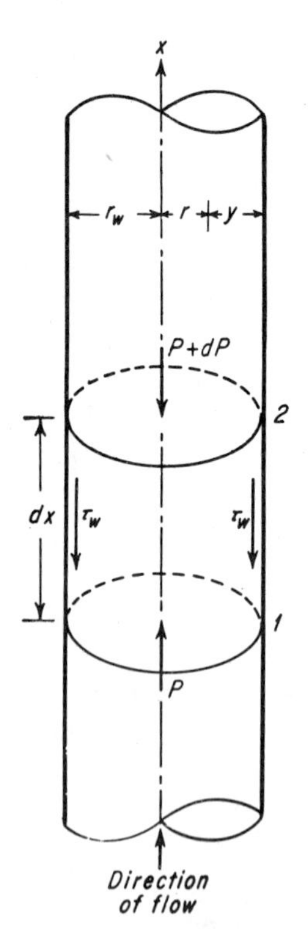

FIG. 4-3. Forces acting on a differential element of fluid flowing steadily in a circular tube.

4-4. Shearing Stresses in a Fluid Flowing Steadily in a Circular Conduit

Shearing stresses are present in the boundary layer. Since these stresses are exerted in a direction opposite to the direction of flow, they represent a force of resistance to the flow. When flow occurs in a closed conduit, the shear stress at the wall τ_w may be expressed in terms of $-dP_f/dx$, the pressure loss due to friction.

Figure 4-3 shows a vertical conduit with fluid flowing upward. The pressure at point 1 is P. At point 2 distance dx downstream from point 1 the pressure is $P + dP$. The shear forces τ_w acting on the fluid at the wall are exerted in a direction opposite to the direction of flow. For steady flow the summation of forces on the cylindrical element must be zero.

$$\Sigma F = \underbrace{\pi r_w^2 P - \pi r_w^2 (P + dP)}_{\text{pressure forces}} - \underbrace{\pi r_w^2\, dx\, \rho \frac{g}{g_c}}_{\text{fluid weight}} - \underbrace{2\pi r_w \tau_w\, dx}_{\text{shear forces}} = 0 \quad (4\text{-}11)$$

Simplifying Eq. (4-11) and dividing by $\rho \pi r_w^2$,

$$-\frac{dP}{\rho} - \frac{g}{g_c}\, dx - \frac{2\tau_w\, dx}{\rho r_w} = 0 \quad (4\text{-}12)$$

Substituting Eq. (4-10) (in its differential form instead of its integrated form) into Eq. (4-12),

$$\tau_w = \frac{r_w}{2} \frac{-dP_f}{dx} \quad (4\text{-}13)$$

where τ_w is the shear stress on the fluid at the wall. Equation (4-13) relates the shear stress to the radius of the tube and the pressure gradient due to friction. At any arbitrary radius the shear stress in the fluid is τ, and by a similar force balance

$$\tau = \frac{r}{2} \frac{-dP_f}{dx} = \tau_w \frac{r}{r_w} \quad (4\text{-}14)$$

The results expressed by Eqs. (4-13) and (4-14) indicate that the shear stresses in the fluid are related to the lost-work terms or friction losses.

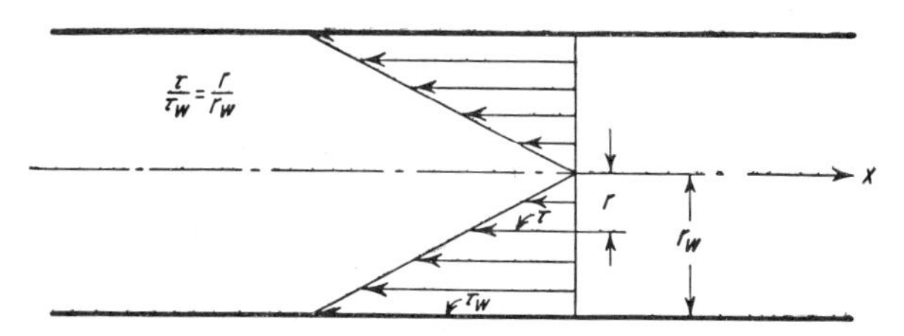

FIG. 4-4. Shear-stress distribution in a fluid flowing in a circular tube.

These relationships may be used to calculate shear stresses when the friction losses are known. Equation (4-14) shows that for steady flow in a circular conduit the shear stress in the fluid is a linear function of the distance from the axis of the tube. This linear distribution is shown in Fig. 4-4.

4-5. The Friction Factor

Numerous studies on the friction losses occurring during turbulent flow have indicated that they are proportional to the kinetic energy of the fluid per unit volume $\rho U^2/2g_c$ and the area A_w of the solid surface in contact with the fluid. This is the basis of the definition of the friction factor. Viscosity has been found to have but slight effect during turbulent flow. The force of resistance F is expressed as

$$F = \frac{f\rho U^2}{2g_c} A_w \tag{4-15}$$

The quantity f is a proportionality factor. Rearranging Eq. (4-15),

$$\frac{F}{A_w} = \frac{f\rho U^2}{2g_c} \tag{4-16}$$

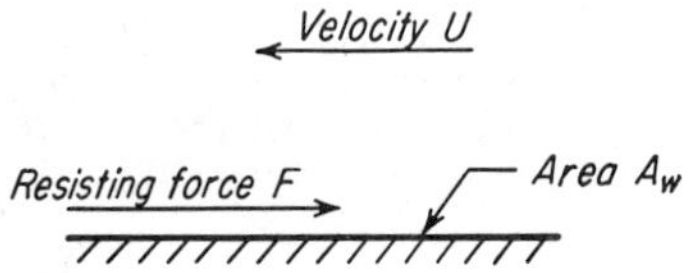

FIG. 4-5. Resistance of a solid surface in a flowing fluid.

The resisting force is opposite to the direction of flow. F/A_w corresponds to shear force at the wall τ_w (see Fig. 4-5.) Hence

$$\tau_w = \frac{f\rho U^2}{2g_c} \tag{4-17}$$

Combining Eqs. (4-17) and (4-13), replacing r_w by $d_w/2$, and solving for f,

$$f = \frac{g_c d_w}{2\rho U^2} \frac{-dP_f}{dx} \tag{4-18}$$

Equation (4-18) defines the *Fanning friction factor* for flow in circular tubes. It is expressed in terms of the tube diameter, fluid density, fluid velocity, and the pressure gradient equivalent to the frictional resistance per unit length of tube.

From Eq. (4-18) the lost work due to friction may be determined in terms of the friction factor. Rearranging Eq. (4-18),

$$\frac{-dP_f}{\rho} = \frac{2fU^2}{g_c d_w} dx \tag{4-19}$$

Integrating Eq. (4-19),

$$\int_1^2 \frac{-dP_f}{\rho} = \overline{lw} = \int_1^2 \frac{2fU^2}{g_c d_w} dx \tag{4-20}$$

Substituting Eq. (4-20) into (4-6),

$$\int_1^2 \frac{dP}{\rho} + \frac{\Delta(U)^2}{\alpha g_c} + \frac{g}{g_c} \Delta Z = -\bar{w} - \int_1^2 \frac{2fU^2}{g_c d_w} dx \tag{4-21}$$

4-6. Other Expressions for the Friction Factor in Tubes

There are several definitions of friction factors for flow in conduits. The friction factor originally defined by Blasius [3] and used by Moody,[15] Rouse,[23] Brown,[5] and others is four times the Fanning friction factor and will be referred to as the Blasius friction factor f_B, where

$$f_B = \frac{2g_c d_w}{\rho U^2} \frac{-dP_f}{dx} \tag{4-22}$$

Thus

$$f_B = 4f \tag{4-23}$$

Other writers define a friction factor which is twice the size of the Fanning friction factor.

4-7. The Hydraulic Radius of a Conduit

Whenever fluid flow in noncircular conduits is considered, it is necessary to determine some characteristic length term to use in the calculation of the Reynolds number and the friction factor. For circular pipes it is the diameter of the pipe, but for conduits having cross sections other than circular a new length term must be defined. This length term is four times the hydraulic radius, which is defined as the ratio of the cross-sectional area of the conduit to the wetted perimeter.

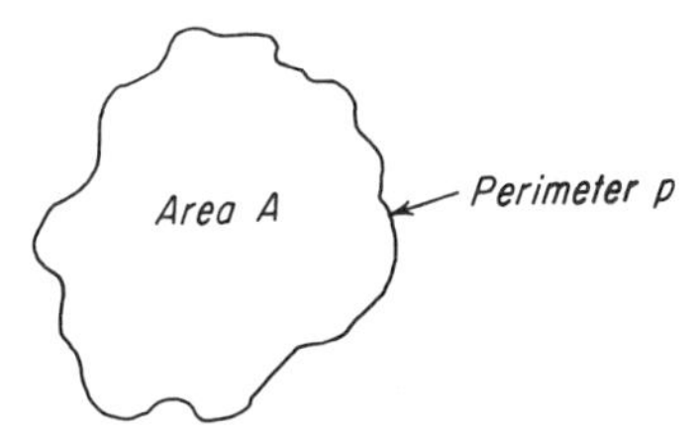

FIG. 4-6. Closed conduit of irregular cross section.

Figure 4-6 shows a noncircular conduit through which a fluid is flowing. The pressure gradient due to friction along the length of the conduit is $-dP_f/dx$. The shear at the solid boundary of the conduit is expressed by Eq. (4-17). Considering a length of conduit dx and a balance of forces like that used in deriving Eq. (4-11) and making use of Eq. (4-10) results in

$$A(-dP_f) = \tau_w p\, dx \tag{4-24}$$

where p is the wetted perimeter. Substituting (4-17) in (4-24) gives

$$A(-dP_f) = \frac{f\rho U^2}{2g_c} p\, dx \tag{4-25}$$

which, upon rearrangement, gives

$$f = \frac{g_c(4A/p)}{2\rho U^2} \frac{-dP_f}{dx} \tag{4-26}$$

Equation (4-26) is the same as Eq. (4-18) except that the diameter d_w has been replaced by the term $4A/p$, which is the equivalent diameter d_e of the noncircular conduit. The expression A/p is the hydraulic radius r_h. For circular conduits the equivalent diameter is the diameter of the tube.

The above definition for the hydraulic radius and equivalent diameter is suitable for conduits which have constant cross section; however, there are conduits in which the cross-sectional area varies in a periodic manner with the length, and for these three types of hydraulic radii have been defined, namely,

$$(r_h)_{\max} = \frac{\text{maximum cross-sectional area}}{\text{wetted perimeter}} \tag{4-27}$$

$$(r_h)_{\min} = \frac{\text{minimum cross-sectional area}}{\text{wetted perimeter}} \tag{4-28}$$

and

$$r_v = \frac{\text{volume of free space per unit length}}{\text{area of wetted surface per unit length}} \tag{4-29}$$

The last term, r_v, is the volumetric hydraulic radius. As before, the equivalent diameter is four times the hydraulic radius.

I. LAMINAR FLOW IN CIRCULAR TUBES

4-8. Laminar and Turbulent Flow in Tubes

In 1883 Osborne Reynolds [20] published a series of papers describing his experiments on the "circumstances which determine whether the motion of water shall be direct or sinuous, and of the law of resistance in parallel channels." This classical experiment consisted of injecting dye into water flowing in glass tubes. Figure 4-7 is a reproduction of the flow patterns published by Reynolds, showing that, under certain conditions of velocity and tube diameter, the streak of dye in the fluid continued at the center of the tube and flowed in a line parallel to the axis of the tube (Fig. 4-7*a*). Under other conditions of velocity and diameter the color band was distributed throughout the cross section of the tube a short distance from the tip of the needle injecting the dye (Fig. 4-7*b*). When the tube was viewed by the light of an electric

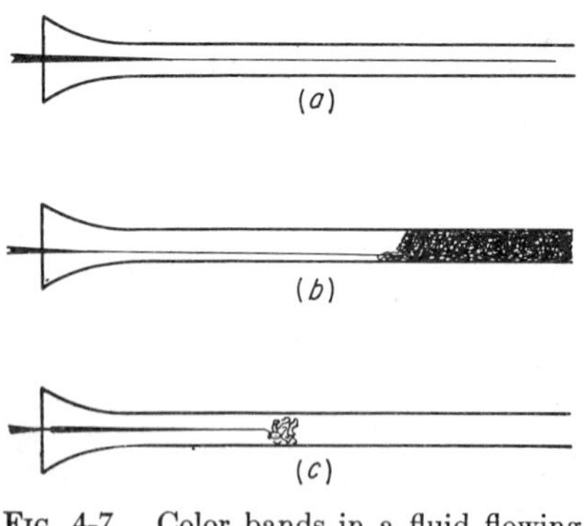

FIG. 4-7. Color bands in a fluid flowing in a circular tube. [*From O. Reynolds, Trans. Roy. Soc. (London)*, **174A**:935 (1883).]

spark, the mass of color was resolved into distinct curls, indicating rapidly changing eddies having circular motion, as shown in Fig. 4-7*c*. Reynolds determined critical velocities at which laminar (direct) flow changed into turbulent (sinuous) flow and found that the critical velocity was proportional to the kinematic viscosity of water and inversely proportional to the diameter of the tube. Combining these two results leads to the equation

$$U_{crit} = \frac{c\nu}{d_w} \tag{4-30}$$

where U_{crit} = critical velocity
ν = kinematic viscosity
d_w = tube ID
c = proportionality constant

The value of the constant c is between 1,900 and 2,000.

Equation (4-30) may be rearranged to give

$$\frac{d_w U_{crit}}{\nu} = c \tag{4-31}$$

The quantity on the left-hand side of Eq. (4-31) is dimensionless if consistent units are used. This dimensionless group is the criterion for determining whether laminar or turbulent flow is the stable form under given conditions. Reynolds showed that, when $d_w U/\nu$ was less than 1,900, the flow was laminar but when the value of the group became greater than 2,000, the flow was turbulent. This dimensionless group is known as the Reynolds modulus or the Reynolds number and is usually written Re (or N_{Re}). It is useful because it is a dimensionless quantity and indicates the manner of flow for any fluid in any size of pipe.

In *laminar flow* (Fig. 4-7*a*) the particles of fluid move in a direction parallel to the solid boundaries, and there is no velocity component normal to the axis of the conduit. In *turbulent flow* (Figs. 4-7*b* and *c*) there is considerable mixing of the fluid during flow. A particle of fluid has an average velocity parallel to the axis of the conduit, but its motion is chaotic, and it has instantaneous velocities both parallel and perpendicular to the axis. The shear stresses in the fluid, which are due to the viscosity, tend to stabilize the unidirectional motion of a fluid particle in flow, while inertia forces, which are due to the weight and velocity of the fluid, tend to disrupt the unidirectional flow. It is these forces which determine the existence of laminar or turbulent flow in a system.

4-9. Velocity Profiles for Laminar Flow in Circular Tubes

Fully developed isothermal laminar flow of an incompressible fluid in a horizontal tube whose axis coincides with the x axis of the rectangular coordinate system may be described by a simplified form of the equations shown in item 3 of Table 2-2. Since the fluid has no velocity components normal to the axis of the tube, v and w are zero; flow is steady, so all derivatives with respect to time become zero; the flow is fully developed, so $\partial u/\partial x$ is zero. Under these conditions the differential equation for such flow is

$$0 = -\frac{g_c}{\rho}\frac{\partial P}{\partial x} + \nu\left(\frac{\partial^2 u}{\partial y^2} + \frac{\partial^2 u}{\partial z^2}\right) \tag{4-32}$$

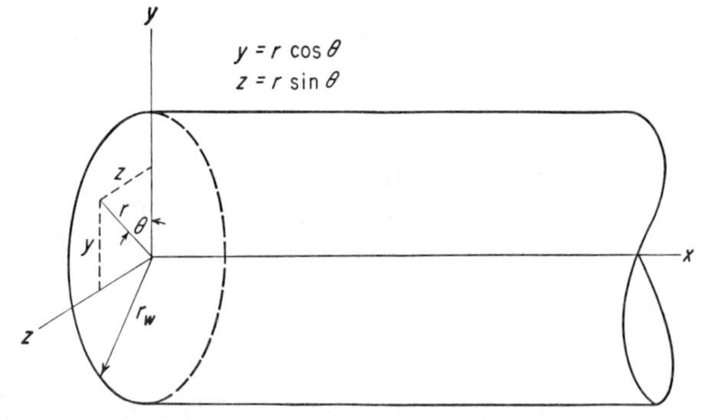

FIG. 4-8. Relation between cylindrical and rectangular coordinate systems.

It is convenient to express Eq. (4-32) in terms of cylindrical coordinates x, r, θ rather than rectangular coordinates x, y, z. The relation between the two coordinate systems is shown in Fig. 4-8.

$$0 = -\frac{g_c}{\rho}\frac{\partial P}{\partial x} + \nu\left(\frac{1}{r}\frac{\partial u}{\partial r} + \frac{\partial^2 u}{\partial r^2} + \frac{1}{r^2}\frac{\partial^2 u}{\partial \theta^2}\right) \tag{4-33}$$

Flow is symmetrical about the x axis, so the derivative with respect to θ becomes zero. Equation (4-33) may be rewritten as

$$\frac{g_c}{\rho}\frac{\partial P}{\partial x} = \frac{\nu}{r}\frac{\partial}{\partial r}\left(r\frac{\partial u}{\partial r}\right) \tag{4-34}$$

Equation (4-34) is a second-order differential equation which may be solved to obtain u in terms of r by integrating with respect to r holding x constant. Two boundary conditions are necessary to evaluate the two arbitrary constants appearing in the solution of a second-order differential equation. These boundary conditions are as follows:

At $r = 0$ axis of the tube

$\frac{\partial u}{\partial r} = 0$ (velocity gradient is zero)

At $r = r_w$ tube wall

$u = 0$ (no slip at the wall)

Integrating Eq. (4-34) and introducing the boundary conditions gives

$$u = -\frac{g_c}{4\mu}\frac{\partial P}{\partial x}(r_w^2 - r^2) \tag{4-35}$$

Equation (4-35) relates the point velocity at any radius to the viscosity, the pressure gradient, and the radius of the tube. If the tube is not horizontal, the point velocity is expressed by

$$u = -\frac{1}{4}\left(\frac{g}{\nu}\frac{\partial Z}{\partial x} + \frac{g_c}{\mu}\frac{\partial P}{\partial x}\right)(r_w^2 - r^2) \tag{4-36}$$

where x is measured along the axis of the pipe and Z is the elevation above a datum plane. For a vertical pipe $\partial Z/\partial x = 1$.

The maximum point velocity u_{max} is obtained from Eq. (4-35) by letting $r = 0$. Hence

$$u_{max} = -\frac{g_c}{4\mu}\frac{\partial P}{\partial x}r_w^2 \tag{4-37}$$

from which it will be seen that

$$-\frac{g_c}{4\mu}\frac{\partial P}{\partial x} = \frac{u_{max}}{r_w^2} \tag{4-38}$$

Substituting Eq. (4-38) into (4-35), the point velocity is obtained as a function of the maximum point velocity.

$$u = u_{max}\left(1 - \frac{r^2}{r_w^2}\right) \tag{4-39}$$

The average velocity is defined by

$$U = \frac{Q}{A} = \frac{Q}{\pi r_w^2} \tag{4-40}$$

where U = average velocity
Q = volumetric rate of flow of fluid in tube
A = cross-sectional area of tube

Q may be expressed in terms of the point velocity u by integration over the cross section from $r = 0$ to $r = r_w$. The incremental discharge at a radius r through an annular ring of thickness dr is

$$dQ = 2\pi r u\, dr \tag{4-41}$$

Substituting the value of u given in Eq. (4-39) and integrating from $r = 0$ to $r = r_w$,

$$Q = 2\pi u_{max}\int_0^{r_w} r\left[1 - \left(\frac{r}{r_w}\right)^2\right]dr = \frac{\pi u_{max} r_w^2}{2} \tag{4-42}$$

The average velocity U may now be found from Eqs. (4-40) and (4-42).

$$U = \frac{\pi u_{\max} r_w^2}{2\pi r_w^2} = \frac{u_{\max}}{2} \tag{4-43}$$

Thus, from Eqs. (4-39) and (4-43), the point velocity is

$$u = 2U\left[1 - \left(\frac{r}{r_w}\right)^2\right] \tag{4-44}$$

Equation (4-44) expresses the point velocity u at any radius r as a function of the average velocity U and the tube radius r_w. The form of the equation indicates that the velocity profile is parabolic in shape. Figure 4-9 shows a velocity profile for incompressible, isothermal, laminar flow in

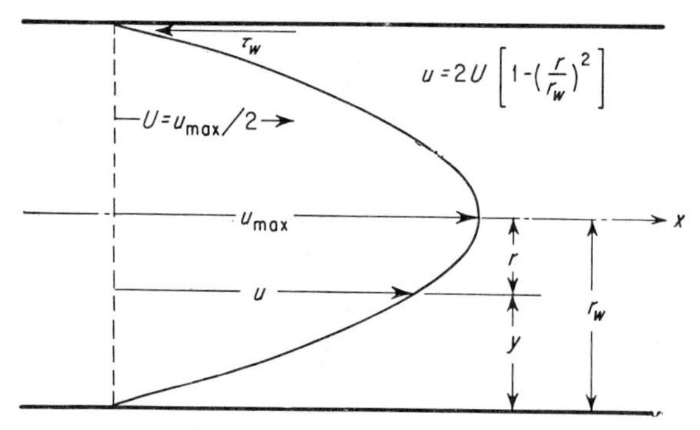

FIG. 4-9. Parabolic velocity distribution for isothermal, incompressible, laminar flow in a circular tube.

circular tubes calculated from Eq. (4-44). Experimental investigations of Senecal and Rothfus [25] and Ferrell, Richardson, and Beatty [10] have essentially substantiated Eq. (4-44). Ferrell and coworkers studied the flow of water and glycerol in a ½-in.-ID glass tube. Point velocities were measured in the region from 0.06 to 0.002 in. from the tube wall, and good agreement with the parabolic velocity-distribution equation was obtained. A dye-displacement technique was used to measure point velocities close to the wall. Senecal and Rothfus measured the velocity distribution for the isothermal flow of air in ½- and ¾-in.-ID smooth tubing. At Reynolds numbers near 1,000 the measured velocity distribution agreed well with Eq. (4-44). Departure from the parabolic distribution was observed as a Reynolds number of 2,000 was approached. The results of these investigations are shown in Fig. 4-10. The points represent experimental data, while the curves are plots of Eq. (4-44).

Equation (4-43) has been verified experimentally by Stanton and Pannell [26] for the flow of air and water in circular tubes and by Senecal and Rothfus [25] for the flow of air. The latter two workers found that in the

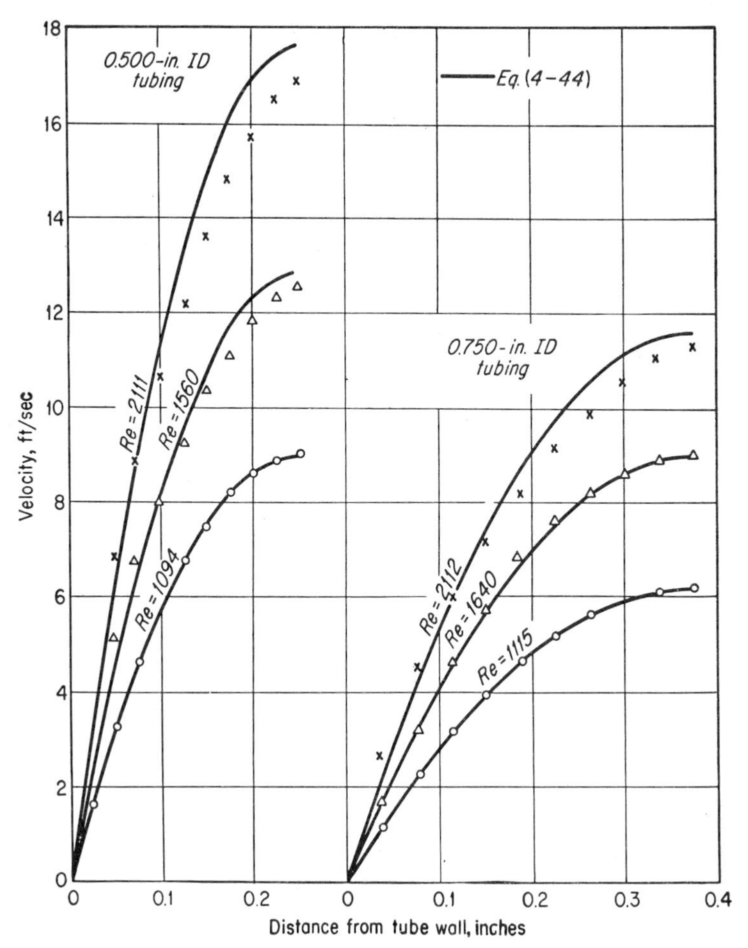

FIG. 4-10. Comparison of experimental and theoretical velocity profiles for laminar flow in circular tubes. Points represent experimental data; curves are plotted from Eq. (4-44). (*From V. E. Senecal, Ph.D. Thesis, Carnegie Institute of Technology,* 1952.)

Reynolds-number range from 500 to 2,000 the ratio of average velocity in the tube to maximum velocity at the center was 0.5 (Figs. 7-3 and 7-4).

4-10. Shear Stresses and Friction Factors for Laminar Flow in Circular Tubes

For laminar flow the shear stress in a fluid is related to the viscosity by Eq. (1-12). For circular tubes, since $r = r_w - y$, the shear stress on the fluid at any radius r is

$$\tau = -\frac{\mu}{g_c}\frac{\partial u}{\partial r} \tag{4-45}$$

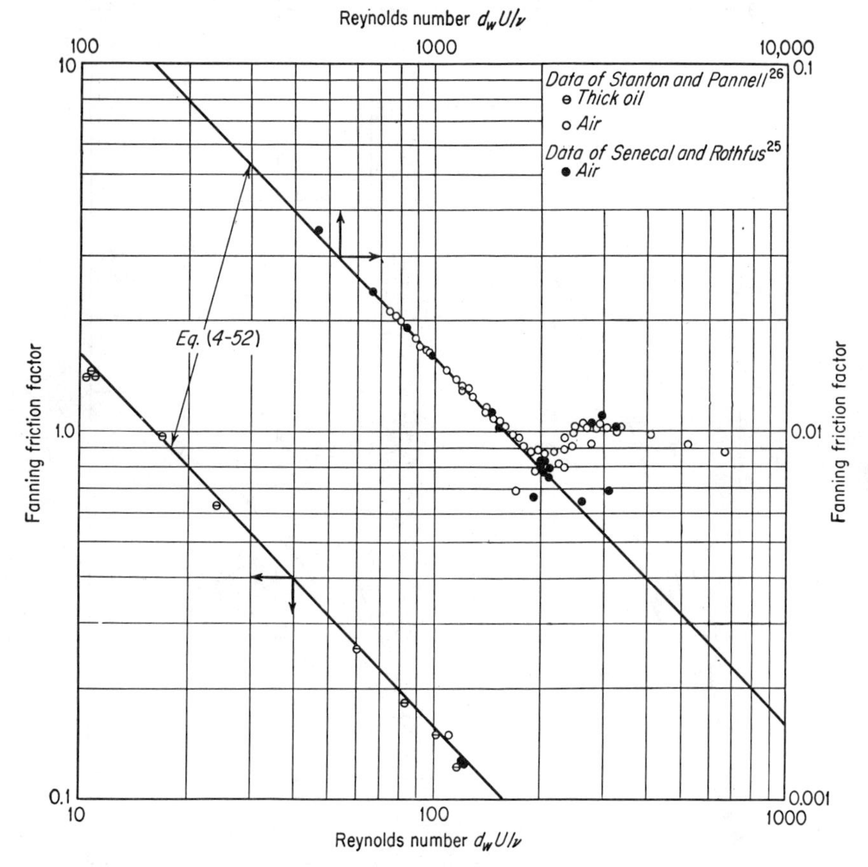

FIG. 4-11. Laminar-flow friction factors in circular tubes.

The shear stress exerted at the wall is

$$\tau_w = -\frac{\mu}{g_c}\left(\frac{\partial u}{\partial r}\right)_{r=r_w} \tag{4-46}$$

From Eq. (4-44)

$$\left(\frac{\partial u}{\partial r}\right)_{r=r_w} = -\frac{4U}{r_w} \tag{4-47}$$

Substituting Eq. (4-47) into (4-46),

$$\tau_w = \frac{4\mu U}{g_c r_w} \tag{4-48}$$

Equation (4-48) gives the shear stress exerted at the wall for laminar flow in terms of the viscosity, the average velocity, and the radius. Combining

Eqs. (4-13) and (4-48) and letting $r_w = d_w/2$, where d_w is the inside diameter of the tube, results in Poiseuille's equation [18] for laminar flow in tubes.

$$\frac{-dP_f}{dx} = \frac{32\mu U}{g_c d_w^2} \tag{4-49}$$

Equation (4-49) relates the pressure loss due to friction to the fluid viscosity, the average velocity in the tube, and the tube diameter. From Eq. (4-18)

$$\frac{-dP_f}{dx} = \frac{2f\rho U^2}{g_c d_w} \tag{4-50}$$

Combining Eqs. (4-49) and (4-50),

$$\frac{32\mu U}{g_c d_w^2} = \frac{2f\rho U^2}{g_c d_w} \tag{4-51}$$

Solving for f,

$$f = \frac{16\mu}{d_w U \rho} = \frac{16}{\text{Re}} \tag{4-52}$$

Equation (4-52) relates the friction factor to the Reynolds number for laminar flow in circular tubes. The data of Stanton and Pannell [26] and Senecal and Rothfus [25] plotted on Fig. 4-11 show agreement with Eq. (4-52) up to a Reynolds number of 2,000.

4-11. Nonisothermal Laminar Flow in Circular Tubes

It was assumed in the integration of Eq. (4-32) that the viscosity and density of the fluid were constant. This is essentially true for isothermal, incompressible flow. If the fluid is being heated or cooled, the properties are also variable, and Eq. (4-34) should be written considering these factors, namely, the variation of viscosity with temperature and the way in which the tube-wall temperature varies with length.

For the heating of liquids during laminar flow in tubes, McAdams [14] points out that the isothermal parabolic velocity profile is distorted and flattened. This is due to the fact there is a radial temperature gradient in the fluid. The high temperature near the wall causes a reduction in the viscosity of the liquid, and the layers of fluid flow at a greater velocity. Conversely, when the liquid is being cooled, the liquid near the wall has a higher viscosity than that near central core, and the velocity profile becomes more pointed than for isothermal laminar flow.

Deissler [9] mathematically analyzed the nonisothermal laminar flow of gases and liquid metals in a circular tube for constant rate of heat transfer

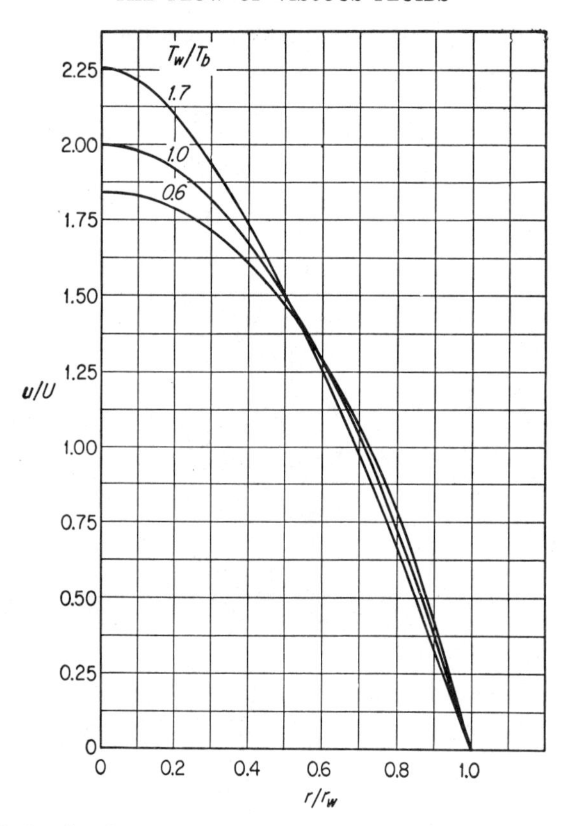

FIG. 4-12. Velocity distribution for laminar flow of the common gases in circular tubes. (*From R. G. Deissler, NACA TN* 2410, *July*, 1951.)

along the tube. For gases the variation of viscosity, thermal conductivity, and density were considered. Figure 4-12 shows the theoretical velocity profiles obtained by Deissler for the common gases. The ratio u/U is plotted versus r/r_w at constant values of T_w/T_b, where T_w is the temperature of the tube wall and T_b is the bulk temperature of the fluid. For heating, T_w/T_b is greater than 1, and the velocity profile is pointed; while for cooling, T_w/T_b is less than 1, and the profile is flattened. These results agree with those of McAdams for liquids, since the viscosity of gases increases with temperature. Figure 4-13 shows Deissler's analytical results for liquid metals, which also agree with those of McAdams.

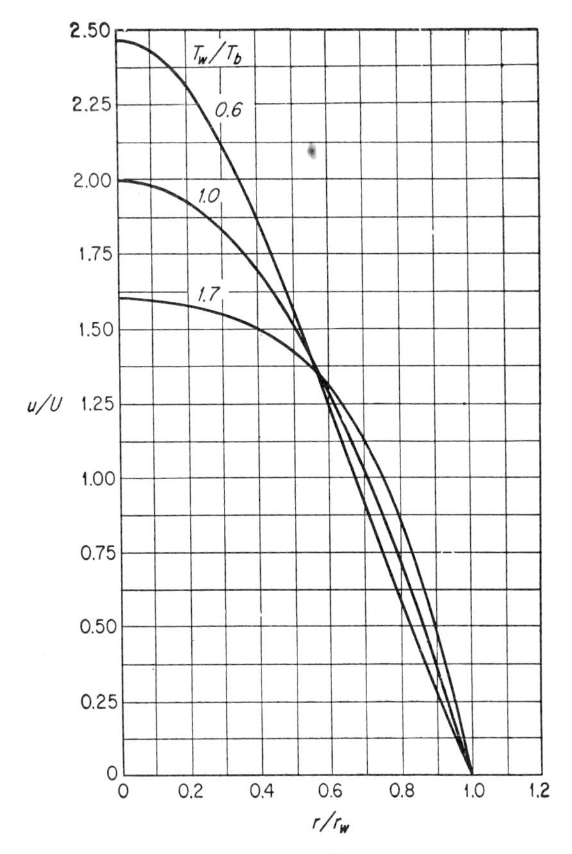

Fig. 4-13. Velocity distribution for laminar flow of liquid metals in circular tubes. (*From R. G. Deissler, NACA TN* 2410, *July,* 1951.)

II. LAMINAR FLOW IN ANNULI

4-12. Flow in Annuli

Flow in annuli is encountered in double-pipe heat exchangers, in which one fluid flows in the inner tube and the other flows in the annular space. Such flow in the annular space may be either laminar or turbulent depending on the Reynolds number. Theoretical equations may be derived for the laminar-flow velocity profiles in annuli, but the geometry of the annulus, although simple, makes the laminar-flow velocity-profile equation somewhat more complicated than that for circular tubes.

Flow in two types of concentric annuli, *plain annuli*, and *modified annuli*, will be considered. Plain annuli consist of two smooth concentric tubes, while modified annuli consist of an outer, smooth tube and an inner, concentric, finned tube, the purpose of the fins on the outer surface of the inner tube being to increase the surface area for heat transfer. The finned tube may have either transverse or longitudinal fins.

4-13. Isothermal Laminar Flow in Plain Annuli

Figure 4-14 shows a cross section of a plain annulus consisting of an outer tube with inside diameter d_2 and an inner tube with outside diameter d_1. The respective radii of the tubes are r_2 and r_1. A velocity profile is shown, with the velocity of the fluid at the boundaries of the annulus assumed to be zero. At some point between the two walls of the annulus

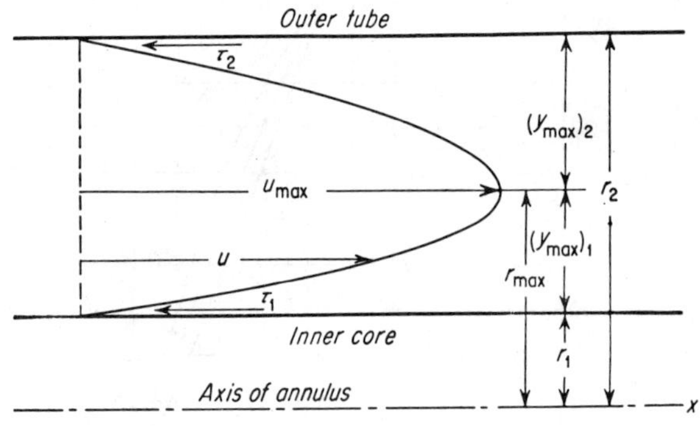

FIG. 4-14. Cross section of a plain concentric annulus showing the velocity profile for laminar flow.

the point velocity reaches a maximum value. Hence, in the derivation of the velocity-profile equation for laminar flow, the following boundary conditions apply:

At $r = r_1$
$u = 0$

At $r = r_2$
$u = 0$

At $r = r_{\max}$
$u = u_{\max}$
$$\frac{\partial u}{\partial r} = 0$$

The differential equation for laminar flow in horizontal annuli is the same as that for circular tubes. The first integration of Eq. (4-34) gives

$$\frac{g_c}{\mu}\frac{\partial P}{\partial x}\frac{r^2}{2} = r\frac{\partial u}{\partial r} + c_1 \tag{4-53}$$

By the boundary condition $r = r_{max}$, $\partial u/\partial r = 0$, the constant c_1 becomes $\frac{g_c}{2\mu}\frac{\partial P}{\partial x} r_{max}^2$. Hence, substituting this value of c_1 in Eq. (4-53),

$$\frac{g_c}{2\mu}\frac{\partial P}{\partial x}(r^2 - r_{max}^2) = r\frac{\partial u}{\partial r} \tag{4-54}$$

Integration of Eq. (4-54) gives

$$\frac{g_c}{2\mu}\frac{\partial P}{\partial x}\left(\frac{r^2}{2} - r_{max}^2 \ln r\right) = u + c_2 \tag{4-55}$$

At $r = r_1$ and $r = r_2$ the point velocity u is zero, which gives the following values of c_2:

$$c_2 = \frac{g_c}{2\mu}\frac{\partial P}{\partial x}\left(\frac{r_2^2}{2} - r_{max}^2 \ln r_2\right) \tag{4-56}$$

or

$$c_2 = \frac{g_c}{2\mu}\frac{\partial P}{\partial x}\left(\frac{r_1^2}{2} - r_{max}^2 \ln r_1\right) \tag{4-57}$$

From Eqs. (4-56) and (4-57) the value of r_{max} is determined in terms of r_1 and r_2.

$$r_{max} = \sqrt{\frac{r_2^2 - r_1^2}{2 \ln (r_2/r_1)}} \tag{4-58}$$

Equation (4-58) is the theoretical position of the point of maximum velocity for laminar flow in an annulus.

Combining Eqs. (4-55) and (4-56), one obtains the expression for the point velocity in an annulus.

$$u = \frac{g_c}{2\mu}\frac{\partial P}{\partial x}\left(\frac{r^2}{2} - \frac{r_2^2}{2} + r_{max}^2 \ln \frac{r_2}{r}\right) \tag{4-59}$$

The average velocity is given by

$$U = \frac{Q}{A} = \frac{\int_{r_1}^{r_2} 2\pi r u\, dr}{\pi(r_2^2 - r_1^2)} \tag{4-60}$$

Integration of Eq. (4-60) results in

$$U = -\frac{g_c}{8\mu}\frac{\partial P}{\partial x}(r_2^2 + r_1^2 - 2r_{max}^2) \tag{4-61}$$

By dividing Eq. (4-59) by (4-61) the point velocity may then be expressed by

$$u = 2U\frac{r_2^2 - r^2 - 2r_{max}^2 \ln (r_2/r)}{r_2^2 + r_1^2 - 2r_{max}^2} \tag{4-62}$$

Equation (4-62) is the relationship between the point velocity at any radius r, the average velocity, and annulus dimensions, for laminar flow in an annulus. It is a function of the radius of both tubes comprising the annulus. As r_1 approaches zero, Eq. (4-62) becomes identical with Eq. (4-44).

A limited number of experimental data are available on laminar velocity profiles in annuli. Rothfus [21] determined velocity profiles for the laminar

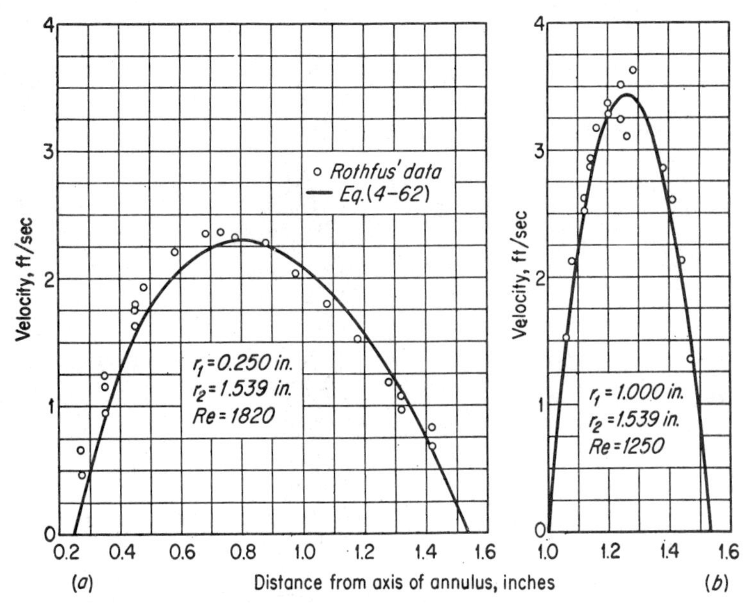

FIG. 4-15. Comparison of experimental and theoretical velocity profiles for laminar flow in plain annuli. (*Data from R. R. Rothfus, Ph.D. Thesis, Carnegie Institute of Technology*, 1948.)

flow of air in two different annuli having the following dimensions: $r_2 =$ 1.539 in., $r_1 = 0.250$ in. and $r_2 = 1.539$ in., $r_1 = 1.000$ in. Rothfus's experimental data are shown in Fig. 4-15, where the point velocity u is plotted versus the radius r. The curves were obtained by calculating the theoretical point velocities from Eq. (4-62). Figure 4-15 shows that Eq. (4-62) agrees closely with the experimental data obtained in the annulus in which $r_1 = 1.000$ in. However, for the annulus in which $r_1 = 0.250$ in., neither the profile nor the theoretical value of r_{max} shows good agreement with the experimental data.

Prengle [19] and Rothfus, Monrad, Sikchi, and Heideger [22] studied laminar and turbulent flow in concentric annuli. Prengle's dye experiments indicated that the laminar velocity profile agreed with Eq. (4-62) up to a Reynolds number of about 700. [For annuli, the diameter term in the Reynolds number is the equivalent diameter $d_2 - d_1$ calculated from Eq.

(4-27).] Above this Reynolds number the point of maximum velocity did not agree with that predicted by Eq. (4-58). Rothfus et al.[22] showed that, for Reynolds numbers ranging from 900 to 1,700, r_{max} was greater than that given by Eq. (4-58) while above a Reynolds number of 1,700 it was less. When Reynolds numbers greater than 10,000 were attained, flow was turbulent, and the point of maximum velocity agreed with that predicted by Eq. (4-58).

4-14. Shear Stresses and Friction Factors for Laminar Flow in Plain Annuli

The rate of shear in the fluid at any radius r during laminar flow in the annulus is given by Eq. (4-45). Differentiating Eq. (4-62) with respect to r gives

$$\frac{\partial u}{\partial r} = \frac{4U(r_{max}^2 - r^2)}{r(r_2^2 + r_1^2 - 2r_{max}^2)} \tag{4-63}$$

From Eqs. (4-45) and (4-63) the shear stress τ at any radius r is given by

$$\tau = \frac{4\mu U(r_{max}^2 - r^2)}{g_c r(r_2^2 + r_1^2 - 2r_{max}^2)} \qquad \text{for } r < r_{max} \tag{4-64}$$

$$\tau = \frac{4\mu U(r^2 - r_{max}^2)}{g_c r(r_2^2 + r_1^2 - 2r_{max}^2)} \qquad \text{for } r > r_{max} \tag{4-65}$$

Letting $r = r_1$ and r_2 in Eqs. (4-64) and (4-65), respectively, τ_1 and τ_2, the shear stresses at the walls of the annulus, may be obtained. The ratio of these two shear stresses is

$$\frac{\tau_1}{\tau_2} = \frac{r_2(r_{max}^2 - r_1^2)}{r_1(r_2^2 - r_{max}^2)} \tag{4-66}$$

Equation (4-66) could also be derived by making a balance of forces on a differential length dx of annular fluid, as was done in the derivation of Eq. (4-13). Although Eq. (4-66) was derived above using relationships applying to laminar flow alone, it also holds for turbulent flow in annuli. Rothfus and coworkers[22] measured shear stresses at the boundaries of two annuli during turbulent flow. Their experimental results agree well with those predicted by Eq. (4-66).

Considering a differential length dx of annular fluid between r_1 and r_2, it may be shown that

$$\frac{-dP_f}{dx} = \frac{2(\tau_2 r_2 + \tau_1 r_1)}{r_2^2 - r_1^2} \tag{4-67}$$

It also may be shown, by considering appropriate portions of the annular fluid, that

$$\frac{-dP_f}{dx} = \frac{2\tau_2 r_2}{r_2{}^2 - r_{\max}^2} \tag{4-68}$$

and

$$\frac{-dP_f}{dx} = \frac{2\tau_1 r_1}{r_{\max}^2 - r_1{}^2} \tag{4-69}$$

Substituting values of τ_1 and τ_2 obtained from Eqs. (4-64) and (4-65) into Eq. (4-67) and replacing the radius terms with diameter terms gives

$$\frac{-dP_f}{dx} = \frac{32\mu U}{g_c(d_2{}^2 + d_1{}^2 - 2d_{\max}^2)} \tag{4-70}$$

Equation (4-70) is Poiseuille's equation for laminar flow in plain annuli and gives the pressure gradient due to friction as a function of the viscosity, the average fluid viscosity, and the diameters of the tubes comprising the annulus.

The equivalent diameter of a plain annulus is

$$d_e = \frac{4\pi(d_2{}^2 - d_1{}^2)}{4\pi(d_2 + d_1)} = d_2 - d_1 \tag{4-71}$$

From Eq. (4-26) a friction factor may be defined for annuli using the equivalent diameter of Eq. (4-71).†

$$f = \frac{g_c(d_2 - d_1)}{2\rho U^2} \frac{-dP_f}{dx} \tag{4-72}$$

Combining Eqs. (4-72) and (4-70),

$$f = \frac{16\mu}{(d_2 - d_1)\rho U} \frac{(d_2 - d_1)^2}{d_2{}^2 + d_1{}^2 - 2d_{\max}^2} \tag{4-73}$$

The first term on the right of Eq. (4-73) is the Reynolds number for the annulus. Replacing $d_{\max}$ by the expression given for it in Eq. (4-58) (with

† The friction factor defined by Eq. (4-72) cannot be related to the wall stresses in annuli as was done for circular tubes [see Eq. (4-17)]. Rothfus et al.[22] define two friction factors for annuli as follows:

$$\tau_1 = \frac{f_1 \rho U^2}{2g_c} \qquad \tau_2 = \frac{f_2 \rho U^2}{2g_c}$$

[see Eqs. (7-94) and (7-95)].

diameter replacing radius) and dividing both numerator and denominator by d_2^2,

$$f = \frac{16}{\text{Re}} \phi \left(\frac{d_1}{d_2}\right) \tag{4-74}$$

where
$$\phi \left(\frac{d_1}{d_2}\right) = \frac{(1 - d_1/d_2)^2}{1 + (d_1/d_2)^2 + \{[1 - (d_1/d_2)^2]/\ln (d_1/d_2)\}} \tag{4-75}$$

The function $\phi(d_1/d_2)$ defined in Eq. (4-75) has a value of unity at $d_1/d_2 = 0$ for a circular tube and a value of 1.5 at $d_1/d_2 = 1$ for parallel planes. In Fig. 4-16 $\phi(d_1/d_2)$ is plotted versus d_1/d_2. Equation (4-74) is compared

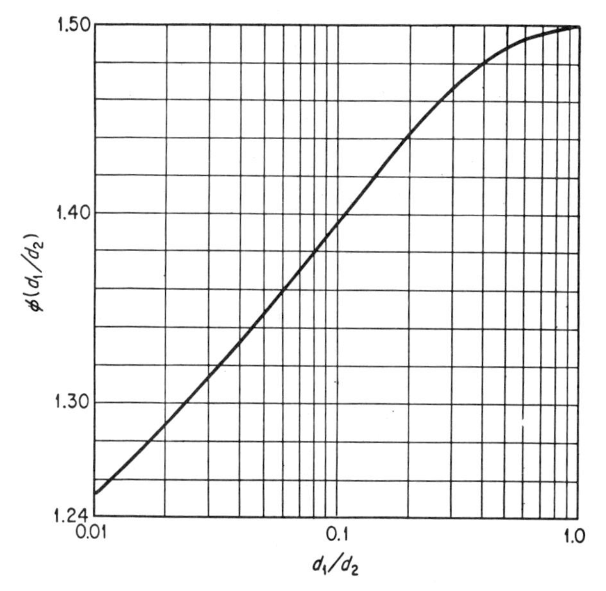

FIG. 4-16. Value of $\phi(d_1/d_2)$ as a function of d_1/d_2 from Eq. (4-75).

with experimental data on laminar flow in annuli in Fig. 4-17, in which $f/\phi(d_1/d_2)$ is plotted versus the annulus Reynolds number. The straight line is Eq. (4-74). Values of d_1/d_2 represented by the data range from 0.0163 to 0.82. Fair agreement is obtained between the theoretical equation and experimental data. An experimental study by Nootbar and Kintner [17] on annuli of small clearances did not show good agreement with Eq. (4-74). With values of d_1/d_2 of 0.64, 0.72, 0.82, and 0.91 these workers obtained a value of $\phi(d_1/d_2)$ of 1.35, compared to a theoretical value of 1.48 to 1.50. They suggest that curvature of the tubes may have an effect that has been neglected in all work on laminar flow in annuli.

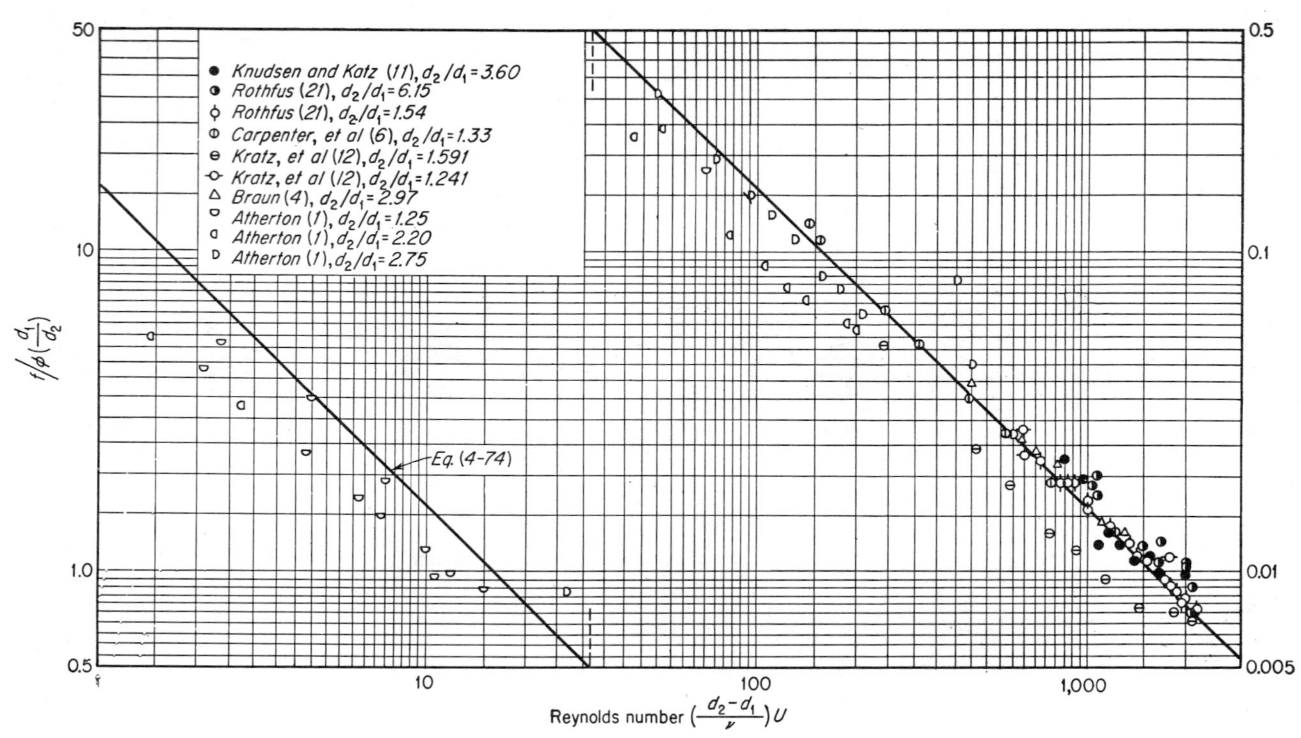

FIG. 4-17. Predicted and experimental friction factors for laminar flow in plain annuli.

4-15. Laminar Flow in Modified Annuli Containing Transverse-finned Tubes

There are indications that laminar flow in annuli containing transverse-finned tubes can be treated similarly to laminar flow in plain annuli. By injecting dye into the stream Knudsen and Katz [11] made visual studies of laminar and turbulent flow in such modified annuli. In the region of laminar flow the dye appeared as a filament for almost the whole length of the annulus. No motion was observed in the space between the fins. This indicated, as did subsequent friction-loss experiments on the same annuli, that for laminar flow the modified annulus behaved like a plain annulus containing a smooth inner core of the same diameter as the fins. A flow pattern for laminar flow in a modified annulus containing a transverse-finned tube is shown in Fig. 7-33.

III. LAMINAR FLOW IN NONCIRCULAR CONDUITS

4-16. Isothermal Laminar Flow between Infinite Parallel Planes

Isothermal laminar flow between two parallel planes of infinite extent is described by the following differential equation:

$$g_c \frac{\partial P}{\partial x} = \mu \frac{\partial^2 u}{\partial {y_c}^2} \tag{4-76}$$

The planes are oriented so that the mid-plane between them coincides with the horizontal xz plane. Flow is in the positive x direction. If the two planes are located a distance b apart, the boundary conditions for Eq. (4-76) are as follows:

At $y_c = 0$

$$\frac{\partial u}{\partial y_c} = 0$$

At $y_c = \frac{b}{2}$

$$u = 0$$

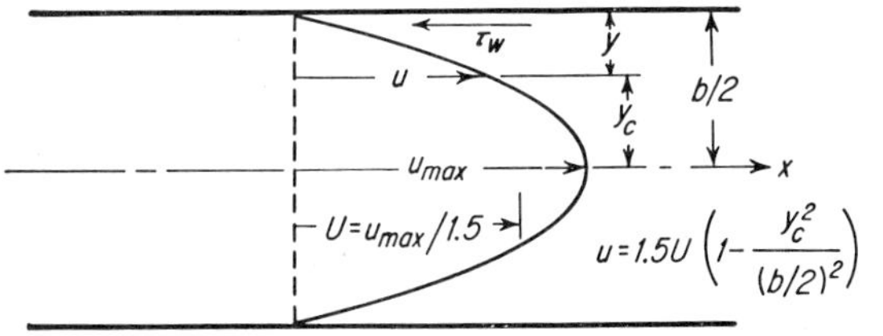

FIG. 4-18. Velocity profile for laminar flow between infinite parallel planes.

Integrating Eq. (4-76) and introducing these boundary conditions gives

$$u = -\frac{g_c}{2\mu}\frac{\partial P}{\partial x}\left[\left(\frac{b}{2}\right)^2 - {y_c}^2\right] \tag{4-77}$$

The average velocity U may be determined by integrating Eq. (4-77) between the limits $y_c = 0$ and $y_c = b/2$ and dividing the result by $b/2$. Thus

$$U = -\frac{g_c}{3\mu}\frac{\partial P}{\partial x}\left(\frac{b}{2}\right)^2 \tag{4-78}$$

Solving for u in Eqs. (4-77) and (4-78),

$$u = \frac{3U}{2}\left[1 - \frac{y_c^{\,2}}{(b/2)^2}\right] \tag{4-79}$$

The maximum point velocity occurs midway between the planes at $y_c = 0$. Hence

$$u_{\text{max}} = \frac{3U}{2} \tag{4-80}$$

Equation (4-79) gives the point velocity between the two planes as a function of the average velocity and the spacing of the planes. Equation (4-80) shows that for laminar flow between parallel planes of infinite extent the maximum velocity is 1.5 times the average velocity. A velocity-distribution curve for laminar flow between parallel planes is shown in Fig. 4-18.

The shear stress on the fluid at any distance y_c from the mid-point may be obtained by differentiating Eq. (4-79) with respect to y_c and making use of Eq. (4-45) with $r = y_c$.

$$\tau = \frac{3\mu U y_c}{g_c(b/2)^2} \tag{4-81}$$

$$\tau_w = \frac{3\mu U}{g_c(b/2)} \tag{4-82}$$

The pressure loss due to friction is related to the shear at the wall by the equation

$$\frac{-dP_f}{dx} = \frac{\tau_w}{b/2} \tag{4-83}$$

Equation (4-83) is derived in the same way as Eq. (4-13). Combining Eqs. (4-82) and (4-83),

$$\frac{-dP_f}{dx} = \frac{12\mu U}{g_c b^2} \tag{4-84}$$

The equivalent diameter of two parallel planes located a distance b apart is $2b$. Using this in the definition of the friction factor and the Reynolds number results in

$$f = \frac{24\mu}{2bU\rho} = \frac{24}{Re} \tag{4-85}$$

Equation (4-85) is the theoretical equation for the friction factor as a function of the Reynolds number for laminar flow between infinite parallel planes.

4-17. Laminar Flow in Noncircular Conduits

Equation (4-32) describes laminar flow in a rectangular conduit oriented with its axis parallel to the horizontal x axis. Rearranging Eq. (4-32),

$$\frac{g_c}{\mu}\frac{\partial P}{\partial x} = \frac{\partial^2 u}{\partial y^2} + \frac{\partial^2 u}{\partial z^2} \tag{4-86}$$

Equation (4-86) is a partial differential equation, and its solution will give u as a function of both y and z. If the axis of the conduit coincides with the x axis and the conduit has dimensions indicated in Fig. 4-19, the boundary conditions for Eq. (4-86) are as follows:

At $z = \pm\frac{a}{2}$ and $y = \pm\frac{b}{2}$

$$u = 0$$

At $y = 0$ and $z = 0$

$$\frac{\partial u}{\partial y} = \frac{\partial u}{\partial z} = 0$$

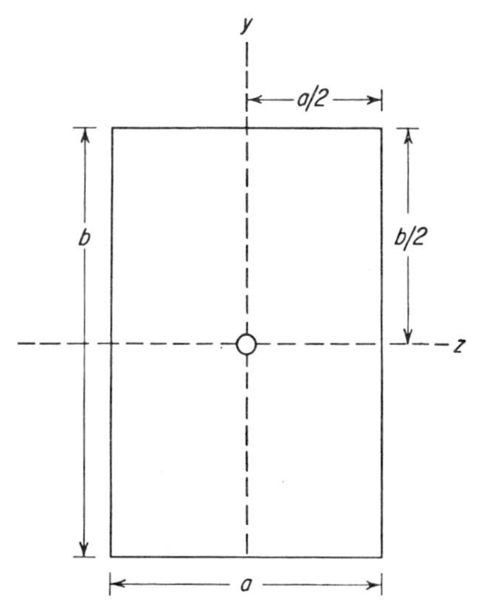

FIG. 4-19. Dimensions of rectangular conduits [Eq. (4-87)].

Timoshenko and Goodier [27] have given a general series solution of Eq. (4-86) in the solving of the analogous equation for torsion in rectangular beams. This series solution is

$$u = \frac{-4a^2 g_c}{\pi^3 \mu}\frac{\partial P}{\partial x}\sum_{n=1,3,5}\frac{1}{n^3}(-1)^{(n-1)/2} \times\left[1 - \frac{\cosh\,(n\pi y/a)}{\cosh\,(n\pi b/2a)}\right]\cos\frac{n\pi z}{a} \tag{4-87}$$

Equation (4-87) shows that u is a function of both y and z and gives the velocity profile for isothermal laminar flow in rectangular conduits. Clark and Kays [7] used a finite-difference approach and a relaxation technique to obtain an approximate solution of Eq. (4-86). The rectangular cross section is divided up into a number of finite increments having dimensions Δy by Δz (see Fig. 4-20*a*). Equation (4-86) in finite-difference form becomes

$$\frac{u_{m-1,n} + u_{m+1,n} - 2u_{m,n}}{(\Delta y)^2} + \frac{u_{m,n-1} + u_{m,n+1} - 2u_{m,n}}{(\Delta z)^2} = \frac{g_c}{\mu}\frac{\partial P}{\partial x} \tag{4-88}$$

The quantities $u_{m-1,n}$, etc., are shown in Fig. 4-20*b*. To solve Eq. (4-88) Δy and Δz are fixed, and, making use of the boundary conditions of the prob-

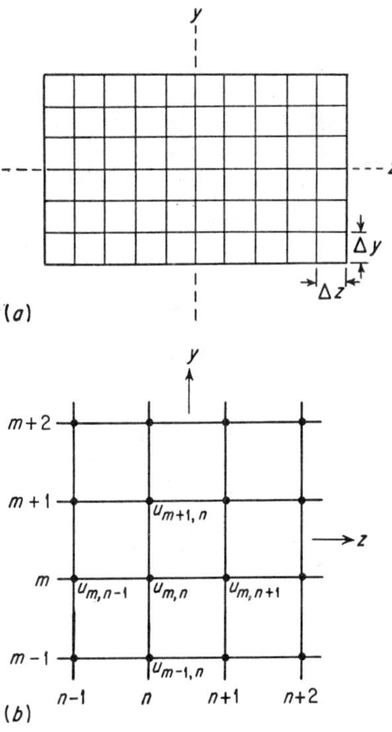

FIG. 4-20. Finite increments of rectangular conduit for solution of Eq. (4-88).

lem, i.e., zero velocity at the wall, values of u are assumed. These values are inserted in Eq. (4-88), and if this equation is not satisfied, new values of u are used. Equation (4-88) is more conveniently solved if brought into dimensionless form by dividing by $\frac{a^2 g_c}{\mu}\frac{\partial P}{\partial x}$, where a is defined in Fig. 4-19. The solution obtained by the finite-difference method was found to be sufficiently accurate by comparing it with the analytical solution represented by Eq. (4-87).

Point velocities for laminar flow in other noncircular horizontal ducts may be predicted by the following:

1. Right isosceles triangle (Fig. 4-21a):

$$u = -\frac{\partial P}{\partial x}\frac{g_c}{2\mu}\left(\tfrac{1}{2}(z^2 + y^2) + zy - \frac{a}{2}(z + y) + \frac{4a^2}{\pi^3}\sum_{n=0}^{\infty}\frac{(-1)^n}{(2n+1)^3 \sinh\{[(2n+1)\pi]/2\}} \times \left[\sinh\frac{(2n+1)\pi y}{a}\cos\frac{(2n+1)\pi z}{a} + \sinh\frac{(2n+1)\pi z}{a}\cos\frac{(2n+1)\pi y}{a}\right]\right) \tag{4-89}$$

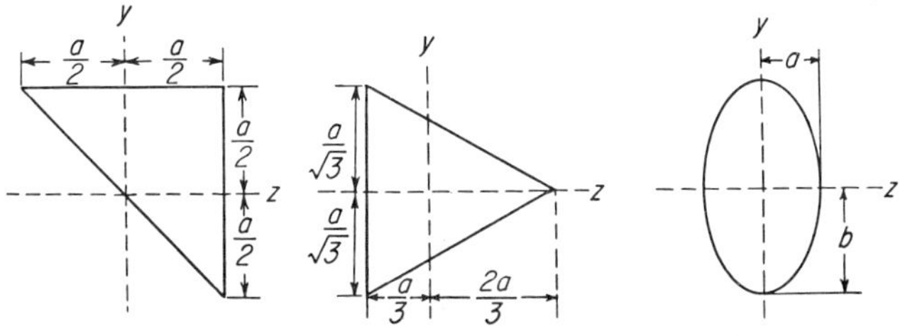

(a) Right isosceles triangle (b) Equilateral triangle (c) Ellipse

FIG. 4-21. Various noncircular conduits.

2. Equilateral triangle (Fig. 4-21b):

$$u = -\frac{\partial P}{\partial x}\frac{g_c}{2\mu}\left[\tfrac{1}{2}(z^2 + y^2) - \frac{1}{2a}(z^3 - 3zy^2) - \frac{2a^2}{27}\right] \tag{4-90}$$

3. Ellipse (Fig. 4-21c):

$$u = -\frac{\partial P}{\partial x}\frac{g_c}{4\mu}\left[(z^2 + y^2) - \frac{a^2 - b^2}{a^2 + b^2}(z^2 - y^2) - \frac{2a^2b^2}{a^2 + b^2}\right] \tag{4-91}$$

4-18. Friction Factors for Laminar Flow in Rectangular Conduits

The equivalent diameter of the rectangular conduit shown in Fig. 4-19 is

$$d_e = \frac{4ab}{2(a + b)} \tag{4-92}$$

Cornish [8] and Lea and Tadros [13] derived a theoretical equation for laminar flow in rectangular conduits expressing the friction factor as a function of the Reynolds number and the conduit dimensions.

$$f\left(\frac{a}{d_e}\right)^2\left[1 - \frac{192}{\pi^5(b/a)}\left(\tanh\frac{\pi b}{2a} - \frac{1}{3^5}\tanh\frac{3\pi b}{2a} + \cdots\right)\right] = \frac{6.0}{\text{Re}} \tag{4-93}$$

Cornish studied laminar flow in a rectangular duct 0.404 by 1.178 cm and obtained results substantiating Eq. (4-93).

Example 4-1

Schiller[24] studied laminar flow in a rectangular conduit measuring 7.9 by 27.8 mm. He reported that the relation between the friction factor and the Reynolds number was $f = 16/\mathrm{Re}$. Using Eq. (4-93), calculate the relationship between the friction factor and the Reynolds number for the conduit studied by Schiller and compare the results with those obtained experimentally.

Solution

$$a = 7.9 \text{ mm}$$

$$b = 27.8 \text{ mm}$$

$$\frac{b}{a} = 3.52$$

$$d_e = \frac{(4)(7.9)(27.8)}{(2)(7.9 + 27.8)} = 12.31 \text{ mm}$$

$$\left(\frac{a}{d_e}\right)^2 = \left(\frac{7.9}{12.31}\right)^2 = 0.412$$

$$\frac{192}{\pi^5(b/a)} = \frac{192}{\pi^5(3.52)} = 0.1785$$

$$\tanh\frac{\pi b}{2a} = \tanh\frac{(\pi)(3.52)}{2} = \tanh 5.52 = 1.000$$

$$1 - \frac{192}{\pi^5(b/a)}\left(\tanh\frac{\pi b}{2a} - \frac{1}{3^5}\tanh\frac{3\pi b}{2a} + \cdots\right) = 0.821$$

Substituting in Eq. (4-93),

$$f(0.412)(0.821) = \frac{6.0}{\mathrm{Re}}$$

from which

$$f = \frac{6.0}{(0.412)(0.821)\mathrm{Re}} = \frac{17.7}{\mathrm{Re}}$$

Fair agreement with Schiller's experimental results is obtained.

4-19. Laminar Flow in Other Noncircular Conduits

Nikuradse[16] reported that for conduits of triangular and trapezoidal cross section the friction factor for laminar flow may be obtained from Eq. (4-52).

$$f = \frac{16}{\mathrm{Re}} \tag{4-52}$$

The equivalent diameter used in determining the Reynolds number is four times the ratio of the cross-sectional area to the wetted perimeter.

Nikuradse also found that the transition from laminar to turbulent flow occurred at a Reynolds number of about 2,000.

BIBLIOGRAPHY

1. Atherton, D. H.: *Trans. ASME*, **48**:145 (1926).
2. Badger, W. L., and J. T. Banchero: "Introduction to Chemical Engineering," McGraw-Hill Book Company, Inc., New York, 1955.
3. Blasius, H.: *Z. Math. u. Phys.*, **56**:1 (1908).
4. Braun, F. W.: Pressure Drop in Annuli Containing Plain and Transverse Fin Tubes, M.S. Thesis, Oregon State College, 1950.
5. Brown, G. G., and associates: "Unit Operations," John Wiley & Sons, Inc., New York, 1950.
6. Carpenter, R. G., A. P. Colburn, E. M. Schoenborn, and A. Wurster: *Trans. AIChE*, **42**:165 (1946).
7. Clark, S. H., and W. M. Kays: *Trans. ASME*, **75**:859 (1953).
8. Cornish, R. J.: *Proc. Roy. Soc.* (*London*), **120A**:691 (1928).
9. Deissler, R. G.: *NACA TN* 2410, 1951.
10. Ferrell, J. K., F. M. Richardson, and K. O. Beatty: *Ind. Eng. Chem.*, **47**:29 (1955).
11. Knudsen, J. G., and D. L. Katz: *Chem. Eng. Progr.*, **46**:490 (1950).
12. Kratz, A. P., H. J. MacIntire, and R. E. Gould: *Univ. Illinois Eng. Expt. Sta. Bull.* 222, 1931.
13. Lea, F. C., and A. G. Tadros: *Phil. Mag.*, **11**:1235 (1931).
14. McAdams, W. H.: "Heat Transmission," 3d ed., McGraw-Hill Book Company, Inc., New York, 1954.
15. Moody, L. F.: *Trans. ASME*, **66**:671 (1944).
16. Nikuradse, J.: *Ing.-Arch.*, **1**:306 (1930).
17. Nootbar, R. F., and R. C. Kintner: *Proc. Midwest. Conf. Fluid Mechanics, 2nd Conf., Ohio State University, 1952*, p. 185.
18. Poiseuille, J. L. M.: *Mém. presentées par divers savants étrangers à l'Acad. Sci. Inst. Nat. France*, **9**:433 (1846).
19. Prengle, R. S.: Ph.D. Dissertation, Department of Chemical Engineering, Carnegie Institute of Technology, 1953.
20. Reynolds, O.: *Trans. Roy. Soc.* (*London*), **174A**:935 (1883).
21. Rothfus, R. R.: Velocity Gradients and Friction in Concentric Annuli, Ph.D. Thesis, Carnegie Institute of Technology, 1948.
22. Rothfus, R. R., C. C. Monrad, K. G. Sikchi, and W. J. Heideger: *Ind. Eng. Chem.*, **47**:913 (1955).
23. Rouse, H.: "Elementary Mechanics of Fluids," John Wiley & Sons, Inc., New York, 1948.
24. Schiller, L.: *Z. angew. Math. u. Mech.*, **3**:2 (1923).
25. Senecal, V. E., and R. R. Rothfus: *Chem. Eng. Progr.*, **49**:533 (1953).
26. Stanton, T. E., and J. R. Pannell: *Trans. Roy. Soc.* (*London*), **214A**:199 (1914).
27. Timoshenko, S. P., and J. N. Goodier: "Theory of Elasticity," 2d ed., McGraw-Hill Book Company, Inc., New York, 1951.
28. Walker, W. H., W. K. Lewis, W. H. McAdams, and E. R. Gilliland: "Principles of Chemical Engineering," 3d ed., McGraw-Hill Book Company, Inc., New York, 1937.

CHAPTER 5

TURBULENCE

5-1. Introduction

In this chapter turbulence is described qualitatively and to some extent quantitatively. The intention is not to present a complete discussion of turbulence but to give the reader a preliminary understanding of turbulence so that he will be familiar with prevailing theories, experimental methods, quantities involved, and application to fluid-flow problems.

Turbulence has been the concern of those interested in fluid mechanics since Reynolds's [20] classical experiment showing that under certain conditions flow in a pipe was direct, or streamlined, and that under other conditions flow was turbulent, or sinuous. The term *sinuous* was applied because the path of fluid particles in turbulent flow was observed to be sinusoidal or irregular. The chaotic nature of fluid motion during turbulent flow has prevented an exact quantitative description of it. Historically, there have been two important theories of turbulence. The earliest theory, or empirical theory, attempted to describe turbulence on the basis of the temporal mean of the flow quantities. Although this theory was highly successful in showing the effect of turbulence on the processes of momentum, heat, and mass transfer, it assumes a mechanism of turbulence which is not realistic. The second theory, or statistical theory of turbulence, considers a more realistic mechanism of turbulent flow, and is based on a knowledge of the fluctuations of the various flow quantities. The development of high-speed electronic equipment has aided the study of turbulence materially from a statistical standpoint.

The empirical theories of turbulence were developed during the period from 1900 to 1915. The outstanding worker in the field at that time was Ludwig Prandtl, whose *mixing-length theory* remained the most important theory of turbulence until about 1935. At present the Prandtl mixing-length theory finds extensive application in turbulent-flow problems where only the temporal mean of the flow quantities is known. Investigators in fluid mechanics from the time of Reynolds realized that a knowledge of the fluctuations occurring during turbulent flow would add materially to the understanding of turbulence. Reynolds [21] modified the Navier-Stokes momentum equations to include the fluctuation of the velocity.

In 1921 Taylor [30] introduced the idea that the velocity of a fluid in turbulent motion was a random continuous function of position and time. He applied statistics to this random continuous function and in 1935 published his classic series of papers on the "Statistical Theory of Turbulence." [31] This represented a new approach to the understanding of turbulence, not previously considered by many workers in the field. The result of Taylor's statistical theory brought renewed effort in experimental work. After 1935 numerous investigations to determine the velocity fluctuations during turbulent flow were made. Such investigations have continued, and today turbulent flow is studied in large wind tunnels with the help of the most modern electronic measuring and control equipment. Some of the prominent workers in the field of the statistical theory of turbulence are von Kármán, who also made important contributions to Prandtl's mixing-length theory, Kolmogoroff, Burgers, Townsend, Dryden, Batchelor, Lin, and Chandrasekhar. The literature on turbulent flow is very extensive, and only a few of the important references will be mentioned here. The reader is referred to them for a more complete bibliography of turbulent flow.

5-2. Stability of Laminar Flow

The Reynolds experiment referred to in Sec. 4-8 showed that laminar flow became unstable as the velocity of flow increased in a given tube. It demonstrated that the transition from laminar to turbulent flow occurred at a value of the Reynolds number in the neighborhood of 2,000. When the Reynolds number was less than 2,000, flow was always laminar. The transition to turbulent flow usually took place in the range of Reynolds numbers from 2,000 to 13,000. There have been instances where laminar flow has existed well above a Reynolds number of 2,000. This is effected by minimizing all disturbances in the flow system.

The actual mechanism of transition from laminar to turbulent flow is not well known. The Reynolds number represents a ratio of the inertia to the viscous forces in the fluid. The viscous forces tend to stabilize laminar flow. At high Reynolds numbers the inertia forces predominate over the viscous forces, and turbulent flow results. With flow in circular tubes it appears that the Reynolds number of 2,000 is the value below which disturbances in the fluid will die out and not produce turbulence. However, the fact that under special conditions laminar flow may be preserved at higher Reynolds numbers indicates that turbulence may be caused by external conditions. The absence of these external conditions allows laminar flow to be maintained even though inertia forces are large.

A number of experimental and theoretical studies on the transition from laminar to turbulent flow [3, 6, 11, 13, 14, 18, 19, 22, 29] have been made for flow over flat plates and flow through circular tubes and between parallel planes. The results for circular tubes have been substantially in agreement. Maurer [13] indicated that turbulence began at a Reynolds number of about 1,500. Senecal and Rothfus [27] found that the major transition from laminar to turbulent flow occurred at a Reynolds number of 2,100, with the remaining transition being complete at about 2,800. Lindgren [12] claims that transition continues up to Reynolds numbers of 3,300.

Gibson [6] and Prengle and Rothfus [18] have made detailed studies of the beginning of transition from laminar to turbulent flow in pipes. Gibson, by means of a special entrance, observed the breakdown of a dye filament at Reynolds numbers as high as 26,000. The dye filament broke down first at a point about $0.6r_w$, where r_w is the tube radius, from the axis of the tube. It was suggested that transition to turbulent flow was related to the energy distribution across the pipe, since for laminar flow the rate of change of kinetic energy with radius is a maximum at $r = 0.577r_w$. Prengle and Rothfus observed that at Reynolds numbers as low as 932 departure from laminar behavior could be observed, and as the Reynolds number increased, the region of turbulent flow spread outward toward the wall of the pipe.

The experimental work of Davies and White [3] on flow between parallel planes indicated a critical Reynolds number of 2,000 ($\mathrm{Re} = 2bU/\nu$, where b is the spacing between the planes). They also showed that for the range of Reynolds numbers from 280 to 1,800 small disturbances die out.

Theoretical studies of the stability of laminar flow have dealt mainly with the effect of disturbances on the vorticity or rate of rotation of the fluid elements. The components of vorticity in the x, y, and z directions are shown in Eqs. (3-11) to (3-13). The vorticity is twice the rate of rotation of the element about the respective axis [see Eq. (3-14)]. The nature of the disturbances which cause transition differs according to the type of flow taking place. For flow in conduits disturbances at the entrance are generally thought to be sufficient to cause turbulent flow. The effect of roughness is considered to be slight. Schiller [23] found that roughness had little effect on the critical Reynolds number and postulated, as Reynolds did in 1883, that transition was a function of the magnitude of the other disturbances, probably those due to the shape of the entrance. For flow over immersed bodies the main source of disturbance is considered to be the turbulence in the main stream. This has been shown experimentally to have an effect on the transition to turbulence in the boundary layer.[4,25] Most theoretical studies have dealt with flow over flat plates and between parallel planes. Disturbances, whatever their nature, are considered to be of a periodic nature, and the effects of the amplitude and frequency of the disturbance have been considered separately.

One of the earliest theoretical investigations on the stability of laminar flow was conducted by Rayleigh.[19] He considered the condition of laminar flow as an equilibrium condition which was subjected to periodic disturbances of a given amplitude. He established the criterion that a continuously increasing or decreasing vorticity in moving from the axis represented a condition for stability. One of the more detailed theoretical works is that by Lin,[11] who determined the effect of the frequency of periodic disturbances on the stability of laminar flow. Lin's work is a revision of the mathematical theory of Tollmien[34] and Schlichting.[24] All these works determine the effect of periodic disturbances on the stability of laminar flow. Lin gives plots for flat plates and infinite parallel planes in which curves relating the frequency of the disturbance to the Reynolds number are shown. The results indicate a region in which the disturbances are damped out and a region in which the disturbances are amplified. In 1947 Schubauer and Skramstad[25] detected periodic disturbances in the boundary layer on a flat plate and essentially substantiated Lin's theoretical results. For flow between parallel planes Lin's results predict a minimum Reynolds number $2bU/\nu$ of about 21,200. Below this Reynolds number disturbances of any frequency are damped. Thomas[33] substantiates these results. Meksyn and Stuart[14] showed the effect of the amplitude of the disturbances on stability of flow between parallel planes. A plot of amplitude versus Reynolds number gives a minimum Reynolds number of 11,600, below which disturbances of all amplitudes die out. For disturbances of infinitesimal amplitude the critical Reynolds number is 20,000. These results are not in good agreement with the experimental critical Reynolds number of 2,000.

5-3. Definition of Various Velocity Terms

Turbulence is characterized by random, chaotic motion of fluid particles. At any point in a three-dimensional turbulent fluid the velocity varies with respect to both time and direction. The *instantaneous velocities* and *pressure* are u_i, v_i, w_i, and p_i, while the *time averages* of these quantities are u, v, w, P, where

$$\begin{aligned} \frac{1}{t}\int_0^t u_i\,dt &= u \\ \frac{1}{t}\int_0^t v_i\,dt &= v \\ \frac{1}{t}\int_0^t w_i\,dt &= w \\ \frac{1}{t}\int_0^t p_i\,dt &= P \end{aligned} \tag{5-1}$$

At any particular instant the difference between u_i and u (v_i and v, w_i and w, and p_i and P) represents the amount of fluctuation of the instantaneous quantity from the average quantity. The *fluctuating components* are u', v', w', and p'.

$$\begin{aligned} u_i &= u + u' \\ v_i &= v + v' \\ w_i &= w + w' \\ p_i &= P + p' \end{aligned} \tag{5-2}$$

From Eqs. (5-1) it is evident that

$$\begin{aligned} \frac{1}{t}\int_0^t u'\,dt &= \overline{u'} = 0 \\ \frac{1}{t}\int_0^t v'\,dt &= \overline{v'} = 0 \\ \frac{1}{t}\int_0^t w'\,dt &= \overline{w'} = 0 \\ \frac{1}{t}\int_0^t p'\,dt &= \overline{p'} = 0 \end{aligned} \tag{5-3}$$

The mean square of the velocity fluctuations is defined as

$$\begin{aligned} \frac{1}{t}\int_0^t u'^2\,dt &= \overline{u'^2} \\ \frac{1}{t}\int_0^t v'^2\,dt &= \overline{v'^2} \\ \frac{1}{t}\int_0^t w'^2\,dt &= \overline{w'^2} \end{aligned} \tag{5-4}$$

The root mean square of the velocity fluctuations is $\sqrt{\overline{u'^2}}$, $\sqrt{\overline{v'^2}}$, $\sqrt{\overline{w'^2}}$. The turbulent shear components are

$$\frac{1}{t}\int_0^t u'v'\,dt = \overline{u'v'}$$
$$\frac{1}{t}\int_0^t v'w'\,dt = \overline{v'w'} \tag{5-5}$$
$$\frac{1}{t}\int_0^t u'w'\,dt = \overline{u'w'}$$

When the mean flow is in the direction of the x axis, v and w are zero; however, the quantities $\overline{u'^2}$, $\overline{v'^2}$, $\overline{w'^2}$, $\overline{u'v'}$, $\overline{v'w'}$, and $\overline{u'w'}$ are all finite.

5-4. Reynolds Equations for Incompressible Turbulent Flow

Even though flow is turbulent, the momentum equations are still applicable if instantaneous values of the velocity and pressure are used. Since the fluctuations are random and chaotic, it is evident that the solution of the momentum equations for turbulent flow is impossible. Reynolds modified the momentum equation by introducing the mean values and fluctuating values of the flow quantities in place of the instantaneous values. In doing so he assumed the fluctuations to be continuous functions of time and space.

The momentum equation for the x direction for incompressible flow is (neglecting field forces)

$$\frac{Du_i}{Dt} = -\frac{g_c}{\rho}\frac{\partial p_i}{\partial x} + \nu\,\nabla^2 u_i \tag{5-6}$$

For incompressible flow, since div $\mathbf{V} = 0$,

$$\frac{Du_i}{Dt} = \frac{\partial u_i}{\partial t} + \frac{\partial(u_i^2)}{\partial x} + \frac{\partial(u_i v_i)}{\partial y} + \frac{\partial(u_i w_i)}{\partial z} \tag{5-7}$$

Substituting Eq. (5-7) into Eq. (5-6) and collecting terms of like derivatives gives

$$\frac{\partial u_i}{\partial t} = -\frac{g_c}{\rho}\frac{\partial p_i}{\partial x} + \frac{\partial}{\partial x}\left(\nu\frac{\partial u_i}{\partial x} - u_i^2\right) + \frac{\partial}{\partial y}\left(\nu\frac{\partial u_i}{\partial y} - u_i v_i\right) + \frac{\partial}{\partial z}\left(\nu\frac{\partial u_i}{\partial z} - u_i w_i\right) \tag{5-8}$$

Each term of Eq. (5-8) is now considered separately. The instantaneous quantities are replaced by their equivalents as given in Eqs. (5-2). After the differentiation is carried out, the fluctuating values of the quantities are replaced by their mean values [these are zero by Eqs. (5-3)].

$$\frac{\partial u_i}{\partial t} = \frac{\partial}{\partial t}(u + u') = \frac{\partial u}{\partial t} + \frac{\partial u'}{\partial t} = \frac{\partial u}{\partial t} \tag{5-9}$$

Similarly

$$\frac{\partial p_i}{\partial x} = \frac{\partial}{\partial x}(P + p') = \frac{\partial P}{\partial x} + \frac{\partial p'}{\partial x} = \frac{\partial P}{\partial x} \tag{5-10}$$

$$\nu \frac{\partial u_i}{\partial x} - u_i^2 = \nu \frac{\partial}{\partial x}(u + u') - (u + u')^2$$

$$= \nu\left(\frac{\partial u}{\partial x} + \frac{\partial u'}{\partial x}\right) - u^2 - 2uu' - u'^2$$

$$= \nu \frac{\partial u}{\partial x} - u^2 - \overline{u'^2} \tag{5-11}$$

Similarly

$$\nu \frac{\partial u_i}{\partial y} - u_i v_i = \nu \frac{\partial u}{\partial y} - uv - \overline{u'v'} \tag{5-12}$$

$$\nu \frac{\partial u_i}{\partial z} - u_i w_i = \nu \frac{\partial u}{\partial z} - uw - \overline{u'w'} \tag{5-13}$$

Substituting Eqs. (5-9) to (5-13) into (5-8) and rearranging,

$$\frac{Du}{Dt} = -\frac{g_c}{\rho}\frac{\partial P}{\partial x} + \nu\,\nabla^2 u - \frac{\partial \overline{u'^2}}{\partial x} - \frac{\partial \overline{u'v'}}{\partial y} - \frac{\partial \overline{u'w'}}{\partial z} \tag{5-14}$$

Similar equations may be obtained for the y and z directions. These are the Reynolds momentum equations for incompressible turbulent flow. Comparing Eq. (5-14) with (5-6), it is seen that the instantaneous quantities in Eq. (5-6) have been replaced by mean quantities and that three additional terms are present which involve the fluctuating components of the velocity.

5-5. The Reynolds Stresses

Multiplying each side of Eq. (5-14) by ρ gives

$$\rho \frac{Du}{Dt} = -g_c \frac{\partial P}{\partial x} + \mu\,\nabla^2 u - \frac{\partial \rho\overline{u'^2}}{\partial x} - \frac{\partial \rho\overline{u'v'}}{\partial y} - \frac{\partial \rho\overline{u'w'}}{\partial z} \tag{5-15}$$

The term $\mu\,\nabla^2 u$ gives the stresses in the fluid due to the mean velocity u and the molecular viscosity μ. The last three terms on the right of Eq.

(5-15) give the stresses in the fluid due to the turbulent fluctuations. The quantities $-\rho\overline{u'^2}$, $-\rho\overline{u'v'}$, and $-\rho\overline{u'w'}$ are called the Reynolds stresses or eddy stresses. In all, there are six independent Reynolds stresses: three normal stresses, $-\rho\overline{u'^2}$, $-\rho\overline{v'^2}$, and $-\rho\overline{w'^2}$, and three shear stresses, $-\rho\overline{u'v'}$, $-\rho\overline{v'w'}$, and $-\rho\overline{u'w'}$.

5-6. Prandtl's Mixing-length Theory

The mixing-length theory arose out of early attempts by Prandtl and von Kármán to describe turbulent flow quantitatively. It was well known that the friction loss occurring during turbulent flow was many times that attributable to molecular viscosity alone. Prandtl [16,17] reasoned that the high friction loss of turbulent flow was due to the exchange of momentum between fluid particles. Consider two adjacent layers of fluid moving at different velocities. If a particle of fluid moves from one layer to the other, a momentum exchange between the two layers results. Slow particles entering the faster layer act as a drag on it. Fast particles entering the slower layer tend to speed it up. The result is the equivalent of shear force between the two layers. Prandtl introduced the concept of *mixing length*, which he described physically as the distance a particle of fluid (containing many molecules) moved transverse to the mean flow before it lost its identity and mingled with other particles.

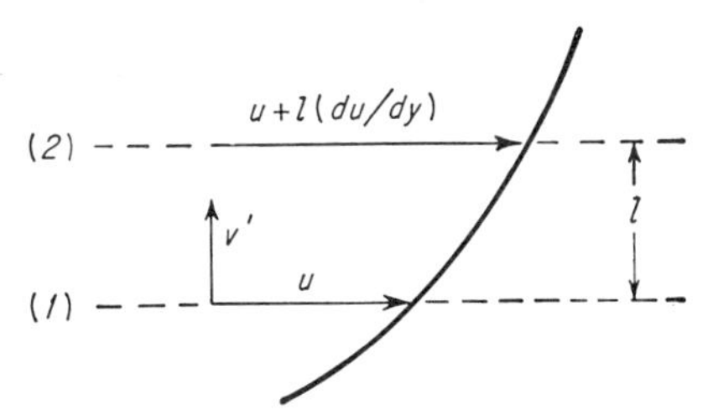

FIG. 5-1. Portion of turbulent-velocity-profile curve showing the Prandtl mixing length.

In Fig. 5-1 the curve represents a portion of a turbulent velocity profile with mean flow in the x direction. Hence v and $w = 0$. At section (1) the mean velocity is u, and at section (2), a distance l away, it is $u + l(du/dy)$. At section (1) the fluctuating velocity in the x direction is u', and the fluctuating cross velocity is v' (in the y direction). Prandtl *defines* the mixing length by the following equation:

$$u' = l_i \frac{du}{dy} \tag{5-16}$$

That is, the x-direction fluctuating velocity is of the order of the difference in the mean velocities of two layers separated by a distance l_i, where l_i is the instantaneous mixing length. The turbulent shear stress between the two layers due to momentum exchange is the instantaneous rate of momentum transfer per unit area. The rate of mass exchange per unit area is $\rho v'$. Thus the turbulent shear stress on the fluid is

$$(\tau_t)_i = \frac{\rho v' l_i}{g_c} \frac{du}{dy} \tag{5-17}$$

where $(\tau_t)_i$ is the instantaneous turbulent shear stress. Substituting Eq. (5-16) into Eq. (5-17) gives an expression for the instantaneous turbulent shear stress, i.e.,

$$(\tau_t)_i = \frac{\rho u' v'}{g_c} \tag{5-18}$$

The mean turbulent shear stress is

$$\tau_t = \frac{\overline{\rho u' v'}}{g_c} \tag{5-19}$$

The shear stress given in Eq. (5-19) is the Reynolds shear stress defined in Sec. 5-5. Prandtl also assumed that v' and u' were of the same order, which results in the following expression for the mean turbulent shear stress:

$$\tau_t = \frac{\rho}{g_c}\left(l \frac{du}{dy}\right)^2 \tag{5-20}$$

where l is the *mean mixing length.*

5-7. Combination of Laminar and Turbulent Shear Stress; Eddy Viscosity

The total shear stress τ on the fluid in Fig. 5-1 consists of a laminar and a turbulent shear stress. The laminar portion is given by Eq. (1-12).

$$\tau_l = \frac{\mu}{g_c} \frac{du}{dy} \tag{5-21}$$

Addition of Eqs. (5-20) and (5-21) gives the total shear stress.

$$\tau = \tau_l + \tau_t = \frac{\mu}{g_c} \frac{du}{dy} + \frac{\rho}{g_c}\left(l \frac{du}{dy}\right)^2 \tag{5-22}$$

Rearranging Eq. (5-22),

$$\tau g_c = \mu \frac{du}{dy} + \rho l^2 \frac{du}{dy} \frac{du}{dy} \tag{5-23}$$

or

$$\tau g_c = (\mu + \mathrm{E}_M) \frac{du}{dy} \tag{5-24}$$

where $\mathrm{E}_M = \rho l^2 (du/dy)$ and is defined as the eddy viscosity. It is analogous to the molecular viscosity μ, since the product of eddy viscosity and velocity gradient gives the turbulent shear stress. Equation (5-24) gives the total shear stress on the fluid at any section as a function of the molecular viscosity, the eddy viscosity, and the mean velocity gradient.

The mixing length is a function of the distance y from the wall, the exact form of this function being unknown. Prandtl and von Kármán postulated relationships between these quantities for turbulent flow in conduits and developed equations for the turbulent velocity profile which agreed well with experiment (see Chap. 7).

5-8. The Statistical Theory of Turbulence; Correlation Coefficients

The mixing-length theory was an attempt to describe turbulence quantitatively. However, it is not realistic to consider discrete fluid particles which retain their identity over a certain distance. If one could follow a particle of fluid starting at a certain point, it is probable that a continuous mixing would be observed. The particle loses some of its identity as soon as it moves an infinitesimal distance from where it was originally observed. It follows that turbulent motion is continuous, a condition which was assumed by Reynolds in deriving the momentum equations for turbulent flow. The velocity is everywhere continuous, and the derivatives of the velocity are continuous. The idea of continuity may be illustrated by Fig. 5-2, where the x direction of the velocity fluctuation at a given point is plotted as a function of time.

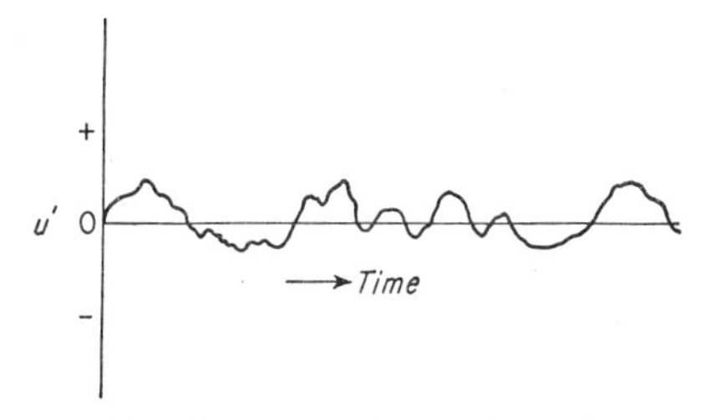

FIG. 5-2. Variation of the velocity fluctuation with time.

In turbulent flow the velocity **V** of the fluid is a random, continuous function of time and space. It is a random function because the instantaneous value of **V** at any point cannot be predicted from the mean value of the flow quantities. The values of **V** at any point are distributed according to the laws of probability. If the instantaneous velocity is considered to be the sum of a mean velocity and a fluctuating velocity, the distribution of the fluctuating velocity follows the gaussian, or normal, distribution curve shown in Fig. 5-3. This curve is a plot of the frequency of occurrence of a fluctuating velocity versus the value of the fluctuating velocity. The most frequent value of the fluctuating velocity is zero. The properties of the normal distribution curves are such that 68.2 per cent of the fluctuating velocities have values in the range $-A$ to A; 95.5 per cent have values in the range $-2A$ to $2A$; and 99.7 per cent have values in the range $-3A$ to $3A$. The value of A depends on the intensity of the turbulence, a term which will be discussed in more detail in succeeding sections.

This statistical concept of the distribution of the fluctuating quantities avoids the difficulties associated with the mixing-length theory. Rather than dealing with frequency distribution of the fluctuating quantities, Taylor [30,31] introduced correlation coefficients between the quantities. A correlation coefficient is defined in the following way:

$$R_t = \frac{\overline{u'_t u'_{t+\Delta t}}}{\overline{u'^2}} \tag{5-25}$$

where u'_t is the fluctuating velocity at time t and $u'_{t+\Delta t}$ is the fluctuating velocity at a time Δt later. Equation (5-25) gives the relation between

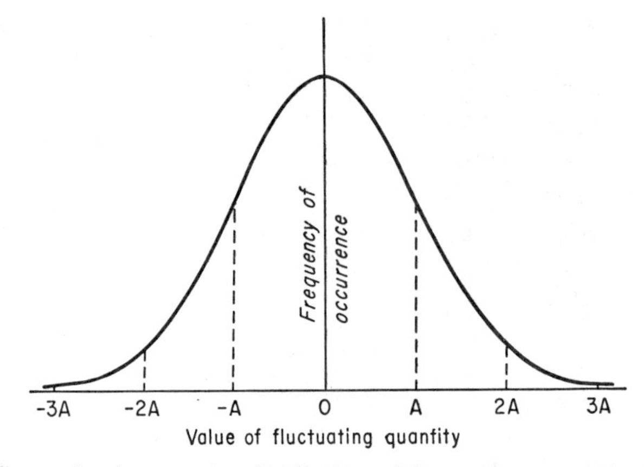

FIG. 5-3. Curve showing gaussian distribution of fluctuating quantities in turbulent flow.

the fluctuating velocities at different times. It gives the effect of u'_t on $u'_{t+\Delta t}$. If Δt is zero, the coefficient is unity. As Δt becomes large, R_t becomes zero, indicating the u'_t has no effect on $u'_{t+\Delta t}$.

The correlation coefficient given in Eq. (5-25) takes the Lagrangian view of flow: the paths of fluid particles are followed. Taylor also defined correlation coefficients based on the Eulerian view, which involves the value of the fluctuating quantity at two different points in space. For example, u' may be measured at two different points in space. The correlation coefficient $R_{xu'}$ is the correlation between values of u' at two points separated by a distance x in the direction of the x coordinate.

$$R_{xu'} = \frac{\overline{u'_1 u'_2}}{\sqrt{\overline{u'^2_1}}\sqrt{\overline{u'^2_2}}} \tag{5-26}$$

where u'_1 and u'_2 are values of u' at points 1 and 2 separated by a distance x.

Similarly

$$R_{yu'} = \frac{\overline{u_1' u_2'}}{\sqrt{\overline{u_1'^2}}\sqrt{\overline{u_2'^2}}} \tag{5-27}$$

where u_1' and u_2' are values of u' at points 1 and 2 separated by a distance y measured in the y direction. Also

$$R_{zu'} = \frac{\overline{u_1' u_2'}}{\sqrt{\overline{u_1'^2}}\sqrt{\overline{u_2'^2}}} \tag{5-28}$$

where u_1' and u_2' are values of u' at points 1 and 2 separated by a distance z measured in the z direction. Similar correlation coefficients may be defined for v' and w'.

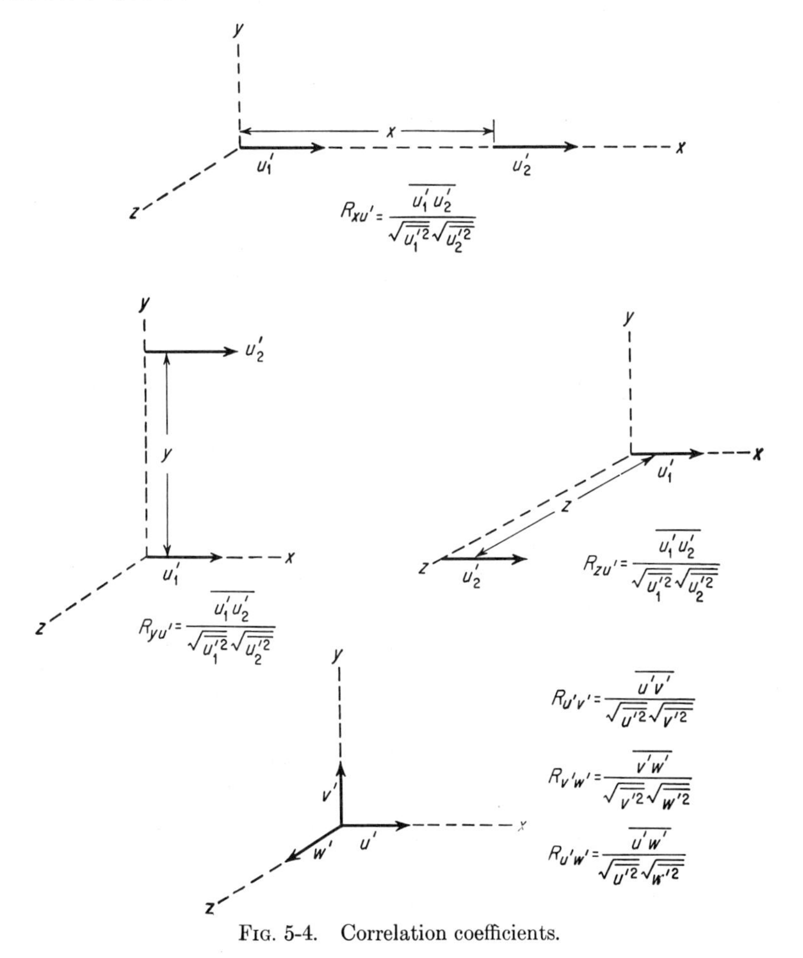

FIG. 5-4. Correlation coefficients.

At a given point correlation coefficients between u' and v', v' and w', and u' and w' are defined as

$$R_{u'v'} = \frac{\overline{u'v'}}{\sqrt{\overline{u'^2}}\sqrt{\overline{v'^2}}} \tag{5-29}$$

$$R_{v'w'} = \frac{\overline{v'w'}}{\sqrt{\overline{v'^2}}\sqrt{\overline{w'^2}}} \tag{5-30}$$

$$R_{u'w'} = \frac{\overline{u'w'}}{\sqrt{\overline{u'^2}}\sqrt{\overline{w'^2}}} \tag{5-31}$$

The correlation coefficients defined in Eqs. (5-26) to (5-31) are shown in Fig. 5-4, along with sketches showing the velocities related by the coefficients.

5-9. The Intensity of Turbulence

The *intensity of turbulence* is a measure of the magnitude of the velocity fluctuations about the mean value. It has been pointed out that the temporal mean values of the fluctuations $\overline{u'}$, $\overline{v'}$, and $\overline{w'}$ are all zero. However, the root mean squares of the fluctuations $\sqrt{\overline{u'^2}}$, $\sqrt{\overline{v'^2}}$, and $\sqrt{\overline{w'^2}}$ are finite and are the components of the intensity of the fluctuations. The intensity of turbulence is often expressed in terms of a percentage of the mean value; i.e., it is given as $100\sqrt{\overline{u'^2}}/u$, $100\sqrt{\overline{v'^2}}/v$, $100\sqrt{\overline{w'^2}}/w$.

In wind tunnels, where the mean flow is in the x direction and the average velocity is U, the intensity of turbulence is expressed as $100\sqrt{\overline{u'^2}}/U$.

5-10. The Scale of Turbulence

The scale of turbulence can be defined by means of the correlation coefficients. Prandtl's mixing length may be considered as representing the scale of turbulence. Using the correlation coefficients, the Lagrangian scale of turbulence is defined as

$$l_1 = \sqrt{\overline{u'^2}} \int_0^\infty R_t \, dt \tag{5-32}$$

where R_t is defined by Eq. (5-25). Using the Eulerian view, the scale is defined as

$$L_y = \int_0^\infty R_{yu'} \, dy \tag{5-33}$$

$$L_x = \int_0^\infty R_{xu'} \, dx \tag{5-34}$$

It is seen that the scale depends on which correlation coefficient is used. Taylor[31] points out that the length L_y may be considered as the *average size of the eddies* in turbulent flow.

5-11. The Reynolds Shear Stresses in Terms of the Correlation Coefficients

The Reynolds, or eddy, shear stresses can be expressed in terms of the correlation coefficients defined in Eqs. (5-29) to (5-31); i.e.,

$$-\rho\overline{u'v'} = -\rho R_{u'v'}\sqrt{\overline{u'^2}}\sqrt{\overline{v'^2}} \tag{5-35}$$

$$-\rho\overline{v'w'} = -\rho R_{v'w'}\sqrt{\overline{v'^2}}\sqrt{\overline{w'^2}} \tag{5-36}$$

$$-\rho\overline{u'w'} = -\rho R_{u'w'}\sqrt{\overline{u'^2}}\sqrt{\overline{w'^2}} \tag{5-37}$$

5-12. Isotropic, Homogeneous Turbulence

Theoretical investigations on the statistical theory of turbulence are greatly simplified by assuming isotropy and homogeneity. Isotropy is that condition wherein the intensity components in all directions are equal. This may be expressed by

$$\overline{u'^2} = \overline{v'^2} = \overline{w'^2} \tag{5-38}$$

Likewise, regardless of the orientation of the coordinate axis in space, the value of the quantities in Eq. (5-38) will always be the same. Isotropy also means that the fluctuations are perfectly random, with the result that there is no correlation between components of the fluctuation in different directions; i.e.,

$$\overline{u'v'} = \overline{v'w'} = \overline{u'w'} = 0 \tag{5-39}$$

Homogeneity in turbulence means that the intensity components are not a function of position in space. The assumption of homogeneity does not lead to so many simplifications as isotropy does. Likewise, it is difficult to produce turbulent flow which is homogeneous, except over short distances.

Isotropic turbulence may be produced in the laboratory, but in many problems of fluid flow isotropy does not exist. In turbulent flow in boundary layers and in closed conduits the turbulence is not isotropic. There appears to be a tendency for turbulence to become isotropic. At the edge of the boundary layer or at the axis of the conduit, where the mean velocity gradient becomes zero, the turbulence is nearly isotropic.

In the laboratory isotropic turbulent flow may be produced in wind tunnels by passing the air stream through a wire grid or screen. Some distance downstream from the grid the irregularities due to the grid disappear, and isotropic turbulence is nearly attained. Much of the experimental work conducted since 1935 has been in flowing streams in which the turbulence is isotropic. The results of these experimental investigations have confirmed the predictions of the statistical theory of turbulence.

Prior to 1941 most theoretical studies concerned isotropic turbulence, since isotropy permitted much simplification of the relationships. No extensive work had been done on the more practical subject of nonisotropic turbulence. The application of the theories of isotropic turbulence to nonisotropic turbulence was greatly aided in 1941 by the theory of locally isotropic turbulence proposed by Kolmogoroff.[8] Although the details of this theory will not be considered here, its physical concepts will be mentioned briefly (it has been described fully by Obukhoff and Yaglom [15] and Batchelor [1]). The theory postulates that turbulent motion consists of eddies of all possible scales. The scale defined in Eqs. (5-33) and (5-34) is the mean scale. The motion of the large eddies produces smaller eddies, which in turn produce yet smaller eddies. The process continues to the smallest eddies. From the standpoint of energy dissipation, it is postulated that the larger eddies lose their energy in forming smaller eddies and that in the process very little energy is dissipated as heat. The motion of the smallest eddies is laminar, and almost all their energy is dissipated as heat because of the molecular viscosity of the fluid. Thus it is seen that the energy dissipated by turbulent flow is dissipated mostly by smaller eddies. Kolmogoroff also states that, although the large eddies of turbulent flow may be nonisotropic, the smaller eddies are isotropic, so most of the energy is thus dissipated in turbulence which is isotropic. This theory of local isotropic turbulence permits the relationships derived for isotropic turbulence to be applied to nonisotropic turbulence.

5-13. The Spectrum of Isotropic Turbulence

Considering turbulence to be made up of eddies of different size, the total kinetic energy of the turbulent fluid may be considered to be distributed among the eddies. The spectrum of turbulence gives the distribution of kinetic energy among the eddies. It is convenient to express the energy distribution in terms of frequency instead of in terms of eddy size. At any point in isotropic turbulence the mean energy of the eddies is proportional to $\overline{u'^2}$. Taylor [32] defined the one-dimensional spectrum function $F(n)\,dn$, where $F(n)$ is the fraction of the total energy which is due to frequencies between n and $n + dn$; thus

$$\int_0^\infty F(n)\,dn = 1 \tag{5-40}$$

Taylor's spectrum function is the Fourier transform of the correlation functions given by Eqs. (5-26) to (5-28). He shows that for isotropic turbulence the spectrum curve may be expressed by the equation

$$\frac{UF(n)}{L_x} = \frac{4}{1 + 4\pi^2 n^2 L_x^2/U^2} \tag{5-41}$$

where L_x is defined by Eq. (5-34) and U is the mean velocity of the turbulent stream.

5-14. Turbulence Measurements; the Hot-wire Anemometer

There are numerous methods of determining point velocities of fluids in space. Where flow is steady and not accompanied by turbulent fluctuations, the pitot tube has proved useful for measuring velocities except in the vicinity close to solid boundaries. In turbulent flow the pitot tube is suitable for measuring only temporal mean velocities and does not react to turbulent fluctuations. The most satisfactory instrument for measuring turbulence and turbulent fluctuations is the hot-wire anemometer. This instrument consists of an electrically heated platinum or tungsten resistance wire mounted between two steel needles. The size of wire ordinarily used ranges from 0.0001 to 0.0003 in. in diameter and 0.025 to 0.06 in. in length. The wire is heated by passing a constant current through the wire, and heat loss from the wire equals the heat generated in it by virtue of its electrical resistance. The heat loss from the wire is a function of its temperature, the stream temperature, and the air velocity. It is given by the following relationship, derived by King:[7]

$$q = (C_1 + C_2\sqrt{u})(T - T_a) \tag{5-42}$$

where q = rate of heat loss from wire
C_1, C_2 = wire constants
T = wire temperature
T_a = air temperature

As the velocity of the air flowing across the wire fluctuates, the temperature of the wire also fluctuates, and so does the wire resistance. Since the wire current is constant, the emf across it fluctuates. Ideally, if there were no lag in the wire, and if the emf were proportional to the velocity fluctuations, the emf fluctuations would give a true picture of the variation of the velocity with time on a cathode-ray oscilloscope. Actually, such a picture is obtained only after suitable electronic equipment is employed to compensate for wire lag and for nonlinear characteristics.

Various arrangements and methods of using hot wires have been developed to measure turbulence. In addition, much specialized electronic equipment is employed along with the hot-wire anemometer to afford a detailed analysis of the quantities which the hot wire measures. In the present discussion various hot-wire arrangements will be described to indicate what quantities they measure. No mention will be made of auxiliary electronic equipment or of the mathematical relationships involved in the measurement. For these, the reader is referred to more detailed articles.[2, 5, 9, 10]

1. *Measurement of Turbulence Intensity.* The measurement of the intensity of turbulence involves the determination of the rms of the x component of the velocity fluctuation $\sqrt{\overline{u'^2}}$, which is measured by inserting a single hot wire in a turbulent stream perpendicular to the direction of the mean flow. Such an arrangement is indicated in Fig. 5-5. A constant cur-

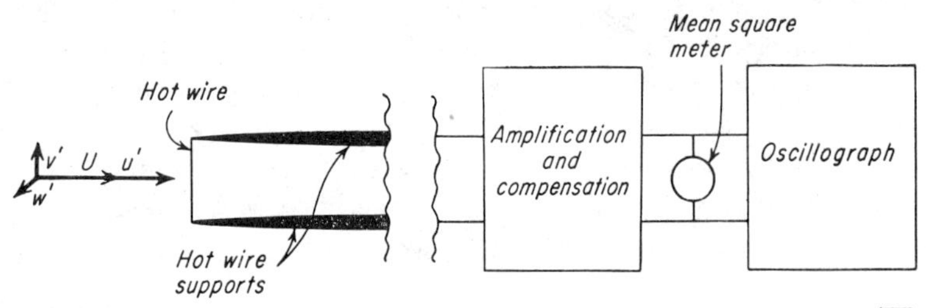

FIG. 5-5. Hot-wire arrangement for the measurement of turbulence intensity $\sqrt{\overline{u'^2}}$ and percentage turbulence $100\sqrt{\overline{u'^2}}/U$.

rent is maintained in the wire by a Wheatstone bridge. When air is flowing past the hot wire, velocity fluctuations cause fluctuations in the emf across the bridge. This fluctuating emf is fed into an amplifier, the output of which may be recorded on an oscillograph or measured by a meter giving the rms values of the amplifier output.

The hot wire is sensitive to u' (x-direction fluctuations) but insensitive to v' and w' (y- and z-component fluctuations). After suitable compensation and correction, the instantaneous amplifier output is proportional to u'. The meter reading is then proportional to $\sqrt{\overline{u'^2}}$, which is the turbulence intensity. The percentage turbulence $100\sqrt{\overline{u'^2}}/U$ may also be determined. The oscillograph record shows the variation of u' with time and looks something like Fig. 5-2.

2. *Measurement of* $\sqrt{\overline{v'^2}}$ *and* $\sqrt{\overline{w'^2}}$. The y and z components of the turbulence intensity are measured by means of a cross-wire anemometer, which consists of two identical wires, one placed at a positive angle to the wind and the other at a negative angle to the wind. The plane of the wires is in the direction of mean flow. Such a probe is shown in Fig. 5-6, oriented in the xy plane. If e_{I} and e_{II} are the instantaneous voltages (suitably

compensated) across wires I and II respectively, then the following relationship holds:

$$\sqrt{\overline{v'^2}} \propto \sqrt{\overline{(e_{\mathrm{I}} - e_{\mathrm{II}})^2}} \tag{5-43}$$

Thus the y component of turbulence intensity $\sqrt{\overline{v'^2}}$ may be obtained by measuring the difference of the voltages on the two wires by an instrument which gives rms readings. In order to obtain $\sqrt{\overline{w'^2}}$ it is necessary to orient the probe in the xz plane.

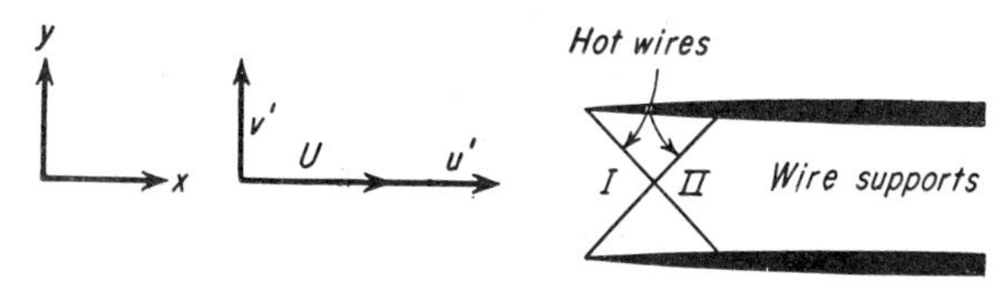

FIG. 5-6. Arrangement of two hot wires in the form of a cross to measure $\sqrt{\overline{v'^2}}$ and $\overline{u'v'}$. Both wires lie in the xy plane.

The intensity of turbulence $\sqrt{\overline{u'^2}}$ may also be measured by this arrangement since

$$\sqrt{\overline{u'^2}} \propto \sqrt{\overline{(e_{\mathrm{I}} + e_{\mathrm{II}})^2}} \tag{5-44}$$

3. *Measurement of Turbulent Shear Components $\overline{u'v'}$ and $\overline{u'w'}$.* The determination of turbulent shear components is also carried out by the use of the cross wires described above and shown in Fig. 5-6. When they are oriented in the xy plane,

$$\overline{u'v'} \propto \overline{e_{\mathrm{I}}^2} - \overline{e_{\mathrm{II}}^2} \tag{5-45}$$

and when oriented in the xz plane,

$$\overline{u'w'} \propto \overline{e_{\mathrm{I}}^2} - \overline{e_{\mathrm{II}}^2} \tag{5-46}$$

Separate mean-square meters may be used to read the rms voltages $\sqrt{\overline{e_{\mathrm{I}}^2}}$ and $\sqrt{\overline{e_{\mathrm{II}}^2}}$, which may then be used to determine $\overline{u'v'}$ or $\overline{u'w'}$, depending on the orientation of the hot-wire probe.

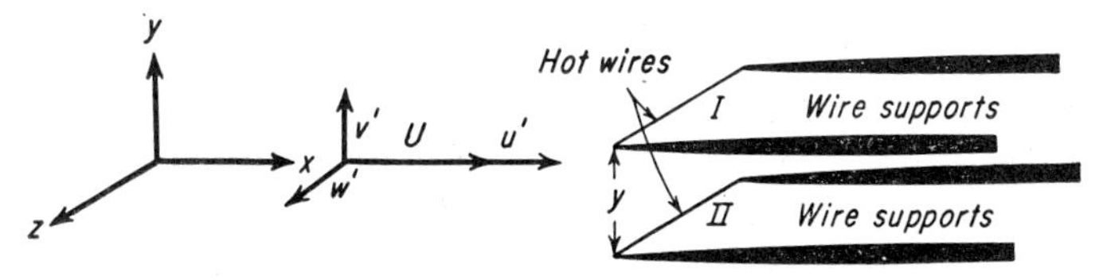

FIG. 5-7. Arrangement of two parallel hot wires to measure the correlation coefficient $R_{yu'}$ and the scale of turbulence L_y.

4. *Measurement of the Correlation Coefficient $R_{yu'}$ and the Scale of Turbulence L_y.* The correlation coefficient $R_{yu'}$ may be measured using two iden-

tical hot wires parallel to each other and placed perpendicular to the direction of mean flow, as shown in Fig. 5-7. If e_{I} and e_{II} are the properly compensated voltages across wires I and II respectively, then

$$R_{yu'} = \frac{\overline{(e_{\text{I}} + e_{\text{II}})^2} - \overline{(e_{\text{I}} - e_{\text{II}})^2}}{\overline{(e_{\text{I}} + e_{\text{II}})^2} + \overline{(e_{\text{I}} - e_{\text{II}})^2}} \tag{5-47}$$

From Eq. (5-47) it is seen that it is necessary to read the rms of the sum of the voltages of the two wires and the rms of the difference in the voltages of the two wires. These two readings may be used to calculate $R_{yu'}$. To calculate L_y, values of $R_{yu'}$ are plotted versus y (the distance between the parallel wires), and the resulting curve is integrated from $y = 0$ to $y = \infty$ according to Eq. (5-33).

5. *Determination of the Energy Spectrum of Turbulence.* The energy spectrum of turbulence may be determined with a single hot-wire probe when use is made of suitable filters and measuring equipment. The single hot wire is placed in the stream perpendicular to the mean flow, as indicated in Fig. 5-5. The probe is first connected to an amplifier which passes voltages of all frequencies. Let the measured voltage be e. Then

$$\overline{u'^2} \propto \overline{e^2} \tag{5-48}$$

A filter is now placed in the system so that voltages of frequencies from 0 to n are measured. All other frequencies above n do not pass through. Let e_n be the voltage indicated by all frequencies between 0 and n. Then

$$\overline{u_n'^2} \propto \overline{e_n{}^2} \tag{5-49}$$

where $\overline{u_n'^2}$ is the mean square of the velocities having frequencies between 0 and n.

For short periods of time the proportionality constants in Eqs. (5-48) and (5-49) are the same, so

$$\frac{\overline{u_n'^2}}{\overline{u'^2}} = \frac{\overline{e_n{}^2}}{\overline{e^2}} \tag{5-50}$$

The ratio on the left side of Eq. (5-50) is the fraction of the total kinetic energy which has frequencies between 0 and n. By varying n from low to high values the energy spectrum may be determined.

5-15. Some Measurements of Intensity, Scale, and Energy Spectrum of Turbulence

Since 1937 a large number of hot-wire-turbulence data have been obtained. There has been extensive investigation of turbulent streams in wind tunnels in which the turbulence is isotropic. Considerable effort, however, has also been put forth in the study of turbulent boundary layers. In this section some measurements of intensity, scale, and energy spectrum of turbulence in wind tunnels are presented to show what types of data are obtainable and to indicate the order of magnitude of these quantities in turbulent streams.

Turbulence is produced in wind tunnels by passing the flowing stream through square-mesh screens placed in the cross section of the tunnel. The

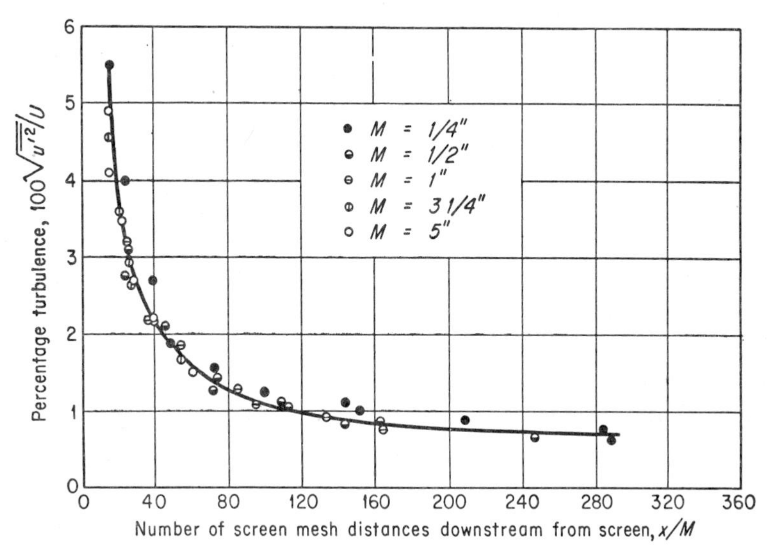

FIG. 5-8. Percentage turbulence downstream from turbulence-producing screens. (*From H. L. Dryden et al., NACA Rept.* 581, 1937.)

intensity and scale of turbulence is a function both of the size of mesh opening in the screen and of the wire diameter. Figure 5-8 shows the percentage turbulence $100\sqrt{\overline{u'^2}}/U$ plotted versus x/M for a wind-tunnel stream ranging in velocity from 20 to 70 ft/sec. The term x/M gives the number of mesh lengths downstream from the turbulence-producing screen, where x is the distance downstream from the screen and M is the mesh size. The data shown were obtained by Dryden and coworkers,[4] using screens having mesh sizes of $\frac{1}{4}$, $\frac{1}{2}$, 1, $3\frac{1}{4}$, and 5 in. in which the ratio of wire size to mesh length was 0.2. The plot indicates that about 20 mesh lengths from the screens the percentage turbulence is about 5 per cent, which means that the rms of the x component of velocity fluctuation is 5 per cent of the mean velocity U. The percentage turbulence decreases rapidly and reaches a value of 0.8 per cent beyond 200 mesh lengths from the screen.

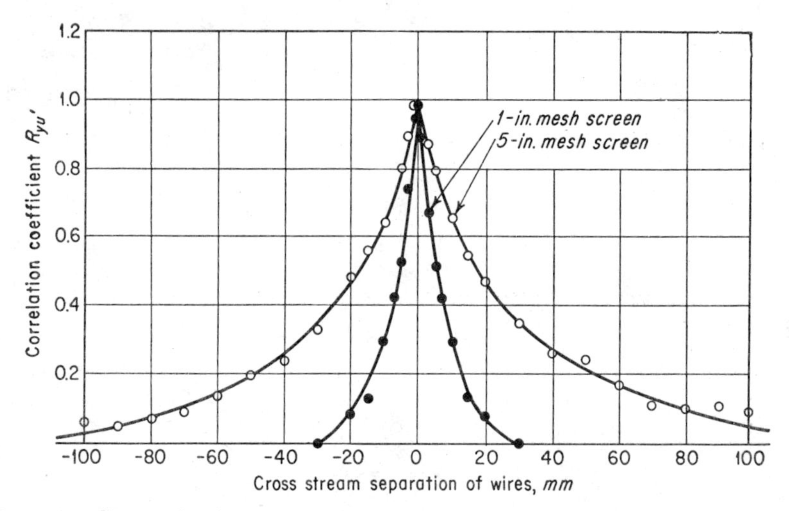

FIG. 5-9. Curves showing the correlation coefficient as a function of the separation of two parallel hot wires. Air speed is 40 ft/sec. Measurements taken 40 mesh lengths from the turbulence-producing screens. (*From H. L. Dryden et al., NACA Rept.* 581, 1937.)

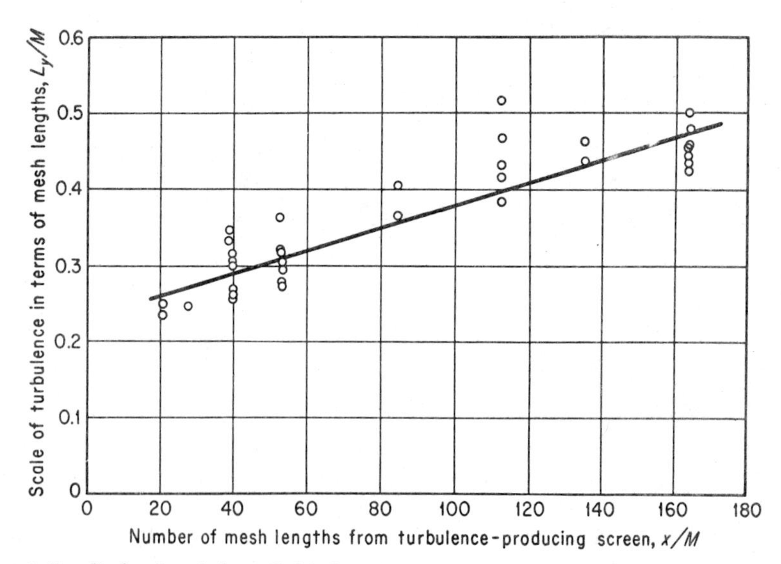

FIG. 5-10. Scale of turbulence behind a 1-in.-mesh screen as a function of the distance downstream from the screen. Air speed about 40 ft/sec. (*From H. L. Dryden et al., NACA Rept.* 581, 1937.)

These workers also determined the scale of turbulence and the correlation coefficient $R_{yu'}$ for the same screens listed above. The correlation coefficient was obtained using two parallel hot wires as described in Sec. 5-14. Figure 5-9 shows $R_{yu'}$ plotted versus y, the distance between the two wires. A symmetrical curve is obtained about $y = 0$. The correlation coefficient has a value of unity at $y = 0$ and eventually becomes zero. These curves of $R_{yu'}$ versus y represent the statistical distribution of u' along the y axis at any instant. They also represent the degree of relationship between u' at $y = 0$ to u' at any other point along the y axis.

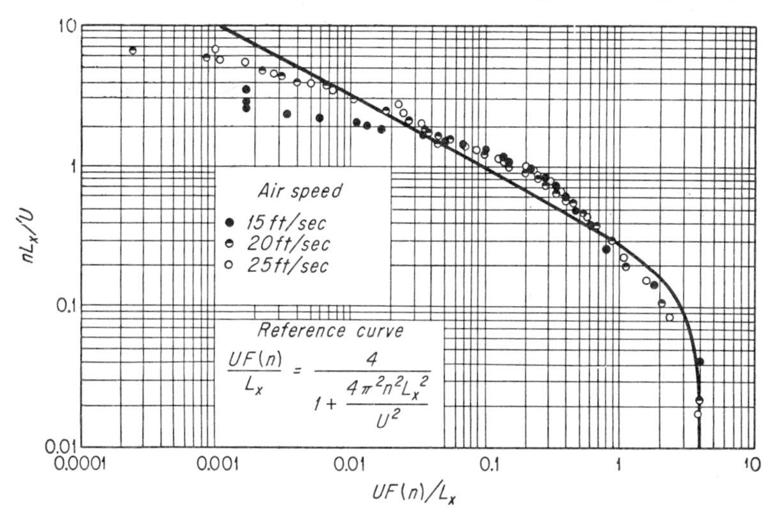

FIG. 5-11. The spectrum of turbulence in a wind tunnel 82 in. downstream from a 3-in.-mesh turbulence-producing screen. [*From L. F. G. Simmons and C. Salter, Proc. Roy. Soc. (London),* **165A**:73 (1938).]

The scale of turbulence, defined by Eq. (5-33), is obtained by determining the area under the curves in Fig. 5-9. Dryden and coworkers also determined the variation of scale with distance downstream from the screen. These data are shown in Fig. 5-10 as a plot of L_y/M versus x/M for the 1-in. mesh size only. The scale increases downstream from the screen, indicating that the smaller eddies die out, leaving the larger ones.

The energy spectrum of isotropic turbulence is shown in Fig. 5-11, where the data of Simmons and Salter [28] and Eq. (5-41) are plotted. These measurements were made on a stream at a point 82 in. downstream from a screen with 3-in. mesh. The data are plotted as $UF(n)/L_x$ against nL_x/U. The reference curve is a plot of Eq. (5-41). This plot shows the relation between $F(n)$ and n, where $F(n)$ is the fraction of the total turbulent kinetic energy arising from frequencies between n and $n + dn$. This analysis of turbulence is similar to that applied to light waves and is physically a consideration of the distribution of turbulence intensity (kinetic energy) as a function of eddy size (frequency).

BIBLIOGRAPHY

1. Batchelor, G. K.: *Proc. Cambridge Phil. Soc.*, **43**:533 (1947).
2. Corrsin, S., and M. S. Uberoi: *NACA Rept.* 1040, 1951.
3. Davies, S. J., and C. M. White: *Proc. Roy. Soc.* (*London*), **119A**:92 (1928).
4. Dryden, H. L., G. B. Schubauer, W. C. Mock, and H. K. Skramstad: *NACA Rept.* 581, 1937.
5. Dryden, H. L., and A. M. Kuethe: *NACA Rept.* 320, 1929.
6. Gibson, A. H.: *Phil. Mag.*, **15**:637 (1933).
7. King, L. V.: *Phil. Mag.*, **29**:556 (1915).
8. Kolmogoroff, A. M.: *Compt. rend. acad. sci. U.R.S.S.*, **30**:301 (1941); **31**:538 (1941); **32**:16 (1941).
9. Kovasznay, L. S. G.: *NACA TN* 2839, 1953.
10. Laufer, J.: *NACA Rept.* 1174, 1954.
11. Lin, C. C.: *Quart. Appl. Math.*, **3**(2):117, **3**(3):218, **3**(4):277 (1946).
12. Lindgren, E. R.: *Arkiv Fysik*, **7**:293 (1953).
13. Maurer, E.: *Z. Physik*, **126**:522 (1949).
14. Meksyn, D., and J. T. Stuart: *Proc. Roy. Soc.* (*London*), **208A**:517 (1951).
15. Obukhoff, A. M., and A. M. Yaglom: *Priklad. Matemat. i Mekh.*, **15**:3 (1951); see also *NACA TM* 1350, 1953.
16. Prandtl, L.: *Z. angew. Math. u. Mech.*, **5**:136 (1925); see also *NACA TM* 1231, 1949.
17. Prandtl, L.: *Z. Ver. deut. Ing.*, **77**:105 (1933); see also *NACA TM* 720, 1933.
18. Prengle, R. S., and R. R. Rothfus: *Ind. Eng. Chem.*, **47**:379 (1955).
19. Rayleigh, Lord: "Scientific Papers," vol. 1, p. 474, Cambridge University Press, London, 1899; see also *Proc. London Math. Soc.*, **11**:57 (1880).
20. Reynolds, O.: *Trans. Roy. Soc.* (*London*), **174A**:935 (1883).
21. Reynolds, O.: *Trans. Roy. Soc.* (*London*), **186A**:123 (1895).
22. Schiller, L.: *Proc. Intern. Congr. Appl. Mech., 3rd Congr., Stockholm, 1930*, **1**:226.
23. Schiller, L.: *Z. angew. Math. u. Mech.*, **2**:96 (1922).
24. Schlichting, H.: *Nachr. Ges. Wiss. Göttingen, Math.-physik. Kl.*, **1933**:181–208.
25. Schubauer, G. B., and H. K. Skramstad: *J. Aeronaut. Sci.*, **14**:69 (1947.)
26. Schubauer, G. B., and P. S. Klebanoff: *NACA WR* W-86 (formerly *ARR* 5K27, March, 1946).
27. Senecal, V. E., and R. R. Rothfus: *Chem. Eng. Progr.*, **49**:533 (1953).
28. Simmons, L. F. G., and C. Salter: *Proc. Roy. Soc.* (*London*), **165A**:73 (1938).
29. Squire, H. B.: *Proc. Roy. Soc.* (*London*), **142A**:621 (1933).
30. Taylor, G. I.: *Proc. London Math. Soc.*, **20**:196 (1921).
31. Taylor, G. I.: *Proc. Roy. Soc.* (*London*), **151A**:421 (I–IV)(1935).
32. Taylor, G. I.: *Proc. Roy. Soc.* (*London*), **164A**:476 (1938).
33. Thomas, L. H.: *Phys. Rev.*, **86**:812 (1952).
34. Tollmien, W.: *Nachr. Ges. Wiss. Göttingen, Math.-physik. Kl.*, **1929**:21–44; see also *NACA TM* 609, 1931.

CHAPTER 6

DIMENSIONAL ANALYSIS AND ITS APPLICATION TO FLUID DYNAMICS

6-1. Dimensional Analysis in Engineering

Dimensional analysis is applied in all fields of engineering. Physical processes may be described by an equation among dimensional physical quantities or variables. Through dimensional analysis these quantities are arranged in dimensionless groups. In applying dimensional analysis, it is necessary that *all* the dimensional variables affecting the process be known. The dimensionless groups obtained provide no information about the mechanism of the process but aid in correlating experimental data and developing functional relationships between dimensional variables. Also, once the functional relationship is experimentally obtained between dimensionless groups, the effect of any dimensional factor may be determined. This is particularly useful where it is difficult, experimentally, to change some variable. In this chapter the various methods of dimensional analysis will be discussed, along with several applications to fluid-flow problems.

6-2. Geometric Similarity

The solid boundaries of any flow system may be adequately described by a number of length dimensions L_1, L_2, L_3, . . ., L_n. If these lengths are divided by L_1, the system may be defined by L_1, r_2', r_3', r_4', . . ., r_n', where

$$r_2' = \frac{L_2}{L_1}$$

$$r_3' = \frac{L_3}{L_1}$$

$$\cdot \quad \cdot \quad \cdot \quad \cdot$$

$$r_n' = \frac{L_n}{L_1}$$

Geometric similarity exists between two solid systems if the ratios r_2', r_3', . . ., r_n' are the same for each system.

6-3. Kinematic Similarity

Kinematic similarity refers to the motion occurring in the system and considers the velocities existing. For kinematic similarity to exist in two geometrically similar systems I and II, the velocities at the same relative point in each system must be related as follows:

$$\frac{u_{\mathrm{I}}}{v_{\mathrm{I}}} = \frac{u_{\mathrm{II}}}{v_{\mathrm{II}}} \qquad \frac{u_{\mathrm{I}}}{w_{\mathrm{I}}} = \frac{u_{\mathrm{II}}}{w_{\mathrm{II}}} \tag{6-1}$$

Likewise, the velocity gradients in each system will bear a similar relationship to each other.

6-4. Dynamic Similarity

Dynamic similarity considers the relationship between the inertial, normal, shear, and field forces acting in systems. In geometrically similar systems dynamic similarity exists at the same relative point in each system if

$$\frac{\text{Inertia force}_{\mathrm{I}}}{\text{Viscous force}_{\mathrm{I}}} = \frac{\text{inertia force}_{\mathrm{II}}}{\text{viscous force}_{\mathrm{II}}} \qquad \frac{\text{Inertia force}_{\mathrm{I}}}{\text{Gravity force}_{\mathrm{I}}} = \frac{\text{inertia force}_{\mathrm{II}}}{\text{gravity force}_{\mathrm{II}}} \tag{6-2}$$

6-5. Dimensional Analysis

1. *Historical.* Dimensional analysis has been used in some form for a long time, particularly in fluid flow. In 1850 Stokes [8] showed that in geometrically similar flow systems the Reynolds number could be used as a criterion of dynamic similarity. Some time later Helmholtz [4] obtained the same result using the differential momentum equations of flow, a method that has been well illustrated by Klinkenberg and Mooy.[5] In 1899 Rayleigh [7] first applied the method of dimensional analysis that is in general use today. In 1914 Buckingham [2] stated the Π theorem, which is the basis of dimensional analysis. Several books [1,3,6] on dimensional analy-

sis have been published, and the reader is referred to them for a detailed mathematical discussion of the subject.

2. *Units and Dimensions.* In dimensional analysis it is necessary to determine the dimensions on physical quantities. The fundamental dimensions of mass, length, time, and temperature are sufficient to express the dimensions of any physical variable. Owing to the common engineering practice of using the terms *pound force* (lb_f) and *pound mass* (lb_m), it is usual to express the dimensions of variables in terms of the fundamental quantities force, mass, length, time, and temperature. In dimensional analysis, if this latter set of fundamental quantities is used, it is necessary to include the conversion factor between poundal and pound force as an additional dimensional quantity in the system.

The introduction of the conversion factor g_c is illustrated using Newton's law for the weight of a body.

$$W = mg \tag{6-3}$$

where g = acceleration of gravity
m = mass
W = weight or force exerted on body

For a unit mass, the weight is g units of force. In the English system, where mass is in pounds mass, this unit of force becomes the poundal. The unit of force called the pound force is the weight of a pound mass at sea level and at 45° latitude, where $g = 32.17$ ft/sec^2. Thus, 1 pound force is 32.17 poundals. Since the dimensions on the poundal are $(lb_m)(ft)/sec^2$, the conversion factor between pound force and poundal becomes 32.17 $(lb_m)(ft)/(lb_f)(sec^2)$ and has the symbol g_c. Thus, if weight is to be expressed in terms of pound force, Eq. (6-3) becomes

$$W = \frac{mg}{g_c} \tag{6-4}$$

3. *Dimensional Homogeneity.* The principle of dimensional homogeneity applies to relationships between dimensional variables. An equation containing dimensional variables is dimensionally homogeneous if each term in the equation has the same dimensions. This may be stated alternatively as follows: a dimensionally homogeneous equation is valid irrespective of the fundamental units used in it. Consider Eq. (3-58) for the pressure distribution for nonviscous flow past a cylinder.

$$P_\theta - P = \frac{\rho U^2}{2g_c}(1 - 4\sin^2\theta) \tag{3-58}$$

The dimensions on each quantity are as indicated in Table 6-1, showing that the equation is dimensionally homogeneous.

TABLE 6-1. DIMENSIONS ON QUANTITIES IN EQ. (3-58)

Quantity	*Dimensions*
ρ	m/L^3
U	L/t
g_c	mL/Ft^2
θ	None
$\rho U^2/2g_c$	F/L^2
P	F/L^2

4. *Methods of Dimensional Analysis.* There are three principal methods of dimensional analysis, all of which yield identical results:

a. The Buckingham method
b. The Rayleigh method
c. The use of differential equations

The basis of all dimensional-analysis theory is Buckingham's theorem,[2] which states that "if an equation is dimensionally homogeneous, it can be reduced to a relationship among a complete set of dimensionless products." This theorem has been proved mathematically by Langhaar.[6]

6-6. The Buckingham Method

Let $Q_1, \ldots, Q_n$ be n dimensional variables upon which a given physical process depends. Let $r'_2, \ldots, r'_n$ (see Sec. 6-2) be the dimensionless length ratios required to geometrically describe the solid boundaries of the system. The functional relation between all these variables may be expressed as

$$F_1(Q_1,Q_2,\ldots,Q_n;r'_2,\ldots,r'_n) = 0 \tag{6-5}$$

If consideration is limited to geometrically similar systems, the relationship becomes

$$F_2(Q_1,Q_2,\ldots,Q_n) = 0 \tag{6-6}$$

Letting j be the number of fundamental dimensions (such as mass, length, time, etc.) necessary to define the dimensions on Q_1, Q_2, $\ldots$, Q_n, and letting i be the number of independent dimensionless products,

$$i = n - j \tag{6-7}$$

Thus every dimensionally homogeneous equation such as Eq. (6-6) may be reduced to

$$F_3(\Pi_1,\Pi_2,\Pi_3,\ldots,\Pi_i) = 0 \tag{6-8}$$

where the dimensionless groups Π are expressed as follows:

$$\begin{aligned} \Pi_1 &= Q_1^{a_1}Q_2^{b_1}\cdots Q_j^{j_1}Q_{j+1} \\ \Pi_2 &= Q_1^{a_2}Q_2^{b_2}\cdots Q_j^{j_2}Q_{j+2} \\ &\cdots\cdots\cdots\cdots \\ \Pi_i &= Q_1^{a_i}Q_2^{b_i}\cdots Q_j^{j_i}Q_{j+i} \end{aligned} \tag{6-9}$$

In Eq. (6-9) the dimensional quantities $Q_1, Q_2, \ldots, Q_j$ must, between them, contain all the j fundamental dimensions. The exponents $a_1, \ldots, j_1$; $a_2, \ldots, j_2$, etc., must be of such value that the Π's are dimensionless.

Example 6-1

The force F exerted on a body immersed in a flowing fluid is a function of fluid velocity U, the fluid density ρ, the fluid viscosity μ, and a characteristic length of the body L. Using Buckingham's method, determine the dimensionless groups in which the dimensional variables may be arranged.

Solution

Table 6-2 shows the dimensions on all quantities.

TABLE 6-2. DIMENSIONS ON QUANTITIES IN EXAMPLE 6-1

Quantity	*Dimensions*
F	F
ρ	m/L^3
U	L/t
L	L
μ	m/Lt
g_c	mL/Ft^2

$$n = 6 \qquad j = 4$$

$$i = n - j = 2$$

Selecting F, ρ, U, and L † as the four quantities which, between them, contain all the fundamental dimensions, the dimensionless groups are

$$\Pi_1 = F^{a_1}\rho^{b_1}U^{c_1}L^{d_1}\mu \tag{6-10}$$

$$\Pi_2 = F^{a_2}\rho^{b_2}U^{c_2}L^{d_2}g_c \tag{6-11}$$

Substituting dimensions on the quantities in Eq. (6-10), the exponents must have values such that

$$F^{a_1}\left(\frac{m}{L^3}\right)^{b_1}\left(\frac{L}{t}\right)^{c_1}L^{d_1}\frac{m}{Lt}$$

is dimensionless. Hence

$$a_1 = 0$$

$$b_1 + 1 = 0$$

$$-c_1 - 1 = 0$$

$$-3b_1 + c_1 + d_1 - 1 = 0$$

from which

$$a_1 = 0$$

$$b_1 = -1$$

$$c_1 = -1$$

$$d_1 = -1$$

† The quantities selected are the ones which one desires to have appear in each dimensionless group.

Similar handling of Eq. (6-11) gives

$$a_2 = 1$$

$$b_2 = -1$$

$$c_2 = -2$$

$$d_2 = -2$$

The two dimensionless groups obtained are

$$\Pi_1 = \frac{\mu}{LU\rho}$$

$$\Pi_2 = \frac{Fg_c}{\rho U^2 L^2}$$

The variables of the system may be related by a functional relationship between these two groups

$$F_3\left(\frac{\mu}{LU\rho}, \frac{Fg_c}{\rho U^2 L^2}\right) = 0 \tag{6-12}$$

Equation (6-12) may be interpreted by stating that for geometrically similar systems there is a functional relationship between the group $\mu/LU\rho$, which is the reciprocal of the Reynolds number, and $Fg_c/\rho U^2L^2$, which is called the Euler number. This also means that in two geometrically similar systems, if the Reynolds numbers are the same, the Euler numbers will also be identical. The functional relationship in Eq. (6-12) must, in most cases, be obtained by experiment.

6-7. The Rayleigh Method

In general terms, the Rayleigh method of dimensional analysis is expressed as Q_1 varies as $Q_2{}^aQ_3{}^bQ_4{}^c$, etc. The dimensionless groups are obtained by evaluating the exponents so that the relationship is dimensionally homogeneous. Like powers of the quantities are grouped together to give dimensionless groups.

Example 6-2

Use the Rayleigh method to make a dimensional analysis of the system considered in Example 6-1. The dimensional quantities to be considered are F, ρ, U, L, μ, and g_c.

Solution

By Rayleigh's method

$$F \propto \rho^a U^b L^c \mu^d g_c{}^e$$

For the relationship to be dimensionally homogeneous $\rho^a U^b L^c \mu^d g_c{}^e$ must have the same dimensions as F; i.e., the expression

$$\left(\frac{m}{L^3}\right)^a \left(\frac{L}{t}\right)^b L^c \left(\frac{m}{Lt}\right)^d \left(\frac{mL}{Ft^2}\right)^e$$

has dimensions of F. Thus

$$\begin{array}{ll} \text{For } F & -e = 1 \\ \text{For } m & a + d + e = 0 \\ \text{For } L & -3a + b + c - d + e = 0 \\ \text{For } t & -b - d - 2e = 0 \end{array}$$

Since there are five unknowns and only four simultaneous equations, four of the unknowns must be determined in terms of the fifth one. The four which are determined must be *on quantities which between them contain all the fundamental dimensions.* It is seen that to determine a, b, c, and d in terms of e would be incorrect, since a, b, c, and d are exponents on quantities which between them do not contain all of the fundamental units (F does not appear on the dimensions of any of these quantities). It is convenient to solve for a, b, c, and e in terms of d. Thus

$$a = 1 - d$$

$$b = 2 - d$$

$$c = 2 - d$$

$$e = -1$$

Therefore

$$F \propto \rho^{1-d} U^{2-d} L^{2-d} \mu^d g_c^{-1}$$

or

$$\frac{F g_c}{\rho U^2 L^2} \propto \left(\frac{\mu}{L U \rho}\right)^d$$

This result implies a functional relationship between $Fg_c/\rho U^2 L^2$ and $\mu/LU\rho$, a result which is the same as that obtained in Example 6-1.

6-8. The Use of Differential Equations

The momentum equations (2-39) are differential equations describing the motion of a fluid. Limiting consideration to incompressible fluids with constant viscosity, the momentum equation for the x direction is

$$\frac{\partial u}{\partial t} + u\frac{\partial u}{\partial x} + v\frac{\partial u}{\partial y} + w\frac{\partial u}{\partial z} = -g_c\frac{\partial \Omega}{\partial x} - \frac{g_c}{\rho}\frac{\partial P}{\partial x} + \frac{\mu}{\rho}\left(\frac{\partial^2 u}{\partial x^2} + \frac{\partial^2 u}{\partial y^2} + \frac{\partial^2 u}{\partial z^2}\right) \tag{6-13}$$

Equation (6-13) is a dimensionally homogeneous equation, and division by one of the terms will yield dimensionless groups. Only the terms which apply to the particular problem need be included in the dimensional analysis. The method of using differential equations to obtain dimensionless groups is well illustrated by Klinkenberg and Mooy.[5]

The variables in Eq. (6-13) are velocity, length, field forces, static pressure, density, and viscosity. In making a dimensional analysis the dimension of each term of the differential equation is expressed in terms of the above variables. The analysis is also restricted to geometrically similar systems.

Table 6-3. Dimensionless Groups Obtainable from the Momentum Equation

Type of flow	Unsteady state term	Inertia terms	Field forces (gravity)	Static-pressure forces	Viscous forces	Boundary condition
	$\frac{\partial u}{\partial t}$	$u \frac{\partial u}{\partial x} + v \frac{\partial u}{\partial y} + w \frac{\partial u}{\partial z}$	$-g_c \frac{\partial \Omega}{\partial x}$	$-\frac{g_c}{\rho} \frac{\partial P}{\partial x}$	$\nu \left(\frac{\partial^2 u}{\partial x^2} + \frac{\partial^2 u}{\partial y^2} + \frac{\partial^2 u}{\partial z^2} \right)$	(a) Wall shear
	$\frac{U}{t}$	$\frac{U^2}{L}$	g	$\frac{g_c P}{\rho L}$	$\frac{\nu U}{L^2}$	$\frac{\tau_w g_c}{\rho L}$
1. Flow of a nonviscous fluid in space or past immersed bodies. Gravitational effects absent		* ⟵	⟶	* $\frac{P g_c}{\rho U^2}$ Euler number		
2. Flow of a nonviscous fluid including gravitational effects. Flow of liquid in open channel, paths of jets, formation of waves on liquid surfaces. Viscosity and surface tension neglected		* ⟵	* ⟶ $\frac{Lg}{U^2}$ Reciprocal of Froude number	*		
		⟵	⟶	$\frac{P g_c}{\rho U^2}$ Euler number		

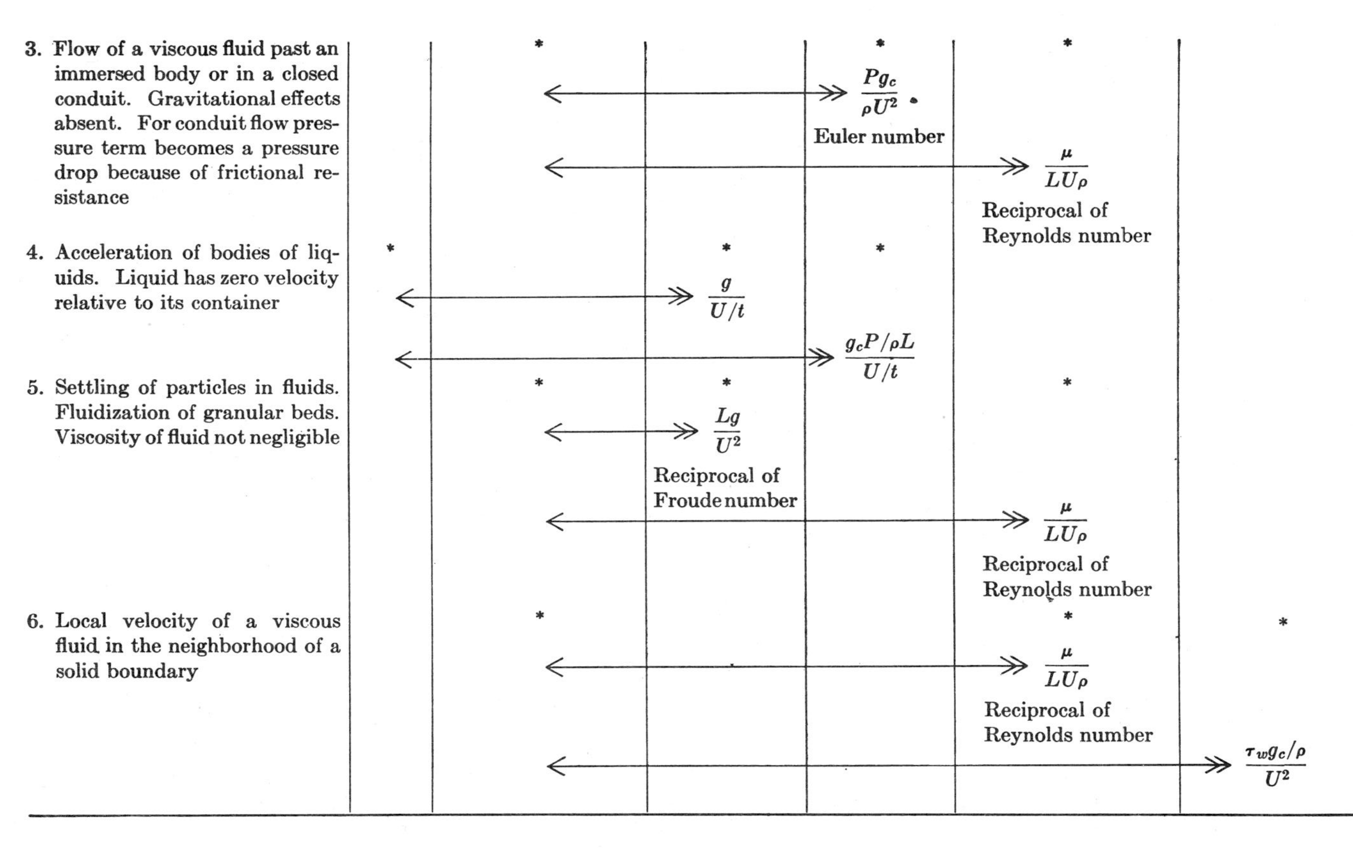
3. Flow of a viscous fluid past an immersed body or in a closed conduit. Gravitational effects absent. For conduit flow pressure term becomes a pressure drop because of frictional resistance
$\frac{Pg_c}{\rho U^2}$
Euler number
$\frac{\mu}{LU\rho}$
Reciprocal of Reynolds number
4. Acceleration of bodies of liquids. Liquid has zero velocity relative to its container
$\frac{g}{U/t}$
$\frac{g_c P/\rho L}{U/t}$
5. Settling of particles in fluids. Fluidization of granular beds. Viscosity of fluid not negligible
$\frac{Lg}{U^2}$
Reciprocal of Froude number
$\frac{\mu}{LU\rho}$
Reciprocal of Reynolds number
6. Local velocity of a viscous fluid in the neighborhood of a solid boundary
$\frac{\mu}{LU\rho}$
Reciprocal of Reynolds number
$\frac{\tau_w g_c/\rho}{U^2}$

TABLE 6-3. DIMENSIONLESS GROUPS OBTAINABLE FROM THE MOMENTUM EQUATION (*Continued*)

Type of flow	Unsteady state term	Inertia terms	Field forces (gravity)	Static-pressure forces	Viscous forces	Boundary condition
	$\frac{\partial u}{\partial t}$	$u\frac{\partial u}{\partial x} + v\frac{\partial u}{\partial y} + w\frac{\partial u}{\partial z}$	$-g_c\frac{\partial \Omega}{\partial x}$	$-\frac{g_c}{\rho}\frac{\partial P}{\partial x}$	$\nu\left(\frac{\partial^2 u}{\partial x^2} + \frac{\partial^2 u}{\partial y^2} + \frac{\partial^2 u}{\partial z^2}\right)$	(*b*) Surface tension
	$\frac{U}{t}$	$\frac{U^2}{L}$	g	$\frac{g_c P}{\rho L}$	$\frac{\nu U}{L^2}$	$\frac{\sigma g_c}{\rho L^2}$
7. Formation of waves on a liquid surface. Surface tension not negligible. Breakup of liquid jets under the influence of gravity		*	*		*	*
		←	→ $\frac{Lg}{U^2}$ Reciprocal of Froude number			
		←			→ $\frac{\mu}{LU\rho}$ Reciprocal of Reynolds number	
		←				→ $\frac{\sigma g_c}{\rho U^2 L}$ Reciprocal of Weber number

Table 6-3 illustrates the use of the differential momentum equation in determining dimensionless groups for a number of flow conditions. The double-headed arrows indicate the terms which are the numerator, and the single-headed arrows indicate the terms which are the denominator in obtaining the dimensionless groups. The asterisks in the columns indicate the terms of the differential equation which are important for the particular flow conditions. The effects of solid boundaries and of liquid surfaces are included in Table 6-3. At a solid boundary the shear force is τ_w. This force per unit area must be divided by the mass of the fluid in order to get the accelerative effect of the wall shear. The product of $\tau_w g_c$ and area gives force, while the product of density and volume gives mass. The ratio of force to mass is acceleration. Hence the boundary condition shown in Table 6-3 for solid surfaces is $[\tau_w g_c \text{ (area)}]/[\rho \text{ (volume)}]$, which has dimensions of $\tau_w g_c/\rho L$. Similarly, when liquid surfaces exist, the force of surface tension at the boundary is important, and the boundary condition is $[\sigma g_c \text{ (length)}]/[\rho \text{ (volume)}]$, which has dimensions $\sigma g_c/\rho L^2$, where σ is the surface tension in force per unit length.

Using differential equations for dimensional analysis has certain advantages. In deriving a differential equation for a process there is little possibility of neglecting some of the variables. Although the resulting differential equation may be too complicated to solve, it is useful in obtaining the dimensionless groups for the system. The physical significance of the dimensionless groups is apparent by this method (see Sec. 6-11).

Example 6-3

As an example of the use of differential equations to obtain dimensionless groups, the steady flow of a nonviscous fluid in the absence of gravitational effects will be considered. One differential equation for this flow is:

$$u\frac{\partial u}{\partial x} + v\frac{\partial u}{\partial y} + w\frac{\partial u}{\partial z} = \frac{-g_c}{\rho}\frac{\partial P}{\partial x} \tag{6-14}$$

$$\frac{U^2}{L} \qquad\qquad \frac{g_c P}{\rho L}$$

The dimensions of the terms using the variables of the system are shown below Eq. (6-14). Dividing the right-hand term by the left-hand term gives the Euler number

$$\frac{g_c P/\rho L}{U^2/L} = \frac{g_c P}{\rho U^2} \tag{6-15}$$

This means that at the same relative position in geometrically similar systems the Euler numbers are identical.

Example 6-4

Determine the value of the Euler number for the flow of a nonviscous fluid past an immersed cylinder. Show that it is the only dimensionless group concerned in the system.

Solution

The pressure distribution at the surface of a cylinder as a nonviscous fluid flows past it is

$$P_\theta - P = \frac{\rho U^2}{2g_c}(1 - 4\sin^2\theta) \tag{3-58}$$

Dividing by $\rho U^2/g_c$.

$$\frac{P_\theta g_c}{\rho U^2} - \frac{P g_c}{\rho U^2} = \frac{1 - 4\sin^2\theta}{2} \tag{6-16}$$

Solving for the Euler number.

$$\frac{P_\theta g_c}{\rho U^2} = \frac{P g_c}{\rho U^2} + \frac{1 - 4\sin^2\theta}{2} \tag{6-17}$$

The first term on the left of Eq. (6-17) is constant, since it includes properties of the undisturbed main stream. The second term is constant at any particular value of θ. Thus, for any value of θ, the Euler number is the same for geometrically similar systems. The angle θ is expressed in radians, which is actually a ratio of two lengths, so the same values of θ in similar systems correspond to the same relative position.

It is seldom that an expression for the Euler number may be found analytically. Whenever two or more dimensionless groups are involved, the relationship between them will usually have to be determined experimentally.

6-9. The Use of Models

The dimensionless groups obtained by dimensional analysis may be used to interpret the data obtained from experiments on models of full-sized equipment. The models studied must be geometrically similar (see Sec. 6-2) to the full-scale prototype. If this condition is satisfied, any functional relationship obtained between the dimensionless groups using the model data may be used on the full-scale equipment.

6-10. Further Examples

Several examples are included in this section illustrating the use of models and showing the procedure when systems are not geometrically similar.

Example 6-5

The inside of a circular pipe may be defined geometrically by three length terms, inside diameter d_w, length L, and wall roughness e. Let the resistance to flow be F. Other variables which affect the resistance are density ρ, viscosity μ, and velocity U. Make a dimensional analysis of these variables.

Solution

By Eq. (6-5)

$$F_1\left(F, d_w, U, \rho, \mu, g_c, \frac{L}{d_w}, \frac{e}{d_w}\right) = 0 \tag{6-18}$$

For geometrically similar systems

$$F_2(F, d_w, U, \rho, \mu, g_c) = 0 \tag{6-19}$$

From Example 6-2, the dimensionless groups obtainable from the quantities are the Reynolds number $d_w U\rho/\mu$ and the Euler number $Fg_c/\rho U^2 d_w^2$. Since F is a force, F/d_w^2 is equivalent to pressure, and the Euler number becomes $\Delta P\, g_c/\rho U^2$. There are four dimensionless groups involving all the variables when geometric similarity does not exist; i.e.,

$$F_3\left(\frac{d_w U\rho}{\mu}, \frac{\Delta P\, g_c}{\rho U^2}, \frac{d_w}{L}, \frac{d_w}{e}\right) = 0 \tag{6-20}$$

It is the practice in correlating resistance data for flow in pipelines to combine the Euler group with the d_w/L ratio, thus giving the dimensionless term $(\Delta P\, g_c d_w)/\rho U^2 L$, which, when divided by 2, is the friction factor f, i.e., for a horizontal pipe

$$f = \frac{\Delta P\, g_c d_w}{2\rho U^2 L} \tag{6-21}$$

Example 6-6 [7]

The mass m of drops formed by a liquid discharging by gravity from a vertical tube is a function of tube diameter d_w, liquid density ρ, surface tension σ, and the acceleration of gravity g. The effect of viscosity is neglected. It is difficult to determine the effect of surface tension on drop size. Make a dimensional analysis using Rayleigh's method. In this way the effect of surface tension may be determined by finding the effect of the dimensionless group that contains it.

Solution

By Rayleigh's method

$$m \propto d_w{}^a \rho^b \sigma^c g^d g_c{}^e$$

Thus

$$L^a \left(\frac{m}{L^3}\right)^b \left(\frac{F}{L}\right)^c \left(\frac{L}{t^2}\right)^d \left(\frac{mL}{Ft^2}\right)^e$$

has dimensions of m.

$$\begin{aligned} &\text{For } L & a - 3b - c + d + e &= 0 \\ &\text{For } m & b + e &= 1 \\ &\text{For } F & c - e &= 0 \\ &\text{For } t & -2d - 2e &= 0 \end{aligned}$$

from which

$$a = 3 - 2e$$

$$b = 1 - e$$

$$c = e$$

$$d = -e$$

Thus

$$m \propto \rho d_w{}^3 \left(\frac{\sigma g_c}{\rho g d_w{}^2}\right)^e$$

Two dimensionless groups are obtained:

$$\frac{m}{\rho d_w{}^3} \quad \text{and} \quad \frac{\sigma g_c}{\rho g d_w{}^2}$$

Experimental data may be obtained by varying ρ and d_w and may be used to give the relationship between the two groups. This relationship may be used to show the effect of all the dimensional variables even though only two were varied in the experiments. This procedure is satisfactory if the quantities which are varied have a significant effect on the droplet size.

Example 6-7

A new type of heat exchanger has been designed for cooling water. No data exist on the resistance to flow in such a heater. It is proposed to study a scale model of the heat exchanger having a length ratio of 1:10 with the commercial prototype. Flow in the prototype will be 200 gal/min of water at an average temperature of 120°F.

(*a*) What flow rate of air at 60°F and 1 atm pressure in the model will give flow conditions similar to those in the prototype?

(*b*) If the pressure drop across the model for the flow rate obtained in (*a*) is 30 in. H_2O, what will be the pressure drop across the prototype?

Solution

For water at 120°F

$$\nu = 0.610 \times 10^{-5}\ \text{ft}^2/\text{sec}$$

$$\rho = 61.7\ \text{lb}_m/\text{ft}^3$$

For air at 60°F

$$\nu = 1.58 \times 10^{-4}\ \text{ft}^2/\text{sec}$$

$$\rho = 0.0763\ \text{lb}_m/\text{ft}^3$$

(*a*) For dynamic similarity to exist in both model and prototype the Reynolds numbers must be equal. The quantities included in the Reynolds number are tabulated in Table 6-4.

TABLE 6-4

	Model	Prototype
Length	L	$10L$
Volume rate of flow	Q ft³/min	$\frac{200}{7.48} = 26.8$ ft³/min
Velocity	$\frac{Q}{L^2}$	$\frac{26.8}{100L^2}$
Kinematic viscosity	1.58×10^{-4} ft²/sec	0.610×10^{-5} ft²/sec
Reynolds number	$L \frac{Q}{L^2} \frac{1}{1.58 \times 10^{-4}}$	$10L \frac{26.8}{100L^2} \frac{1}{0.610 \times 10^{-5}}$

The Reynolds numbers are equal. Thus

$$Q = \frac{(10)(26.8)(1.58 \times 10^{-4})}{(100)(0.610 \times 10^{-5})} = 69.4 \text{ ft}^3/\text{min}$$

The air flow rate in the model must be 69.4 ft³/min.

(*b*) If dynamic similarity exists between model and prototype, the Euler numbers must be equal.

TABLE 6-5

	Model	Prototype
Length	L	$10L$
Pressure drop	30 in. H_2O	ΔP
Velocity	$\frac{69.4}{L^2}$	$\frac{26.8}{100L^2}$
Density	0.0763 lb_m/ft^3	61.7 lb_m/ft^3
Euler number	$\frac{30g_c}{0.0763(69.4/L^2)^2}$	$\frac{\Delta P \, g_c}{61.7(26.8/100L^2)^2}$

Since the Euler numbers are equal,

$$\Delta P = \frac{(30)(61.7)(26.8)^2}{(0.0763)(69.4)^2(100)^2} = 0.36 \text{ in. } H_2O$$

This will be the pressure drop across the prototype corresponding to the 30-in. pressure drop across the model.

6-11. Significance of Dimensionless Groups

The use of the Buckingham and Rayleigh methods of dimensional analysis will yield dimensionless groups, but their physical significance is not evident. The use of differential equations allows one to interpret physically the dimensionless groups thus derived but still provides no information on the fundamental mechanism of the process. The Reynolds numbers derived in Examples 6-1 and 6-2 have no obvious significance, but it is apparent from item 3 of Table 6-3 that the Reynolds number represents the ratio of viscous forces to inertia forces in the system. Similarly, the Euler number is the ratio of pressure forces to inertia forces. Table 6-6 shows the physical significance of the common dimensionless groups encountered in fluid flow.

TABLE 6-6. PHYSICAL SIGNIFICANCE OF SOME DIMENSIONLESS GROUPS IN FLUID FLOW

Group	Name	Significance
$\frac{LU\rho}{\mu}$	Reynolds number (Re)	$\frac{\text{Inertia forces}}{\text{Viscous forces}}$
$\frac{Pg_c}{\rho U^2}$	Euler number (Eu)	$\frac{\text{Pressure forces}}{\text{Inertia forces}}$
$\frac{U^2}{Lg}$	Froude number (Fr)	$\frac{\text{Inertia forces}}{\text{Gravity forces}}$
$\frac{\rho U^2 L}{\sigma g_c}$	Weber number (We)	$\frac{\text{Inertia forces}}{\text{Surface-tension forces}}$
$\frac{U}{U_c}$	Mach number (Ma)	$\frac{\text{Velocity}}{\text{Sonic velocity}}$
$\frac{\tau_w g_c/\rho}{u^2}$	$(u^+)^{-}$	$\frac{\text{Wall shear forces}}{\text{Inertia forces}}$

6-12. Uses and Limitations of Dimensional Analysis

It has been pointed out that dimensional analysis gives no indications of the fundamental mechanism of a process. This constitutes one of the serious limitations of the method. Likewise, a dimensional analysis of any process is invalid if any significant variable has been neglected. It is extremely useful, however, in correlating experimental data, and there are numerous relationships in the field of fluid flow and heat transfer which bear out this fact. Even though no information is provided about the fundamental mechanism, the groups obtained by means of dimensional analysis are useful in any basic study of a process.

BIBLIOGRAPHY

1. Bridgman, P. W.: "Dimensional Analysis," Yale University Press, New Haven, Conn., 1946.
2. Buckingham, E.: *Phys. Rev.*, **4**:345 (1914).
3. Duncan, W. J.: "Physical Similarity and Dimensional Analysis," Edward Arnold & Co., London, 1953.
4. Helmholtz, H.: *Charlottenburg Physik.-tech. Reichsanstalt, Wiss. Abhandl.*, **1**:158 (1894).

5. Klinkenberg, A., and H. H. Mooy: *Chem. Eng. Progr.*, **44**:17 (1948).
6. Langhaar, H. L.: "Dimensional Analysis and Theory of Models," John Wiley & Sons, Inc., New York, 1951.
7. Rayleigh, Lord, *Phil. Mag.*, **48**:321 (1899).
8. Stokes, G. G.: "Mathematical and Physical Papers," vol. 3, p. 1, Cambridge University Press, London, 1922; see also *Trans. Cambridge Phil. Soc.*, **9**:8 (1856).

CHAPTER 7

TURBULENT FLOW IN CLOSED CONDUITS

7-1. Introduction

In Chap. 4 theoretical relationships were presented which predicted both velocity profiles and friction factors for laminar flow in closed conduits. Turbulent flow occurs much more frequently than laminar flow, but it cannot be analyzed satisfactorily from a theoretical standpoint. Relationships for predicting mean point velocities have been derived using Prandtl's mixing-length theories as a basis. These theories predict velocities quite accurately, but they do not represent the physical case at certain points in the stream. Some studies have been made with hot-wire anemometers to determine mean velocity fluctuations, correlation coefficients, and the spectrum of turbulence, but they have found limited application.

In the present chapter velocity distributions and friction factors for turbulent flow in closed conduits are considered. For turbulent flow, the wall roughness of the conduit influences both velocity distribution and the friction factor. This is not the case in laminar flow. The effect of wall roughness in circular tubes and annuli is considered also.

I. TURBULENT FLOW IN CIRCULAR TUBES

7-2. The Transition from Laminar to Turbulent Flow

Since turbulent flow in circular tubes is the type usually encountered in practice, it has been studied extensively. Numerous velocity profiles have been determined experimentally, and attempts have been made to determine a universal relationship to express the velocity distribution for flow in tubes. The work of Stanton et al.,[56] Nikuradse,[34] Reichardt,[45] Deissler,[6] and Rothfus and Monrad[51] has provided the data for the study of turbulent flow in circular tubes.

The nature of the velocity profile in turbulent flow is quite different from that in streamline flow. A comparison of the two profiles is shown in Fig.

7-1, in which the point velocity is plotted as a function of the tube diameter. The open circles represent experimental data obtained by Nikuradse for water flowing in a tube 0.394 in. in diameter. The average velocity is

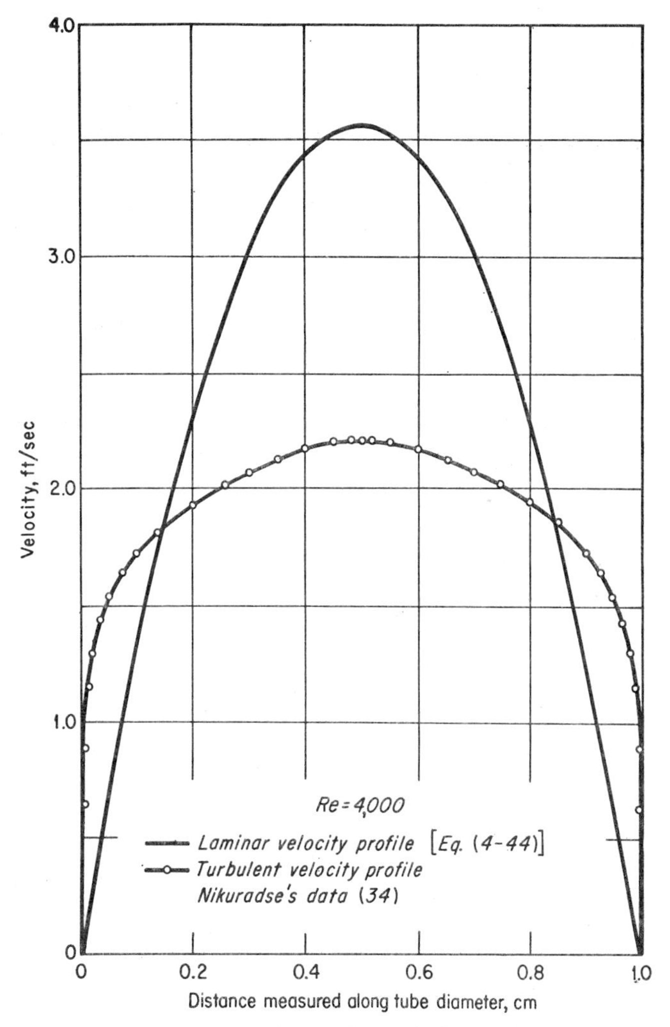

FIG. 7-1. Comparison of laminar and turbulent velocity profiles at the same average velocity.

1.79 ft/sec, and the Reynolds number is 4,000. Ordinarily, laminar flow will not exist at this Reynolds number except when all possible disturbances have been eliminated. The velocity distribution curve for laminar flow was calculated from Eq. (4-44). The difference between the two velocity profiles is considerable, although the average velocity in the tube is the same for both cases. The form of the laminar-flow profile is due to viscous forces between adjacent layers of fluid. The interface between layers diminishes toward the center of the tube and causes the velocity to change gradually. However, when turbulence exists, most of the velocity gradient occurs in the fluid adjacent to the wall, where the fluid retains some degree of streamline motion. The profile in the center core is almost flat, since in this region the motion is turbulent and inertia forces are high and have an effect over the major portion of the tube cross section.

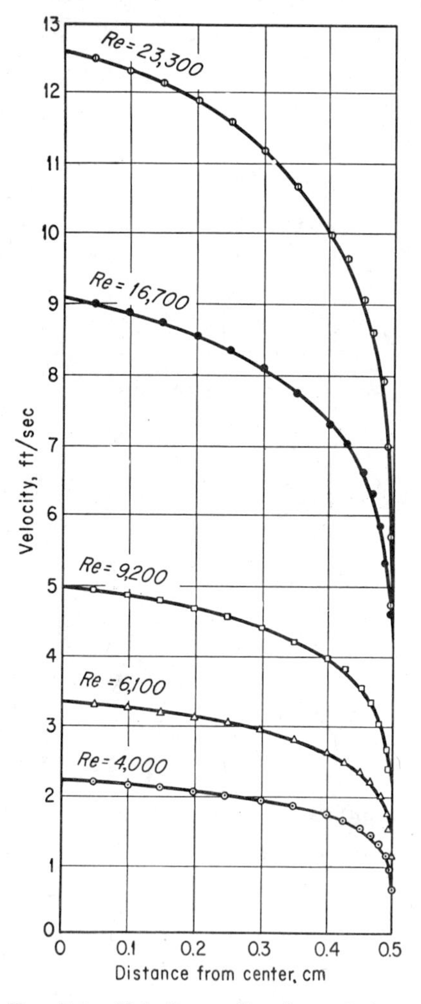

Fig. 7-2. Velocity profiles for turbulent flow in smooth circular tubes. (*From J. Nikuradse, VDI-Forschungsheft* 356, 1932.)

Nikuradse determined the velocity distribution for turbulent flow in smooth circular tubes with diameters of 1.0, 2.0, 3.0, 5.0, and 10.0 cm and at Reynolds numbers ranging from 4×10^3 to 3.24×10^6. Figure 7-2 is a plot of some of the experimental velocity profiles obtained; the distance from the axis r is plotted versus the measured lineal velocity u. The five curves shown are velocity profiles determined in a tube 1 cm in diameter and at values of Reynolds numbers ranging from 4×10^3 to 2.33×10^4. All the curves have the same general shape and indicate a maximum velocity at the center of the tube. There are indications that the velocity becomes zero at the tube wall, and in all cases there is a very high velocity gradient du/dy in the vicinity of the tube wall.

In both laminar and turbulent flow in tubes the point of maximum velocity is at the axis of the tube. Equation (4-43) represents a theoretical relationship between the average velocity and the maximum velocity for laminar flow and indicates that the average velocity is one-half of the maximum velocity. Figure 7-1 shows that for turbulent flow the ratio of the average velocity to the maximum velocity is much greater than 0.5, attaining a value slightly higher than 0.8.

Figure 7-3 is a plot of $U/u_{\max}$ versus the Reynolds number showing the data of Nikuradse [34] and Stanton and Pannell.[57] Figure 7-4 is a similar

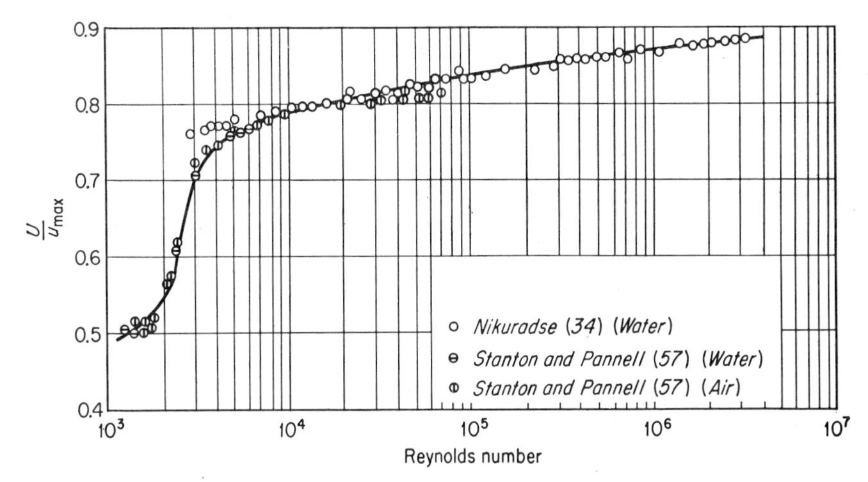

FIG. 7-3. Variation of axial velocity with the Reynolds number for flow in smooth circular tubes.

plot of the data of Senecal and Rothfus.[55] In the laminar region (below Re = 2,000) $U/u_{\max}$ has a value of 0.5, which is in agreement with Eq. (4-43). At Reynolds numbers slightly greater than 2,000, $U/u_{\max}$ increases sharply, reaching a value near 0.8 and thereafter increasing very slowly

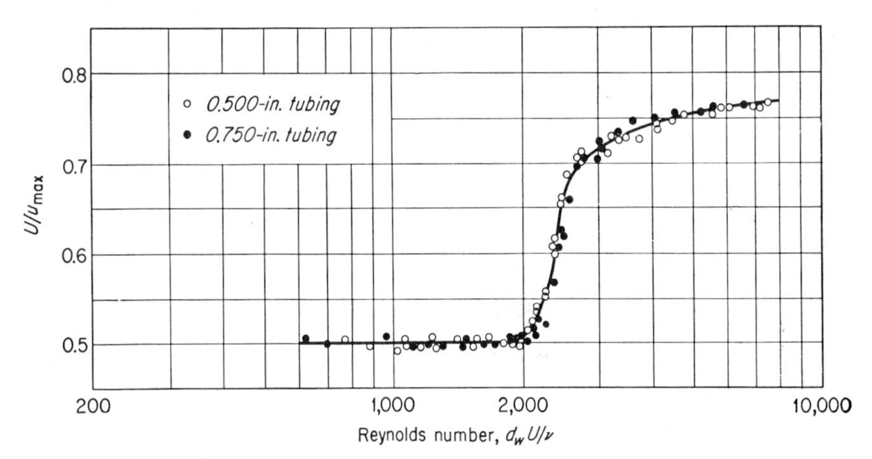

FIG. 7-4. Ratio of average to maximum velocity as a function of the Reynolds number for transition flow in tubes. [*From V. E. Senecal and R. R. Rothfus, Chem. Eng. Progr.*, **49**:533 (1953).]

over the range of Reynolds numbers studied. Figure 7-3 provides a means of determining the average velocity in pipelines by measuring the velocity at the pipe axis. Using the measured value of u_{max}, an approximate Reynolds number may be calculated from which a value of U/u_{max} may be obtained from Fig. 7-3. From this a value of U may be calculated from which a second Reynolds number is calculated. A second determination of U/u_{max} should be sufficient to give a correct average velocity.

Example 7-1

Air is flowing at 100°F in a circular duct having an ID of 12.0 in. A differential manometer placed across a pitot tube reads 0.85 in. H_2O. A manometer also shows that the pressure of the flowing air is 2.0 in. H_2O above atmospheric pressure. The barometer reads 29.43 in. Hg. Determine the mass rate of flow in the duct.

Solution

Applying Eq. (4-6) between points 1 and 2 in Fig. 7-5,

$$\int_1^2 \frac{dP}{\rho} + \frac{\Delta u^2}{\alpha g_c} = 0$$

since ΔZ, $\bar{w}$, and $\overline{lw}$ are zero. The pressure change is small between points 1 and 2,

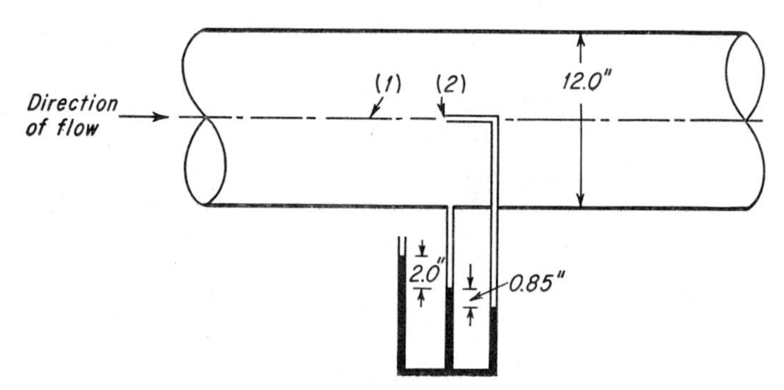

FIG. 7-5. Example 7-1.

so average density may be used. At point 1 the velocity is u_{max}; at point 2 the velocity is zero:

$$P_2 - P_1 = 0.85 \text{ in. } H_2O$$

$$\text{Average air density} = 0.0702 \text{ lb}_m/\text{ft}^3$$

$$\text{Viscosity of air at 100°F} = 12.84 \times 10^{-6} \text{ lb}_m/(\text{ft})(\text{sec})$$

Substituting in equation, taking $\alpha = 2$,

$$\frac{(0.85 \text{ in.})(62.4 \text{ lb}_m/\text{ft}^3)(32.2 \text{ ft/sec}^2)}{(12 \text{ in./ft})(0.0702 \text{ lb}_m/\text{ft}^3)[32.2\ (\text{lb}_m)(\text{ft})/(\text{lb}_f)(\text{sec}^2)]} - \frac{u_{max}^2 \text{ ft}^2/\text{sec}^2}{(2)[32.2\ (\text{lb}_m)(\text{ft})/(\text{lb}_f)(\text{sec}^2)]} = 0$$

from which
$$u_{\max}^2 = \frac{(0.85)(62.4)(2)(32.2)}{(12)(0.0702)} = 4{,}035 \text{ ft}^2/\text{sec}^2$$
giving
$$u_{\max} = 63.6 \text{ ft/sec}$$

Now calculate the Reynolds number based on $u_{\max}$.

$$\text{Re}_{\max} = \frac{d_w u_{\max} \rho}{\mu} = \frac{(1 \text{ ft})(63.6 \text{ ft/sec})(0.0702 \text{ lb}_m/\text{ft}^3)}{12.84 \times 10^{-6} \text{ lb}_m/(\text{ft})(\text{sec})} = 3.48 \times 10$$

At this Reynolds number, from Fig. 7-3

$$\frac{U}{u_{\max}} = 0.855$$

Thus
$$U = (0.855)(63.6) = 54.5 \text{ ft/sec}$$

$$\text{Re} = \frac{(1)(54.5)(0.0702)}{12.84 \times 10^{-6}} = 2.98 \times 10^5$$

At this Reynolds number

$$\frac{U}{u_{\max}} = 0.85$$

Thus
$$U = (0.85)(63.6) = 54.1 \text{ ft/sec}$$

This is the average velocity in the duct. The mass rate of air flow is

$$\left(\frac{\pi}{4} \text{ ft}^2\right)(54.1 \text{ ft/sec})(0.0702 \text{ lb}_m/\text{ft}^3) = 2.98 \text{ lb}_m/\text{sec}$$

7-3. Prandtl's Power Law of Velocity Distribution in Tubes

The power law of velocity distribution for turbulent flow in circular tubes represents one of the earliest attempts to correlate velocity data. Nikuradse [34] plotted his data as shown in the upper graph of Fig. 7-6, where $u/u_{\max}$ (the ratio of the point velocity at a position y from the wall to the maximum velocity) is plotted versus y/r_w (the ratio of the distance from the wall to the radius of the tube). Such a plot succeeds in bringing the extremities of the velocity-profile curves into coincidence; however, the middle portions of the curves are still separated, and it is obvious that such a dimensionless plot is not universal in character.

The most extensive work on the study of turbulent velocity profiles was done by Prandtl.[39,40,43] By means of his theories it is possible to obtain relationships which succeed in predicting turbulent velocity profiles quite accurately.

Prandtl's ideas regarding the mechanism of fluid flow were associated with the laws of resistance to flow. He postulated [39] that during turbulent flow the velocity of the fluid adjacent to wall was zero and that there was a layer of fluid close to the wall which was in laminar motion. The existence of this laminar film has been widely accepted, although its actual presence has not been detected experimentally with any degree of satisfaction. In 1920 Stanton et al.[56] claimed to have measured the velocity profile in the laminar layer adjacent to the pipe wall. Since the laminar layer is ex-

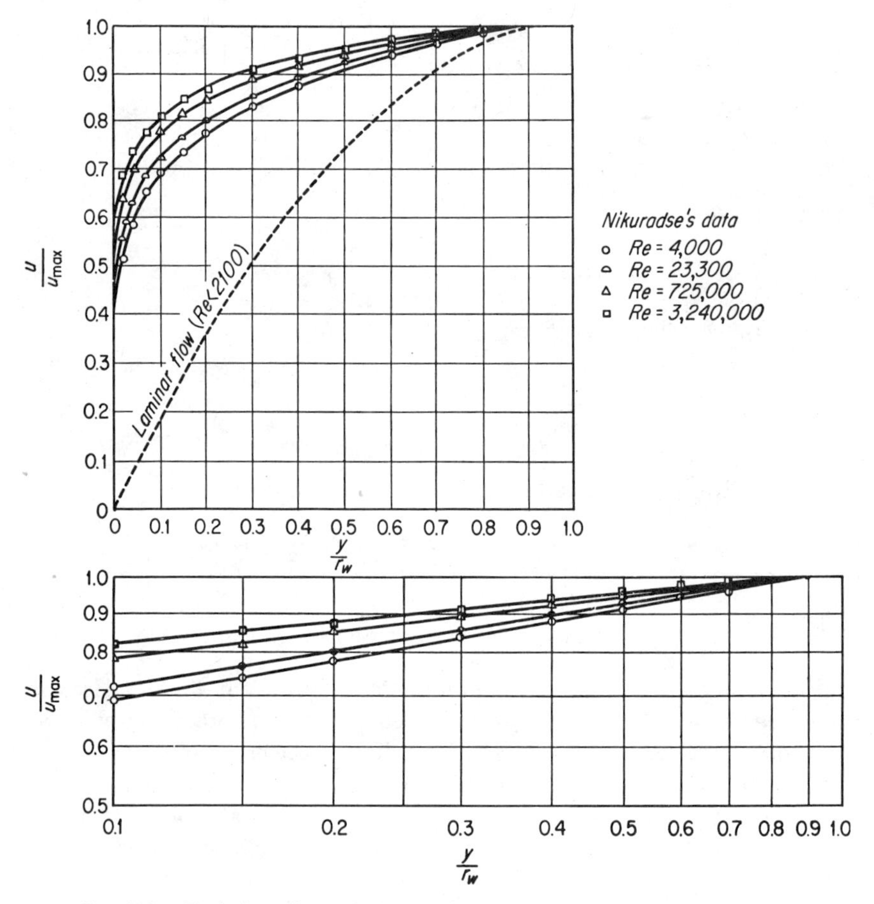

FIG. 7-6. Turbulent-flow velocity profiles plotted on a dimensionless basis.

tremely thin, its detection by any velocity-measuring device is difficult, particularly since the presence of the measuring equipment disturbs the laminar layer. In 1949 Miller [29] stated that the existence of the laminar layer has not been proved and that he will question whether it exists until conclusive proof is obtained. In this regard, however, it may be stated that present theories of heat and mass transfer depend largely on the belief that such a laminar film does exist, and their successful application constitutes at least a partial proof that Prandtl's theory is correct.

If such a laminar layer is present, the only shearing force acting in this layer would be a viscous force, in which case the velocity profile in the neighborhood of the wall could be expressed by

$$\left(\frac{du}{dy}\right)_{y=0} = \frac{\tau_w g_c}{\mu} \tag{7-1}$$

Thus, knowing τ_w, which is related to the pressure loss due to friction by Eq. (4-13), the velocity profile in the laminar layer can be obtained for any flow and for any fluid. This can be done for both laminar and turbulent flow, since in both types a laminar layer adjacent to the tube wall is assumed.

For turbulent flow, however, at positions in the main stream beyond the laminar layer the flow is turbulent, and the shear stress cannot be expressed by Eq. (7-1). Prandtl [41] derived a preliminary relationship for velocity profiles in turbulent flow from Blasius' [2] friction-factor equation for the flow of fluids in tubes. The form of the relationship which Prandtl derived is

$$\frac{u}{u_{\max}} = c\left(\frac{y}{r_w}\right)^{1/7} \tag{7-2}$$

which is known as the power law for the velocity distribution in turbulent flow. In order to test this law, Nikuradse's data [34] are plotted in Fig. 7-6 as log $(u/u_{\max})$ versus log (y/r_w) for various values of the Reynolds number. The slopes of the lines vary from a value of 1/7 at Re $= 4 \times 10^3$ to 1/10 at Re $= 3.24 \times 10^6$.

The Blasius friction-factor equation relates the friction factor to the Reynolds number

$$f = 0.079(\mathrm{Re})^{-1/4} \tag{7-3}$$

where f is the friction factor. Combining Eqs. (7-3) and (4-17) and noting that $d_w = 2r_w$ and Re $= 2r_wU/\nu$,

$$\frac{2\tau_w g_c}{\rho U^2} = 0.079\left(\frac{2r_wU}{\nu}\right)^{-1/4} \tag{7-4}$$

From Eq. (7-4), since τ_w, g_c, ρ, and ν are constant,

$$U = c_1 r_w^{1/7} \tag{7-5}$$

Since the ratio of the average velocity to the maximum velocity is approximately constant,

$$u_{\max} = c_2 r_w^{1/7} \tag{7-6}$$

Equation (7-6) may be written to express the velocity at any radius r

$$u = c_3 r^{1/7} \tag{7-7}$$

and since the distance from the tube wall y varies in the same way as the radius r, Eq. (7-7) may be written

$$u = c_4 y^{1/7} \tag{7-8}$$

Dividing Eq. (7-8) by Eq. (7-6) gives the power law for velocity distribution

$$\frac{u}{u_{\max}} = c\left(\frac{y}{r_w}\right)^{1/7} \tag{7-2}$$

It is evident from Fig. 7-6 that the velocity distribution in the turbulent core can be expressed in some form of a power law. However, Eq. (7-2) is based on Blasius' friction-factor equation (7-3), which is valid only up to a Reynolds number of 100,000. It would be expected that Eq. (7-2) would be valid up to this Reynolds number. The curves in Fig. 7-6 bear out this conclusion.

7-4. Logarithmic Velocity-distribution Equations

A logarithmic velocity-distribution equation for turbulent flow in smooth tubes was derived by Prandtl,[43] using his mixing-length theory as a basis. This theory was referred to in Sec. 5-6, where it was shown that

$$\tau = \frac{\mu}{g_c}\frac{du}{dy} + \frac{\rho}{g_c}\left(l\frac{du}{dy}\right)^2 \tag{5-22}$$

where τ is the shear stress at any distance y from the tube wall. The turbulent portion of the shear stress is contained in the last term of Eq. (5-22). If viscous shear is neglected,

$$\tau = \frac{\rho}{g_c}\left(l\frac{du}{dy}\right)^2 \tag{7-9}$$

This equation expresses the turbulent shear between any two layers of fluid in terms of the Prandtl mixing length, the density, and the velocity gradient. Expressing the shear at any point in terms of the shear at the wall by the use of Eq. (4-14) gives the following equation, in which r/r_w is replaced by the the term $1 - y/r_w$:

$$\tau_w\left(1 - \frac{y}{r_w}\right) = \frac{\rho}{g_c}\left(l\frac{du}{dy}\right)^2 \tag{7-10}$$

from which

$$l\frac{du}{dy} = \sqrt{\frac{\tau_w g_c}{\rho}}\sqrt{1 - \frac{y}{r_w}} \tag{7-11}$$

The term $\sqrt{\tau_w g_c/\rho}$ appears frequently in analyzing velocity profiles. In dimensional analysis of the problem (see Table 6-3) this term appeared in one of the dimensionless groups obtained. It has dimensions of velocity and is called the friction velocity or the shear velocity. It is given the symbol u^*.

$$u^* = \sqrt{\frac{\tau_w g_c}{\rho}} \tag{7-12}$$

The friction velocity is constant for a given set of flow conditions.

Prandtl proceeded to develop his velocity-distribution equation by assuming that near the wall the term $1 - y/r_w$ is very nearly equal to 1, and thus from Eqs. (7-11) and (7-12) one obtains

$$u^* = l\frac{du}{dy} \tag{7-13}$$

It is further assumed that neither the viscosity nor the roughness of the wall has any appreciable effect on the position being considered. Therefore, for a point at the distance y from the wall there is no other characteristic length than the distance y. Hence follows the relationship

$$l = \kappa y \tag{7-14}$$

where κ is a universal constant. Thus, substituting Eq. (7-14) into (7-13),

$$\frac{du}{dy} = \frac{u^*}{\kappa y} \tag{7-15}$$

Integration yields

$$u = \frac{u^*}{\kappa}\ln y + c \tag{7-16}$$

The universal constant κ was determined by Prandtl [42] and Nikuradse [34] from the data which the latter obtained on turbulent velocity profiles. The constant was evaluated by means of Eqs. (7-11) and (7-15) and was found to have a value of 0.4.

The eddy viscosity is defined in terms of the mixing length [see Eqs. (5-23) and (5-24)] as

$$\mathrm{E}_M = \rho l^2 \frac{du}{dy} \tag{7-17}$$

The eddy diffusivity of momentum ϵ_M is

$$\epsilon_M = \frac{\mathrm{E}_M}{\rho} = l^2 \frac{du}{dy} \tag{7-18}$$

Combining Eq. (7-14) with (7-18),

$$\epsilon_M = \kappa^2 y^2 \frac{du}{dy} \tag{7-19}$$

By dimensional analysis, if it is assumed that ϵ_M is a function of y and du/dy, one dimensionless group $\dfrac{\epsilon_M}{y^2(du/dy)}$ is obtained. Prandtl's relationship given in Eq. (7-14) is equivalent to assuming a constant value κ^2 for this dimensionless group.

The velocity distribution of Prandtl can be obtained by solving for the constant of integration in Eq. (7-16) by making use of the boundary condition that $u = u_{\max}$ at $y = r_w$. [NOTE: This use of the boundary condition is not entirely appropriate because Eq. (7-13) applies strictly to the vicinity near the tube wall. However, the resulting velocity-distribution equation represents experimental data very well.]

$$u = u_{\max} + 2.5u^* \ln \frac{y}{r_w} \tag{7-20}$$

Equation (7-20) is known as the Prandtl velocity-distribution equation. It is plotted in Fig. 7-7 as $(u_{\max} - u)/u^*$ versus y/r_w and is compared with the actual experimental data of Nikuradse.[34]

Von Kármán [19] obtained a different expression for the point velocity. He shows [19,20] by his similarity theory that the turbulent shear stress is given by

$$\tau = \frac{\rho\kappa^2(du/dy)^4}{g_c(d^2u/dy^2)^2} \tag{7-21}$$

from which

$$\epsilon_M = \frac{\kappa^2(du/dy)^3}{(d^2u/dy^2)^2} \tag{7-22}$$

and from Eq. (7-18)

$$l = \frac{\kappa(du/dy)}{d^2u/dy^2} \tag{7-23}$$

The result given in Eq. (7-22) could be obtained from dimensional analysis by assuming ϵ_M to be a function of du/dy and d^2u/dy^2.

Substitution of Eq. (7-23) into Eq. (7-11) and subsequent integration gives

$$u = u_{\max} + \frac{1}{\kappa} u^* \left[\ln\left(1 - \sqrt{1 - \frac{y}{r_w}}\right) + \sqrt{1 - \frac{y}{r_w}} \right] \tag{7-24}$$

Conditions:

At $y = 0$

$$\frac{du}{dy} = \infty$$

At $y = r_w$

$$u = u_{\max}$$

The value of κ is taken as 0.4. Equation (7-24) is plotted in Fig. 7-7.

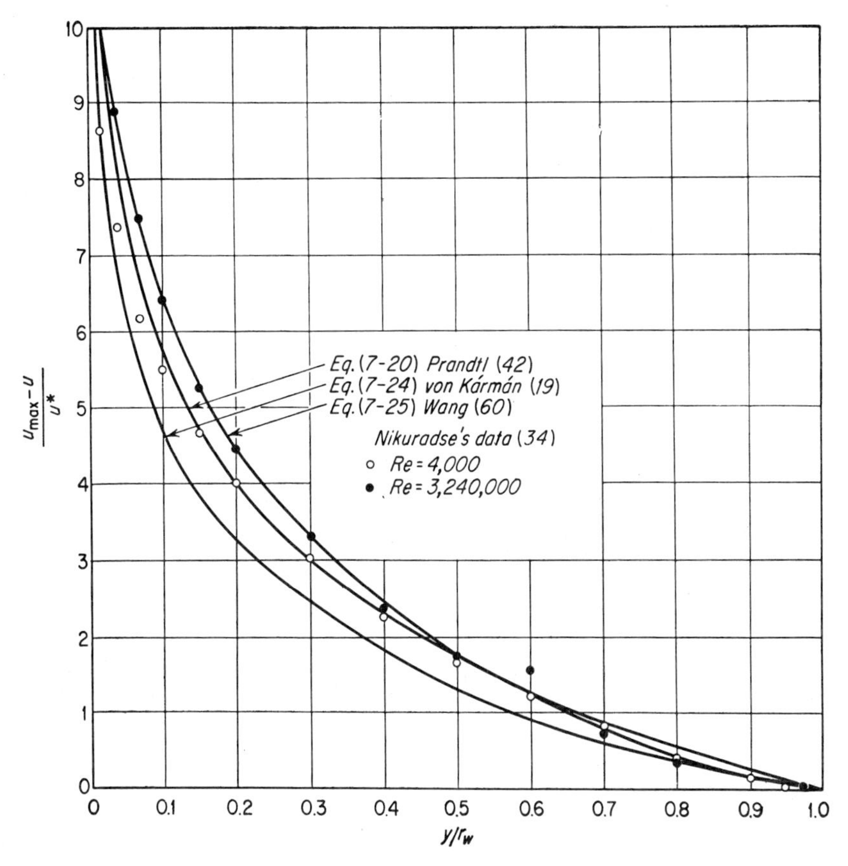

FIG. 7-7. Comparison of the various velocity-distribution equations.

Wang[60] has developed another velocity-distribution equation which agrees with experimental data better than the equations of Prandtl or von Kármán do. Wang develops an expression for the mixing length from the actual velocity distribution, substitutes this expression into Eq. (7-11), and, on integration and solving for the constant, derives the following equation:

$$\frac{u_{\max} - u}{u^*} = 2.5 \left(\ln \frac{1 + \sqrt{1 - y/r_w}}{1 - \sqrt{1 - y/r_w}} - 2 \tan^{-1} \sqrt{1 - \frac{y}{r_w}} - 0.572 \ln \frac{1 - y/r_w + 1.75\sqrt{1 - y/r_w} + 1.53}{1 - y/r_w - 1.75\sqrt{1 - y/r_w} + 1.53} + 1.14 \tan^{-1} \frac{1.75\sqrt{1 - y/r_w}}{1.53 - (1 - y/r_w)} \right) \tag{7-25}$$

Wang's equation is also plotted in Fig. 7-7. The agreement of Wang's equation with the experimental data is very good. However, considering its complexity, it is less useful than either Prandtl's or von Kármán's equation. In fact, the Prandtl equation agrees very well with the data, and its simplicity makes it very convenient to use.

7-5. The Universal Velocity Distribution for Smooth Tubes

Equation (7-20) may be taken as a basis for the development of a still more general equation for the velocity distribution in circular tubes. One must consider the flow pattern in the vicinity of the wall, where both laminar and turbulent flow exist, as shown in Fig. 7-8. Although the transition from laminar to turbulent flow is gradual, it will be assumed that the transition takes place at a distance δ_1 from the wall and that beyond this point only fully developed turbulence exists. Because the velocity gradient is assumed to be uniform throughout the laminar layer, it can be expressed as

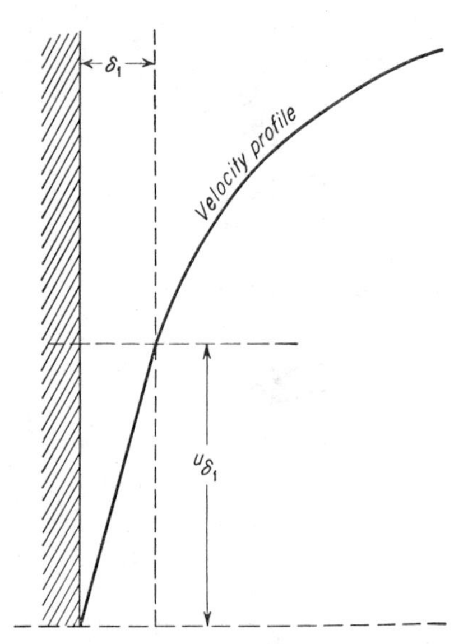

FIG. 7-8. Velocity distribution near the wall of a smooth tube.

$$\left(\frac{du}{dy}\right)_{y=\delta_1} = \frac{u_{\delta_1}}{\delta_1} \tag{7-26}$$

Since only viscous forces are acting in the laminar layer, the shear at the wall may be expressed in terms of the velocity and the viscosity, giving

$$g_c\tau_w = \mu\frac{u_{\delta_1}}{\delta_1} \tag{7-27}$$

Equation (7-27) may be rearranged by using Eq. (7-12).

$$\frac{u_{\delta_1}}{u^*} = \frac{\delta_1 u^*}{\nu} \tag{7-28}$$

In Eq. (7-28) $\delta_1 u^*/\nu$ is a dimensionless quantity having the same form as the Reynolds number. This equation states that the ratio of the velocity at the edge of the laminar layer to the friction velocity is equal to some modified Reynolds number. By letting u_{δ_1} be a function of τ_w, ρ, μ, and δ_1 the groups appearing in Eq. (7-28) could be obtained by dimensional analysis. It is assumed that this ratio is constant for all turbulent flow in all sizes of tubes. Thus Eq. (7-28) is a dimensionless equation, each side of which is constant; i.e.,

$$\frac{u_{\delta_1}}{u^*} = \frac{\delta_1 u^*}{\nu} = c \tag{7-29}$$

Rearranging Eq. (7-20) to

$$\frac{u}{u^*} = \frac{u_{max}}{u^*} + 2.5 \ln \frac{y}{r_w} \tag{7-30}$$

and substituting $u = u_{\delta_1}$ and $y = \delta_1$ gives

$$c = \frac{u_{\delta_1}}{u^*} = \frac{u_{max}}{u^*} - 2.5 \ln \frac{r_w u^*}{c\nu} \tag{7-31}$$

Collecting all the constant terms in a constant c_1 results in

$$\frac{u_{max}}{u^*} = c_1 + 2.5 \ln \frac{r_w u^*}{\nu} \tag{7-32}$$

Substituting Eq. (7-32) into (7-30) gives a dimensionless equation for the turbulent velocity distribution in smooth circular tubes.

$$\frac{u}{u^*} = c_1 + 2.5 \ln \frac{yu^*}{\nu} \tag{7-33}$$

The terms u/u^* and u^*y/ν are dimensionless terms. The former is the ratio of the point velocity to the friction velocity; the latter is a modified Reynolds number involving the friction velocity, the distance from the wall, and the kinematic viscosity of the fluid. Thus Eq. (7-33) may be written

$$u^+ = c_1 + 2.5 \ln y^+ \tag{7-34}$$

where $u^+ = \dfrac{u}{u^*}$

$y^+ = \dfrac{yu^*}{\nu}$

Equation (7-34) is the universal velocity-distribution equation for turbulent flow in circular tubes. The dimensionless groups u^+ and y^+ could have been obtained by dimensional analysis; however, a further consideration of the mechanism of turbulent flow involving an assumption of the variation of the mixing length [Eq. (7-14)] provided the functional relationship between the dimensionless groups.

Nikuradse [34] plotted his experimental data to obtain the constant in Eq. (7-34). In Fig. 7-9 Nikuradse's experimental data are plotted, and the value of c_1 is 5.5 for curve I. Thus

$$u^+ = 5.5 + 2.5 \ln y^+ \tag{7-35}$$

An investigation of Nikuradse's original data indicates that his calculated values of y^+ are different by an amount equal to seven units from those determined from tables containing his original data. It appears that

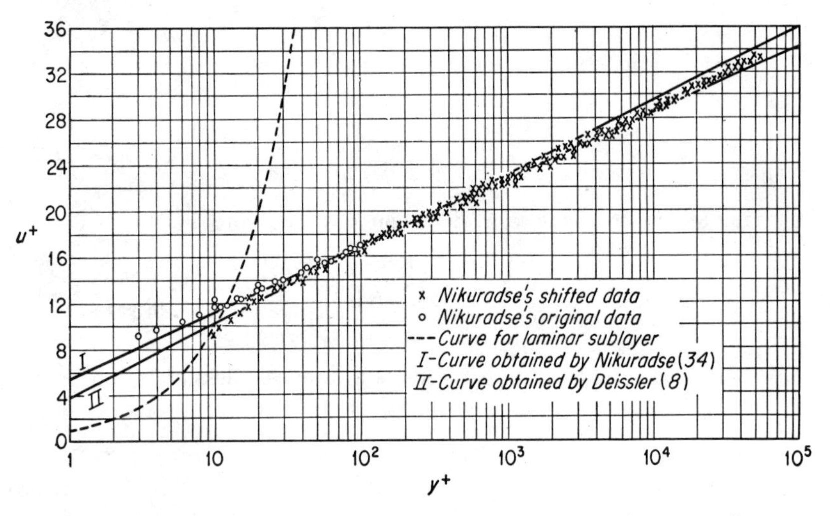

FIG. 7-9. Experimental velocity-distribution data plotted as u^+ versus y^+. (*From J. Nikuradse, VDI-Forschungsheft* 356, 1932.)

Nikuradse shifted his original data [29] in order that his velocity-distribution data near the wall of the tube would agree with Prandtl's laminar-sublayer theory, which is represented on Fig. 7-9 by the broken curve. Both the shifted and original data of Nikuradse are plotted in Fig. 7-9, and the difference is apparent only at values of y^+ less than 50.

Curve II in Fig. 7-9 represents the equation

$$u^+ = 3.8 + 2.78 \ln y^+ \tag{7-36}$$

which was obtained by Deissler [8] for the flow of air in a smooth circular tube. Deissler investigated values of y^+ up to 5,000. The difference between curves I and II in the range of y^+ from 30 to 5,000 is very small. Since Nikuradse [34] investigated a very large range of y^+, it appears that the equation represented by curve I in Fig. 7-9 is probably the more dependable over the whole range of y^+.

Reichardt [45] obtained some velocity-distribution data for flow in an open channel, and, using his own results and Nikuradse's, he obtained the plot in Fig. 7-10, which shows u^+ as a function of y^+ throughout the cross section of the tube. The curve may be divided into three distinct parts. In the region adjacent to the wall the fluid motion is laminar, according to Prandtl's theory. Therefore, in this region

$$u^+ = y^+ \tag{7-37}$$

The relation given in Eq. (7-37) for the laminar layer may be obtained from Eq. (7-28), which was derived for the edge of the laminar layer but applies throughout when δ_1 is replaced by y and u_{δ_1} is replaced by u

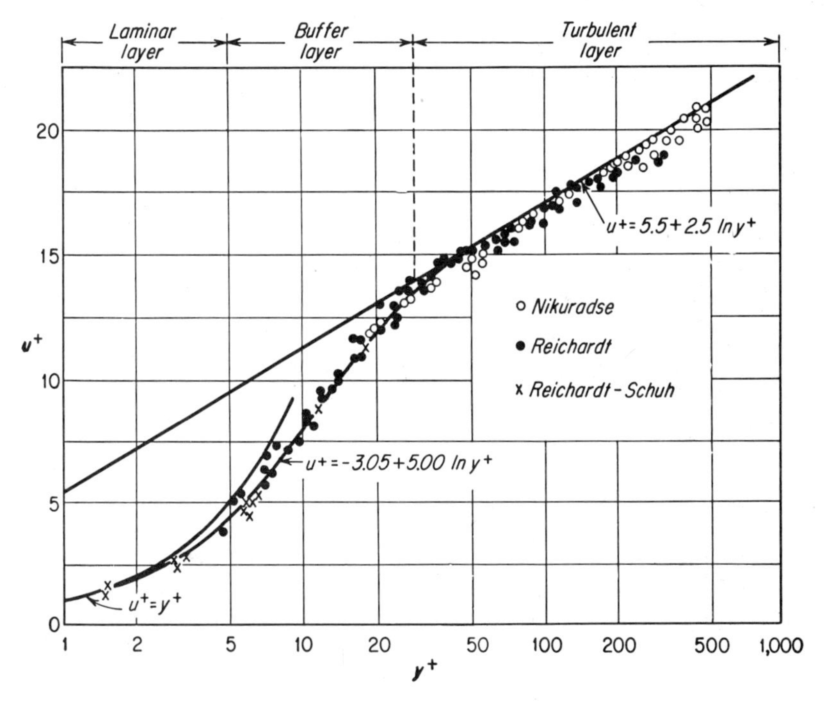

FIG. 7-10. Universal velocity distribution for turbulent flow in circular tubes. (*From H. Reichardt, NACA TM* 1047, 1943.)

Von Kármán [20] analyzed the curve in Fig. 7-10 and found the extent of the laminar boundary layer to be between $y^+ = 0$ and $y^+ = 5$. Between $y^+ = 5$ and $y^+ = 30$ is a buffer layer, in which the relation between y^+ and u^+ can be expressed as a logarithmic relationship.

Thus the velocity distribution for the whole cross section of the circular tube is expressed by the following three relationships:

Laminar layer:

$$u^+ = y^+ \qquad y^+ < 5 \tag{7-37}$$

Buffer layer:

$$u^+ = -3.05 + 5.0 \ln y^+ \qquad 5 < y^+ < 30 \tag{7-38}$$

Turbulent core:

$$u^+ = 5.5 + 2.5 \ln y^+ \qquad y^+ > 30 \tag{7-35}$$

Equations (7-35), (7-37), and (7-38) may be used to calculate the complete velocity profile for turbulent flow in smooth circular tubes.

7-6. Inconsistencies in the Universal Velocity-distribution Equations

It is apparent that experimental velocity data are in good agreement with the curves given by Eqs. (7-35), (7-37), and (7-38). The relationships, however, are not consistent with physical reality. The change from laminar to turbulent motion is gradual, and the transition probably starts at values of y^+ less than 5. It is therefore not correct to have three different curves (u^+ versus y^+) representing the complete velocity distribution when a single curve should be used. A second, more serious inconsistency is the fact that at the axis of the tube Eq. (7-35) does not give a zero velocity gradient. For this reason a velocity profile calculated from Eqs. (7-35), (7-37), and (7-38) has an average velocity somewhat higher than it should. Despite these inconsistencies, however, the universal velocity-distribution equations have been used extensively in studying the relation between momentum and heat transfer.

7-7. Improvements in the Universal Velocity-distribution Equations

Aware of the inconsistencies mentioned above, a number of workers have developed relationships to avoid them if possible. Deissler [6] developed a single equation relating u^+ to y^+ up to $y^+ = 26$. He derived an expression for the velocity profile in the vicinity of the wall, neglecting the effect of kinematic viscosity, i.e., assuming

$$\epsilon_M = n^2 uy \tag{7-39}$$

where n is a constant determined experimentally,

$$y^+ = \frac{1}{n} \frac{1/\sqrt{2\pi} \int_0^{nu^+} e^{-[(nu^+)^2/2]} d(nu^+)}{(1/\sqrt{2\pi}\,) e^{-[(nu^+)^2/2]}} \tag{7-40}$$

The constant n has a value of 0.109.

A subsequent analysis by Deissler [10] taking the effect of kinematic viscosity into account gives a relationship differing only slightly from Eq.

(7-40). For the turbulent core Deissler obtained Eq. (7-36). Equations (7-36) and (7-40) are plotted in Fig. 7-11, along with Deissler's experimental data. Equation (7-40) applies for $0 < y^+ < 26$, while Eq. (7-36) applies for $y^+ > 26$. Equation (7-40) is distinctive in that it represents by a single curve the velocity distribution in the laminar layer and buffer layer. It does not assume the existence of a laminar layer, but in the limit as y^+ approaches zero, Eq. (7-40) approaches Eq. (7-37), which would indicate that laminar motion exists immediately adjacent to the pipe wall.

Nikuradse's velocity data were critically reanalyzed by Ross,[47] who considered only that part of Nikuradse's data which appeared most reliable. He found that in the turbulent core in the region close to the wall (not including the laminar layer and buffer layer) the equation

$$u^+ = 5.6 + 5.6 \log y^+ \tag{7-41}$$

best represented the data up to a value of $y/r_w = 0.13$, while Eq. (7-24) with $\kappa = 0.3$ represented the data in the range $0.13 < y/r_w < 1.0$. Equations (7-41) and (7-24) (with $\kappa = 0.30$) are plotted in Fig. 7-12 as $(u_{\text{max}} - u)/u^*$ versus y/r_w. In plotting Eq. (7-41), $u = u_{\text{max}}$ at a value of $y/r_w = 1.38$. This value was determined from Nikuradse's data. The resulting curve shown in Fig. 7-12 agrees with velocity data very well and gives a zero velocity gradient at the axis of the tube.

A study by Rothfus and Monrad [51] indicates that u^+ and y^+ are not the only dimensionless groups involved in a universal relationship, particularly in the region just beyond the buffer layer. These workers obtained better correlation of data by plotting $u^+(U/u_{\text{max}})$ versus $y^+(u_{\text{max}}/U)$. Pai [36] derived an equation for the velocity distribution from the Reynolds momentum equations (5-14). The theoretical relationship obtained is shown in Eq. (7-42), the numerical constants of which were obtained from experimental data.

$$\frac{u}{u_{\text{max}}} = 1 - 0.204\left(\frac{r}{r_w}\right)^2 - 0.796\left(\frac{r}{r_w}\right)^{32} \tag{7-42}$$

Up to a value of $r/r_w = 0.9$ Eq. (7-42) agrees with Nikuradse's [34] velocity data for the flow of water in a tube at a Reynolds number of 3.24×10^6. Pai found that the following equation agreed with Nikuradse's data over the whole tube cross section:

$$\frac{u}{u_{\text{max}}} = 1 - 0.204\left(\frac{r}{r_w}\right)^2 - 0.250\left(\frac{r}{r_w}\right)^{32} \tag{7-43}$$

In view of the fact that Nikuradse's data in the vicinity of the wall are questionable, further investigation in this region to check Eq. (7-42) is necessary.

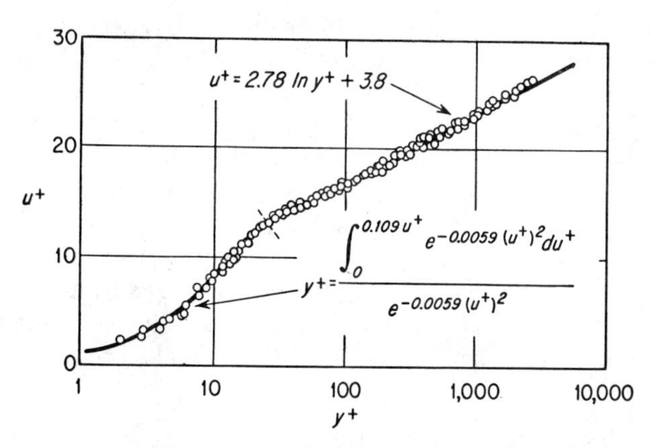

FIG. 7-11. Deissler's circular-tube velocity-distribution data. [*From R. G. Deissler, Trans. ASME*, **73**:101 (1951).]

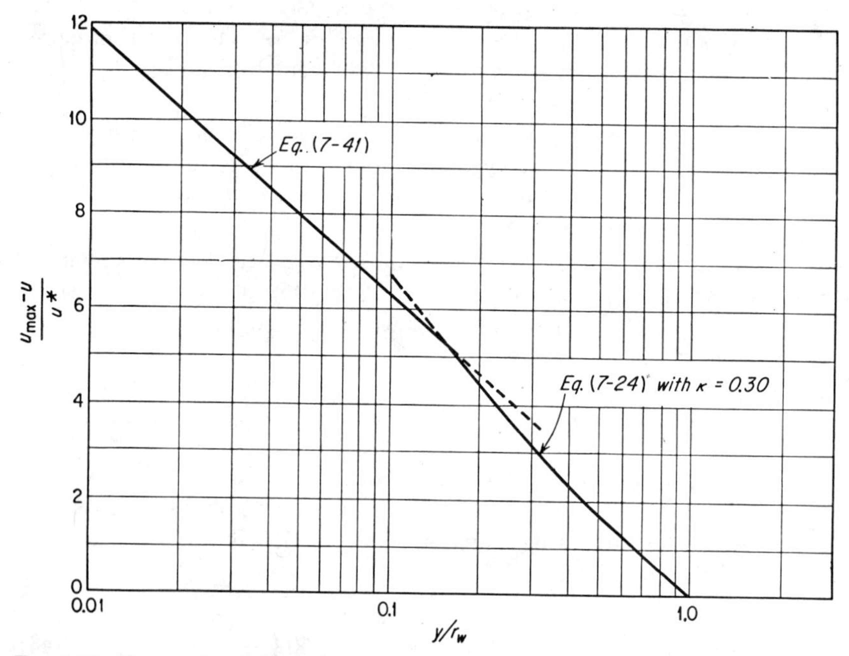

FIG. 7-12. Proposed turbulent-core velocity-distribution curve for flow in circular tubes. (*From D. Ross, Proc. Midwest. Conf. Fluid Mechanics, 3rd Conf., Univ. Minn., 1953*, p. 651.)

7-8. Effect of Variable Fluid Properties on Turbulent Velocity Profiles

The equations given above for turbulent-flow velocity profiles apply to incompressible fluids in the absence of heat transfer between fluid and surroundings. Deissler [6,7] showed that for adiabatic flow the effects of compressibility on the velocity distribution were small up to the sonic velocity.

Deissler [7] developed velocity-distribution relationships for the case of heat transfer to the fluid flowing in a tube. He considered only variation of fluid viscosity and assumed density to be constant.

For gases in the region near the wall ($0 < y^+ < 26$)

$$y^+ = \exp\left(-\frac{n^2u^+}{\omega}\right)(1 - \omega u^+)^{-\frac{n^2}{\omega^2}} \int_0^{u^+} \exp\left(\frac{n^2u^+}{\omega}\right)(1 - \omega u^+)^{\frac{n^2}{\omega^2}+d'}\, du^+ \tag{7-44}$$

and in the turbulent core

$$y^+ = \frac{c_1\omega}{2\kappa^2} \exp\left(-\frac{2\kappa}{\omega}\sqrt{1 - \omega u^+}\right)\left(\frac{2\kappa}{\omega}\sqrt{1 - \omega u^+} + 1\right) \tag{7-45}$$

where $n = 0.109$

$\kappa = 0.36$

c_1 = constant of integration evaluated at $y^+ = 26$

ω = dimensionless heat-transfer factor:

$$\omega = \frac{q_w\sqrt{\tau_w g_c/\rho_w}}{A_w C_p g_c T_w \tau_w} \tag{7-46}$$

d' = exponent giving relation of gas viscosity to temperature, i.e.,

$$\frac{\mu}{\mu_w} = \left(\frac{T}{T_w}\right)^{d'} \tag{7-47}$$

Equation (7-45) reduces to Eq. (7-36) for the case of no heat transfer (i.e., $\omega = 0$). Equations (7-44) and (7-45) are plotted in Fig. 7-13 for a gas with $d' = 0.68$ and a Prandtl number $C_p\mu/k = 1.0$. These curves are applicable without great error to the common gases.[7] In using Eqs. (7-44) and (7-45) the fluid properties are evaluated at the tube-wall temperature.

For liquids [10] in the region near the wall ($0 < y^+ < 26$)

$$u^+ = \int_0^{y^+} \frac{dy^+}{\mu/\mu_w + n^2u^+y^+\left[1 - \exp\left(-\frac{n^2u^+y^+}{\mu/\mu_w}\right)\right]} \tag{7-48}$$

and for the turbulent core

$$u^+ = c_1 + \frac{1}{\kappa}\ln y^+ \tag{7-49}$$

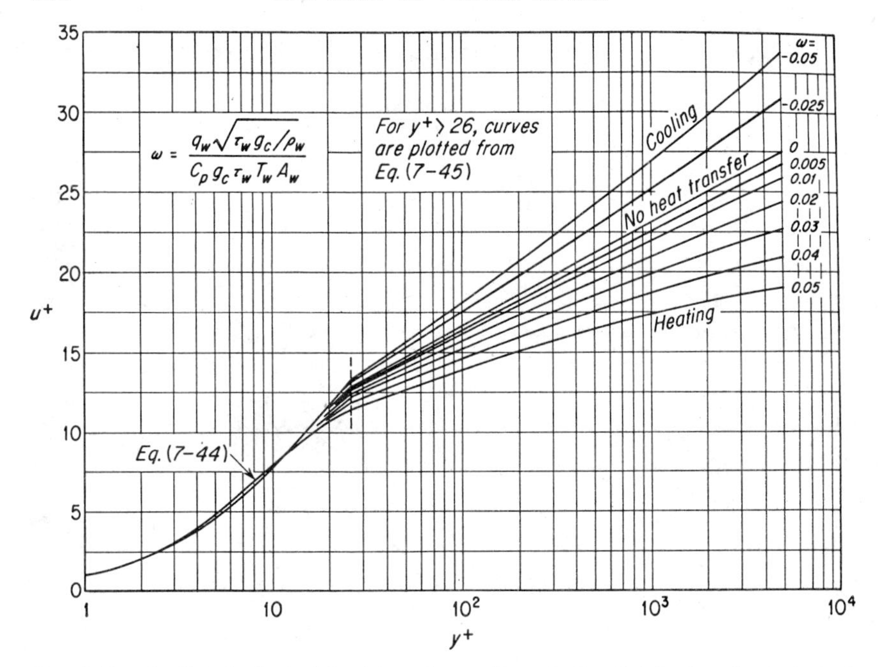

FIG. 7-13. Predicted effect of heat transfer on the velocity distribution for turbulent flow of a gas having a Prandtl number of 1. All properties evaluated at T_w. (*From R. G. Deissler, NACA TN* 2242, 1950.)

where $n = 0.109$

$$\frac{\mu}{\mu_w} = \left(\frac{T}{T_w}\right)^{d'} \quad (d' \text{ has values between } -1 \text{ and } -4 \text{ for liquids})$$

$\kappa = 0.36$

c_1 = constant evaluated at $y^+ = 26$

7-9. Relation of u^+ and y^+ to the Friction Factor and Reynolds Number

The friction velocity is defined as

$$u^* = \sqrt{\frac{\tau_w g_c}{\rho}} \tag{7-12}$$

and the shear stress exerted on the wall is

$$\tau_w = \frac{f \rho U^2}{2g_c} \tag{4-17}$$

Combining Eqs. (4-17) and (7-12),

$$u^* = U\sqrt{\frac{f}{2}} \tag{7-50}$$

Thus from Eq. (7-34)

$$u^+ = \frac{u}{u^*} = \frac{u/U}{\sqrt{f/2}} \tag{7-51}$$

Equation (7-51) provides a means of calculating the dimensionless quantity u^+ from the friction factor and the average velocity in the tube.

Likewise the relation

$$y^+ = \frac{yu^*}{\nu} = \frac{y}{r_w}\frac{\text{Re}}{2}\sqrt{\frac{f}{2}} \tag{7-52}$$

allows the calculation of y^+ from the Reynolds number and the friction factor.

7-10. Velocity Distribution for Turbulent Flow in Rough Tubes

The internal surface of tubes is usually quite rough, especially after the tubes have been in service for some time. Drawn tubing is considered to be smooth when it is new. Standard pipe generally is quite rough, even when new. Such boundary roughness is shown schematically in Fig. 7-14. The average height of the roughness projections is expressed by the quantity e, while the radius of the tube extends from the center of the tube to the average distance of the wall. The relative roughness of the surface is expressed by the relation e/r_w. This dimensionless quantity must be considered also when correlating friction and velocity data for turbulent flow in tubes. For laminar flow the relative roughness has no effect on velocity distribution or friction loss.

Nikuradse [32,35] studied artificially roughened pipes and determined velocity profiles for values of r_w/e from 15.0 to 507.0. The tubes were artificially roughened by cementing sand grains of diameter e to the inside surface.

The difference between velocity profiles in smooth tubes and rough tubes is shown in Fig. 7-15, which is a plot of the point velocity u versus y/r_w for a constant volumetric flow rate in tubes having different surface conditions.

In deriving a general correlation for velocity distribution in rough tubes Prandtl [41] modified Eq. (7-16) by replacing the distance from the wall y with the relative roughness y/e, giving

$$\frac{u}{u^*} = \frac{1}{\kappa}\ln\frac{y}{e} + c_1 \tag{7-53}$$

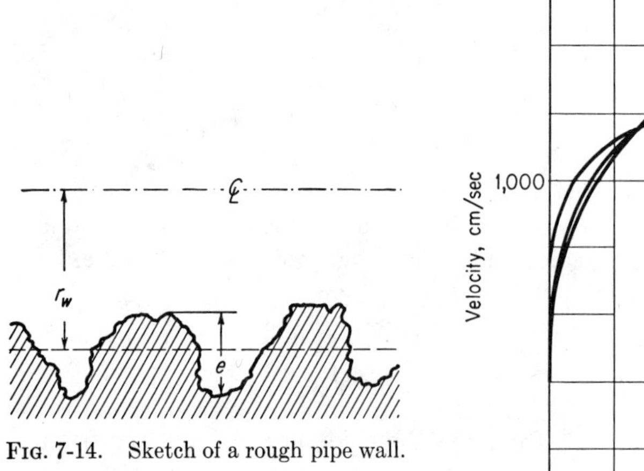

FIG. 7-14. Sketch of a rough pipe wall.

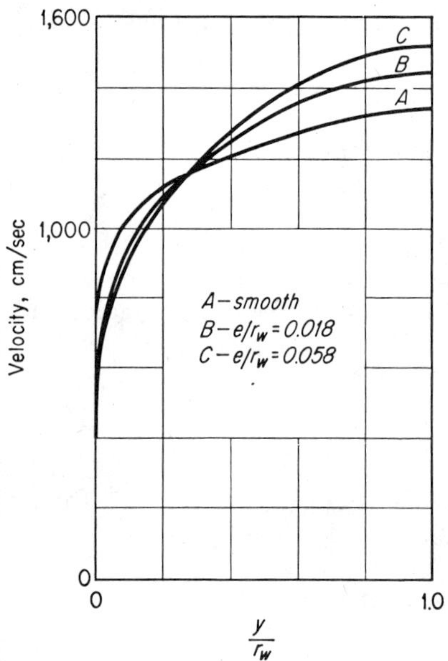

FIG. 7-15. Comparison of velocity profiles in rough and smooth tubes.

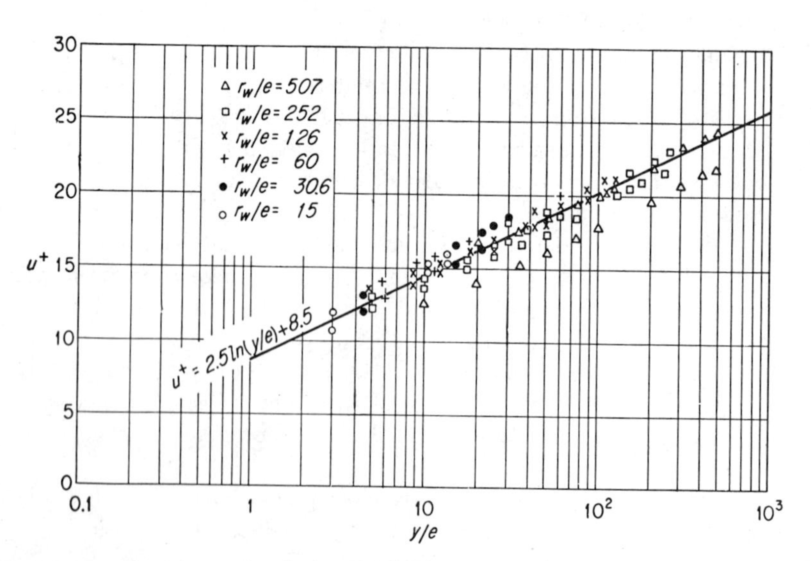

FIG. 7-16. Experimental velocity-distribution data in rough tubes. (*Data from J. Nikuradse, VDI-Forschungsheft* 361, 1933.)

In Fig. 7-16 Nikuradse's experimental velocity data are plotted in the form u^+ versus y/e. The data shown are for the turbulent core only. The equation of the line is

$$u^+ = 2.5 \ln \frac{y}{e} + 8.5 \tag{7-54}$$

and this relationship fits the data quite well. Equation (7-54) indicates that the value of κ in Eq. (7-53) is 0.4, which is the same value as that obtained by Nikuradse for turbulent flow through smooth tubes. The relationship given in Eq. (7-54) may be used to calculate velocity distributions in the turbulent core of rough tubes.

7-11. A Velocity-distribution Equation for Both Rough and Smooth Tubes

Rouse [52] derived a general equation to predict the velocity distribution in the turbulent core for either rough or smooth tubes. The average velocity may be expressed by the following equation:

$$U = \frac{Q}{A} = \frac{\int_0^{r_w} 2\pi r u \, dr}{\pi r_w^2} \tag{7-55}$$

From Eq. (7-35) the point velocity u in smooth tubes is given by

$$u = u^* \left(5.5 + 2.5 \ln \frac{u^* y}{\nu}\right) \tag{7-56}$$

Combining Eqs. (7-55) and (7-56) and carrying out the integration between the specified limits results in

$$\frac{U}{u^*} = 2.5 \ln \frac{u^* r_w}{\nu} + 1.75 \tag{7-57}$$

A similar procedure may be carried out for the velocity profiles in rough pipes, and the result is

$$\frac{U}{u^*} = 2.5 \ln \frac{r_w}{e} + 4.75 \tag{7-58}$$

By subtracting Eq. (7-57) from (7-35) or Eq. (7-58) from (7-54) one obtains the following relationship in both cases:

$$u^+ = 2.5 \ln \frac{y}{r_w} + 3.75 + \frac{U}{u^*} \tag{7-59}$$

Hence, using the average velocity, the equations become identical for both smooth and rough pipes. This provides evidence that the turbulence mechanism is independent of the condition of the solid boundary.

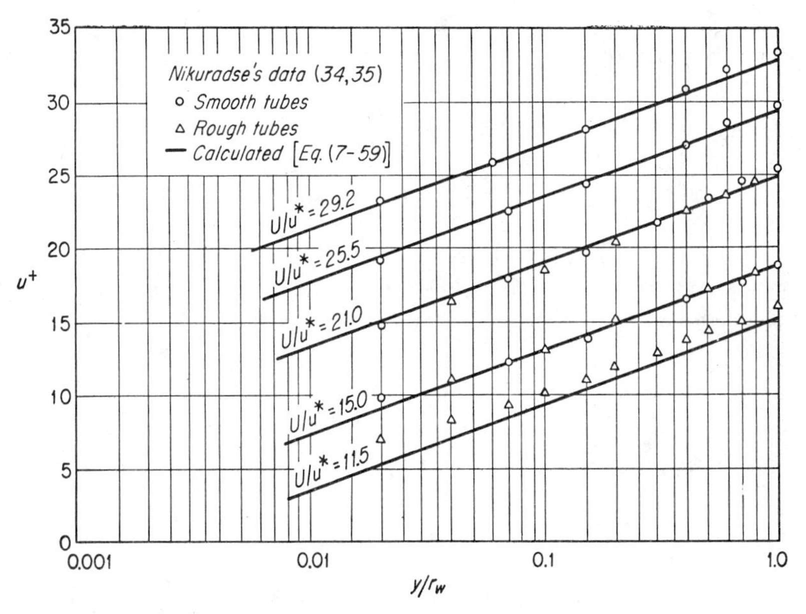

FIG. 7-17. Combined velocity distribution for both rough and smooth tubes plotted according to Eq. (7-59).

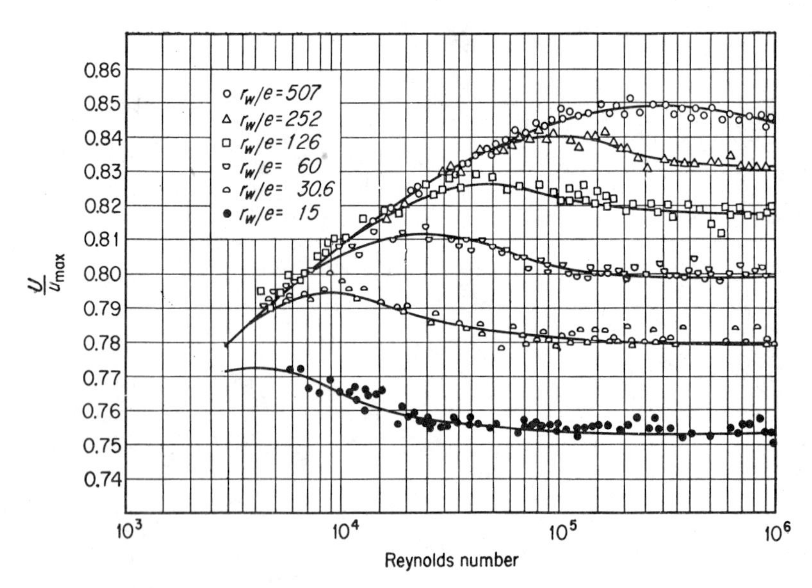

FIG. 7-18. Effect of the Reynolds number on axial velocity for flow in rough tubes. (*From J. Nikuradse, VDI-Forschungsheft* 361, 1933.)

Equation (7-59) is plotted in Fig. 7-17, where u^+ is plotted versus y/r_w for various values of the parameter U/u^*. The parameter is constant if the rate of flow in the pipe is constant. Figure 7-17 may be used for the determination of point velocities for turbulent flow in either smooth or rough tubes. The experimental data of Nikuradse [34,35] are included in Fig. 7-17 to show that Eq. (7-59) has experimental verification. The plot has the advantage that when it is used to determine point velocities in rough tubes, the magnitude of the wall roughness need not be known.

Nikuradse also presented data which related the value of $U/u_{\max}$ to the Reynolds number for the rough tubes he studied. These data are plotted in Fig. 7-18, and the resulting curves, at various roughnesses, are useful for determining the maximum velocity in rough pipes.

7-12. Friction Factors for Turbulent Flow in Smooth Tubes

It is shown by the dimensional analysis in Example 6-5 that for geometrically similar systems the friction factor and Reynolds number are the two dimensionless groups obtainable from the variables of the system. When wall roughness has a significant effect, an additional dimensionless group, the relative roughness, must be considered. All friction data for turbulent flow in tubes have been correlated by plotting the friction factor versus the Reynolds numbers for geometrically similar systems. Various friction-factor–Reynolds-number relationships have been obtained, all of which predict the friction factor with a fair degree of accuracy.

Some of the first friction data for turbulent flow in smooth tubes were obtained by Reynolds; [46] however, he determined the pressure gradient along the tube due to friction and did not report friction factors. Extensive data on various fluids and various sizes of tubes have been obtained since Reynolds's experiments, and in 1913 Blasius [2] analyzed all the data and presented a correlation between the friction factor and the Reynolds number. Most workers had investigated only a relatively short range of Reynolds numbers, and the relation obtained was applicable only over the range studied. Blasius suggested the following expression for the friction factor:

$$f = 0.079(\mathrm{Re})^{-1/4} \qquad (7\text{-}3)$$

Equation (7-3) may be used to predict friction factors with excellent accuracy for Reynolds numbers from 3,000 to 100,000. Subsequent experiments covering wider ranges of Reynolds numbers have shown that the Blasius equation does not predict friction factors accurately for Reynolds numbers above 100,000.

In 1914 Stanton and Pannell [57] conducted extensive experiments in which they investigated the flow of air, water, and oil, covering a range of Reynolds numbers from 10 to 500,000. Following these workers, Nikuradse [34] conducted experiments on the flow of water in smooth pipes for Reynolds numbers ranging from 4,000 to 3,240,000. The data of these investigators are shown in Fig. 7-19, where the friction factor is plotted versus the Reynolds number. For comparison, Eq. (7-3) is plotted as a broken line. The deviation of the Blasius equation from the experimental data at values of the Reynolds number greater than 100,000 is evident. By means of his own data and those of Stanton and Pannell and other investigators

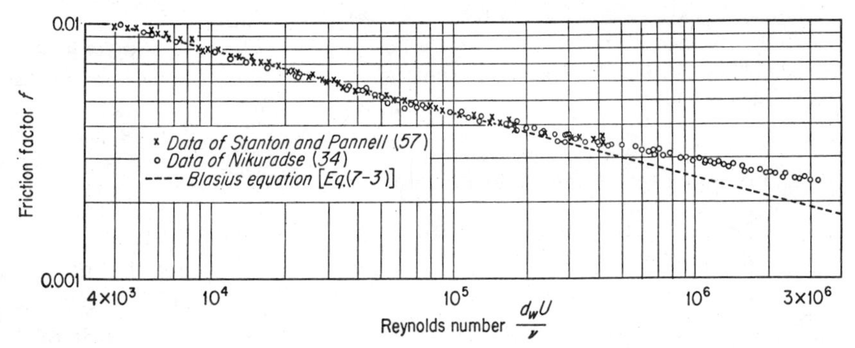

FIG. 7-19. Friction factors for turbulent flow in smooth circular tubes.

Nikuradse obtained the following relationship between f and Re, which is applicable over the whole range of Reynolds numbers investigated:

$$\frac{1}{\sqrt{f}} = 4.0 \log (\mathrm{Re}\sqrt{f}) - 0.40 \tag{7-60}$$

Equation (7-60) is the friction-factor equation recommended for determining friction factors in smooth tubes.

Von Kármán [19] derived a theoretical equation which had the same form as Eq. (7-60) and differed only slightly from it in the value of the constants. The universal velocity-distribution equation for the turbulent core was used. The value of u^* given in Eq. (7-50) was substituted in Eq. (7-57), resulting in an expression relating the average velocity to the friction factor and the tube radius.

$$\frac{1}{\sqrt{f/2}} = 5.75 \log \left(U \sqrt{\frac{f}{2}} \frac{r_w}{\nu} \right) + 1.75 \tag{7-61}$$

Equation (7-61) reduces to

$$\frac{1}{\sqrt{f}} = 4.06 \log (\mathrm{Re}\sqrt{f}) - 0.60 \tag{7-62}$$

The difference between Eqs. (7-62) and (7-60) is small.

An empirical equation relating the friction factor and the Reynolds number was presented by Drew et al.[13]

$$f = 0.00140 + 0.125(\text{Re})^{-0.32} \tag{7-63}$$

Equation (7-63) is somewhat simpler than Nikuradse's equation and is based on 1,310 experiments covering a Reynolds-number range from 3,000 to 3,000,000. The friction-factor plot based on this equation has been used extensively for determining friction factors in smooth tubes.[38, 59]

A simpler relationship between the Reynolds number and the friction factor is given in Eq. (7-64). This relationship is used in heat-transfer calculations which make use of the analogy between the transfer of momentum and the transfer of heat.

$$f = 0.046(\text{Re})^{-0.2} \tag{7-64}$$

TABLE 7-1. COMPARISON OF THE FRICTION-FACTOR EQUATIONS AT VARIOUS VALUES OF THE REYNOLDS NUMBER

Re	Eq. (7-3)	Eq. (7-64)	Eq. (7-60)	Eq. (7-62)	Eq. (7-63)
3,000	0.0107	0.00930	0.0109	0.0109	0.0110
10,000	0.00790	0.00730	0.00772	0.00774	0.00797
100,000	0.00443	0.0046	0.00448	0.00449	0.00456
1,000,000	0.00250	0.00289	0.00291	0.00292	0.00290
10,000,000	0.00140	0.00183	0.00204	0.00202	0.00218

In Table 7-1, Eqs. (7-3), (7-64), (7-60), (7-62), and (7-63) are compared with each other at various Reynolds numbers. Good agreement is obtained at Reynolds numbers less than 100,000, but at greater values of the Reynolds number the Blasius equation deviates somewhat from the others. The von Kármán equation, which was derived from the velocity-distribution equation, shows excellent agreement with Nikuradse's empirical equation. Moody [30] and Rouse [52] have published friction-factor charts based on Eq. (7-60).

7-13. Friction Factors for Turbulent Flow in Rough Tubes

In general, the friction factor for turbulent flow in circular conduits is a function of two dimensionless quantities, the Reynolds number and the relative roughness e/d_w. For smooth tubes the relative roughness may be neglected. Commercial pipe is usually quite rough, while glass tubing and drawn-metal tubing are considered to be smooth. For rough tubes new friction-factor equations are required which take into account the roughness e/d_w. Drew and Genereaux [14] have given an equation representing the bulk of the data on various types of commercial pipe made of various materials. The data for all commercial pipe were quite scattered, and the following equation represents average values:

$$\frac{1}{\sqrt{f}} = 3.2 \log (\text{Re}\sqrt{f}) + 1.2 \tag{7-65}$$

A number of investigators have studied the resistance to flow in rough pipes. The method of investigation consists in producing some artificial roughness on the inner surface. The degree of this artificial roughness is thus known accurately, and the friction factors are found as a function of the roughness. Schiller [53] made rough pipes by cutting threads of varying height and pitch in smooth pipes. He concluded that the Reynolds number at which transition from laminar to turbulent flow occurred was independent of the nature of the wall. Hopf [17] and Fromm [15] studied the resistance to flow in rough channels in which the roughness was caused by wire net, serrated zinc sheet, and corrugated steel. Hopf found that for rough surfaces the friction factor in the region of flow well past the critical point was independent of the Reynolds number. He expressed the friction factor for rough surfaces as follows:

$$f = 0.01 \left(\frac{e}{r_w}\right)^{0.314} \tag{7-66}$$

The experiments of these workers cover a number of roughness types but only a narrow range of Reynolds numbers. Nikuradse [35] undertook the determination of friction factors along with his studies of the velocity profiles in artificially roughened pipes. The range of values of r_w/e covered

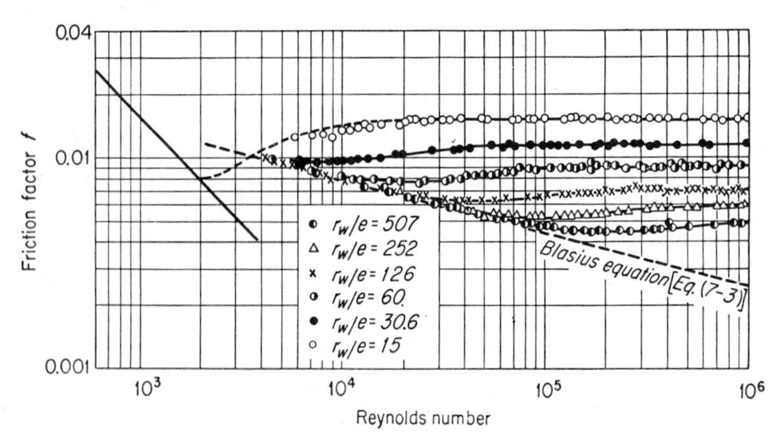

FIG. 7-20. Friction factors for flow in artificially roughened tubes. (*From J. Nikuradse, VDI-Forschungsheft* 361, 1933.)

was from 15 to 507, while the Reynolds numbers varied from 1,000 to 1,000,000. The results Nikuradse obtained are shown in Fig. 7-20, where the friction factor is plotted versus the Reynolds number. It is seen that for each value of r_w/e an individual friction-factor curve is obtained. The transition from laminar to turbulent flow is unaffected by the roughness of the pipe and occurs at the same point as in smooth tubes. In the turbulent region of flow it is seen that each friction-factor curve eventually becomes horizontal; i.e., the friction factor becomes independent of the Reynolds number and dependent only on the roughness. This region of horizontal friction-factor curves is referred to as the region of fully turbulent flow. Von Kármán [19] determined the resistance equation for flow in rough pipes in a manner similar to that used for smooth tubes. Substitution of Eq. (7-50) into Eq. (7-58) gives Eq. (7-67), which is von Kármán's theoretical equation relating the friction factor and the roughness in the region of fully turbulent flow.

$$\frac{1}{\sqrt{f}} = 4.06 \log \frac{r_w}{e} + 3.36 \tag{7-67}$$

The corresponding equation which Nikuradse obtained experimentally is

$$\frac{1}{\sqrt{f}} = 4 \log \frac{r_w}{e} + 3.48 \tag{7-68}$$

Equations (7-67) and (7-68) are very nearly identical.

It is seen from Fig. 7-20 that the friction-factor curves for rough pipes follow the smooth-tube friction-factor equation for a short range of Reynolds numbers. Beyond this region there is a transition region to fully turbulent flow.

For the transition region Colebrook [4] proposed the following empirical equation:

$$\frac{1}{\sqrt{f}} = 4 \log \frac{r_w}{e} + 3.48 - 4 \log \left(1 + 9.35 \frac{r_w/e}{\mathrm{Re}\sqrt{f}}\right) \tag{7-69}$$

In the transition region the friction factor is a function of both the roughness r_w/e and the Reynolds number. This equation applies to the transition region up to a value of $(r_w/e)/\mathrm{Re}\sqrt{f} = 0.005$. At values greater than this the flow is fully turbulent, and the friction factor is then independent of the Reynolds number.

7-14. Complete Friction-factor Plot for Both Rough and Smooth Tubes

A summary of the equations available for the prediction of friction factors for laminar and turbulent flow in both smooth and rough pipes is given below.

Laminar flow (smooth or rough pipe):

$$f = \frac{16}{\text{Re}} \qquad \text{Re} < 2{,}000 \tag{4-52}$$

Turbulent flow (smooth pipe):

$$\frac{1}{\sqrt{f}} = 4.0 \log (\text{Re}\sqrt{f}) - 0.40 \qquad \text{Re} > 3{,}000 \tag{7-60}$$

Fully turbulent flow (rough pipe):

$$\frac{1}{\sqrt{f}} = 4 \log \frac{d_w}{e} + 2.28 \qquad \frac{r_w/e}{\text{Re}\sqrt{f}} > 0.005 \tag{7-70}$$

and for transition flow

$$\frac{1}{\sqrt{f}} = 4 \log \frac{d_w}{e} + 2.28 - 4 \log \left(1 + 4.67 \frac{d_w/e}{\text{Re}\sqrt{f}}\right) \tag{7-71}$$

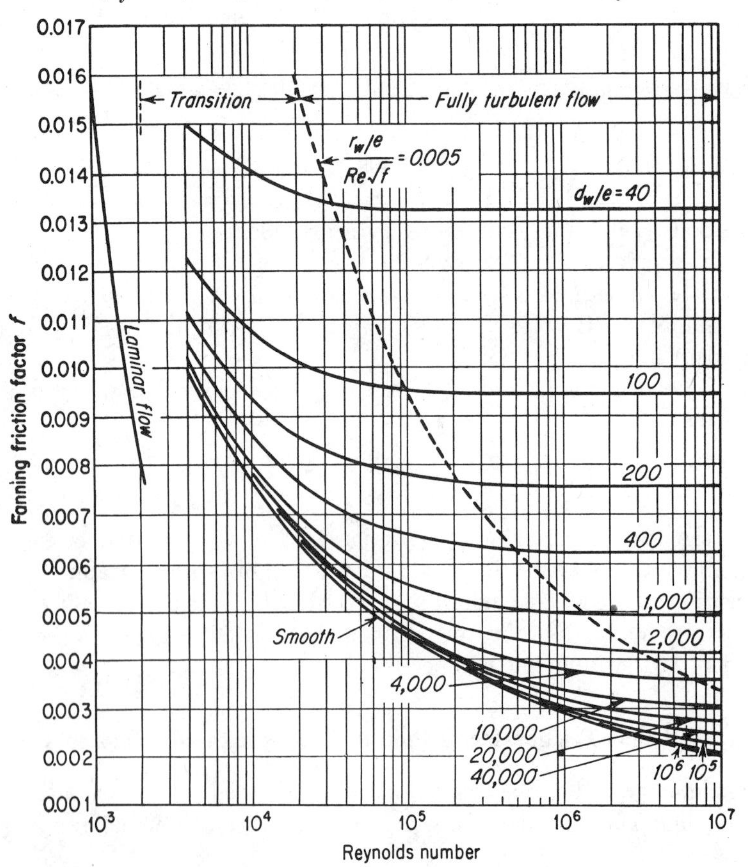

FIG. 7-21. Complete friction-factor plot for both rough and smooth tubes.

Moody[30] and Rouse[52] have presented identical friction-factor charts for the determination of friction factors in either smooth or rough pipes. The charts are based on Eq. (4-52) for laminar flow, Eq. (7-60) for turbulent flow in smooth tubes, Eq. (7-70) for fully turbulent flow in rough pipes, and Eq. (7-71) for the transition region, where the friction factor is a function of both the roughness and the Reynolds number. A complete friction-factor plot based on Moody's chart is given in Fig. 7-21, where the Fanning

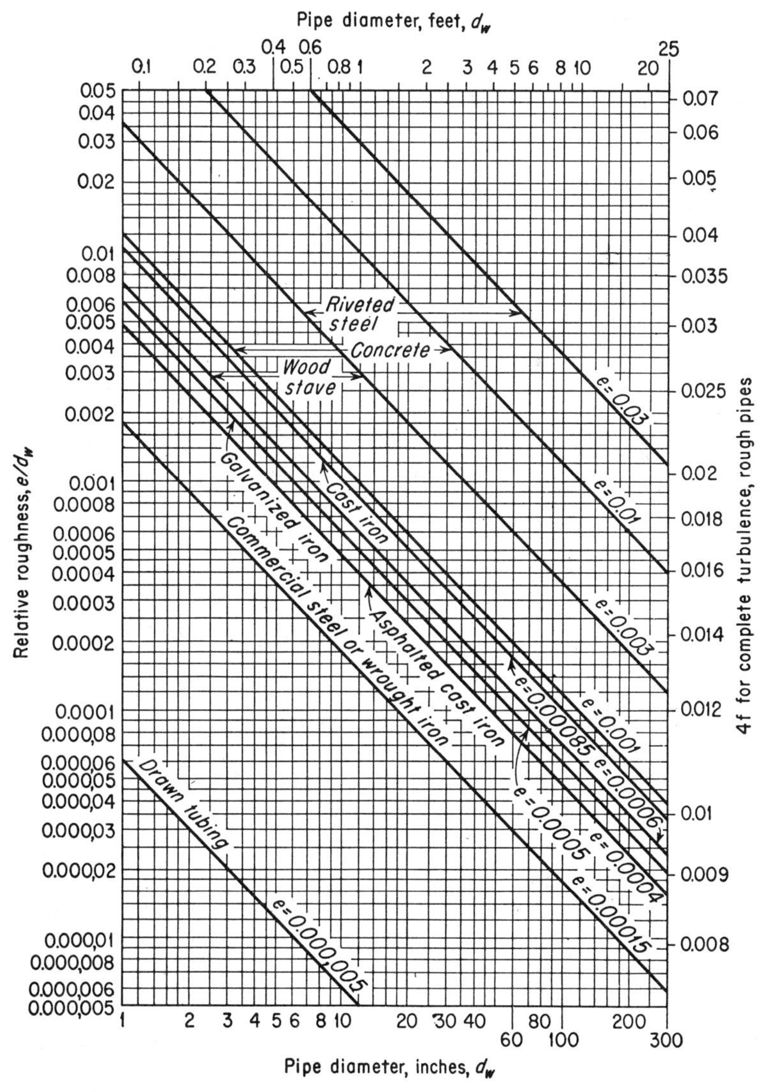

FIG. 7-22. Relative roughness of pipes made of different materials. [*From L. F. Moody, Trans. ASME,* **66**:671 (1944).]

friction factor is plotted versus the Reynolds number with relative roughness d_w/e as the parameter. Moody presented the chart shown in Fig. 7-22, giving the roughness of commercial pipe as a function of the diameter. Various materials are considered. From Fig. 7-22 it is possible to obtain the value of the term d_w/e, and then from Fig. 7-21 the friction factor may be obtained at any Reynolds number.

7-15. Friction Factors for Nonisothermal Flow

The friction-factor plot of Fig. 7-21 may be used to obtain friction factors for nonisothermal flow if the fluid properties are evaluated at a film temperature which may be calculated from the bulk temperature of the fluid T_b and the wall temperature T_w by the relation

$$T_f = 0.4(T_w - T_b) + T_b \tag{7-72}$$

Deissler and Eian [9] showed that the effects of heat-transfer rate on the friction factor are essentially eliminated by evaluating fluid properties at the film temperature defined in Eq. (7-72).

7-16. Turbulence Measurements for Flow in Tubes

Turbulent flow in circular tubes has been studied with hot-wire turbulence-measuring equipment to determine values of velocity fluctuations, turbulent shear stresses, and the spectrum of turbulence. An investigation by Laufer [26] covers the structure of turbulent pipe flow quite thoroughly. Laufer used a hot-wire anemometer to measure velocity profiles, the rms of the turbulent velocity fluctuations $\sqrt{\overline{u'^2}}$, $\sqrt{\overline{v'^2}}$, and $\sqrt{\overline{w'^2}}$, shear stresses expressed as $\overline{u'v'}/u^{*2}$, and turbulent energy spectra.

The general variations of the velocity fluctuations from the tube wall to the axis are shown in Fig. 7-23 for air flowing in a 9.72-in.-ID tube at a Reynolds number of 50,000. This plot shows $\sqrt{\overline{u'^2}}/u^*$, $\sqrt{\overline{v'^2}}/u^*$, and $\sqrt{\overline{w'^2}}/u^*$ as a function of y/r_w. The velocity fluctuations u', v', and w' are in the direction of the cylindrical coordinates x, r, and θ respectively, the tube axis coinciding with the x coordinate.

The rms velocity fluctuations approach zero at the pipe wall, reach maximum values at $y/r_w = 0.02$ to 0.2, then all decrease to the same value at the tube axis. The terms $\sqrt{\overline{u'^2}}/u^*$, $\sqrt{\overline{v'^2}}/u^*$, $\sqrt{\overline{w'^2}}/u^*$ all have values slightly less than 0.8 at the tube axis, indicating that in this vicinity the turbulence becomes isotropic. The axial component of the fluctuation $\sqrt{\overline{u'^2}}/u^*$ has higher values than the others over the whole range of y/r_w from 0 to 1. The variation of the Reynolds shearing stress $\overline{u'v'}/u^{*2}$ from

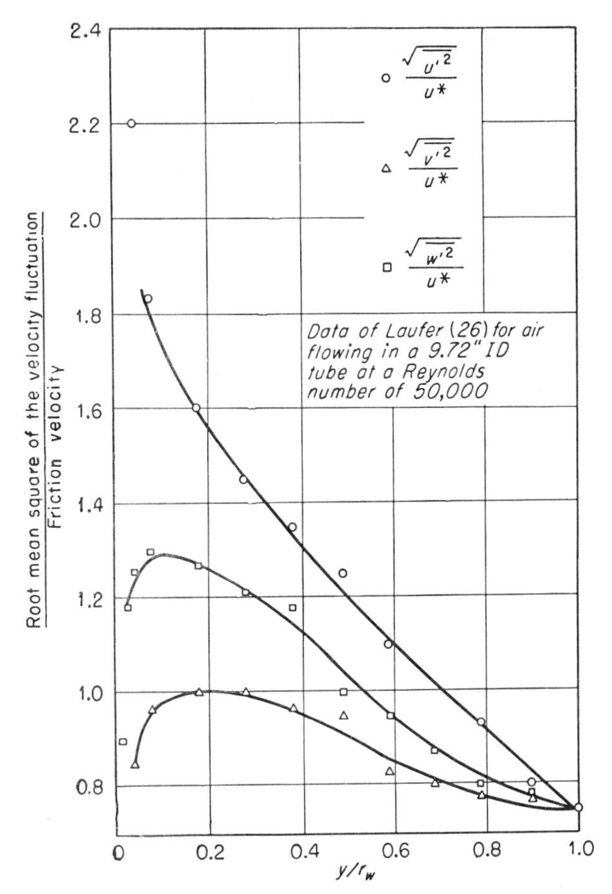

FIG. 7-23. Distribution of rms velocity fluctuation for turbulent flow in a circular tube. (*Data from J. Laufer, NACA Rept.* 1174, 1954.)

tube wall to tube axis is shown in Fig. 7-24. The data points are those obtained by Laufer [26] for $\overline{u'v'}/u^{*2}$, while the curve is a plot of the equation

$$\frac{\overline{u'v'}}{u^{*2}} = 0.9835\left(1 - \frac{y}{r_w}\right)\left[1 - \left(1 - \frac{y}{r_w}\right)^{30}\right] \tag{7-73}$$

Equation (7-73) is a theoretical equation Pai [37] derived using the Reynolds equations (5-14). The constants in Eq. (7-73) were evaluated by means of Nikuradse's experimental velocity-distribution data at a Reynolds number of 3.24×10^6. Figure 7-24 shows that the turbulent shear stresses are zero at the tube wall, reach a maximum value very close to the wall, then decrease almost linearly to zero at the tube axis.

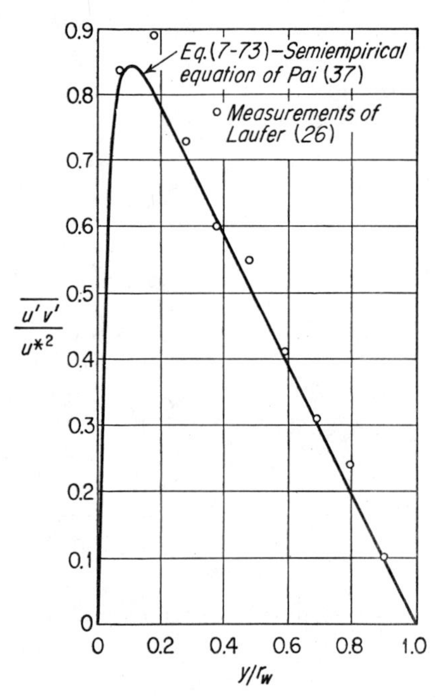

FIG. 7-24. Reynolds shearing stress for the flow of air in a circular tube.

Laufer's measurements of the turbulent energy spectrum for flow in tubes show the same general distribution of turbulent kinetic energy as in Fig. 5-11, particularly near the tube axis, where turbulence is nearly isotropic.

7-17. Illustrative Problems

The following examples involve the use of some of the relationships presented during the discussion of turbulent flow in tubes.

Example 7-2

In Example 7-1 the average velocity of the air flowing in a 12-in.-ID duct was found to be 54.1 ft/sec. Consider the tube wall to be smooth.

(*a*) Using Eqs. (7-20), (7-24), and (7-35), calculate the fluid velocity at a point 3 in. from the tube wall.

(*b*) Calculate the thickness of the laminar layer.

(*c*) Using Eq. (7-37), calculate the velocity of the fluid at the edge of the laminar layer.

(*d*) Using Eq. (7-40), calculate the value of y^+ for a value of $u^+ = 5$. Compare it with Eq. (7-37).

(*e*) Calculate the shear stress at the wall.

Solution

(*a*) Properties of air:

$$\rho = 0.0702 \text{ lb}_m/\text{ft}^3$$

$$\mu = 12.84 \times 10^{-6} \text{ lb}_m/(\text{ft})(\text{sec})$$

$$\text{Re} = \frac{12}{12}\left|\frac{54.1}{\ }\right|\frac{0.0702}{12.84 \times 10^{-6}} = 2.96 \times 10^5$$

$$u_{\max} = 63.6 \qquad \text{(from Example 7-1)}$$

The friction factor is calculated from Eq. (7-60).

$$\frac{1}{\sqrt{f}} = 4.0 \log (2.96 \times 10^5 \sqrt{f}) - 0.40$$

Solving for f by trial and error,

$$f = 0.0036$$

From Eq. (7-50)

$$u^* = U\sqrt{\frac{f}{2}} = 54.1\sqrt{\frac{0.0036}{2}} = 2.30 \text{ ft/sec}$$

At a point 3 in. from the wall

$$\frac{y}{r_w} = 0.5$$

From Eq. (7-20)

$$u = 63.6 + (2.5)(2.30) \ln 0.5 = 59.6 \text{ ft/sec}$$

From Eq. (7-24)

$$u = 63.6 + (2.5)(2.30)[\ln (1 - \sqrt{0.5}) + \sqrt{0.5}] = 60.6 \text{ ft/sec}$$

By Eq. (7-52)

$$y^+ = 0.5 \frac{2.96 \times 10^5}{2} \sqrt{\frac{0.0036}{2}} = 3{,}150$$

By Eq. (7-35)

$$u^+ = 5.5 + 2.5 \ln 3{,}150 = 25.8$$

from which $\qquad u = (25.8)(2.30) = 59.3 \text{ ft/sec}$

Summary of calculated point velocities at $y = 3.0$ in.:

By Eq. (7-20)	$u = 59.6$ ft/sec
By Eq. (7-24)	$u = 60.6$ ft/sec
By Eq. (7-35)	$u = 59.3$ ft/sec

(*b*) The laminar layer extends to a value of $y^+ = 5$. Thus, letting δ_1 = thickness of laminar layer,

$$5 = \frac{\delta_1}{r_w} \frac{\text{Re}}{2} \sqrt{\frac{f}{2}} = \frac{\delta_1}{6} \frac{2.96 \times 10^5}{2} \sqrt{\frac{0.0036}{2}}$$

giving $\qquad \delta_1 = 0.00477$ in.

(*c*) At $u^+ = 5$

$$u = (5)(2.30) = 11.50 \text{ ft/sec}$$

The velocity at the edge of the laminar layer is 11.50 ft/sec.

(*d*) Equation (7-40) may be rearranged by letting

$$\sqrt{2}\,x = nu^+$$

then

$$\sqrt{2}\,dx = d(nu^+)$$

so

$$y^+ = \frac{1}{n}\,\frac{1/\sqrt{\pi}\int_0^{\sqrt{2}x} e^{-x^2}\,dx}{(1/\sqrt{2\pi})e^{-x^2}}$$

The integral may be evaluated from any standard table of integrals, with $n =$ 0.109 and $u^+ = 5.0$. Thus

$$\sqrt{2}\,x = (5.0)(0.109) = 0.545$$

and

$$x^2 = 0.1485$$

$$\frac{1}{\sqrt{\pi}}\int_0^{0.545} e^{-x^2}\,dx = 0.2796$$

Thus at $u^+ = 5$

$$y^+ = \frac{\sqrt{2\pi}\,(0.2796)}{0.109e^{-0.1485}} = 7.4$$

which is the value of y^+ where $u^+ = 5$.

(*e*) From Eq. (7-12)

$$u^* = \sqrt{\frac{\tau_w g_c}{\rho}}$$

from which

$$\tau_w = \frac{u^{*2}\rho}{g_c} = \frac{(2.30 \text{ ft/sec})^2(0.0702 \text{ lb}_m/\text{ft}^3)}{32.2\ (\text{lb}_m)(\text{ft})/(\text{lb}_f)(\text{sec}^2)} = 0.0115 \text{ lb}_f/\text{ft}^2$$

Example 7-3

An 8-in.-ID cast-iron pipeline has been in service for 10 years. The maximum water (at 60°F) flow rate possible in the line is 950 gal/min with the present pumps and motors. Past experience with this type of pipe indicates that the relative roughness is a function of time according to the relation

$$\frac{e}{d_w} = \left(\frac{e}{d_w}\right)_0\left(1 + \frac{t}{5}\right)$$

where $(e/d_w)_0$ is the relative roughness of new cast-iron pipe and t is the time of service in years.

(*a*) Calculate the velocity of the fluid at a point 1 in. from the pipe wall by Eqs. (7-54) and (7-59).

(*b*) Calculate the pressure loss per 100 ft of horizontal pipe at the maximum flow rate.

Solution

(*a*) From Fig. 7-22 for new 8-in.-ID cast-iron pipe

$$(e/d_w)_0 = 0.00125$$

After 10 years of service

$$\frac{e}{d_w} = 0.00125(1 + {}^{10}\!/_{5}) = 0.00375$$

Thus

$$e = (0.00375)(8) = 0.03 \text{ in.}$$

$$\text{Average velocity in pipe} = \frac{950 \text{ gal/min}}{(60 \text{ min/sec})(7.48 \text{ gal/ft}^3)[(\pi/4)(8/12)^2 \text{ ft}^2]} = 6.06 \text{ ft/sec}$$

For water at 60°F, the kinematic viscosity is 1.21×10^{-5} ft²/sec.

$$\text{Re} = \frac{(8)(6.06)}{(12)(1.21 \times 10^{-5})} = 3.34 \times 10^5$$

From Fig. 7-21 at $d_w/e = 267$

Thus

$$f = 0.0069$$

$$u^* = 6.06\sqrt{\frac{0.0069}{2}} = 0.354 \text{ ft/sec}$$

By Eq. (7-54)

$$u^+ = \frac{u}{u^*} = 2.5 \ln \frac{1}{0.03} + 8.5 = 17.27$$

Thus

$$u = (17.27)(0.354) = 6.11 \text{ ft/sec}$$

By Eq. (7-59)

$$u^+ = 5.75 \log {}^1\!/_4 + \left(3.75 + \frac{6.06}{0.354}\right) = 17.39$$

and

$$u = (17.39)(0.354) = 6.15 \text{ ft/sec}$$

(*b*) By the energy equation for horizontal pipe and incompressible fluids

$$\frac{\Delta P}{\rho} = -\int_0^{100} \frac{2fU^2\,dx}{g_c d_w}$$

$$\Delta P = -\frac{(2)(0.0069)(6.06)^2(100)(62.4)(12)}{(32.2)(8)(144)}$$

$$= -1.02 \text{ psi}$$

The pressure decrease per 100 ft of pipe is 1.02 psi.

Example 7-4

Natural gas (assume pure methane) is to be pumped through a 100-mile section of 40-in.-ID pipeline at the rate of 1.5×10^8 ft³ per 24 hr measured at 60°F and standard atmospheric pressure. The gas is to be at a pressure of 10 psig at the discharge end of the line. What is the gas pressure at the beginning of the 100-mile section? The average temperature of the flowing gas is 60°F.

Solution

For a differential length of pipe Eq. (4-21) becomes (assuming $\bar{w} = 0$ and neglecting elevation effects)

$$\frac{dP}{\rho} + \frac{U\,dU}{g_c} = -\frac{2fU^2\,dx}{g_c d_w}$$

Multiplying by $g_c/2U^2$.

$$\frac{g_c\,dP}{2\rho U^2} + \frac{dU}{2U} = -\frac{f}{d_w}\,dx$$

Letting w = the mass rate of flow,

$$w = \rho A U$$

or

$$U = \frac{w}{\rho A} \quad \text{and} \quad U^2 = \left(\frac{w}{\rho A}\right)^2$$

Thus

$$\frac{\rho A^2 g_c\,dP}{2w^2} + \frac{dU}{2U} = -\frac{f}{d_w}\,dx$$

but

$$\frac{P}{\rho} = ZnRT$$

where n = moles per unit mass $= \dfrac{1}{29(Gr)}$

Gr = gas gravity (air = 1)

Hence

$$\rho = \frac{29(Gr)P}{ZRT}$$

So

$$\frac{29(Gr)A^2 g_c P\,dP}{2w^2 ZRT} + \frac{dU}{2U} = -\frac{f}{d_w}\,dx$$

Integrating over a length of pipe L, assuming Z, T, and f constant at average values,

$$\frac{29(Gr)A^2 g_c(P_1^2 - P_2^2)}{4w^2 ZRT} - \tfrac{1}{2}\ln\frac{U_2}{U_1} = \frac{fL}{d_w}$$

Neglecting $\ln(U_2/U_1)$ and solving for w.

$$w = \left[\frac{29(Gr)A^2 g_c(P_1^2 - P_2^2)d_w}{4ZRTfL}\right]^{1/2}$$

and since $A = \pi d_w^2/4$.

$$w\,\dagger = \left[\frac{29(Gr)\pi^2 g_c(P_1^2 - P_2^2)d_w^5}{64ZRTfL}\right]^{1/2}$$

† This equation may be used to derive the Weymouth equation for gas flow in pipelines:

$$Q = 1.6156\,\frac{T_0}{P_0}\left[\frac{(P_1^2 - P_2^2)d_w^5}{Z(Gr)TfL}\right]^{1/2}$$

where Q = ft³ of gas per hour measured at T_0 and P_0, °R and psia
P_1, P_2 = inlet and outlet pressure, respectively, psia
d_w = pipe diameter, in.
T = temperature of flowing gas, °R
L = pipe length, miles

Volume of 1 lb mole of gas at 60°F, 1 atm pressure = 379 ft^3

$$w = \text{mass rate of flow} = \frac{(1.5 \times 10^8)(16)}{(379)(24)(3{,}600)} = 73.4 \text{ lb}_m/\text{sec}$$

$$\text{Gravity of methane} = {}^{16}\!/\!_{29} = 0.552$$

Viscosity of methane at 60°F, 1 atm = 0.0108 centipoise

$$\text{Reynolds number} = \frac{(4)(73.4)(12)}{(\pi)(40)(0.0108)(0.000672)}$$

$$= 3.87 \times 10^6$$

From Fig. 7-21

$$f = 0.0023$$

Assuming $Z = 1$,

$$73.4 = \left[\frac{(29)(0.552)(\pi^2)(32.2)(P_1^2 - P_2^2)(^{40}\!/\!_{12})^5}{(64)(1)(1{,}543)(520)(0.0023)(100)(5{,}280)}\right]^{1/2}$$

from which

$$(P_1^2 - P_2^2)^{1/2} = 12{,}680$$

$$P_1^2 - P_2^2 = 160.5 \times 10^6$$

$$P_2 = 10 \text{ psig} = 24.7 \text{ psia}$$

$$P_2^2 = [(24.7)(144)]^2 = 12.7 \times 10^6$$

$$P_1^2 = 160.5 \times 10^6 + 12.7 \times 10^6 = 173.2 \times 10^6$$

and

$$P_1 = 13{,}160 \text{ psf}$$

$$= 91.4 \text{ psia}$$

This is the pressure at the beginning of the 100-mile section. At this pressure the compressibility factor is unity (Fig. 1-3), and the viscosity is essentially the same as at atmospheric pressure (Fig. 1-10).

II. TURBULENT FLOW IN ANNULI

7-18. The Transition from Laminar to Turbulent Flow in Plain Annuli

Only limited investigation of the laminar-turbulent transition in plain annuli has been made. The results of most friction-loss and velocity-distribution experiments have indicated that transition occurs at a Reynolds number (based on equivalent diameter) of about 2,000. However, the transition is gradual and covers a fairly wide range of Reynolds numbers.

Prengle and Rothfus [44] made a detailed study of the transition in plain annuli by injecting dye into the stream. Transition first occurred at the point of maximum velocity. These workers defined a Reynolds number based on the point of maximum velocity, i.e.,

$$\mathrm{Re}_2 = \frac{4r_H U\rho}{\mu} \tag{7-74}$$

$$4r_H = \frac{2(r_2{}^2 - r_{\max}^2)}{r_2} \tag{7-75}$$

where $4r_H$ is the equivalent diameter of the portion of the annulus between the outer tube and the point of maximum velocity, which is calculated from Eq. (4-58). At values of Re_2 near 700, deviations from laminar flow occurred and became progressively greater up to values of Re_2 about 2,200. Above Re_2 = 2,200, turbulent flow existed in the annuli studied.

7-19. Velocity Distribution for Turbulent Flow in Plain Annuli

Few studies have been made on the turbulent velocity profiles in annuli. Rothfus[48,49] determined point velocities for air flowing in turbulent flow in

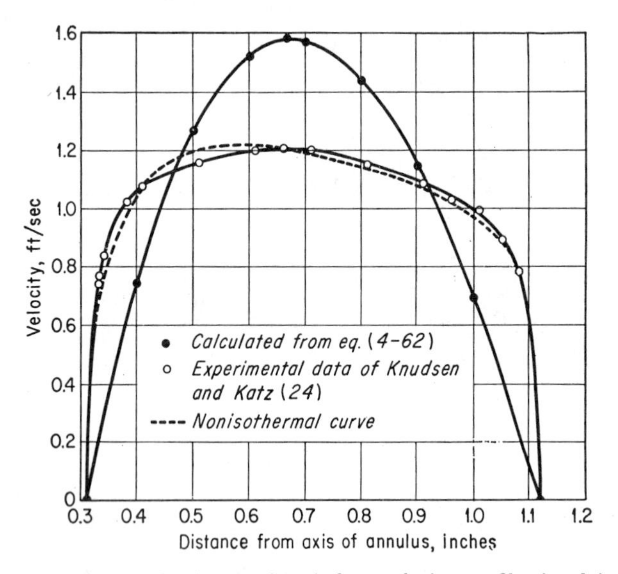

FIG. 7-25. Comparison of laminar and turbulent velocity profiles in plain annuli for the same average velocity.

two different annuli. Knudsen and Katz[23] have also determined velocity profiles in an annulus for water flowing both isothermally and with heat being transferred through the inner wall of the annulus.

The difference between laminar and turbulent velocity profiles in annuli is indicated in Fig. 7-25, where the open circles represent experimental data on the velocity profile for water flowing in an annulus for which $r_2 = 1.120$ in. and $r_1 = 0.311$ in. The Reynolds number is 9,190. The solid circles represent calculated point velocities [from Eq. (4-62)] for laminar flow in the same annulus at the same velocity. The value of r_{max} for both curves is very nearly the same. The broken curve in Fig. 7-25 is the turbulent velocity profile for the case where heat is being transferred through the wall of the inner tube. The point of maximum velocity is shifted somewhat for the nonisothermal profile. The experimental data of Knudsen and Katz on water and of Rothfus on air are plotted in Fig. 7-26. These investigators studied different annuli, and in order to plot all the curves on the same basis the abscissa of the curves in Fig. 7-26 is $y_1/(r_2 - r_1)$, where y_1 is the distance from the outer surface of the inner tube and $r_2 - r_1$ is the difference of the radii of the outer and the inner tube. This dimensionless ratio varies from 0 to 1 for any annulus. The ordinate in Fig. 7-26 is the point velocity. Also shown is the point where the maximum velocity

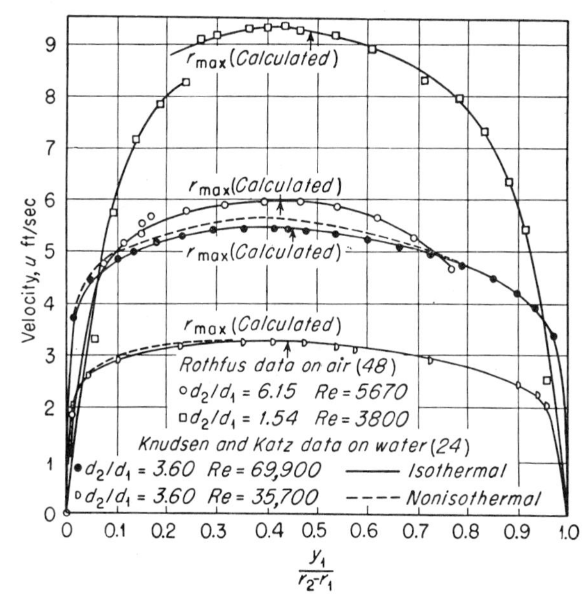

FIG. 7-26. Experimental turbulent-flow velocity profiles in plain annuli.

would be if the flow were laminar. For each annulus the positions indicated by the arrows represent the value of

$$\frac{\sqrt{(r_2^2 - r_1^2)/2 \ln (r_2/r_1)} - r_1}{r_2 - r_1}$$

[calculated from Eq. (4-58)]. This is the point of maximum velocity for laminar flow in the annuli, and for all turbulent velocity profiles shown in Fig. 7-26 the point of maximum velocity lies very close to the arrows. However, these curves are only for values of r_2/r_1 less than 6.5. The turbulent velocity profile is very flat in the vicinity of the maximum velocity, and experimental determination of r_{max} is quite difficult.

During the transition from laminar to turbulent flow, the point of maximum velocity shifts from its position for laminar flow. However, for fully turbulent flow its position is the same as for laminar flow. This shift during the transition is shown in Fig. 7-27, which illustrates the data of Rothfus

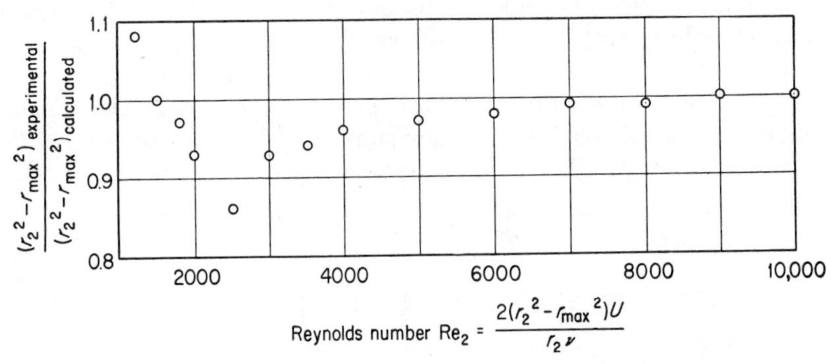

FIG. 7-27. Dependence of the point of maximum velocity on the Reynolds number for the transition region of flow in plain annuli. [*Data from R. R. Rothfus, C. C. Monrad, K. G. Sikchi, and W. J. Heideger, Ind. Eng. Chem.*, **47**:913 (1955).]

et al.[50] The ratio of the observed value of $r_2^2 - r_{max}^2$ to the calculated value of $r_2^2 - r_{max}^2$ [from Eq. (4-58)] is plotted versus Re_2 [see Eq. (7-74)].

It is important to know the position of the point of maximum velocity since it is used to calculate shear stresses at both the inner and outer walls of the annulus. The ratio of shear stresses at the inner and outer walls of the annulus is given by Eq. (4-66).

$$\frac{\tau_1}{\tau_2} = \frac{r_2(r_{max}^2 - r_1^2)}{r_1(r_2^2 - r_{max}^2)} \tag{4-66}$$

Using

$$\frac{-dP_f}{dx} = \frac{2(\tau_2 r_2 + \tau_1 r_1)}{r_2^2 - r_1^2} \tag{4-67}$$

gives

$$\tau_2 = \frac{r_2^2 - r_{max}^2}{2r_2}\frac{-dP_f}{dx} \tag{7-76}$$

and

$$\tau_1 = \frac{r_{max}^2 - r_1^2}{2r_1}\frac{-dP_f}{dx} \tag{7-77}$$

where τ_1 and τ_2 are the shear stresses at the walls of the annulus.

Since the position of the point of maximum velocity is not midway between the walls of the annulus, the velocity-distribution curve is not symmetrical. It is convenient to livide the velocity profile into an outer portion, extending from the outer-tube wall to the point of maximum velocity, and an inner portion, extending from the point of maximum velocity to the wall of the inner tube. In dealing with quantities in the annulus the subscript 1 refers to quantities at the inner wall or in the inner portion of the velocity profile, and the subscript 2 refers to quantities at the outer wall or in the outer portion of the profile.

For a given steady flow in an annulus τ_1 and τ_2 are constant. The friction velocities for the inner and outer portion of the stream are, respectively,

$$u_1^* = \sqrt{\frac{\tau_1 g_c}{\rho}} \tag{7-78}$$

$$u_2^* = \sqrt{\frac{\tau_2 g_c}{\rho}} \tag{7-79}$$

from which may be defined

$$u_1^+ = \frac{u_1}{u_1^*} \tag{7-80}$$

$$u_2^+ = \frac{u_2}{u_2^*} \tag{7-81}$$

$$y_1^+ = \frac{u_1^* y_1}{\nu} \tag{7-82}$$

$$y_2^+ = \frac{u_2^* y_2}{\nu} \tag{7-83}$$

where u_1, u_2 = point velocities in inner and outer profiles respectively
y_1 = distance from inner tube
y_2 = distance from outer tube

The experimental annular velocity data of Knudsen and Katz [24] and Rothfus [48] are plotted in Figs. 7-28 and 7-29. In Fig. 7-28 u_2^+ is plotted versus y_2^+ for the outer portion of the velocity curve, and in Fig. 7-29 u_1^+ is plotted versus y_1^+ for the inner portion of the velocity curve. For values of y_2^+ greater than 30 the following equation represents most of the data of Knudsen and Katz:

$$u_2^+ = 6.1 \log y_2^+ + 3.0 \tag{7-84}$$

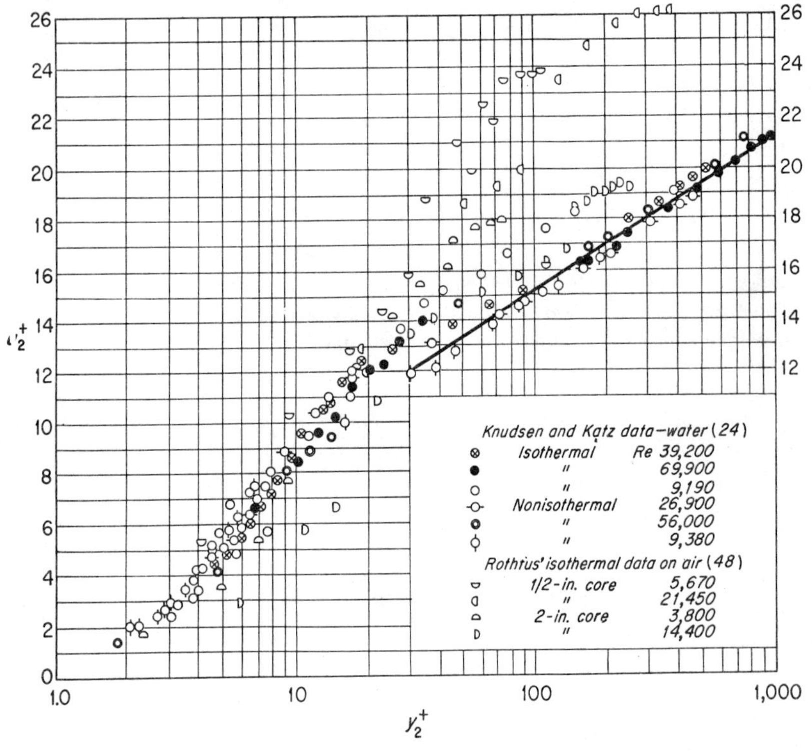

FIG. 7-28. Experimental velocity-distribution data for the outer portion of the annular velocity profile.

In Fig. 7-29, where data for the inner portion of the velocity curve are plotted, the bulk of the data of Knudsen and Katz can be represented by the equation

$$u_1^+ = 4.4 \log y_1^+ + 6.2 \tag{7-85}$$

In this case considerably more scattering of the data is noted. Equation (7-85) applies for the turbulent portion of the velocity-profile curve at values of y_1^+ greater than 40 (approximately). The data for the isothermal profiles agree more closely with Eq. (7-85) than the data for the nonisothermal profiles do. This is to be expected, since heat was being transferred into the annulus through the inner tube, and it is probable that the inner portion of the velocity-profile curve would be more affected by the transference of heat. Equations (7-84) and (7-85) represent at least a partial correlation of the turbulent-velocity data in annuli. Since very few data are available, it is unlikely that these equations represent a final correlation.

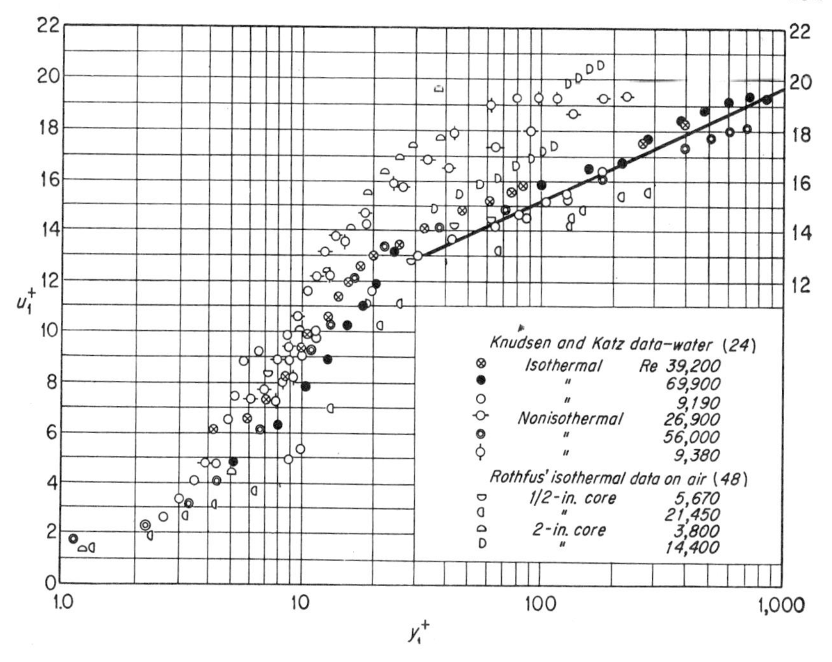

FIG. 7-29. Experimental velocity-distribution data for the inner portion of the annular velocity profile.

Bailey [1] investigated the data of Knudsen and Katz and reported the following relationships as holding within experimental error for the inner and outer velocity profiles respectively:

$$\frac{u_{\max} - u_1}{u_1^*} = -2.5 \ln \left(1 - \frac{r_{\max}^2 - r^2}{r_{\max}^2 - r_1{}^2} \frac{r_1}{r}\right) \tag{7-86}$$

and

$$\frac{u_{\max} - u_2}{u_2^*} = -2.5 \ln \left(1 - \frac{r^2 - r_{\max}^2}{r_2{}^2 - r_{\max}^2} \frac{r_2}{r}\right) \tag{7-87}$$

In Fig. 7-30 these equations are represented by the straight line, and in the same figure the experimental velocity-profile data are plotted. There is considerable scattering of the data about the solid line. The data for the outer portion of the annulus, in general, correlate better with Eq. (7-87) than the data for the inner portion correlate with Eq. (7-86).

A further attempt at correlating the velocity data in annuli is shown in Fig. 7-31, in which the data of Knudsen and Katz are plotted as $u/u_{\max}$ versus $y/y_{\max}$, where $y_{\max}$ is the distance from the point of maximum velocity to the wall. The data are represented quite well by the following equa-

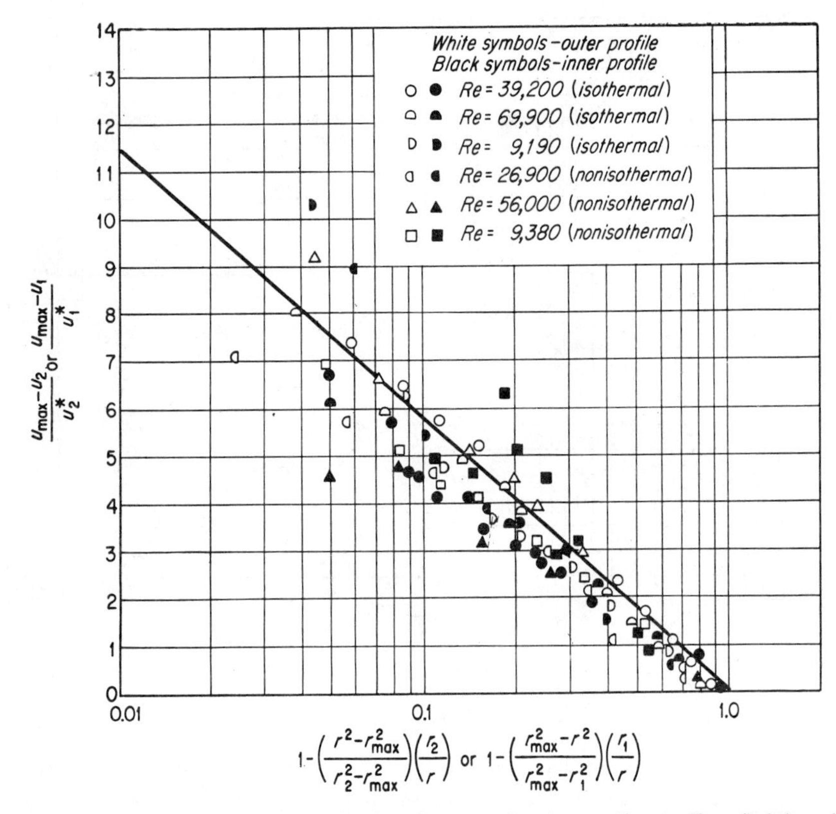

FIG. 7-30. Annular velocity-distribution data correlated according to Eqs. (7-86) and (7-87). (*From R. V. Bailey, Oak Ridge Natl. Lab., Tech. Div. Eng. Research Sec. ORNL 521, 1950.*)

tions for the outer and inner profiles respectively:

$$\frac{u_2}{u_{max}} = \left[\frac{y_2}{(y_{max})_2}\right]^{0.142} \tag{7-88}$$

and

$$\frac{u_1}{u_{max}} = \left[\frac{y_1}{(y_{max})_1}\right]^{0.102} \tag{7-89}$$

Experimentally it was found[24] that U/u_{max} had an average value of 0.876 ± 1.8 per cent. Thus Eqs. (7-88) and (7-89) could be expressed by the following relationships, in which the average velocity U is used:

$$u_2 = 1.14U\left[\frac{y_2}{(y_{max})_2}\right]^{0.142} = 1.14U\left(\frac{r_2 - r}{r_2 - r_{max}}\right)^{0.142} \tag{7-90}$$

$$u_1 = 1.14U\left[\frac{y_1}{(y_{\max})_1}\right]^{0.102} = 1.14U\left(\frac{r - r_1}{r_{\max} - r_1}\right)^{0.102} \tag{7-91}$$

There are indications that the above equations are not valid below Reynolds numbers of 10,000. However, they do express quite adequately the point velocity for both turbulent portions of the velocity-profile curve and for both isothermal and nonisothermal flow.

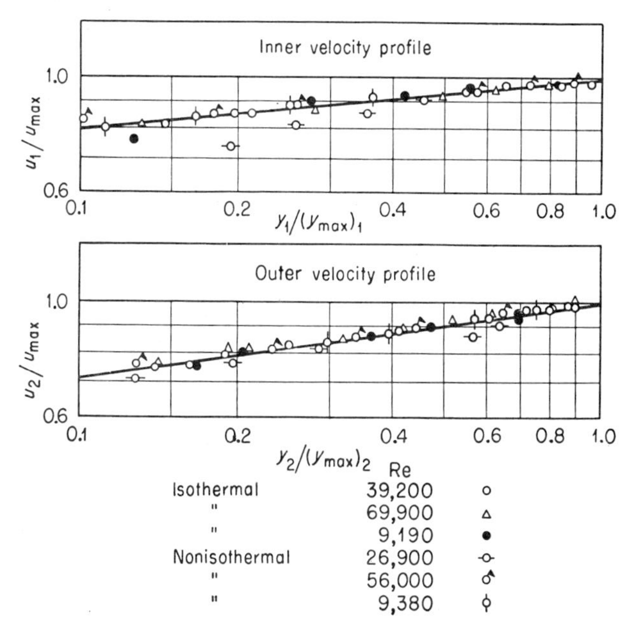

FIG. 7-31. Annular velocity-distribution data plotted according to Eqs. (7-88) and (7-89).

It may be shown theoretically, as it was for turbulent flow in circular tubes, that for the laminar film in the vicinity of the wall of the annulus $y_1^+ = u_1^+$ and $y_2^+ = u_2^+$ for the inner and outer portions of the profile, respectively. Figures 7-28 and 7-29 indicate that this is true for values of y_1^+ and y_2^+ up to 7. These conclusions are not too definite, since the impact tube used in determining the velocities was not calibrated, and there is always some question regarding velocity data in the region close to the wall, where there is a rapid change in velocity with radius.

7-20. Turbulent Velocity Distribution in Modified Annuli

Modified annuli which contain inner fin tubes are used as extended-surface heat exchangers. The purpose of increasing the surface area of the inner tube is to increase the rate of heat transfer in the annulus, and thus, for a given space occupied by the heat exchanger, a larger amount of heat may be transferred. Some types of fin tubes have longitudinal fins parallel to the axis of the tube; other types have fins in the form of flat, circular disks mounted perpendicular to the axis of the tube. The latter are called transverse-fin tubes. Very often the transverse fins are in the form of a helix, and the tubes are called transverse-helical-fin tubes. Other types of extended-surface tubes used in annuli are spine tubes, which have small spines or needles closely spaced over the whole surface, and serrated-fin tubes, which have transverse disk fins with segments cut from them, making them approximately star-shaped.

Little investigation has been made of the mechanism of the flow of fluids in modified annuli, although considerable work has been done on the study of the pressure drop in such annuli. Knudsen [22] carried out a study of the mechanism of flow in five modified annuli containing transverse-helical-fin tubes. The investigation covered the determination of velocity profiles in these annuli and the visual study of the flow patterns occurring in the spaces between the fins with water flowing in the modified annulus. The five different annuli studied by Knudsen are described dimensionally in Table 7-2.

TABLE 7-2. MODIFIED ANNULI STUDIED BY KNUDSEN †

Tube no.	Fins per inch	Diam. at base of fins, in.	Diam. of fins, in.	Fin spacing, in.	Fin height, in.	Ratio of fin spacing to fin height
A	3.98	0.664	1.050	0.222	0.193	1.15
B	8.12	0.648	1.031	0.098	0.191	0.51
C	4.19	0.655	1.241	0.214	0.293	0.73
D	5.85	0.649	1.295	0.146	0.323	0.45
E	8.02	0.639	1.319	0.102	0.341	0.30

† J. G. Knudsen, Heat Transfer, Friction, and Velocity Gradients in Annuli Containing Plain and Transverse Fin Tubes, Ph.D. Thesis, University of Michigan, 1949; J. G. Knudsen and D. L. Katz, *Chem. Eng. Progr.*, **46**:490 (1950).

In an annulus with transverse fins on the inner tube the cross-sectional area for flow varies. At a fin it is equal to $\pi(r_2^2 - r_f^2)$, but between fins it is equal to $\pi(r_2^2 - r_1^2)$, where r_f is the outside radius of the fin and r_1 is the root radius. Velocity profiles were studied only at the fins, i.e., at the points of minimum cross section. Figure 7-32 shows the nine velocity profiles obtained. In order to represent annuli of different dimensions on the same basis the point velocity has been plotted as a function of $y_2/(r_2 - r_f)$.

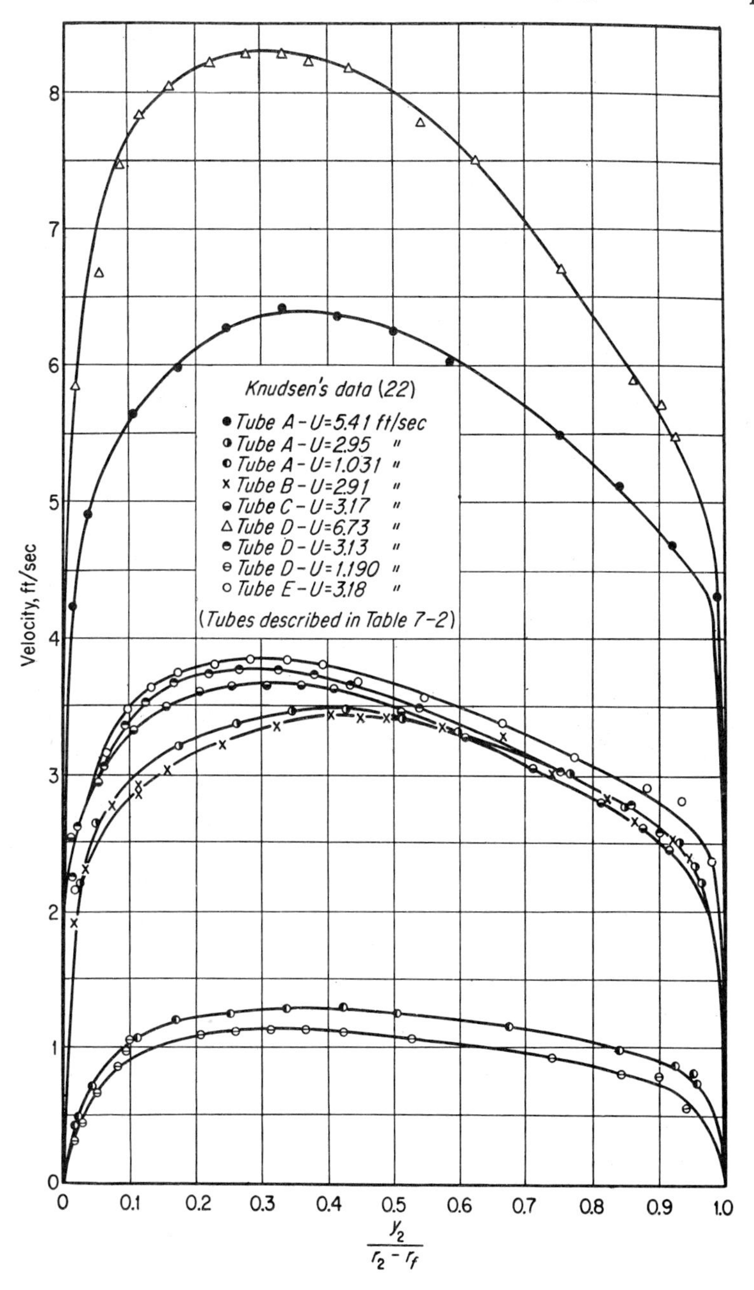

FIG. 7-32. Turbulent-flow velocity profiles in modified annuli.

All the velocity profiles shown in Fig. 7-32 have the same general shape. The profile is unsymmetrical, and the point of maximum velocity lies closer to the outer wall of the annulus than it would for a plain annulus containing an inner tube of the same diameter as the fins.

Knudsen and Katz [23] made visual studies of the flow patterns occurring in modified annuli. The investigation was carried out by injecting colored

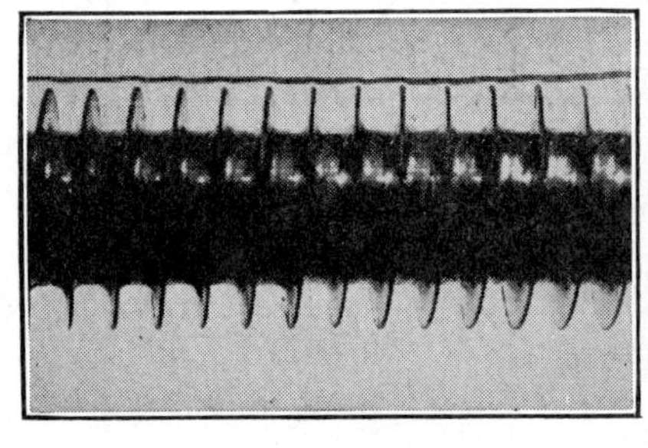

(*a*) $U = 0.09$ ft/sec
Re = 632

(*b*) $U = 0.211$ ft/sec
Re = 1,480

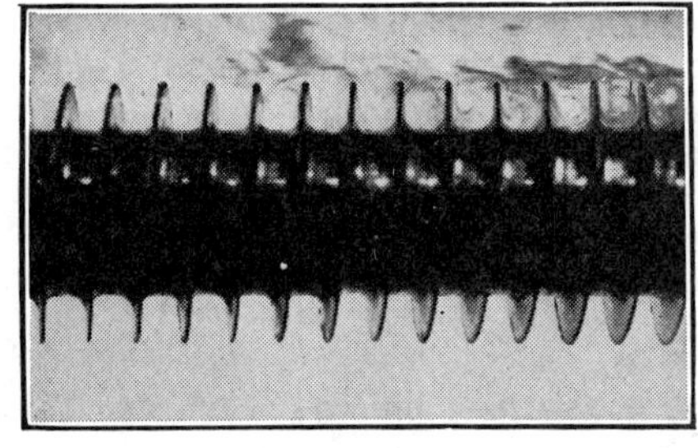

(*c*) $U = 0.724$ ft/sec
Re = 5,070

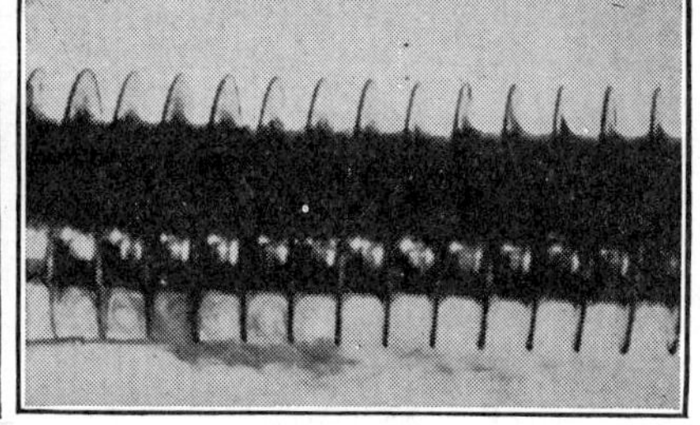

(*d*) $U = 1.24$ ft/sec
Re = 10,400

FIG. 7-33. Flow patterns in modified annuli. [*From J. G. Knudsen and D. L. Katz, Chem. Eng. Progr.*, **46**:490 (1950).]

fluid into the flowing annular stream at a point just above the edge of the fin. It was possible to see the path of the flowing fluid in the fin spaces. This visual study was made both for laminar and turbulent flow and covered a Reynolds-number range of 500 to 20,000.

In the region of laminar flow the dye appeared as a ribbon and maintained this shape for almost the whole length of the annulus. For turbulent flow there was considerable motion of the fluid between the fins. Eddies appeared between the fins and were observed to rotate at a rate which was a function of the velocity of fluid in the open cross section of the annulus. Figure 7-33 shows a series of flow patterns photographed for tube *A* of Table 7-2. The laminar region of flow is depicted in Fig. 7-33*a*, while in Fig. 7-33*b*, *c*, and *d* turbulent flow is evident. The circular eddies may be observed for turbulent flow. Sketches of flow patterns for all tubes described in Table 7-2 are shown in Fig. 7-34, from which it can be seen that the flow patterns change as the Reynolds number increases and that

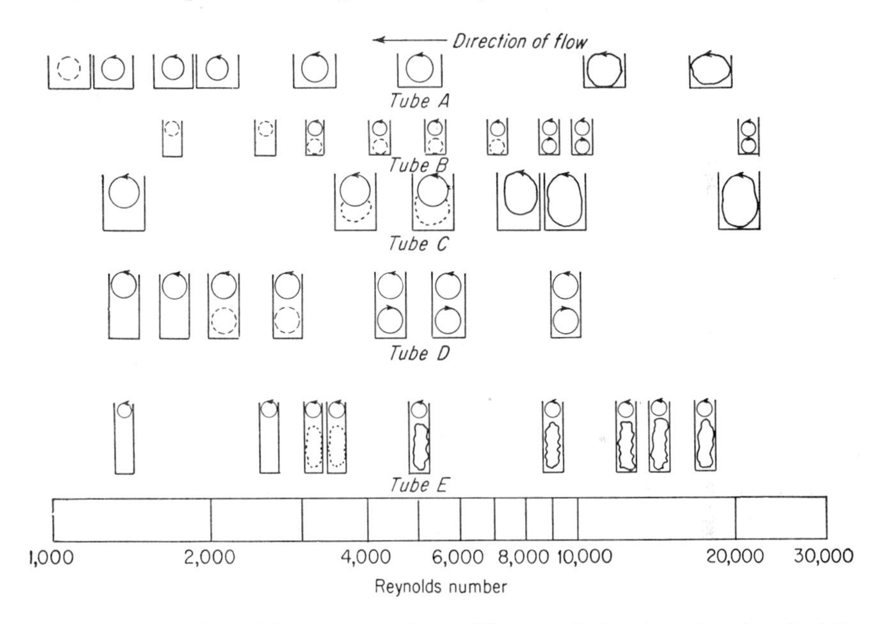

FIG. 7-34. Sketches of flow patterns in modified annuli for the tubes described in Table 7-2. [*From J. G. Knudsen and D. L. Katz, Chem. Eng. Progr.*, **46**:490 (1950).]

the fin spacing and the fin height have an effect on the type of flow pattern which occurs between the fins.

The results depicted in Figs. 7-33 and 7-34 indicate that it is possible to predict the flow patterns which occur between transverse fins during turbulent flow. The prediction of the flow pattern is based on a dimensionless quantity which is the ratio of the fin spacing to the fin height. For values of this ratio between 1.15 and 0.73 the flow pattern is characterized by one circular eddy between the fins, which becomes slightly elongated when the ratio reaches a value near 0.73. When the ratio ranges from 0.51 to 0.45, two circular eddies form between the fins, and they rotate in opposite directions. When the ratio reaches a value of 0.31, a somewhat different pattern is observed. In this case, a circular eddy forms at the outer edge of the fin space, but in the space between this eddy and the tube wall no steady circular eddies are observed.

7-21. Friction Factors in Plain Annuli

In order to describe an annulus geometrically, it is necessary to use the diameters of both tubes making up the annulus. Thus it would be expected that the friction factor is a function of the Reynolds number and the diameter ratio d_2/d_1, and this can be shown by dimensional analysis when wall roughness is assumed to have no effect. Extensive investigation has failed to produce a satisfactory relationship involving the diameter ratio.

The friction factor for annuli is expressed by Eq. (4-72).

$$f = \frac{(d_2 - d_1)g_c}{2\rho U^2} \frac{-dP_f}{dx} \tag{4-72}$$

The Reynolds number for annuli is based on the equivalent diameter $d_2 - d_1$.

A large number of experimental data have been obtained on turbulent friction factors in annuli involving various diameter ratios, fluids, and types of tube surfaces. Eq. (7-65) has been recommended for use with annuli. Davis [5] made a comprehensive study of all existing annular friction-factor data and proposed the following equation:

$$f\left(\frac{d_2/d_1 - 1}{d_2/d_1}\right)^{-0.1} = 0.055(\mathrm{Re})^{-0.2} \tag{7-92}$$

Davis's equation for turbulent annular friction factors involves the diameter ratio d_2/d_1. Empirical correlations of most investigators express the friction factor in terms of the Reynolds number alone. All data vary considerably because some annuli investigated do not have tubes with smooth surfaces. The average of most experimental results can be represented by the following equation:

$$f = 0.076(\mathrm{Re})^{-0.25} \tag{7-93}$$

Experimental data deviate as much as 35 per cent from Eq. (7-93).

Equations (7-92) and (7-93) are proposed relationships for the turbulent friction factors in smooth annuli. They are compared with each other in Table 7-3 at Reynolds numbers of 3,000, 10,000, 100,000, and 1,000,000. With the Davis equation, three diameter ratios are considered.

Table 7-3 indicates that the friction-factor equations agree quite well, and they may be used to predict turbulent friction factors in smooth annuli. Davis's equation shows that the effect of the diameter ratio is small at values of d_2/d_1 greater than approximately 3. At low values of d_2/d_1 the effect of the diameter ratio is large. Rothfus, Monrad, Sikchi, and Heideger [50] defined inner- and outer-wall friction factors for annuli and corre-

TABLE 7-3. COMPARISON OF FRICTION-FACTOR EQUATIONS FOR ANNULI

Re	Eq. (7-92)			Eq. (7-93)
	$d_2/d_1 = 1.5$	$d_2/d_1 = 4$	$d_2/d_1 = 7$	
3,000	0.00958	0.0108	0.0109	0.0103
10,000	0.00745	0.00850	0.00860	0.0076
100,000	0.00494	0.00535	0.00542	0.00426
1,000,000	0.00292	0.00336	0.00341	0.00239

lated friction-loss data over a Reynolds-number range from 10,000 to 45,000. The inner- and outer-wall friction factors are, respectively,

$$f_1 = \frac{2\tau_1 g_c}{\rho U^2} \tag{7-94}$$

$$f_2 = \frac{2\tau_2 g_c}{\rho U^2} \tag{7-95}$$

These relationships are similar to Eq. (4-17).

Combining Eqs. (7-76) and (7-94),

$$f_2 = \frac{(r_2{}^2 - r_{\max}^2)g_c}{r_2 \rho U^2} \frac{-dP_f}{dx} \tag{7-96}$$

From Eq. (7-96) the pressure drop due to friction may be calculated when f_2, the dimensions of the annulus, the average velocity, and the fluid density are known. Rothfus and coworkers report that for $10{,}000 < \mathrm{Re}_2 < 45{,}000$ and for long annuli

$$\frac{1}{\sqrt{f_2}} = 4.0 \log (\mathrm{Re}_2 \sqrt{f_2}) - 0.40 \tag{7-97}$$

where Re_2 is the Reynolds number defined in Eq. (7-74). Since Eq. (7-97) is the same as (7-60), the friction-factor curve for smooth tubes on Fig. 7-21 may be used to obtain f_2 from Re_2. For short annuli, entrance conditions have a significant effect on the outer-wall friction factor. Equation (7-97) applies for annuli in which $L/(d_2 - d_1)$ is greater than 250. At values of this ratio less than 250 plots are available to determine f_2 (Sec. 9-7).

7-22. Effect of Inner-tube Eccentricity in Plain Annuli

Deissler and Taylor [11] made a theoretical study of flow in eccentric and concentric plain annuli and showed that the friction factor decreases as the eccentricity of the annulus is increased. Figure 7-35 shows the results reported by Deissler and Taylor, in which is plotted the friction factor versus the Reynolds number at various eccentricities for an annulus with $d_2/d_1 = 3.5$. Experimental friction-loss data of Stein and others [58] on eccentric

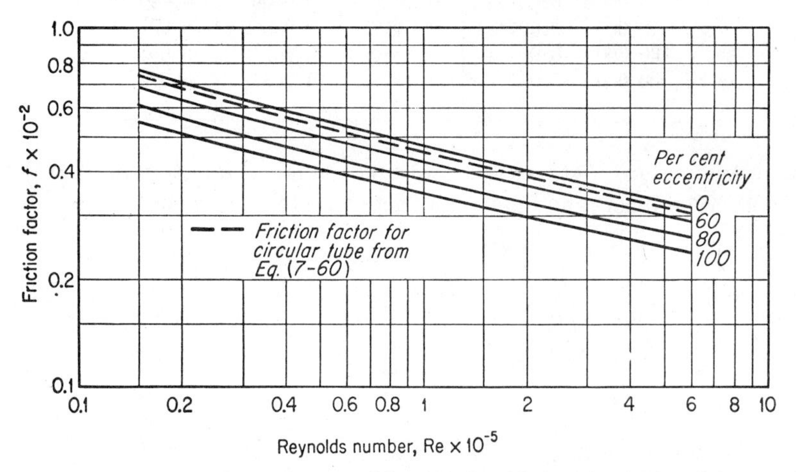

FIG. 7-35. Predicted effect of eccentricity on the friction factor in plain annuli ($r_2/r_1 = 3.5$). (*From R. G. Deissler and M. F. Taylor, NACA TN 3451, 1955.*)

annuli agree approximately with Fig. 7-35. The per cent eccentricity is defined as the per cent of annulus width by which the inner tube is eccentric.

Example 7-5

A double-tube heat exchanger consists of an inner tube 1.00 in. ID and an outer tube 2.00 in. ID. Carbon tetrachloride at an average temperature of 120°F flows in the annular space of the heat exchanger at a rate of 15 gal/min. The heat exchanger is installed in a vertical position, and flow is downward.

(*a*) Determine the point of maximum velocity in the annulus.

(*b*) Calculate the lost work due to friction per pound mass of fluid per foot of annulus (*i*) using Eq. (7-65) for the friction factor and (*ii*) using Eq. (7-97) for the friction factor.

(*c*) Determine the shear stress at the inner and outer wall of the annulus using the friction factor calculated from Eq. (7-97).

Solution

For carbon tetrachloride at 120°F

$$\rho = 98.0 \text{ lb}_m/\text{ft}^3$$

$$\mu = 0.72 \text{ centipoise}$$

Equivalent diameter of annulus = 2.0 − 1.0 = 1.0 in.

$$\text{Average liquid velocity in annulus} = \frac{(15)(576)}{(7.48)(60)(\pi)[(2)^2 - (1)^2]} = 2.04 \text{ ft/sec}$$

$$\text{Re (based on equivalent diameter)} = \frac{1}{12}\frac{(2.04)(98.0)}{(0.72)(0.000672)} = 3.44 \times 10^4$$

(*a*) The position of maximum velocity is calculated from Eq. (4-58).

$$r_{\max} = \sqrt{\frac{(1.0)^2 - (0.5)^2}{2 \ln (1.0/0.5)}} = 0.735 \text{ in.}$$

The point of maximum velocity is 0.235 in. from the inner wall of the annulus.

(*b*) From Eq. (7-65) at Re = 3.44×10^4

$$f = 0.0066$$

From Eq. (4-20), using the equivalent diameter of the annulus,

$$\overline{lw} = \frac{(2)(0.0066)(2.04)^2(1)(12)}{(32.2)(1.0)} = 0.0204 \text{ (ft)(lb}_f)/(\text{lb}_m)(\text{ft of length})$$

$$\text{Re}_2 \text{ [defined by Eq. (7-74)]} = \frac{(2)[(1.0)^2 - (0.735)^2](2.04)(98.0)}{(1.0)(12)(0.72)(0.000672)}$$

$$= 3.16 \times 10$$

From Eq. (7-97)

$$f_2 = 0.0058$$

From Eq. (7-96) the lost work per foot of annulus is

$$\overline{lw} = \frac{1}{\rho}\frac{-dP_f}{dx} = \frac{(0.0058)(1.0)(2.04)^2(12)}{[(1.0)^2 - (0.735)^2](32.2)}$$

$$= 0.0197 \text{ (ft)(lb}_f)/(\text{lb}_m)(\text{ft of length})$$

Equations (7-65) and (7-97) give almost the same value of the lost work per foot of annulus.

(*c*) In part (*b*)

$$f_2 = 0.0058$$

From Eq. (7-95)

$$\tau_2 = \frac{(0.0058)(98.0)(2.04)^2}{(2)(32.2)} = 0.0367 \text{ lb}_f/\text{ft}^2$$

This is the shear stress at the outer wall.

$$\tau_1 = \frac{(0.0367)(1.0)[(0.735)^2 - (0.5)^2]}{(0.5)[(1.0)^2 - (0.735)^2]} = 0.0465 \text{ lb}_f/\text{ft}^2$$

Thus $\tau_1 = 0.0465 \text{ lb}_f/\text{ft}^2$, which is the shear stress exerted at the inner wall of the annulus.

7-23. Friction Factors for Turbulent Flow in Annuli Containing Transverse-fin Tubes

To describe the geometry of an annulus containing a transverse-fin tube a number of dimensions are necessary, including the outside-tube diameter d_2, the inside-tube diameter, i.e., the root diameter d_1, the fin diameter d_f, the fin height e, the fin spacing s, and the fin thickness. The friction factor for such annuli is a function not only of the Reynolds number but also of all the above dimensions.

There is no empirical relationship involving all these quantities which will predict the friction factor in these modified annuli since existing experimental data are not sufficiently extensive to show the effect of each variable. Knudsen and Katz [23] determined the friction factors for water flowing in the annuli containing the transverse-helical-fin tubes described in Table 7-2. A relatively small range of fin spacings was investigated, and the diameter of the outside tube was not varied. Braun and Knudsen [3] determined friction factors for similar annuli. They constructed fin tubes by placing circular metal disks on a solid rod, the distance between the disks being maintained by spacers, which also fitted the rod. In this way it was possible to investigate a wide range of fin spacings. However, the diameter of the outside tube was again held constant for all tests. Pressure-drop data in annuli containing transverse-fin tubes indicate that a friction-factor curve is obtained for each tube, and charts relating the friction factor and the Reynolds number have been presented which are similar to the plot obtained by Nikuradse [35] in his study of rough pipes (Fig. 7-20).

The equivalent diameter of a modified annulus containing a transverse-fin tube is difficult to determine because of the varying cross-sectional area. Perhaps the best equivalent diameter would be the volumetric diameter based on the volume of free flow per unit length and the wetted surface per unit length. This volumetric diameter is four times the volumetric hydraulic radius defined in Eq. (4-29). However, it is not readily calculated. Knudsen and Katz have used as the equivalent diameter the difference between the outside-tube diameter and the fin diameter, i.e., $d_2 - d_f$. This is the equivalent diameter at the minimum cross section defined in Eq. (4-28).

The friction-factor chart obtained by Braun and Knudsen [3] on modified annuli is shown in Fig. 7-36. The plot was obtained from cross plots of the friction-factor–Reynolds-number chart. In Fig. 7-36 the friction factor is plotted versus the dimensionless quantity $(d_2 - d_f)/(d_2 - d_r)$ at constant values of s/e, the ratio of fin spacing to fin height. Each set of curves is for a constant Reynolds number. Figure 7-36 may be used to determine the friction-factor curve for any annulus containing a transverse-fin tube. For a given annulus the value of s/e and $(d_2 - d_f)/(d_2 - d_r)$ may be easily calculated, and from Fig. 7-36 the friction factors at the four given Reynolds numbers may be obtained. The friction-factor–Reynolds-number curve may then be drawn for the given annulus.

The fin thickness is not considered in Fig. 7-36, but its effect is assumed to be negligible for the finned tubes investigated.

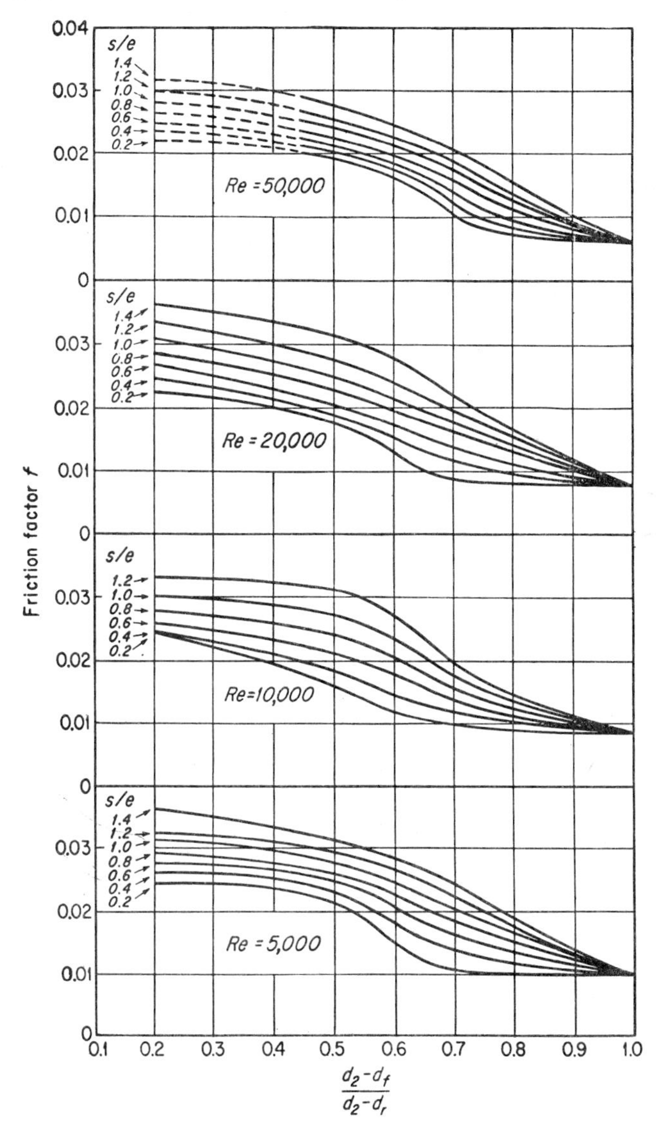

FIG. 7-36. Friction factors in annuli containing transverse-fin tubes. [*From F. W. Braun and J. G. Knudsen, Chem. Eng. Progr.*, **48**:517 (1952).]

Winkel[61] studied several kinds of modified annuli in which the inner tube had rings on its surface. The clearance between the rings and the outside tube was quite small. No general correlation can be obtained from Winkel's investigation, but his results may be of some use if annuli of this type are encountered. A summary of Winkel's work is given in Fig. 7-37.

Annulus tested	Laminar-flow friction factor	Turbulent-flow friction factor	Transition Reynolds number
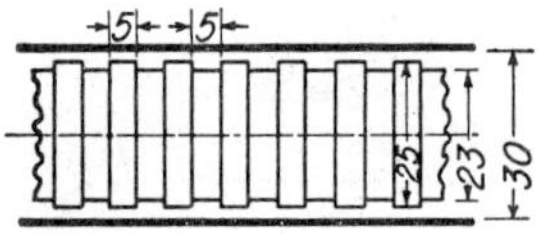	$f = 24(\mathrm{Re})^{-1}$	$f = 0.118(\mathrm{Re})^{-0.2}$	2,000–2,700
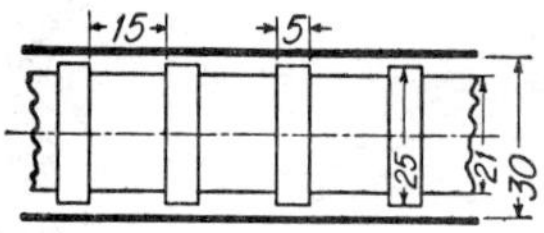	$f = 15.4(\mathrm{Re})^{-1}$	$f = 0.165(\mathrm{Re})^{-0.2}$	500–700
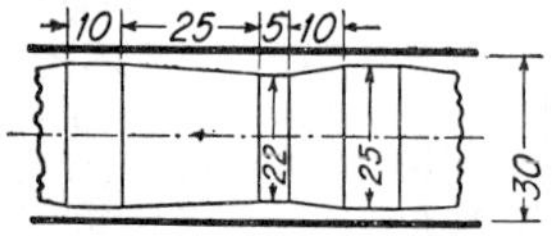	$f = 17.1(\mathrm{Re})^{-1}$	$f = 0.327(\mathrm{Re})^{-0.46}$	1,800
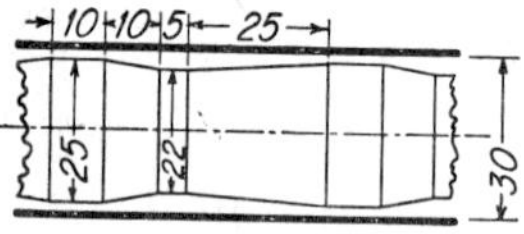	$f = 17.9(\mathrm{Re})^{-1}$	$f = 2.24(\mathrm{Re})^{-0.646}$	1,600–5,000
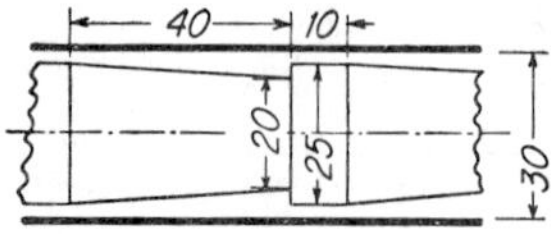	$f = 13.7(\mathrm{Re})^{-1}$	$f = 0.0935(\mathrm{Re})^{-0.25}$	1,000–1,600

FIG. 7-37. Friction factors in modified annuli. [*From R. Winkel, Z. angew. Math. u. Mech.*, **3**:251 (1923).]

The left column of Fig. 7-37 shows the annuli studied and their dimensions; the second column gives the expressions for the laminar friction factors, while the third column gives the expressions for the turbulent friction factors. The fourth column gives the range of Reynolds numbers in which transition from laminar to turbulent flow occurred and in which the friction-factor formulas given in the second and third columns do not apply. The turbulent friction-factor equations shown in Fig. 7-37 apply for Reynolds numbers up to 16,000. The equivalent diameter used in the equations is the equivalent diameter at the minimum cross section.

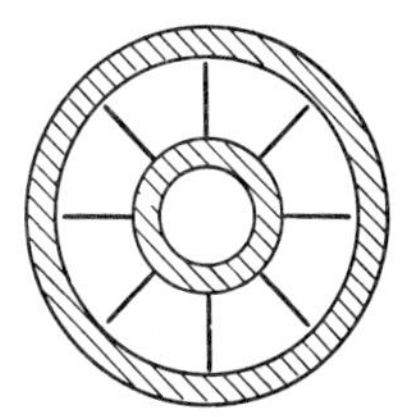

FIG. 7-38. Cross section of an annulus containing a longitudinal-fin tube.

7-24. Friction Factors in Annuli Containing Longitudinal-fin Tubes

Friction factors in annuli containing longitudinal-fin tubes having a transverse cross section as shown in Fig. 7-38 have been investigated by de Lorenzo and Anderson.[27] They recommend the friction-factor curve shown in Fig. 7-39, where the friction factor is plotted versus the Reynolds number for both laminar and turbulent flow. The equivalent diameter of annuli containing longitudinal-fin tubes is four times the ratio of the free cross-sectional area to the wetted perimeter.

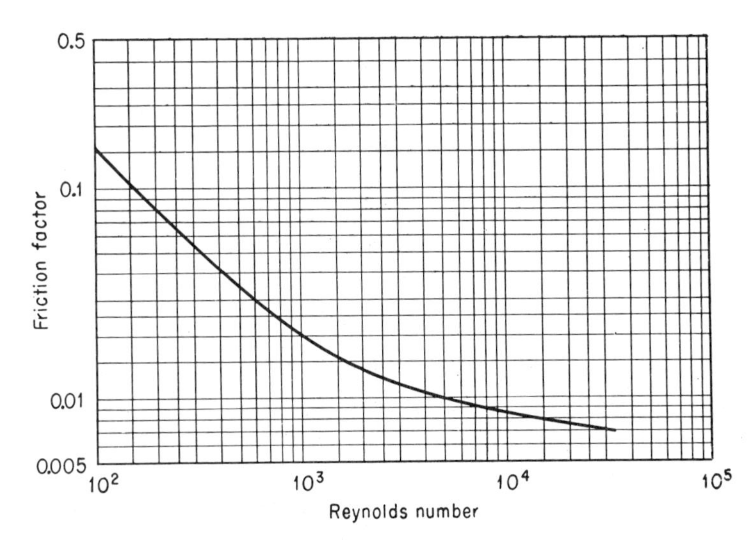

FIG. 7-39. Friction factors for annuli containing longitudinal-fin tubes. [*From B. de Lorenzo and E. D. Anderson, Trans. ASME,* **67**:697 (1945).]

III. TURBULENT FLOW IN NONCIRCULAR CONDUITS

7-25. Velocity Distribution for Turbulent Flow between Infinite Parallel Planes

Donch [12] and Nikuradse [31] measured respective point velocities for the turbulent flow of air and water between parallel planes. Goldstein [16] analyzed their data and obtained the following relationship for the turbulent core:

$$\frac{u_{\max} - u}{u^*} = -3.385 \left[\ln \left(1 - \sqrt{\frac{y_c}{b/2}} \right) + \sqrt{\frac{y_c}{b/2}} \right] - 0.172 \qquad (7\text{-}98)$$

where $u_{\max}$ is the maximum velocity in the stream and u^* is the friction velocity.

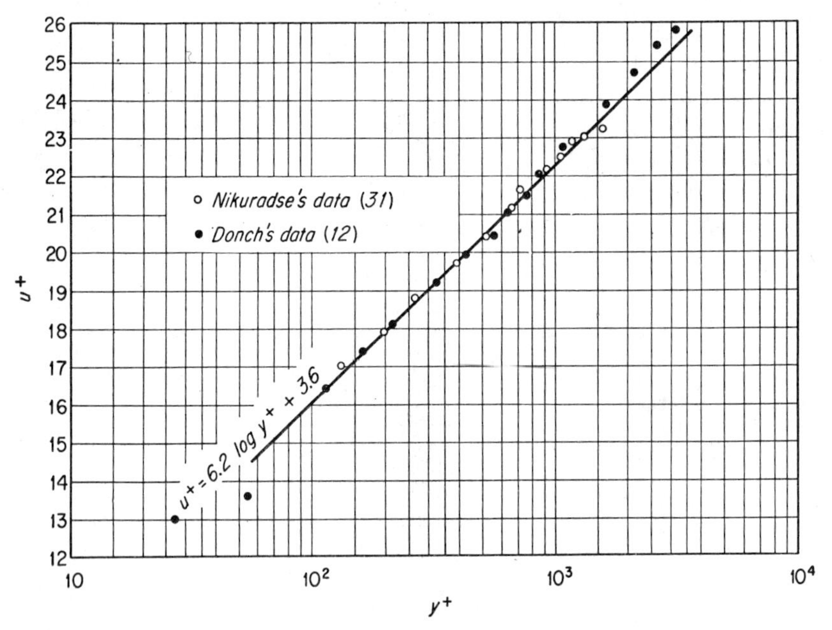

Fig. 7-40. Velocity distribution for turbulent flow between parallel planes.

Donch's [12] and Nikuradse's [31] velocity-distribution data are plotted in Fig. 7-40 as u^+ versus y^+, where $u^+ = u/u^*$ and $y^+ = yu^*/\nu$. The data agree with the equation

$$u^+ = 6.2 \log y^+ + 3.6 \qquad (7\text{-}99)$$

Schlinger and Sage [54] report velocity-distribution data for flow between parallel planes up to a value of $y^+ = 750$. Their data agree with Eq. (7-99) and also coincide quite well with the circular-tube equations (7-35) and (7-36).

Laufer [25] studied turbulent flow between parallel planes, and by means of a hot-wire anemometer he was able to obtain point velocities at values of y^+ as low as 2. His results for small values of y^+ indicate the relation

$$u^+ = y^+ \qquad 0 < y^+ < 5 \tag{7-37}$$

Laufer's turbulent-core data lie considerably above those given by Eq. (7-99).

Pai [36] used Reynolds's momentum equations (5-14) to derive the following semiempirical velocity distribution relation:

$$\frac{u}{u_{\max}} = 1 - 0.3293\left(\frac{y_c}{b/2}\right)^2 - 0.6707\left(\frac{y_c}{b/2}\right)^{32} \tag{7-100}$$

where y_c is the distance from the center line and b is the distance between the planes. The exponents in Eq. (7-100) are determined from Laufer's experimental data; the form of the relationship is determined from theoretical considerations. Equation (7-100) applies over the whole distance between the parallel planes.

7-26. Friction Factors for Turbulent Flow between Parallel Planes

The friction-factor–Reynolds-number relation for turbulent flow between parallel planes is the same as that for circular tubes.

$$\frac{1}{\sqrt{f}} = 4.0 \log (\mathrm{Re}\sqrt{f}) - 0.40 \tag{7-60}$$

where $f = \dfrac{bg_c}{\rho U^2}\dfrac{-dP_f}{dx}$

$\mathrm{Re} = \dfrac{2bU}{\nu}$

b = spacing of the planes

7-27. Velocity Distribution for Turbulent Flow in Rectangular, Triangular, and Trapezoidal Conduits

No extensive studies have been conducted on the velocity distribution in conduits having noncircular cross sections. The most thorough investigations were made by Prandtl [42] and Nikuradse,[33] who, in addition to determining point velocities in such conduits, also carried out visual studies of the flow patterns.

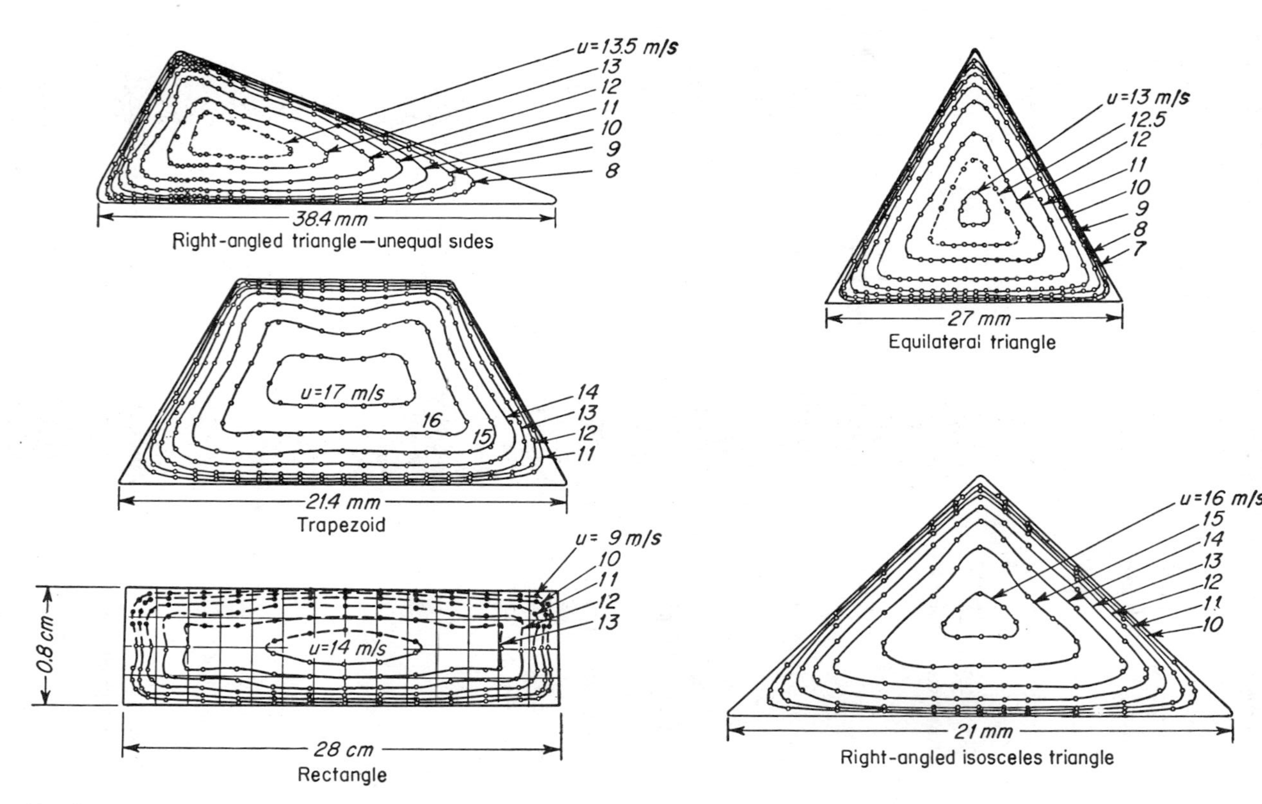

FIG. 7-41. Point velocities in noncircular conduits. [*From L. Prandtl, Proc. Intern. Congr. Appl. Mech., 2nd Congr., Zurich, 1927, p. 62; J. Nikuradse, Ing.-Arch.,* **1**:306 (1930).]

The velocity data Prandtl and Nikuradse obtained for water flowing in noncircular conduits are shown in Fig. 7-41. The various conduits studied are shown in cross section, and the experimentally determined velocities are plotted. Lines joining points of equal velocity give the complete velocity distribution.

Of considerable interest are the results of Prandtl's and Nikuradse's visual studies of the flow patterns in these conduits. By injecting a milky fluid at various points in the cross section the direction of flow at these

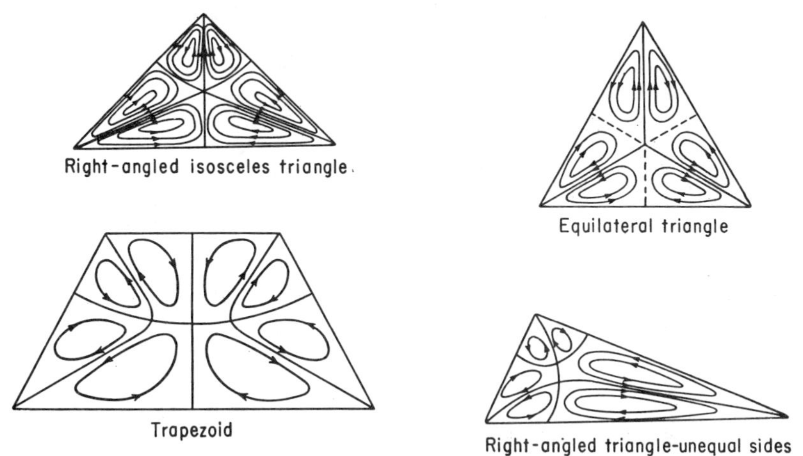

FIG. 7-42. Secondary flow in noncircular conduits. [*From L. Prandtl, Proc. Intern. Congr. Appl. Mech., 2nd Congr., Zurich, 1927, p.* 62; *J. Nikuradse, Ing.-Arch.*, **1**:306 (1930).]

points could be determined. It was found that there was flow toward the corners of the conduit and away from the sides of the conduit. This motion was superimposed on the longitudinal motion of the fluid particles. The patterns of this secondary motion are reproduced in Fig. 7-42. The effect of the secondary flow on the lines of constant velocity is to transpose these lines in the direction of the secondary flow. Hence, in the corners these lines are pushed toward the corner, and in the vicinity of the wall they are pushed away from the wall.

7-28. Friction Factors for Turbulent Flow in Rectangular, Triangular, and Trapezoidal Conduits

Experimental data on the pressure drop during turbulent flow in conduits of noncircular cross section indicate that the Blasius equation (7-3) will suitably predict friction factors from the point of transition to a Reynolds number of 100,000.

$$f = 0.079(\mathrm{Re})^{-1/4} \tag{7-3}$$

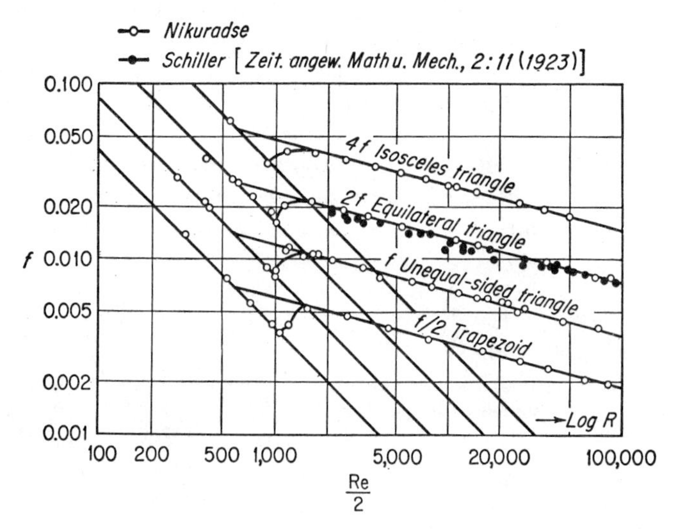

FIG. 7-43. Turbulent-flow friction factors in noncircular conduits. [*From J. Nikuradse, Ing.-Arch.*, **1**:306 (1930).]

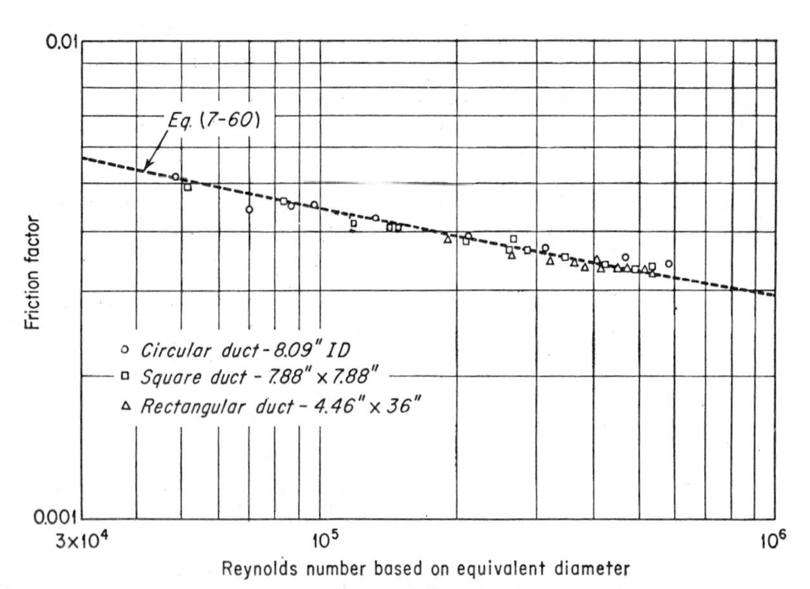

FIG. 7-44. Friction factors in rectangular, square, and circular ducts. [*Data from P. G. Huebscher, Heating, Piping, Air Conditioning*, **19**:127 (1947).]

Nikuradse[33] determined friction factors for noncircular conduits, and his data covering the laminar- and turbulent-flow regions are shown in Fig. 7-43. Nikuradse's results on all conduits may be reduced to Eq. (7-3).

Huebscher[18] reported turbulent friction-factor data for air flowing in noncircular conduits in the Reynolds-number range from 20,000 to 500,000. These data, shown in Fig. 7-44, agree better with Eq. (7-60) than with Eq. (7-3). Lowdermilk and others[28] also report that friction-factor data for noncircular conduits can be correlated with Eq. (7-60). The same workers also studied nonisothermal flow in noncircular ducts and found that Eq. (7-60) could also be used to predict nonisothermal friction factors if the fluid properties were evaluated at $(T_w + T_b)/2$, which is arithmetic average of the wall and bulk temperatures.

BIBLIOGRAPHY

1. Bailey, R. V.: *Oak Ridge Natl. Lab. Tech. Div. Eng. Research Sec., ORNL* 521, 1950.
2. Blasius, H.: *Mitt. Forschungsarb.* 131, pp. 1–40, 1913.
3. Braun, F. W., and J. G. Knudsen: *Chem. Eng. Progr.*, **48**:517 (1952).
4. Colebrook, C. F.: *J. Inst. Civil Engrs. (London)*, **11**:133 (1938–1939).
5. Davis, E. S.: *Trans. ASME*, **65**:755 (1943).
6. Deissler, R. G.: *NACA TN* 2138, 1950.
7. Deissler, R. G.: *NACA TN* 2242, 1950.
8. Deissler, R. G.: *Trans. ASME*, **73**:101 (1951).
9. Deissler, R. G., and C. S. Eian: *NACA TN* 2629, 1952.
10. Deissler, R. G.: *NACA TN* 3145, 1954.
11. Deissler, R. G., and M. F. Taylor: *NACA TN* 3451, 1955.
12. Donch, F.: *VDI-Forschungsheft* 282, 1926.
13. Drew, T. B., E. C. Koo, and W. H. McAdams: *Trans. AIChE*, **28**:56 (1932).
14. Drew, T. B., and R. P. Genereaux: *Trans. AIChE*, **32**:17 (1936).
15. Fromm, K.: *Z. angew. Math. u. Mech.*, **3**:339 (1923).
16. Goldstein, S.: *Proc. Roy. Soc. (London)*, **159A**:473 (1937).
17. Hopf, L.: *Z. angew. Math. u. Mech.*, **3**:329 (1923).
18. Huebscher, P. G.: *Heating, Piping, Air Conditioning*, **19**:127 (1947).
19. Kármán, T. von: *NACA TM* 611, 1931.
20. Kármán, T. von: *J. Aeronaut. Sci.*, **1**:1 (1934).
21. Kármán, T. von: *Trans. ASME*, **61**:705 (1939).
22. Knudsen, J. G.: Heat Transfer, Friction, and Velocity Gradients in Annuli Containing Plain and Transverse Fin Tubes, Ph.D. Thesis, University of Michigan, 1949.
23. Knudsen, J. G., and D. L. Katz: *Chem. Eng. Progr.*, **46**:490 (1950).
24. Knudsen, J. G., and D. L. Katz: *Proc. Midwest. Conf. Fluid Dynamics, 1st Conf., Univ. Illinois, 1950*, p. 175.
25. Laufer, J.: *NACA Rept.* 1053, 1951.
26. Laufer, J.: *NACA Rept.* 1174, 1954.
27. Lorenzo, B. de, and E. D. Anderson: *Trans. ASME*, **67**:697 (1945).
28. Lowdermilk, W. H., W. F. Weiland, and J. N. B. Livingood: *NACA RM* E53J07, 1954.
29. Miller, B.: *Trans. ASME*, **71**:357 (1949).
30. Moody, L. F.: *Trans. ASME*, **66**:671 (1944).

31. Nikuradse, J.: *VDI-Forschungsheft* 289, 1929.
32. Nikuradse, J.: *Proc. Intern. Congr. Appl. Mech., 3rd Congr., Stockholm, 1930,* **1**:239.
33. Nikuradse, J.: *Ing.-Arch.,* **1**:306 (1930).
34. Nikuradse, J.: *VDI-Forschungsheft* 356, 1932.
35. Nikuradse, J.: *VDI-Forschungsheft* 361, 1933.
36. Pai, S. I.: *J. Appl. Mechanics,* **20**:109 (1953).
37. Pai, S. I.: *J. Franklin Inst.,* **256**:337 (1953).
38. Perry, J. H.: "Chemical Engineers' Handbook," 3d ed., McGraw-Hill Book Company, Inc., New York, 1950.
39. Prandtl, L.: *Z. angew. Math. u. Mech.,* **1**:431 (1921).
40. Prandtl, L.: *Z. angew. Math. u. Mech.,* **5**:136 (1925); see also *NACA TM* 1231, 1949.
41. Prandtl, L.: *Ergeb. Aerodyn. Versuchanstalt Göttingen* 3, p. 1, 1927.
42. Prandtl, L.: *Proc. Intern. Congr. Appl. Mech., 2nd Congr., Zurich, 1927,* p. 62
43. Prandtl, L.: *Z. Ver. deut. Ing.,* **77**:105 (1933); see also *NACA TM* 720, 1933
44. Prengle, R. S., and R. R. Rothfus: *Ind. Eng. Chem.,* **47**:379 (1955).
45. Reichardt, H.: *NACA TM* 1047, 1943.
46. Reynolds, O.: *Trans. Roy. Soc. (London),* **174A**:935 (1883).
47. Ross, D.: *Proc. Midwest. Conf. Fluid Mechanics, 3rd Conf., Univ. Minn., 1953,* p. 651.
48. Rothfus, R. R.: Velocity Gradients and Friction in Concentric Annuli, Ph.D. Thesis, Carnegie Institute of Technology, 1948.
49. Rothfus, R. R., C. C. Monrad, and V. E. Senecal: *Ind. Eng. Chem.,* **42**:2511 (1950).
50. Rothfus, R. R., C. C. Monrad, K. G. Sikchi, and W. J. Heideger: *Ind. Eng. Chem.,* **47**:913 (1955).
51. Rothfus, R. R., and C. C. Monrad: *Ind. Eng. Chem.,* **47**:1144 (1955).
52. Rouse, H.: "Elementary Mechanics of Fluids," John Wiley & Sons, Inc., New York, 1948.
53. Schiller, L.: *Z. angew. Math. u. Mech.,* **3**:2 (1923).
54. Schlinger, W. G., and B. H. Sage: *Ind. Eng. Chem.,* **45**:2636 (1953).
55. Senecal, V. E., and R. R. Rothfus: *Chem. Eng. Progr.,* **49**:533 (1953).
56. Stanton, T. E., D. Marshall, and C. N. Bryant: *Proc. Roy. Soc. (London),* **97A**:413 (1920).
57. Stanton, T. E., and J. R. Pannell: *Trans. Roy. Soc. (London),* **214A**:199 (1914).
58. Stein, R. P., J. W. Hoopes, M. Markels, W. A. Selke, A. J. Bendler, and C. F. Bonilla: *Chem. Eng. Progr., Symposium Ser.* [11], **50**:115 (1954).
59. Walker, W. H., W. K. Lewis, W. H. McAdams, and E. R. Gilliland: "Principles of Chemical Engineering," 3d ed., McGraw-Hill Book Company, Inc., New York, 1937.
60. Wang, Chi-teh: *J. Appl. Mechanics,* **68**:A-85 (1946).
61. Winkel, R.: *Z. angew. Math. u. Mech.,* **3**:251 (1923).

CHAPTER 8

THE LAMINAR SUBLAYER

8-1. The Laminar-film Hypothesis

In early studies of the transfer of heat between a fluid flowing turbulently and a smooth solid wall the idea was put forth that the heat flows by conduction through a film immediately adjacent to the solid boundary. This implies the existence of a layer of fluid which is in laminar motion. It has been pointed out that the thickness of this laminar film could be calculated from heat-transfer measurements on the basis that the distance over which laminar flow occurs is exactly the same as the distance over which the heat is transferred by molecular conduction. Miller,[7] in a discussion of the laminar-film hypothesis, states that this need not be the case. Prandtl's [9] two theories regarding the laminar film adjacent to a solid boundary have been mentioned in Chap. 7. These theories are (1) that the fluid touching the solid boundary has zero velocity with respect to the wall and (2) that a layer of fluid adjacent to the solid boundary moves in laminar motion. Miller pointed out that the experimental proof of such a laminar layer was lacking. As a basis, he used the velocity-distribution data available at that time and indicated that measured velocity gradients adjacent to solid boundaries are different from those calculated from friction-loss data when the existence of laminar motion is assumed. From Eqs. (4-46) and (4-17) the velocity gradient in the laminar layer may be calculated from the relation

$$\left(\frac{du}{dy}\right)_{y=0} = \frac{f\rho U^2}{2\mu} \tag{8-1}$$

In the following sections the difficulties associated with the detection of a laminar boundary layer are discussed, and the significant studies on flow adjacent to a solid boundary are mentioned.

8-2. The Thickness of the Laminar Sublayer

It was shown in Chap. 7 that if laminar flow exists adjacent to the solid boundary, the relation between u^+ and y^+ is

$$u^+ = y^+ \tag{7-37}$$

Von Kármán determined that this relation holds up to a value of $y^+ = 5$, which implies that complete laminar flow exists up to this distance from

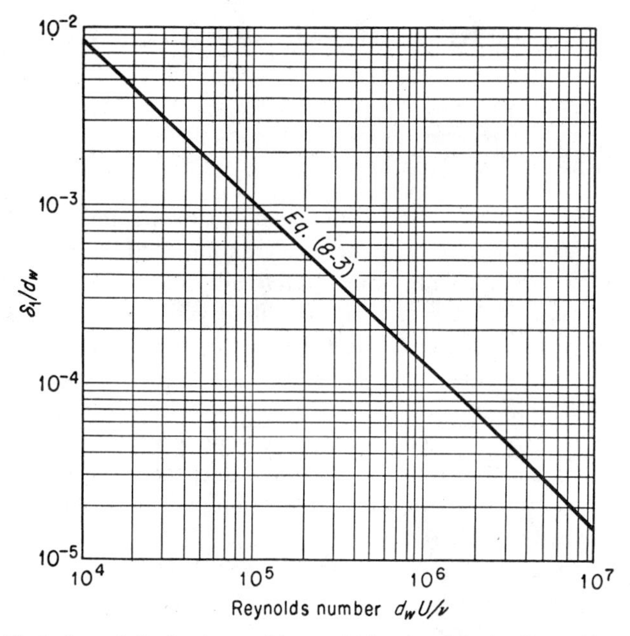

FIG. 8-1. Variation of the laminar-sublayer thickness with the Reynolds number for turbulent flow in tubes (based on $y^+ = 5$ at the edge of the laminar layer).

the wall. Letting δ_1 be the thickness of the sublayer in which flow is completely laminar and expressing y^+ by Eq. (7-52),

$$\frac{\delta_1}{r_w}\frac{\text{Re}}{2}\sqrt{\frac{f}{2}} = 5 \tag{8-2}$$

Therefore

$$\frac{\delta_1}{2r_w} = \frac{\delta_1}{d_w} = \frac{5}{\text{Re}}\sqrt{\frac{2}{f}} \tag{8-3}$$

The relation between δ_1/d_w and the Reynolds number is shown in Fig. 8-1, where the curve represents Eq. (8-3). The friction factor in Eq. (8-3) is calculated from Eq. (7-60). These results indicate that for turbulent flow in tubes δ_1 is less than 1 per cent of the tube diameter at a Reynolds number of 10,000 and decreases rapidly as the Reynolds number increases. The actual thickness of the laminar sublayer is very small except for very large pipes and low Reynolds numbers. It is extremely difficult to measure a detailed velocity profile in this small distance.

The velocity at the edge of the laminar sublayer (at $y^+ = 5$) may be obtained from the relation

$$u_{\delta_1}^+ = \frac{u_{\delta_1}}{U\sqrt{f/2}} = 5 \tag{8-4}$$

or

$$\frac{u_{\delta_1}}{U} = 5\sqrt{\frac{f}{2}} \tag{8-5}$$

where u_{δ_1} is the velocity at $y^+ = 5$. Equation (8-5) gives the ratio of the velocity at the edge of the laminar sublayer to the average velocity in the

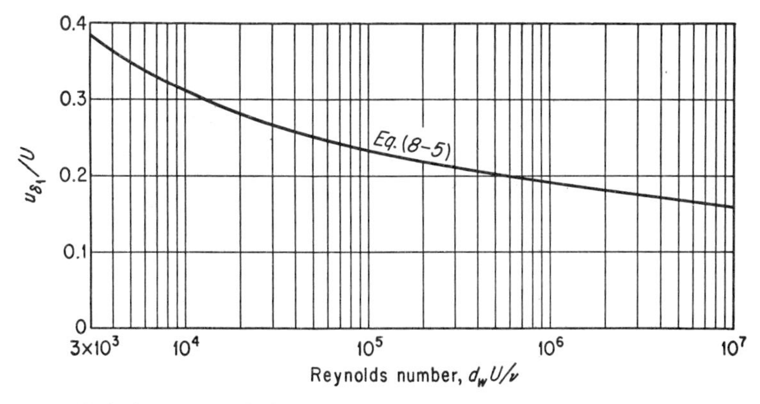

FIG. 8-2. Variation of the velocity at the edge of the laminar sublayer with the Reynolds number for turbulent flow in circular tubes (based on $u^+ = 5$ at the edge of the laminar layer).

pipe as a function of the friction factor, which, in turn, is a function of the Reynolds number. In Fig. 8-2, u_{δ_1}/U is plotted versus the Reynolds number.

8-3. Measurement of Point Velocities near a Solid Boundary

The actual measurement of point velocities (time-average velocities) may be carried out with a pitot tube or a hot-wire anemometer. Since each of these instruments must be of finite size, their very presence disturbs the flow to be measured. The hot-wire anemometer has been used to a greater extent since the development of electronic accessory equipment to go with it. It has proved to be more useful than the pitot tube in determining point velocities near solid boundaries. Laufer [6] obtained very good profiles near a solid wall using a hot-wire probe with a wire diameter of 0.00024 in.

Pitot tubes with very small openings have been used, but they are still large compared to the probable thickness of the laminar sublayer. Stanton, Marshall, and Bryant [14] used a pitot tube which was part of the pipe wall, thereby obtaining pitot-tube openings as small as 0.001 in. Rothfus, Monrad, and Senecal [12] used a pitot tube having an opening 0.0015 by 0.024 in. However, the opening was in the side of hypodermic-needle tubing, and in the vicinity of the wall the opening became smaller as the tube was moved out. In this way very small openings could be obtained. Deissler [1] used a pitot tube having width of opening of 0.005 in. Most pitot tubes which have been used have a width of opening almost as great as the thickness of the laminar sublayer.

8-4. Disadvantages of the Pitot Tube in Measuring Velocities Close to a Solid Wall

The impact pressure exerted when an incompressible fluid impinges on a pitot-tube opening is a measure of the velocity at that point according to the following relation, which was derived from Eq. (4-8):

$$P_I = \frac{\rho u^2}{2g_c} \tag{8-6}$$

where P_I is the impact pressure and is the difference between static pressure of the stream and the static pressure at the impact-tube opening. As the fluid impinges on the impact tube, its kinetic energy is converted to pressure energy, and Eq. (8-6) holds for complete conversion. For actual pitot tubes the right side of Eq. (8-6) is usually multiplied by a coefficient of 0.98 or 0.99.

The opening of the pitot tube must be of finite size, and the point velocity of the fluid will vary across the tube opening. Usually the velocity calculated by means of Eq. (8-6) is considered to be at the center of the tube opening. This is true if point velocities are being measured in the turbulent portion of the flowing stream; however, when point velocities are being measured near a solid boundary, Eq. (8-6) does not give an accurate value of the velocity at the center of the pitot-tube opening.

Figure 8-3*a* shows an impact tube with a square opening of width b placed in a turbulent stream so that the center of the tube opening is a distance y_0 from the solid boundary. The point velocity in the stream is

$$u = cy^{1/7} \tag{8-7}$$

The actual impact pressure $(P_I)_c$ corresponding to the point velocity at the center of the tube opening is obtained from Eqs. (8-6) and (8-7).

$$(P_I)_c = \frac{c^2\rho}{2g_c} y_0^{2/7} \tag{8-8}$$

The impact pressure at any point in the pitot-tube opening is $\rho c^2 y^{2/7}/2g_c$. The average pressure $(P_I)_{av}$ over the tube opening is the measured pressure and is expressed as

$$(P_I)_{av} = \frac{\displaystyle\int_{y_0-b/2}^{y_0+b/2} (c^2 b y^{2/7} \rho/2g_c)\,dy}{b^2}$$

$$= \frac{7}{9b}\frac{c^2\rho}{2g_c}\left[\left(y_0 + \frac{b}{2}\right)^{9/7} - \left(y_0 - \frac{b}{2}\right)^{9/7}\right] \tag{8-9}$$

$(P_I)_c$ and $(P_I)_{av}$ may be compared with each other only when the value of y_0 is known relative to b. If $y_0 = 2b$, which means that the pitot tube is

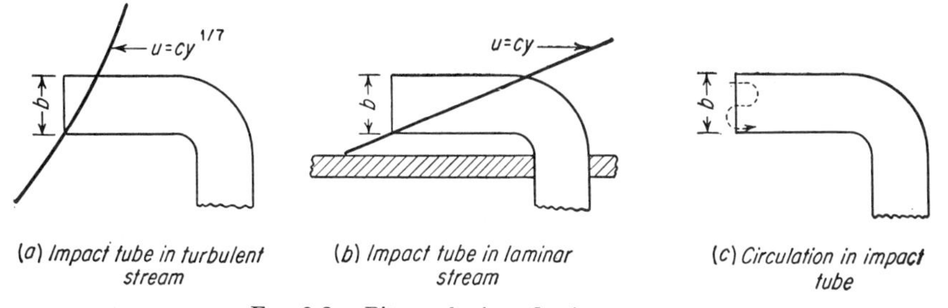

(a) Impact tube in turbulent stream (b) Impact tube in laminar stream (c) Circulation in impact tube

FIG. 8-3. Pitot tube in a flowing stream.

relatively close to the solid boundary, $(P_I)_c$ and $(P_I)_{av}$ are within 1 per cent of each other. Thus it may be assumed that a small pitot tube measures accurately the point velocity at the center of its opening when the flow is turbulent.

Figure 8-3*b* shows the same pitot tube located very close to the solid boundary, where the point velocity changes very rapidly, i.e.,

$$u = cy \tag{8-10}$$

The actual impact pressure at the center of the tube opening is

$$(P_I)_c = \frac{c^2\rho}{2g_c} y_0^2 \tag{8-11}$$

The pressure at any point in the tube opening is $\rho c^2 y^2 / 2g_c$.

$$(P_I)_{av} = \frac{\int_{y_0 - b/2}^{y_0 + b/2} (bc^2 y^2 \rho / 2g_c)\, dy}{b^2}$$

$$= \frac{1}{3b} \frac{c^2 \rho}{2g_c} \left[\left(y_0 + \frac{b}{2} \right)^3 - \left(y_0 - \frac{b}{2} \right)^3 \right] \qquad (8\text{-}12)$$

If the pitot tube is immediately adjacent to the solid wall, $y_0 = b/2$ (approximately), and from Eqs. (8-11) and (8-12) it may be shown that $(P_I)_{av}$ is 1.33 times as great as $(P_I)_c$. In this case the measured pressure $(P_I)_{av}$ actually corresponds to the point velocity at a position which is 58 per cent of the distance b from the edge of the pitot tube.

Another complication which arises in the measurement of point velocities close to a solid wall is illustrated in Fig. 8-3*c*. Because of the rapidly changing pressure across the pitot-tube opening there is a possibility that circulation will occur in the tube as indicated by the broken arrow; i.e., flow occurs from the high- to the low-pressure area. If this should happen, the actual measured impact pressure would be greater than that given by Eq. (8-12) and would correspond to a point velocity even farther out in the stream and perhaps beyond the pitot tube.

Example 8-1

A smooth tube is 1.00 in. OD and has a wall thickness of 0.049 in. Benzene is flowing in the tube at the rate of 2 gal/min. A pitot tube is located in the laminar sublayer so that the pitot-tube opening is 0.005 by 0.005 in. The center of the pitot-tube opening is 0.0025 in. from the wall. The benzene temperature is 80°F.

(*a*) Calculate the measured impact pressure.
(*b*) Calculate the velocity corresponding to the measured impact pressure.
(*c*) Calculate the velocity at the center of the pitot-tube opening.

Solution

$$\text{Average velocity} = \frac{2}{60} \Big| \frac{}{7.48} \Big| \frac{4}{\pi} \Big| \frac{144}{0.902} \Big| \frac{}{0.902} = 1.0 \text{ ft/sec}$$

At 80°F

$$\text{Viscosity of benzene} = 0.6 \text{ centipoise} \qquad \text{(Fig. 1-8)}$$

$$\text{Density of benzene} = 7.25 \text{ lb}_m\text{/gal} \qquad \text{(Fig. 1-2)}$$

$$\text{Re} = \frac{0.902}{12} \Big| \frac{1.0}{} \Big| \frac{7.25}{0.6} \Big| \frac{7.48}{0.000672} = 1.01 \times 10^4$$

From Fig. 7-21

$$f = 0.0077$$

From Eq. (8-1)

$$\left(\frac{du}{dy}\right)_{y=0} = \frac{0.0077}{0.000672}\left|\frac{54.2}{0.6}\right|\frac{1.0}{}\left|\frac{1.0}{}\right.$$

$$= 10.35 \times 10^2 \text{ ft/(sec)(ft)}$$

$$= 86.3 \text{ ft/(sec)(in.)}$$

(*a*)

$$y_0 = 0.0025 \text{ in.}$$

$$b = 0.005 \text{ in.}$$

$$y_0 - \frac{b}{2} = 0$$

$$y_0 + \frac{b}{2} = 0.005 \text{ in.}$$

$$\left(y_0 + \frac{b}{2}\right)^3 = 125 \times 10^{-9} \text{ in.}^3$$

From Eq. (8-12)

$$(P_I)_{av} = \frac{86.3}{3}\left|\frac{86.3}{0.005}\right|\frac{54.2}{2}\left|\frac{125 \times 10^{-9}}{32.2}\right.$$

$$= 5.22 \times 10^{-2} \text{ lb}_f/\text{ft}^2$$

(*b*) The velocity corresponding to 5.22×10^{-2} lb$_f$/ft^2 is

$$\sqrt{\frac{(5.22 \times 10^{-2})(2)(32.2)}{54.2}} = 0.249 \text{ ft/sec}$$

(*c*) At the center of the pitot-tube opening the point velocity is

$$(86.3)(0.0025) = 0.216 \text{ ft/sec}$$

The velocity corresponding to the measured impact pressure is about 15 per cent above the point velocity of the center of the pitot-tube opening. The calculations neglect effects due to circulation, surface tension, or other disturbances.

8-5. Experimental Work on Velocity Distribution near Solid Boundaries

Some of the earliest work on velocity profiles close to solid boundaries was conducted by Stanton et al.,[14] who studied the flow of air in a circular tube and used a very small pitot tube, one side of which was the tube wall. Velocities were measured with the center of the pitot-tube opening about 0.05 mm from the tube wall. Stanton and his associates calibrated this pitot tube for the case when the flow was laminar in the main tube. They obtained the calibration curve shown in Fig. 8-4, where the pitot-tube opening is plotted versus the effective distance. The calibration assumes that the laminar-flow velocity-distribution equation (4-44) is true. The effective distance means that the pressure measured at a certain pitot-tube opening corresponds to a velocity at some effective distance from the wall.

In Fig. 8-4 the effective distance becomes even greater than the pitot-tube opening when the opening becomes very small. By means of this calibration Stanton tentatively proved the existence of a laminar sublayer at the tube wall.

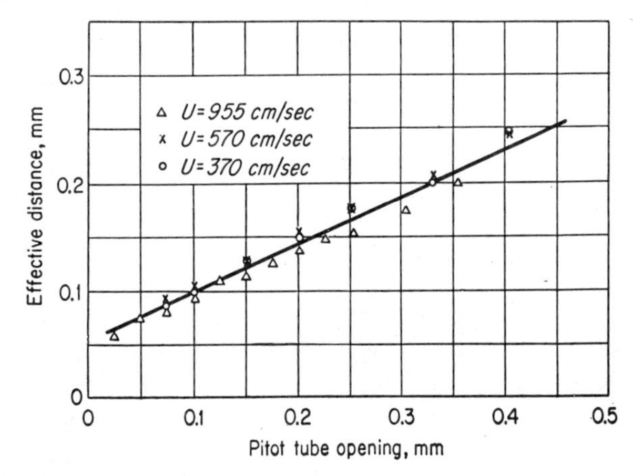

FIG. 8-4. Pitot-tube calibration. [*From T. E. Stanton et al., Proc. Roy. Soc. (London),* **97A**:413 (1920).]

If the laminar flow exists at the tube wall, then Eq. (8-1) gives the velocity distribution in this layer; i.e.,

$$\left(\frac{du}{dy}\right)_{y=0} = \frac{f\rho U^2}{2\mu} \tag{8-1}$$

Thus, for any set of flow conditions, the velocity distribution may be calculated in the laminar sublayer. Stanton did this for all his tests. He then plotted his velocity-distribution data both with and without the pitot-tube calibration. Figure 8-5 is a sketch of the type of curve obtained, showing a plot of the point velocity versus the distance from the wall. Curve I is the velocity distribution calculated for the laminar layer. Curve II shows the measured velocity distribution plotted as the actual distance from the center of the pitot tube versus the velocity. Curve III is obtained using the pitot-tube calibration; it is a plot of the effective distance versus the point velocity. Curve III is shifted to the right of curve II and approaches curve I asymptotically. On the basis of this corrected curve Stanton concluded that laminar flow exists in a layer of fluid adjacent to the tube wall.

Other investigators have encountered the same difficulty with regard to the determination of the point velocities in the laminar layer. It has been shown (Fig. 7-9) how Nikuradse [8] shifted his velocity data to conform with the theory that laminar flow occurs near the tube wall. The data of Reichardt [10] shown in Fig. 7-10 indicate that he measured velocities in the region where the flow is laminar.

Velocity-distribution studies using pitot tubes have been made by Rothfus [11] and by Knudsen and Katz.[4] Rothfus calibrated pitot tubes and obtained a calibration similar to Stanton's. Rothfus was unable to come to a definite conclusion regarding the existence of a laminar layer. Knudsen and Katz measured velocity profiles close to the tube walls of an annulus but used an uncalibrated impact tube. Velocity profiles were obtained close to the wall by first determining curve I in Fig. 8-5 for the particular flow conditions under consideration. Then, since curve III in Fig. 8-5 is very similar to curve II, though shifted to the right, Knudsen and Katz plotted their data directly so that it would fit curve I and thereby obtained velocity profiles near the solid wall. Their studies also indicated that laminar flow exists near the tube wall. However, calculations of heat-transfer coefficients from their velocity data showed that the thickness of the laminar layer which they determined was too great.

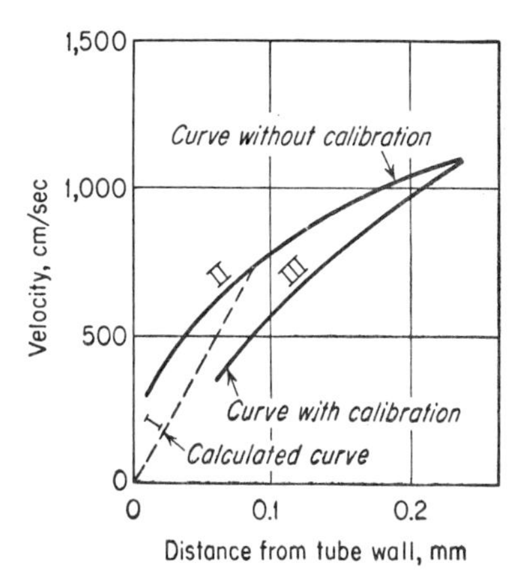

FIG. 8-5. Type of velocity profiles obtained by Stanton et al. [*From T. E. Stanton et al., Proc. Roy. Soc. (London),* **97A**:413 (1920).]

Deissler [1] also obtained velocity data adjacent to the wall of smooth pipe for air flowing at Reynolds numbers between 16,000 and 25,000. His data extend to values of y^+ as low as 2.0 and in the region $2.0 < y^+ < 5.0$ the data agree well with Eqs. (7-37) and (7-40). To determine the position of the pitot tube, the data for a number of velocity traverses were plotted against the scale reading on the pitot-tube holder. All velocity curves were extrapolated to zero velocity, where all curves intersected. This point corresponded to zero distance from the wall.

Using a pitot tube for low Reynolds numbers and a hot-wire anemometer for high Reynolds numbers, Laufer [5,6] obtained velocity distributions in a tube of 9.72 in. ID and in a rectangular duct 60 by 5 in. He measured velocities in the vicinity of 0.0025 in. from the solid boundary and obtained detailed data in the range of $2.0 < y^+ < 5.0$. Again the data agree well with Eqs. (7-37) and (7-40).

In Fig. 8-6 Laufer's velocity-distribution data near the wall for air flowing in a circular tube are shown. The dashed line is calculated from friction-loss measurements using Eq. (8-1) and assuming laminar flow in the sublayer. The measured velocity distribution adjacent to the wall indicates that motion in the sublayer is laminar.

A dye-displacement technique for measuring velocity distribution near solid boundaries was employed by Ferrell, Richardson, and Beatty,[3] who measured velocities within 0.002 in. of the tube wall for laminar flow and obtained excellent agreement with Eq. (4-44). These authors indicate that measurements closer to the wall could be obtained if molecular diffusion

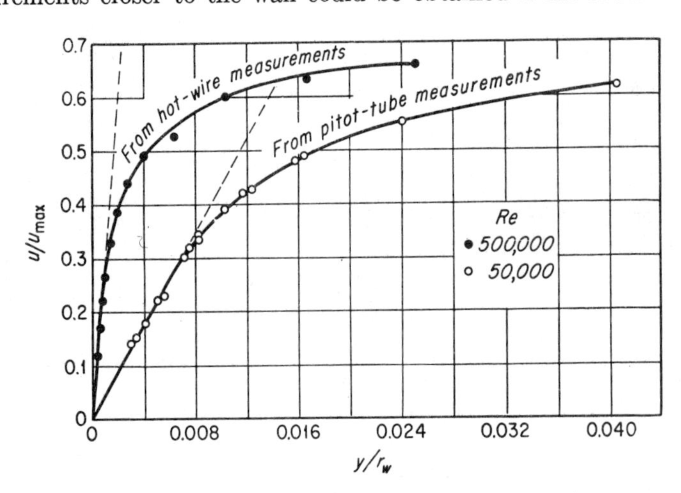

FIG. 8-6. Mean velocity distribution near the wall for air flowing in a 9.72-in.-ID circular tube. (*From J. Laufer, NACA Rept.* 1174, 1954.) (Reynolds number is based on u_{max}.)

were reduced. It is suggested that the dye-displacement technique might possibly be used for measuring velocity distribution during turbulent flow.

8-6. Visual Studies of the Laminar Sublayer

In a series of investigations Fage and Townend [2] observed the motion of small particles in a flowing fluid by means of an electron microscope. They noted that the colloidal particles normally present in tap water were suitable for observation. For turbulent flow in a tube the movement of these particles was observed at a distance of 0.000025 in. from the wall. A sketch given by Fage and Townend of the movement of the particles near the wall is shown in Fig. 8-7. At this distance from the wall it was found that there was no fluctuation of the particles in the radial direction; so the flow was of a laminar type. However, the motion of the particles in the laminar layer was sinuous, and no particle was observed to move in a straight path.

Rothfus and Prengle [13] made a detailed study of the laminar-turbulent

FIG. 8-7. Movement of particles near a solid boundary observed under an electron microscope. [*From A. Fage and H. C. H. Townend, Proc. Roy. Soc. (London),* **135A**:656 (1932).]

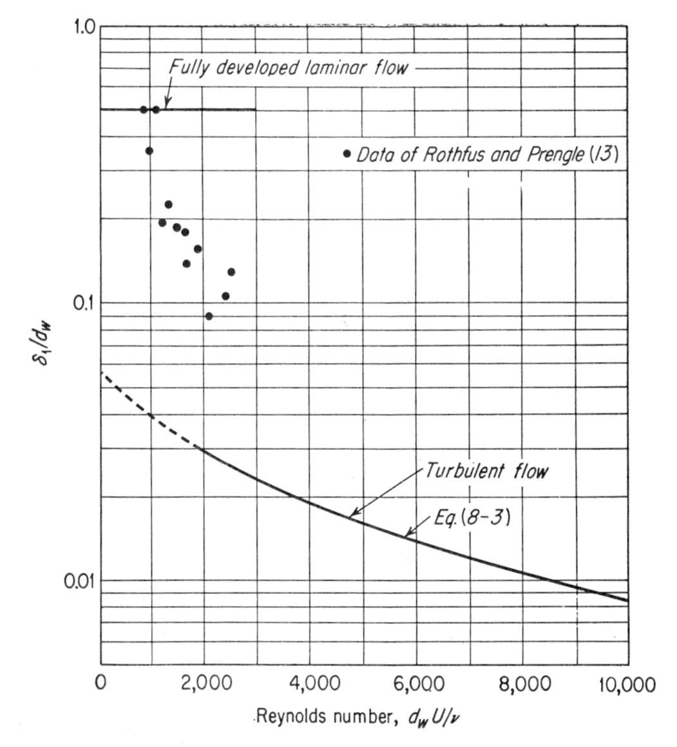

FIG. 8-8. Variation of the laminar-sublayer thickness with the Reynolds number for flow in circular tubes. [*Data from R. R. Rothfus and R. S. Prengle, Ind. Eng. Chem.,* **44**:1683 (1952).]

transition in tubes by studying the behavior of a thin filament of dye and were able to measure the thickness of the laminar sublayer from the behavior of the dye adjacent to the wall. Their results are plotted in Fig. 8-8, where δ_1/d_w is plotted versus the Reynolds number. When δ_1/d_w is 0.5, flow is completely laminar in the tube. These results also indicate that departure from laminar flow may occur at a Reynolds number of about 1,000, which is considerably below the 2,000 where the transition from laminar to turbulent flow is usually assumed to take place.

8-7. The Possibility of the Existence of the Laminar Sublayer

The difficulties encountered in the detection of a laminar sublayer have been mentioned. Early investigations on velocity distribution near solid boundaries and extensive work on heat transfer between solid surfaces and turbulent fluids provided evidence for a laminar layer but did not prove its existence. More recent work of Deissler, Laufer, Rothfus, and Ferrell et al., using fine pitot tubes, hot-wire anemometers, and dye techniques, have provided velocity data nearer the wall than ever before. The results of these workers are additional evidence to support the theory that a laminar sublayer exists adjacent to the wall, provided that all their calibration procedures are correct.

On the theoretical side, the approach made by Deissler [1] appears to be the most realistic. Basing his analysis on the assumption that in the vicinity of the wall

$$\epsilon_M = n^2 u y \tag{7-39}$$

he obtained the equation

$$y^+ = \frac{1}{n} \frac{1/\sqrt{2\pi} \int_0^{nu^+} e^{-[(nu^+)^2/2]} \, d(nu^+)}{(1/\sqrt{2\pi}\,) e^{-[(nu^+)^2/2]}} \tag{7-40}$$

where $n = 0.109$ as determined experimentally. As y^+ approaches zero, it becomes equal to u^+, which indicates that laminar motion exists as the wall is approached.

It is probably best to think of the laminar sublayer from the statistical standpoint. All investigations of velocity distribution near solid boundaries have involved the measurement of temporal-mean velocities. The effect of turbulent velocity fluctuations has not been determined. It is likely that the laminar sublayer is a layer having the most probable thickness of y^+ in the vicinity of 5.0, and it is reasonable to expect that the thickness would, in some instances, be nearly zero ($y^+ < 1.0$) or relatively large (y^+ up to 10.0). The instantaneous thickness of the laminar sublayer is probably a function of time in the same way that turbulent velocity fluctuations are a function of time. From this point of view, the laminar sublayer is not an entity apart from the turbulent core but exists in conjunction with it, and a continuous transition from one to the other occurs. The location of this transition is a fluctuating quantity, which varies with time.

BIBLIOGRAPHY

1. Deissler, R. G.: *NACA TN* 2138, 1950.
2. Fage, A., and H. C. H. Townend: *Proc. Roy. Soc. (London)*, **135A**:656 (1932).
3. Ferrell, J. K., F. M. Richardson, and K. O. Beatty: *Ind. Eng. Chem.*, **47**:29 (1955).
4. Knudsen, J. G., and D. L. Katz: *Proc. Midwest. Conf. Fluid Dynamics, 1st Conf., Univ. Illinois, 1950*, p. 175.
5. Laufer, J.: *NACA Rept.* 1053, 1951.
6. Laufer, J.: *NACA Rept.* 1174, 1954.
7. Miller, B.: *Trans. ASME*, **71**:357, 1949.
8. Nikuradse, J.: *VDI-Forschungsheft* 356, 1932.
9. Prandtl, L.: *Z. angew. Math. u. Mech.*, **1**:431 (1921).
10. Reichardt, H.: *NACA TM* 1047, 1943.
11. Rothfus, R. R.: Velocity Gradients and Friction in Concentric Annuli, Ph.D. Thesis, Carnegie Institute of Technology, 1948.
12. Rothfus, R. R., C. C. Monrad, and V. E. Senecal: *Ind. Eng. Chem.*, **42**:2511 (1950).
13. Rothfus, R. R., and R. S. Prengle: *Ind. Eng. Chem.*, **44**:1683 (1952).
14. Stanton, T. E., D. Marshall, and C. N. Bryant: *Proc. Roy. Soc. (London)*, **97A**:413 (1920).

CHAPTER 9

FLOW IN THE ENTRANCE SECTION OF CLOSED CONDUITS

9-1. Introduction

In previous chapters considerations of flow of Newtonian fluids in tubes, annuli, and other conduits have assumed that the flow pattern depended only on the fluid properties and the boundary of the system and not on the prior history of the fluid. In laboratory apparatus and in much commercial equipment the dynamics of flow in the entrance section depends on disturbances created before or at the entrance. Such disturbances in the flow are dissipated as the fluid flows through the equipment. The reliability of standard coefficients for orifice meters is known to be dependent on the absence of upstream disturbances, and recommended installations call for a straight tube length at least 50 diameters ahead of the orifice. Straightening vanes, in the form of bundles of small tubes, are often inserted in the conduit in order to dampen the disturbances which may be present.

When a fluid enters a conduit, the boundary layer begins forming at the entrance. The fully developed velocity profile exists after the edge of the boundary layer coincides with the axis of the duct. The fluid-dynamical conditions at the entrance of the conduit greatly influence the length required for the fully developed velocity profile to form. The entrance to a conduit involves either a sudden expansion or contraction in the cross-sectional area of flow, and for this reason the configuration of the entrance is an important consideration in studying the flow downstream. When the rate is such that the fully developed flow will be laminar, the configuration of the conduit entrance has slight effect on the subsequent flow. For this case, flow will usually be laminar in the boundary layer even though the entering fluid may be turbulent. When the fully developed flow is turbulent in the conduit, the configuration of the entrance is of prime importance in determining the dynamics of flow downstream. If the entrance is abrupt, the boundary layer will probably be turbulent from the beginning.

If the entrance is rounded, the boundary layer may be laminar immediately beyond the entrance and then become turbulent some distance downstream. Figure 9-1 illustrates schematically the formation of the boundary layer at the entrance to a circular tube. In Fig. 9-1*a* flow is laminar in the boundary layer and in the tube. In Fig. 9-1*b* flow is turbulent in the tube

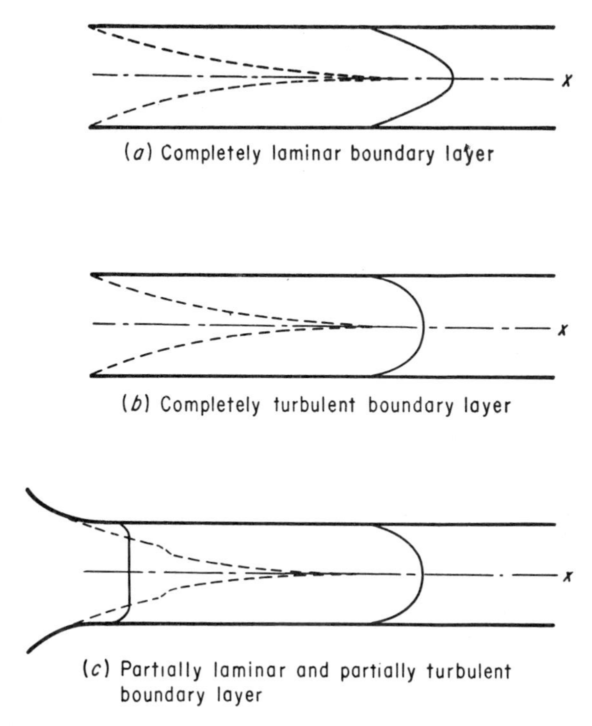

Fig. 9-1. Formation of boundary layers at the entrance to smooth circular tubes.

and in the whole boundary layer. In Fig. 9-1*c*, which illustrates a rounded entrance, flow is turbulent in the tube but is first laminar and then turbulent in the boundary layer.

In considering flow in the entrance section of conduits, two entrance lengths are involved, the length required for the velocity profile to develop and the length required for the shear stress at the wall to reach its fully developed value. As the fluid enters the conduit, the velocity gradient at the wall is theoretically infinite. In a relatively short distance the velocity profile adjacent to the wall will have developed to a steady value. Formation of the fully developed velocity profile, however, requires a considerable length.

9-2. Velocity Distribution for Laminar Flow at the Entrance of a Pipe

It is commonly assumed that the velocity profiles in the inlet section following a sharp-edged entrance have the form shown in Fig. 4-2. The profile is flat at the entrance. The boundary layer increases in thickness, but the central core beyond the boundary layer maintains its flat profile and is accelerated in order to satisfy the continuity equation. The various theoretical analyses of laminar flow in entrance sections have been based on the assumption that the fully developed velocity profile forms in this manner.

Schiller [10] predicted theoretically that

$$\frac{L_e}{d_w} = 0.02875\text{Re} \tag{9-1}$$

where L_e is the axial distance from the entrance required for a fully developed parabolic velocity profile [as given by Eq. (4-44)] to form. Prandtl and Tietjens [8] indicate that the velocity profile is developed in a length given by

$$\frac{L_e}{d_w} = 0.05\text{Re} \tag{9-2}$$

Langhaar's [6] analysis of laminar flow at the entrance of a circular tube produced useful relationships. He solved the momentum equations, linearizing them by assuming that the acceleration of the fluid along the conduit was a function of x alone (where the conduit axis coincides with the x axis). Langhaar shows that

$$\frac{L_e}{d_w} = 0.0575\text{Re} \tag{9-3}$$

where L_e is now the length required for the center-line velocity to reach 99 per cent of its fully developed value. This result is in fair agreement with Eq. (9-2).

The following relation given by Langhaar permits the calculation of the complete laminar velocity profile in the entrance section of a tube:

$$\frac{u}{U} = \frac{I_0[\phi(x)] - I_0[\phi(x)r/r_w]}{I_2[\phi(x)]} \tag{9-4}$$

where I_0 and I_2 are modified Bessel functions of the first kind and are functions of x. Values of $\phi(x)$ are given in Table 9-1 as a function of $(x/d_w)/\text{Re}$, and these are also plotted in Fig. 9-2.

Example 9-1

A fluid is flowing in a 2.1-in.-ID tube at a Reynolds number of 1,000. Using Eq. (9-4), calculate the complete velocity profile a distance 7½ in. from the entrance of the tube. Plot the velocity profile as r/r_w versus u/U, and compare with the fully developed laminar velocity profile as calculated by Eq. (4-44).

Solution

$$x = 7.5 \text{ in.}$$

$$d_w = 2.1 \text{ in.}$$

$$\text{Re} = 1{,}000$$

Thus

$$\frac{x/d_w}{\text{Re}} = \frac{7.5}{(2.1)(1{,}000)} = 3.58 \times 10^{-3}$$

From Fig. 9-2

$$\phi(x) = 6.0$$

TABLE 9-1. VALUES OF $\phi(x)$ TO USE IN EQ. (9-4) †

$\phi(x)$	$\frac{x/d_w}{\text{Re}}$
20.0	0.000205
11.0	0.00083
8.0	0.001805
6.0	0.003575
5.0	0.00535
4.0	0.00838
3.0	0.01373
2.5	0.01788
2.0	0.02368
1.4	0.0341
1.0	0.04488
0.6	0.06198
0.4	0.0760

† H. L. Langhaar, *Trans. ASME*, **64**:A-55 (1942).

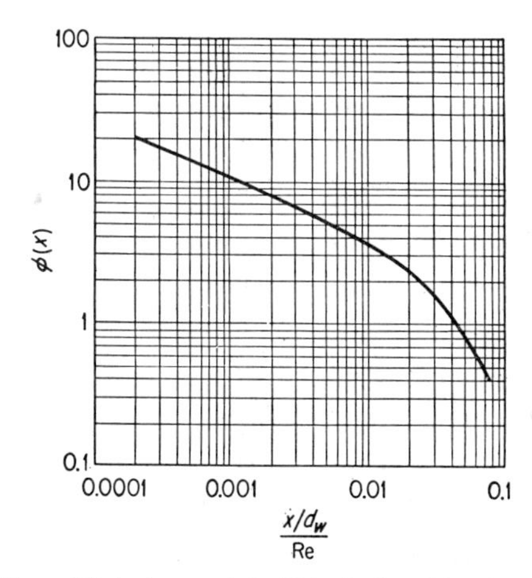

FIG. 9-2. The function $\phi(x)$ to be used in Eq. (9-4). [*Results from H. L. Langhaar, Trans. ASME*, **64**:A-55 (1942).]

From any standard table of Bessel functions

$$I_0(6.0) = 67.234$$

$$I_2(6.0) = 46.787$$

At various values of r/r_w, $I_0(6.0r/r_w)$ is determined, and u/U is calculated from Eq. (9-4).

At $r/r_w = 0.95$

$$(6.0)(0.95) = 5.70$$

$$I_0(5.70) = 51.17$$

From Eq. (9-4)

$$\frac{u}{U} = \frac{67.23 - 51.17}{46.787} = 0.344$$

The results are tabulated in Table 9-2 and plotted in Fig. 9-3.

TABLE 9-2

$\frac{r}{r_w}$	$I_0\left(\frac{6.0r}{r_w}\right)$	$\frac{u}{U}$	
		Eq. (9-4)	Fully developed flow Eq. (4-44)
1.0	67.23	0	0
0.95	51.17	0.344	0.195
0.90	39.01	0.604	0.38
0.85	29.79	0.779	0.555
0.80	22.79	0.949	0.72
0.75	17.48	1.065	0.875
0.70	13.44	1.150	1.02
0.65	10.70	1.210	1.155
0.60	8.03	1.268	1.28
0.55	6.24	1.304	1.345
0.50	4.88	1.337	1.50
0.40	3.05	1.372	1.68
0.30	1.99	1.397	1.82
0.20	1.39	1.408	1.92
0.10	1.09	1.413	1.98
0	1.0	1.415	2.0

Figure 9-4 shows the development of the velocity at the axis of the tube as determined theoretically by Langhaar. Here u_{max}/U is plotted versus $(x/d_w)/\text{Re}$.

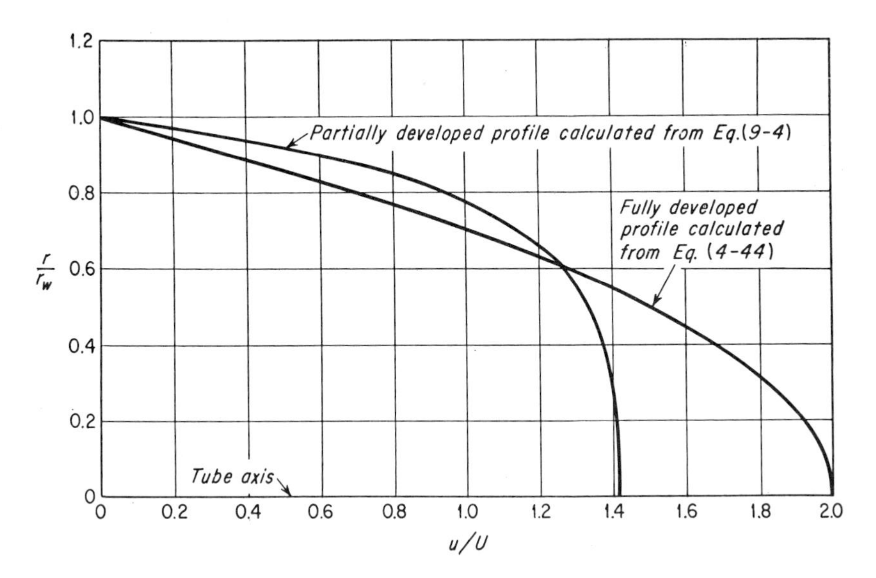

FIG. 9-3. Partially developed and fully developed velocity profiles in circular tubes (Re = 1,000; partially developed profile is 3.57 diameters from the tube entrance).

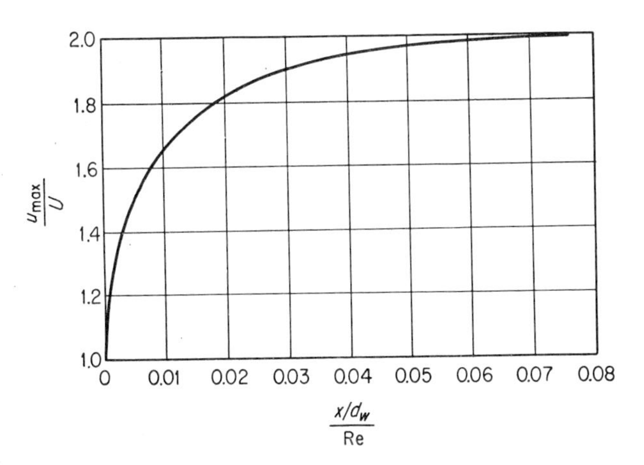

FIG. 9-4. Development of axial velocity for laminar flow in the entrance of a circular tube. [*Results from H. L. Langhaar, Trans. ASME*, **64**:A-55 (1942).]

9-3. Friction Factors for Laminar Flow in the Inlet Section of a Circular Tube

The friction factor for laminar flow in the entrance section of circular conduits has been determined by Langhaar,[6] whose theoretical results are shown in Table 9-3 and by curve I of Fig. 9-5, in which $f_{av}(x/d_w)$ is plotted

TABLE 9-3. ANALYTICAL RESULTS OF LANGHAAR †

$\dfrac{x/d_w}{\text{Re}}$	$f_{av}\left(\dfrac{x}{d_w}\right)$	$\dfrac{2(-\Delta P)g_c}{\rho U^2}$ ‡
0.000205	0.0530	1.2122
0.00083	0.0965	1.3866
0.001805	0.1413	1.5732
0.003575	0.2075	1.8300
0.00535	0.2605	2.0442
0.00838	0.340	2.3614
0.01373	0.461	2.8474
0.01788	0.547	3.1898
0.02368	0.659	3.6354
0.0341	0.845	4.3840
0.04488	1.028	5.1146
0.06198	1.308	6.2326
0.0760	1.538	7.1478

† H. L. Langhaar, *Trans. ASME*, **64**:A-55 (1942).
‡ When flow is not horizontal, replace $(-\Delta P)/\rho$ by $(-\Delta P)/\rho - \Delta Z(g/g_c)$.

versus $(x/d_w)/\text{Re}$. The friction factor f_{av} is the average friction factor over the length of conduit from $x = 0$ to $x = x$, thus:

$$f_{av}\left(\frac{x}{d_w}\right) = \frac{-\Delta P_f\, g_c}{2\rho U^2} \tag{9-5}$$

where $-\Delta P_f$ is the pressure loss due to friction over the length from 0 to x.

Curve IV in Fig. 9-5 shows the variation of $(-\Delta P\, g_c)/(\rho U^2/2)$ † with $(x/d_w)/\text{Re}$ as determined theoretically by Langhaar, where ΔP is the difference between the pressure at point x in the conduit and the pressure in the reservoir where the velocity is zero. Beyond the transition region, i.e., for $(x/d_w)/\text{Re} > 0.1$, Langhaar showed that

$$\frac{-\Delta P\, g_c}{\rho U^2/2} = \frac{64(x/d_w)}{\text{Re}} + 2.28 \tag{9-6}$$ †

† See second footnote in Table 9-3.

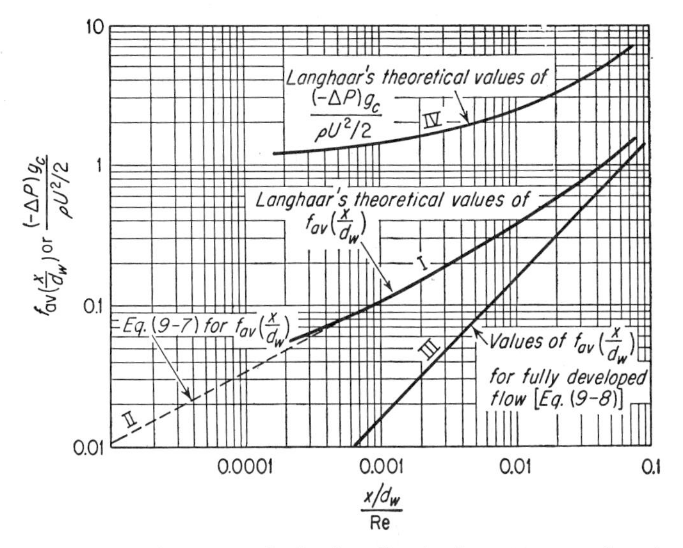

FIG. 9-5. Average friction factors for laminar flow in the entrance region of a circular tube.

where ΔP is now the difference between the pressure at the end of the capillary and the pressure in the reservoir of fluid before the capillary. Equation (9-6) is the relationship used to calculate viscosity from measurements made on a capillary viscometer. When $(x/d_w)/\text{Re}$ is less than 0.01, an exit loss correction must be made.†

Example 9-2

One means of determining the radius of the capillary tube in a capillary viscometer is to determine the time of efflux for a certain quantity of a liquid of known density and viscosity.

In one calibration test the amount of liquid discharged from the capillary was 9 in.³/min. The capillary was located at the bottom of a vertical cylindrical chamber. During the test the average height of liquid above the exit of the capillary was

† An energy balance between the reservoir and the exit of the capillary gives

$$\frac{\Delta P}{\rho} + \frac{\Delta Z g}{g_c} = -\frac{2f_{av}U^2x}{g_c d_w} - \frac{U^2}{g_c}$$

where U^2/g_c is the exit loss for laminar flow (recommended by J. H. Perry, "Chemical Engineers Handbook," p. 388, McGraw-Hill Book Company, Inc., New York, 1950). This equation reduces to

$$\frac{1}{U^2/2g_c}\left[-\frac{\Delta P}{\rho} - \frac{\Delta Z g}{g_c}\right] = \frac{4f_{av}x}{d_w} + 2$$

where ΔP and ΔZ are differences in pressure and elevation between a point beyond the exit of the capillary and a point in the reservoir.

1 ft, and the pressure above the liquid in the chamber was 150 psia. The capillary was 0.5 in. long. The liquid had a viscosity of 56 centipoises and a density of 59.1 lb_m/ft^3. Calculate the diameter of the capillary tube.

Solution

It will be assumed that Eq. (9-6) is applicable (this assumption must be checked). Letting d_w = capillary diameter,

$$U = \frac{(9)(4)}{(60)(1{,}728)(\pi)(d_w^2)} = \frac{1.11 \times 10^{-4}}{d_w^2} \text{ ft/sec}$$

$$-\Delta P = 150 \text{ psi}$$

$$-\Delta Z = 1 \text{ ft}$$

Thus

$$-\frac{\Delta P}{\rho} - \frac{\Delta Z g}{g_c} = \frac{(+150)(144)}{59.1} + 1 = 367$$

$$\frac{2g_c}{U^2}\left(-\frac{\Delta P}{\rho} - \frac{\Delta Z g}{g_c}\right) = \frac{(2)(32.2)(367)(d_w)^4}{(1.11 \times 10^{-4})^2}$$

$$= 1.918 \times 10^{12} d_w^4$$

$$\frac{x/d_w}{\text{Re}} = \frac{(0.5)(d_w)^2(56)(6.72 \times 10^{-4})}{(12)(d_w)(d_w)(1.11 \times 10^{-4})(59.1)} = 0.238$$

The above result justifies the assumption made at the beginning of the solution. Substituting in Eq. (9-6),

$$(1.918 \times 10^{12})(d_w)^4 = (64)(0.238) + 2.28$$

$$= 17.51$$

Thus

$$d_w^4 = 9.13 \times 10^{-12}$$

$$d_w = 1.74 \times 10^{-3} \text{ ft}$$

$$= 0.0209 \text{ in.}$$

The capillary diameter is 0.0209 in.

Curve II of Fig. 9-5 represents the experimental results of Kline and Shapiro,[4] which are expressed by the relation

$$f_{\text{av}}\left(\frac{x}{d_w}\right) = 3.435\sqrt{\frac{x/d_w}{\text{Re}}} \qquad 10^{-5} < \frac{x/d_w}{\text{Re}} < 10^{-3} \tag{9-7}$$

For comparison, curve III in Fig. 9-5 is included to show how the friction factor is affected by the entrance. Curve III shows what the value of the friction factor would be if fully developed laminar flow existed at the entrance. It is based on Eq. (4-52) and represents the relation

$$f_{\text{av}}\left(\frac{x}{d_w}\right) = \frac{16(x/d_w)}{\text{Re}} \tag{9-8}$$

The comparison shows that the friction factor in the entrance section is considerably higher than that for fully developed flow. This is due to two factors, namely, the very high velocity gradients which exist at the tube entrance and the fact that the core of fluid beyond the boundary layer is being accelerated in accordance with the continuity equation.

9-4. Turbulent Velocity Profiles in the Entrance Section of Circular Tubes

When a fluid enters a tube at such a rate that the fully developed flow will be turbulent in the tube, the entrance configuration and turbulence of

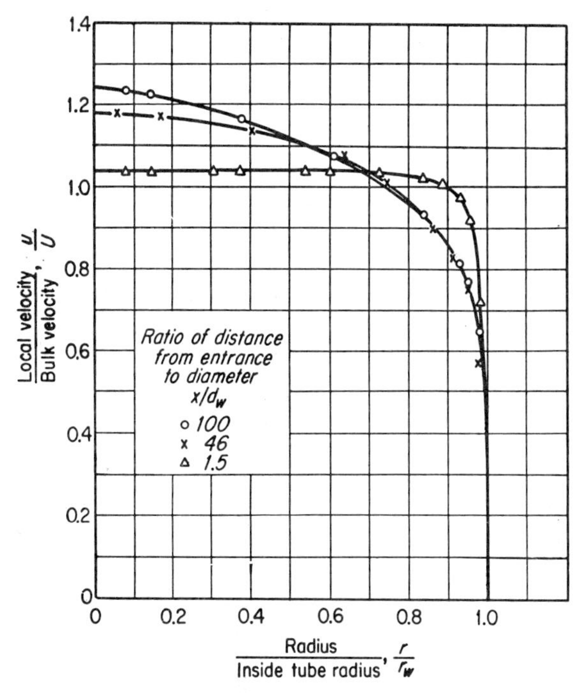

FIG. 9-6. Turbulent velocity profiles for flow through a tube with rounded entrance at various distances from the entrance. Reynolds number = 46,000. (*From R. G. Deissler, NACA TN* 2138, 1950.)

the entering stream are factors which determine the magnitude of the entrance length. No general relationship is available to predict the distance required for a fully developed turbulent velocity profile. In general, a greater distance is required for a rounded entrance than for a sharp-edged entrance,[3] since in the former the initial portion of the boundary layer is laminar and a laminar boundary layer increases in thickness more slowly than a turbulent one.

For turbulent flow at Reynolds numbers greater than 10,000, Schiller and Kirsten [11] observed that entrance lengths greater than 50 tube diameters were generally necessary for the formation of a fully developed turbulent velocity profile. These workers used a rounded entrance, as did Deissler,[1] who obtained the velocity profiles reproduced in Fig. 9-6. Three profiles are shown for the flow of air in a 0.87-in.-ID tube at a Reynolds number of 46,000. The velocity profiles were measured at 1.5, 46, and 0 diameters downstream from the entrance. The results indicate that a distance of more than 50 diameters is needed to form the fully developed profile.

9-5. Turbulent Friction Factors for Flow in the Entrance Section of Circular Tubes

For turbulent flow the distance required for the local friction factor to equal the fully developed friction factor is considerably less than that re-

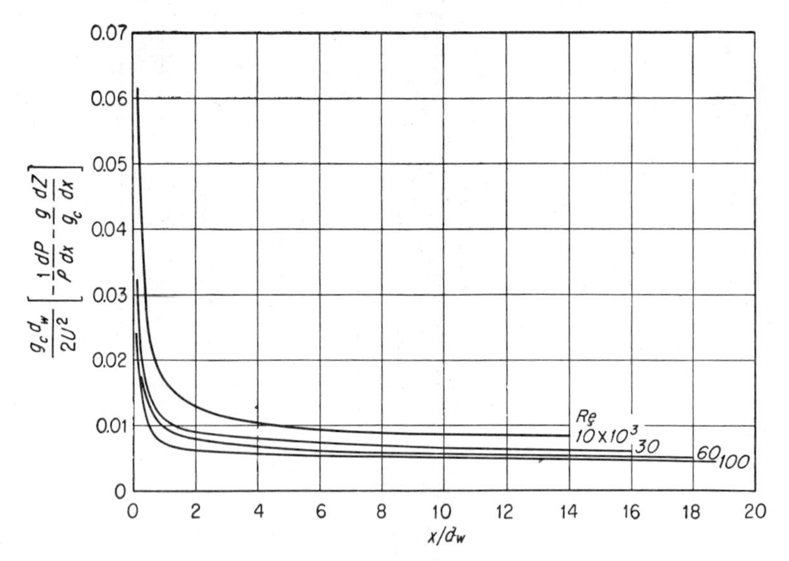

FIG. 9-7. Predicted local static-pressure gradient in the entrance of a circular tube. (*From R. G. Deissler, NACA TN* 3016, 1953.)

quired for the development of the velocity profile. Figure 9-7 shows Deissler's [2] predicted static-pressure gradient

$$\frac{g_c d_w}{2U^2}\left(-\frac{1}{\rho}\frac{dP}{dx}-\frac{g}{g_c}\frac{dZ}{dx}\right)$$

for a tube as a function of the distance from the entrance. The curves were derived assuming a completely turbulent boundary layer beginning at the tube entrance and show that at a distance of about 10 diameters from the entrance the static-pressure gradient becomes constant. This constant value is equal to the fully developed friction factor in the tube [given by Eq. (7-60)]. Figure 9-8 shows Deissler's predicted values of the local wall shear stress. It is seen that the term $2\tau_w g_c/\rho U^2$, which is the local friction

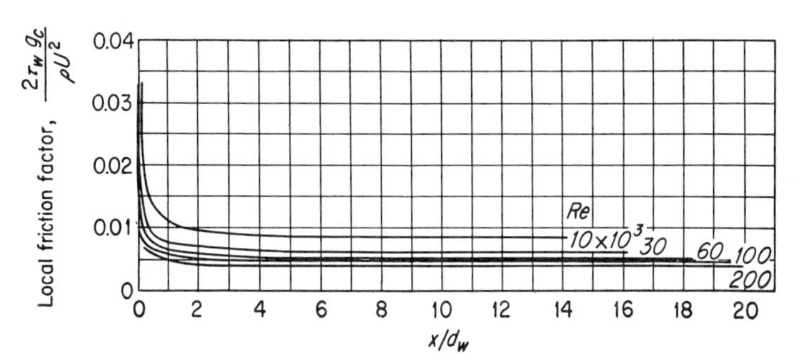

FIG. 9-8. Predicted wall shear stress in the entrance section of a circular tube. (*From R. G. Deissler, NACA TN* 3016, 1953.)

factor, becomes constant in a distance of about 6 diameters, indicating that the velocity profile adjacent to the wall becomes established in a very short distance. At positions close to the entrance the static-pressure gradient, since it includes the kinetic-energy effect due to the formation of the

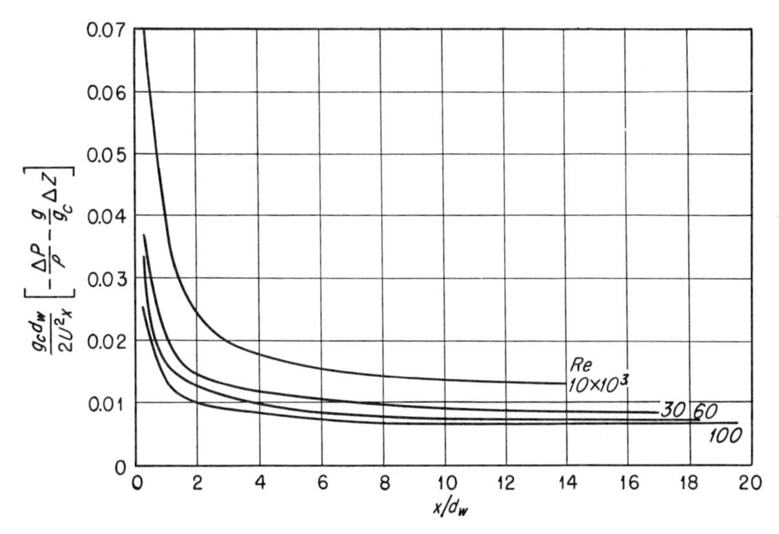

FIG. 9-9. Predicted static-pressure drop in the entrance section of a circular tube (*From R. G. Deissler, NACA TN* 3016, 1953.)

velocity profile, is higher than $2\tau_w g_c/\rho U^2$. Figure 9-9 shows a plot of $\frac{g_c d_w}{2U^2 x}\left(-\frac{\Delta P}{\rho} - \frac{\Delta Z g}{g_c}\right)$ as a function of x/d_w, where $-\Delta P$ is the pressure drop and $-\Delta Z$ the drop in elevation between $x = 0$ and $x = x$.

Latzko[3,7] predicted the entrance length required for the friction factor to become constant as follows:

$$\frac{L_e}{d_w} = 0.623(\text{Re})^{1/4} \tag{9-9}$$

which appears to agree with Deissler's theoretical results.

Example 9-3

Compare the static-pressure drop over the first foot of a pipe with that over 1 ft of pipe 100 diameters from the entrance.

Flow conditions:

Re = 100,000

Tube diameter: 3-in.-ID smooth tubing

Fluid: water at 60°F

Solution

Let $(-\Delta P)_\infty$ be the pressure drop per foot at a point 100 diameters from the entrance. At Re = 100,000

$$f = 0.00448$$

From Eq. (4-21)

$$\frac{(\Delta P)_\infty}{\rho} = \frac{-2fU^2}{g_c d_w}$$

from which

$$\frac{(-\Delta P)_\infty g_c d_w}{2\rho U^2} = 0.00448$$

Let $(-\Delta P)_1$ be the pressure drop for the first foot of pipe.

$$\frac{x}{d_w} = \frac{1}{0.25} = 4$$

From Fig. 9-9

$$\frac{g_c d_w}{2U^2 x}\left[\frac{(-\Delta P)_1}{\rho} - \frac{\Delta Z g}{g_c}\right] = 0.008$$

For $x = 1$ and $Z = 0$

$$\frac{(-\Delta P)_1 g_c d_w}{2\rho U^2} = 0.008$$

and

$$\frac{(-\Delta P)_1}{(-\Delta P)_\infty} = \frac{0.008}{0.00448} = 1.79$$

The pressure drop over the first foot of pipe is 1.8 times the pressure drop over 1 ft of pipe after the velocity profile is fully developed.

9-6. Effect of Entrance Geometry on Turbulent Velocity Profiles in Circular Tubes

The influence of entrance configuration on turbulent velocity profiles was observed by Knudsen and Katz.[5] The entrance configuration was varied, and the effect on the velocity profile was determined. The velocity profile for water flowing in a circular tube was determined at 44 diameters from the entrance, and the point velocity, instead of being a maximum at the axis of the tube, was found to be a maximum some distance from the axis. The types of velocity profiles obtained are shown in Fig. 9-10, along with a cross section of the tube and entrance. The fluid entered the tube through

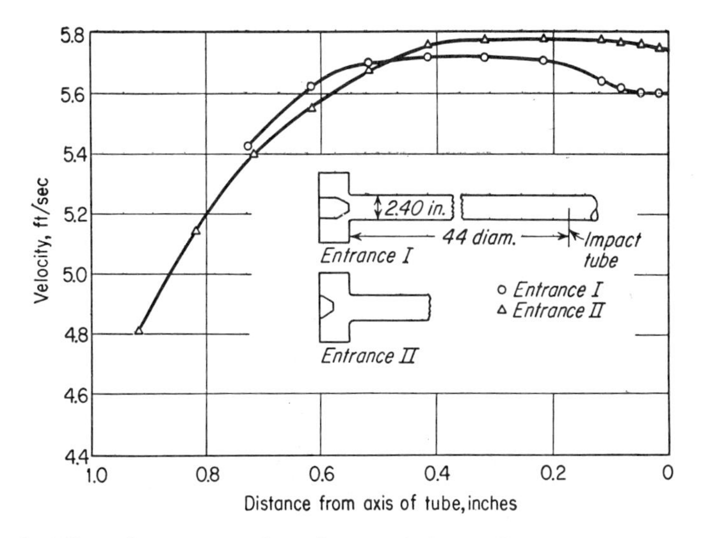

FIG. 9-10. Effect of entrance configuration on velocity profiles in circular tubes. (*From J. G. Knudsen and D. L. Katz, unpublished work.*)

an enlarged section, the purpose of which was to distribute the fluid evenly. However, a tapered plug in the enlarged section protruded nearly to the tube entrance, so that, as the fluid entered the tube, it flowed first over the plug. The effect was to reduce the velocity of the fluid at the tube axis, and this effect was still evident 44 diameters downstream. The open circles in Fig. 9-10 represent data obtained when the end of the plug was even with the tube entrance, and the triangles represent data obtained when the end of the plug protruded only slightly into the enlarged section. There is a difference in the two curves, indicating that the velocity profile was affected by the entrance disturbances.

9-7. Turbulent Friction Factors in the Inlet Section of Concentric Annuli

The effect of the entrance conditions on the shear stress on the outer wall of a concentric annulus is depicted very well in Fig. 9-11, which shows the results of Rothfus, Monrad, Sikchi, and Heideger.[9] The ratio

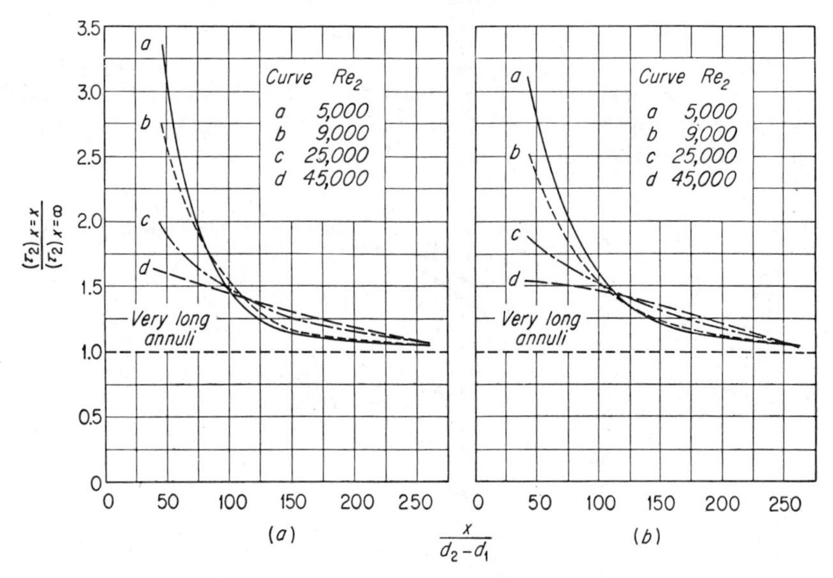

Fig. 9-11. The variation of shear stress on the outer wall of a plain annulus as a function of distance from the entrance. (a) High turbulence; (b) low turbulence. [*From R. R. Rothfus, C. C. Monrad, K. G. Sikchi, and W. J. Heideger, Ind. Eng. Chem.*, **47**:913 (1955).]

$(\tau_2)_{x=x}/(\tau_2)_{x=\infty}$ is plotted versus $x/(d_2 - d_1)$. Figure 9-11*a* shows results when the entering fluid has high turbulence, and Fig. 9-11*b* is for low turbulence in the entering stream. It is noted that the ratio $(\tau_2)_{x=x}/(\tau_2)_{x=\infty}$ does not approach unity until values of $x/(d_2 - d_1)$ of 200 and above are reached.

9-8. Laminar Velocity Profiles in the Entrance Section of Parallel Planes

Sparrow [13] studied laminar flow in the inlet section of two parallel planes. He assumed that the velocity in the boundary layer was expressed by the relation

$$\frac{u}{u_\delta} = 2\frac{y}{\delta} - \left(\frac{y}{\delta}\right)^2 \tag{9-10}$$

where u_δ = velocity at edge of boundary layer
δ = thickness of boundary layer
y = distance from wall where velocity is u (see Fig. 9-12)

The following approximate relationships were obtained:

$$\frac{dx}{du_\delta} = \frac{3b^2}{40\nu} \frac{(u_\delta - U)(9u_\delta - 7U)}{u_\delta{}^2} \tag{9-11}$$

and

$$\frac{\delta}{b} = 1.5\left(1 - \frac{U}{u_\delta}\right) \tag{9-12}$$

where b is the distance between the parallel planes. Equation (9-11) relates the velocity u_δ at the edge of the boundary layer to the distance x from the entrance. This is a first-order differential equation which must be integrated. Equation (9-12) gives the thickness of the boundary layer

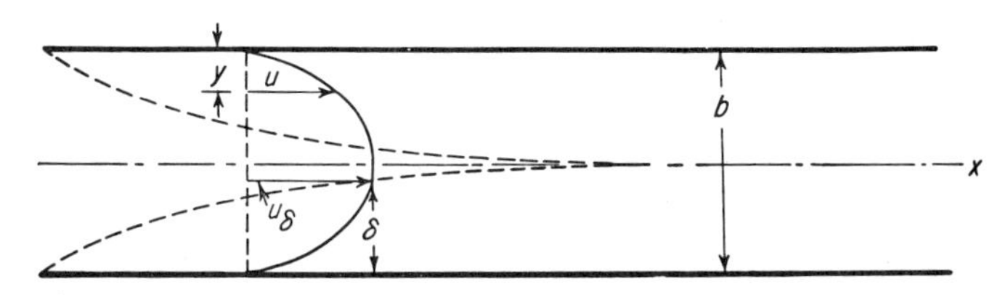

FIG. 9-12. Development of velocity profile for laminar flow between parallel planes.

once u_δ is known from Eq. (9-11). When both δ and u_δ are known, the velocity profile in the boundary layer may be obtained from Eq. (9-10). Rearranging Eq. (9-11) and integrating,

$$\frac{x/2b}{\text{Re}} = \frac{3}{160}\int_1^{u_\delta/U} \frac{(u_\delta/U - 1)(9u_\delta/U - 7)}{(u_\delta/U)^2}\, d\frac{u_\delta}{U} \tag{9-13}$$

where $\text{Re} = 2bU/\nu$. Equation (9-13) has been integrated graphically, and the results are shown in Fig. 9-13, where $(x/2b)/\text{Re}$ is plotted against u_δ/U from $u_\delta/U = 1$ up to a value of $u_\delta/U = 1.5$. At this value fully developed laminar flow is established. At $u_\delta/U = 1.5$, $(x/2b)/\text{Re}$ has a value of 0.00648, thus giving

$$\frac{L_e}{2b} = 0.00648\text{Re} \tag{9-14}$$

where L_e is the length required for a fully developed laminar velocity profile to form between two parallel planes.

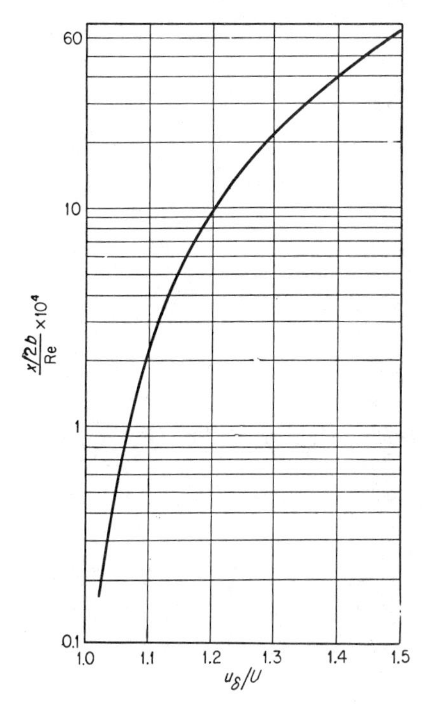

FIG. 9-13. Variation of velocity at edge of boundary layer as a function of distance from the entrance to two parallel planes [obtained from the integration of Eq. (9-13)].

Equations (9-11) and (9-12) give velocity profiles in the entrance section between parallel planes which agree fairly well with the experimental results of Schlichting[12] except at relatively large values of $(x/2b)/\text{Re}$. Schlichting's result gives an entrance length almost twice that predicted by Eq. (9-14).

9-9. Turbulent Friction Factors and Velocity Profiles in the Entrance Section of Two Parallel Planes

It is probable that the distance required for the formation of a fully developed velocity profile between parallel planes is similar to that for circular tubes, i.e., about 50 equivalent diameters from the entrance.

The wall shear stress and pressure drop in the entrance section of parallel planes have been predicted by Deissler[2] and are shown in Figs. 9-14 and 9-15.

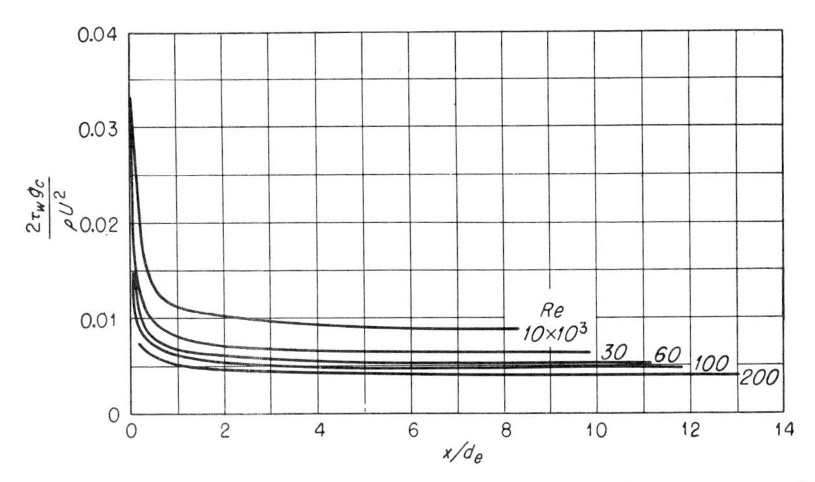

FIG. 9-14. Predicted wall shear stress in the entrance section between two parallel planes. (*From R. G. Deissler, NACA TN* 3016, 1953.)

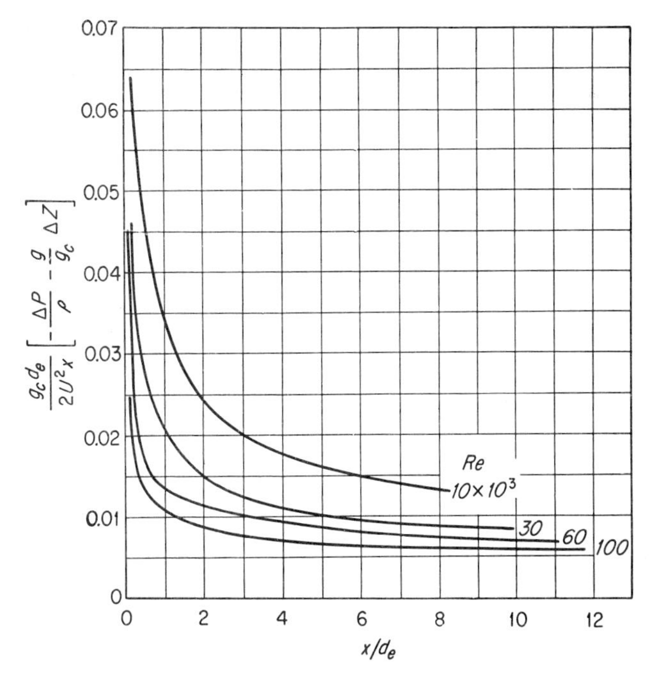

FIG. 9-15. Predicted static-pressure drop in the entrance section between two parallel planes. (*From R. G. Deissler, NACA TN* 3016, 1953.)

Example 9-4

Air is flowing through a bank of flat parallel plates at a Reynolds number of 1,000. The plates have negligible thickness and are spaced 0.125 in. apart. They act as cooling fins on a hot-water radiator. Determine the length of flow required for the formation of a fully developed velocity profile. Calculate the velocity profile at $x = L_e/3$, where L_e is the entrance length.

Solution

$$b = 0.125 \text{ in.}$$

From Eq. (9-14)

$$\frac{L_e}{(2)(0.125)} = (0.00648)(1{,}000)$$

from which

$$L_e = 1.62 \text{ in.}$$

At $x = 1.62/3 = 0.54$ in.

$$\frac{x/2b}{\text{Re}} = \frac{0.54/0.250}{1{,}000} = 0.00216$$

From Fig. 9-13.

$$\frac{u_\delta}{U} = 1.29$$

From Eq. (9-12)

$$\frac{\delta}{b} = 1.5\left(1 - \frac{1}{1.29}\right) = 0.337$$

Giving

$$\delta = (0.337)(0.125) = 0.0422 \text{ in.}$$

The point velocity is calculated from Eq. (9-10) as u/u_δ for various values of y/δ. Results are tabulated in Table 9-4 and plotted in Fig. 9-16.

TABLE 9-4

$\frac{y}{b/2}$	$\frac{y}{\delta}$	y, in.	$\frac{u}{u_\delta}$	$\frac{u}{U}$
0	0	0	0	0
0.135	0.2	0.00845	0.36	0.464
0.271	0.4	0.0169	0.64	0.825
0.407	0.6	0.0254	0.84	1.082
0.541	0.8	0.0338	0.96	1.239
0.675	1.0 (edge of boundary layer)	0.0422	1.00	1.29

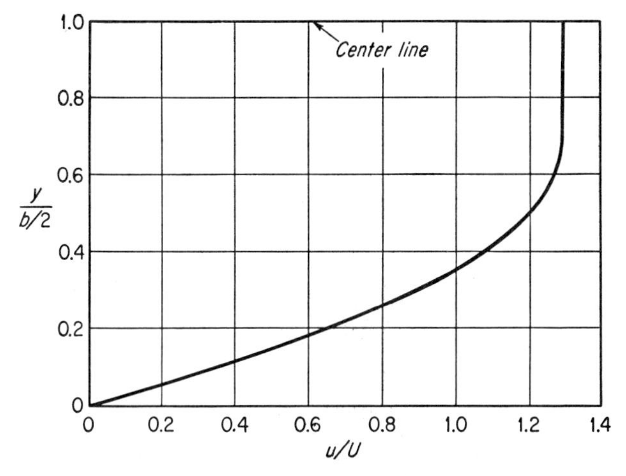

FIG. 9-16. Partially developed laminar velocity profile between parallel planes calculated from Example 9-4 (Re = 1,000; profile calculated for a value of $x/2b = 2.16$).

BIBLIOGRAPHY

1. Deissler, R. G.: *NACA TN* 2138, 1950.
2. Deissler, R. G.: *NACA TN* 3016, 1953.
3. Deissler, R. G.: *Trans. ASME*, **77**:1221 (1955).
4. Kline, S. J., and A. H. Shapiro: *NACA TN* 3048, 1953.
5. Knudsen, J. G., and D. L. Katz: unpublished work.
6. Langhaar, H. L.: *Trans. ASME*, **64**:A-55 (1942).
7. Latzko, H.: *Z. angew. Math. u. Mech.*, **1**:268 (1921).
8. Prandtl, L., and O. Tietjens: "Hydro- und Aeromechanik," vol. 2, p. 28, Springer-Verlag OHG, Berlin, 1931.
9. Rothfus, R. R., C. C. Monrad, K. G. Sikchi, and W. J. Heideger: *Ind. Eng. Chem.*, **47**:913 (1955).
10. Schiller, L.: *Z. angew. Math. u. Mech.*, **2**:96 (1922).
11. Schiller, L., and H. Kirsten: *Z. tech. Phys.*, **10**:268 (1929).
12. Schlichting, H.: *Z. angew. Math. u. Mech.*, **14**:368 (1934).
13. Sparrow, E. M.: *NACA TN* 3331, 1955.

CHAPTER 10

FLOW OF INCOMPRESSIBLE FLUIDS PAST IMMERSED BODIES

10-1. The Boundary Layer on Immersed Bodies

Flow past immersed bodies differs from flow in closed conduits in that the fluid is considered to have no outside boundaries. Prior to encountering an immersed body the undisturbed fluid has uniform mean velocity in the x direction and zero mean velocity in the y and z directions. The flowing stream is usually turbulent. As the fluid flows past the immersed body, a boundary layer forms adjacent to the solid surface. This type of flow is encountered in the flight of aircraft through the atmosphere. Flow around solid objects is not only important in aerodynamics but also of interest in the process industries. In the shell side of tubular heat exchangers flow takes place parallel and perpendicular to circular cylinders. The transfer of heat from extended surfaces depends on the flow of fluids past flat surfaces. Numerous industrial processes depend on the fluidization of granular solids, and the mechanism of fluid flow past the particles governs the rates of heat and mass transfer.

The formation of a boundary layer on a flat plate is shown schematically in Fig. 10-1. The undisturbed fluid has a velocity U in the x direction.

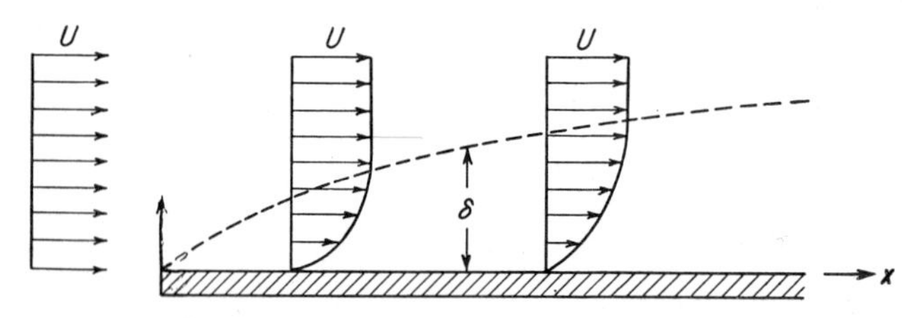

FIG. 10-1. Formation of a boundary layer on a flat plate.

The fluid in contact with the surface is assumed to have zero velocity, and frictional (viscous) resistance retards the moving fluid in a thin layer near the wall. The boundary layer extends from the surface, where the velocity is zero, to the point where the velocity is that of the undisturbed main stream. Theoretically, the edge of the boundary layer is an infinite distance from the wall. However, most of the velocity change occurs in a very small distance. Usually the edge of the boundary layer is defined as the point at which the velocity reaches some fraction of the velocity in the undisturbed stream. The thickness δ of the boundary layer is variable along the surface and is zero at the leading edge. The broken line in Fig. 10-1 indicates the edge of the boundary layer. The thickness is measured normal to the surface.

10-2. Equations of Two-dimensional Flow in the Boundary Layer

The complete momentum equations of flow in two dimensions are given in Table 2-2, items 3 and 6, for an incompressible fluid with constant viscosity.

$$\frac{\partial u}{\partial t} + u\frac{\partial u}{\partial x} + v\frac{\partial u}{\partial y} = -g_c\frac{\partial \Omega}{\partial x} - \frac{g_c}{\rho}\frac{\partial P}{\partial x} + \nu\left(\frac{\partial^2 u}{\partial x^2} + \frac{\partial^2 u}{\partial y^2}\right) \tag{10-1}$$

$$\frac{\partial v}{\partial t} + u\frac{\partial v}{\partial x} + v\frac{\partial v}{\partial y} = -g_c\frac{\partial \Omega}{\partial y} - \frac{g_c}{\rho}\frac{\partial P}{\partial y} + \nu\left(\frac{\partial^2 v}{\partial x^2} + \frac{\partial^2 v}{\partial y^2}\right) \tag{10-2}$$

These equations were considerably simplified by Prandtl[44] on the basis that frictional effects of viscosity are important only in the immediate vicinity of the surface and can be considered negligible beyond this thin layer of fluid. The simplifications obtained by Prandtl's order-of-magnitude analysis of the terms in Eqs. (10-1) and (10-2) are summarized below.†

(1) Flow is steady, i.e., $\partial u/\partial t = 0$, $\partial v/\partial t = 0$. This is strictly applicable for laminar flow but applies to turbulent flow if mean velocities are considered.
(2) The boundary layer is thin compared to the distance measured from the leading edge.
(3) As a result of (2), $\partial^2 u/\partial x^2$ is much smaller than $\partial^2 u/\partial y^2$.
(4) The velocity v (in the y direction) is much smaller than the velocity u (in the x direction), so all terms in Eq. (10-2) which involve v are of such an order of magnitude that they may be neglected.
(5) $\partial P/\partial y \simeq 0$. The pressure across the boundary layer is assumed constant.
(6) The field force Ω is neglected.

† For a detailed order-of-magnitude analysis of Eqs. (10-1) and (10-2) see ref. 48 or ref. 2 of Chap. 2.

The simplified two-dimensional equation of motion becomes

$$u \frac{\partial u}{\partial x} + v \frac{\partial u}{\partial y} = -\frac{g_c}{\rho} \frac{\partial P}{\partial x} + \frac{\mu}{\rho} \frac{\partial^2 u}{\partial y^2} \tag{10-3}$$

In addition, the two-dimensional continuity equation for an incompressible fluid describes the law of conservation of mass in the boundary layer, i.e.,

$$\frac{\partial u}{\partial x} + \frac{\partial v}{\partial y} = 0 \tag{10-4}$$

Equation (10-3) is the simplified momentum equation for the boundary layer. It is applicable for flow past flat or curved surfaces. For two-dimensional flow along a curved surface, the distance x for Eq. (10-3) is measured along the surface from the leading edge, and y is measured normal to the curved surface.

The number of unknowns in Eqs. (10-3) and (10-4) is three, u, v, and $\partial P/\partial x$. If $\partial P/\partial x$ is specified, the two equations are sufficient to give u and v as functions of x and y.

10-3. Local and Total Drag Coefficients for Flow over Immersed Bodies

As a solid body moves through a fluid, resistance is encountered by the body, and, conversely, as a fluid flows past a solid body, the fluid encounters resistance. This resistance to the movement of a solid in a fluid is known as drag, and it is caused by the shear stresses exerted in the boundary layer of the fluid next to the solid surface. If the boundary layer is completely laminar in character, the shear forces are viscous alone, while if the boundary layer is turbulent, the resistance results from velocity fluctuations of the fluid in the boundary layer. Drag is an energy loss, and, if caused by stresses in the boundary layer, it is surface drag. However, as will be seen in the discussion of the mechanism of flow around cylinders, separation of the boundary layer occurs, and the fluid immediately behind the solid body is in turbulent motion. This turbulent wake results in loss of energy in addition to that lost because of surface drag. The energy loss due to the turbulent wake is form drag and is a function of the form or shape of the body past which the fluid is flowing.

The coefficient of friction, or the coefficient of drag, for flow past solid bodies is defined in the same way as the friction factor for circular tubes; i.e.,

$$F = \frac{f \rho U^2 A_w}{2 g_c} \tag{10-5}$$

where F is the force of resistance exerted on the solid body by the fluid flowing past it and A_w is the wetted area of the immersed body. Another drag coefficient for flow past solid bodies is defined as follows:

$$F = \frac{f_D \rho U^2 A_P}{2g_c} \tag{10-6}$$

where A_P is the projected area of the body on a plane normal to the direction of flow. It is necessary to distinguish between these two drag coefficients. The coefficient f is based on the wetted area and applies mainly for flat plates and thin streamlined struts, while the coefficient f_D is based on the projected area of the body and applies for cylinders and bodies of revolution. When the force is calculated from the coefficient of drag, it includes both the surface drag and the form drag.

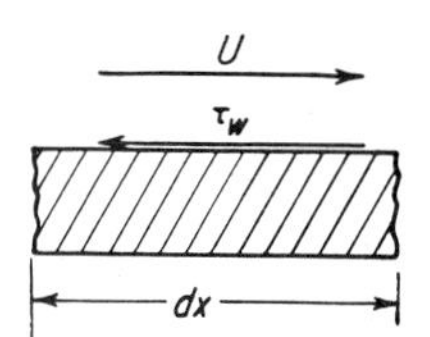

FIG. 10-2. Fluid flowing over element of solid surface.

The local coefficient of drag f' is obtained from a knowledge of the conditions at the surface of the body. In Fig. 10-2 an element of a solid surface of width dz and length dx is shown. The shear at the surface of this element is τ_w, so that the force acting on the element is $\tau_w\, dz\, dx$. This force may be expressed in the same way as the total force in Eq. (10-5) by making use of the local drag coefficient.

$$\tau_w\, dx\, dz = f' \frac{\rho U^2}{2g_c}\, dx\, dz \tag{10-7}$$

This equation expresses the point-resisting force in terms of the local drag coefficient and the average velocity in the main stream. The total force F acting on a surface of length L and width z is

$$F = \int_0^z \int_0^L \tau_w\, dx\, dz \tag{10-8}$$

From Eqs. (10-5), (10-7), and (10-8) it may be shown (assuming f' to be independent of z) that

$$f = \frac{z}{A_w} \int_0^L f'\, dx = \frac{1}{L} \int_0^L f'\, dx \tag{10-9}$$

Equation (10-9) gives the relation between the local drag coefficient and the total drag coefficient when there is no separation of the boundary layer from the surface of the immersed body. If separation occurs, the total drag coefficient will also include form drag, which cannot be taken into account by the integral on the right of Eq. (10-9).

I. FLOW PAST FLAT PLATES

10-4. Blasius' Exact Solution for the Laminar Boundary Layer on a Flat Plate

The simplest example of flow past immersed bodies is the flow past a thin, flat plate oriented parallel to the direction of flow. The flow in the forward portion of the boundary layer is laminar, even though the main-stream flow may be turbulent. The length of the boundary layer in which laminar flow exists depends on the main-stream turbulence and mean velocity, the distance from the leading edge, and the kinematic viscosity of the fluid.

Blasius[1] obtained an exact solution of Eqs. (10-3) and (10-4) for the laminar boundary layer on a flat plate immersed in a fluid with uniform velocity. At the edge of the boundary layer the velocity is U. Substituting this in Eq. (10-3) (since the momentum equation applies over the whole thickness of the boundary layer) gives

$$U \frac{\partial U}{\partial x} + v \frac{\partial U}{\partial y} = -\frac{g_c}{\rho} \frac{\partial P}{\partial x} + \frac{\mu}{\rho} \frac{\partial^2 U}{\partial y^2} \tag{10-10}$$

Since U is the velocity in the x direction only and does not vary in the y direction,

$$U \frac{\partial U}{\partial x} = -\frac{g_c}{\rho} \frac{\partial P}{\partial x} \tag{10-11}$$

Substituting Eq. (10-11) in (10-3),

$$u \frac{\partial u}{\partial x} + v \frac{\partial u}{\partial y} = U \frac{\partial U}{\partial x} + \frac{\mu}{\rho} \frac{\partial^2 u}{\partial y^2} \tag{10-12}$$

For flow past a thin, flat plate in a uniform stream, the velocities at the edge of the boundary layer and in the undisturbed stream are identical. Thus, since U is independent of x,

$$\frac{\partial U}{\partial x} = 0 \tag{10-13}$$

Thus

$$u \frac{\partial u}{\partial x} + v \frac{\partial u}{\partial y} = \frac{\mu}{\rho} \frac{\partial^2 u}{\partial y^2} \tag{10-14}$$

Equation (10-14) and the continuity equation

$$\frac{\partial u}{\partial x} + \frac{\partial v}{\partial y} = 0 \tag{10-4}$$

are two partial differential equations describing laminar flow past a thin, flat plate. Since Eq. (10-14) is a third-order equation, three boundary conditions must be specified in order to obtain a particular solution for u and v in terms of x and y. These boundary conditions are as follows:

At $y = 0$ surface of plate
$u = 0$
$v = 0$

At $y = \infty$ edge of boundary layer
$u = U$

Blasius' solution of Eqs. (10-14) and (10-4) involves the introduction of the stream function ψ, defined by Eqs. (3-16) and (3-17),

$$\frac{\partial \psi}{\partial y} = u \tag{3-16}$$

$$\frac{\partial \psi}{\partial x} = -v \tag{3-17}$$

which satisfies the two-dimensional continuity equation (10-4).

Equation (10-14) is reduced to an ordinary differential equation by defining a variable η which is a function of x and y and a function ϕ which is a function of η alone. The variable η is defined by

$$\eta = \frac{y}{2}\left(\frac{U\rho}{\mu x}\right)^{1/2} \tag{10-15}$$

and the function ϕ is defined by

$$\psi = \left(\frac{\mu U x}{\rho}\right)^{1/2} \phi \tag{10-16}$$

Without showing the details of the mathematics involved, Eqs. (10-15) and (10-16) may be used to obtain the following expressions for the terms appearing in Eq. (10-14):

$$u = \frac{U}{2}\phi' \tag{10-17}$$

$$\frac{\partial u}{\partial x} = -\frac{U\eta}{4x}\phi'' \tag{10-18}$$

$$\frac{\partial u}{\partial y} = \frac{U}{4}\left(\frac{U\rho}{\mu x}\right)^{1/2}\phi'' \tag{10-19}$$

$$\frac{\partial^2 u}{\partial y^2} = \frac{U}{8}\frac{U\rho}{\mu x}\phi''' \tag{10-20}$$

$$v = \frac{1}{2}\left(\frac{\mu U}{\rho x}\right)^{1/2}(\eta\phi' - \phi) \tag{10-21}$$

where ϕ', ϕ'', ϕ''' are first, second, and third total derivatives of ϕ with respect to η.

Substituting Eqs. (10-17) to (10-21) into Eq. (10-14) gives

$$\phi''' + \phi\phi'' = 0 \tag{10-22}$$

Equation (10-22) is a nonlinear ordinary differential equation of third order. The boundary conditions for Eq. (10-22) are as follows:

At $\eta = 0$ corresponding to $y = 0$
$\phi = 0$
$\phi' = 0$

At $\eta = \infty$ corresponding to $y = \infty$
$\phi' = 2$

The above three boundary conditions are sufficient to permit a solution of Eq. (10-22).

No solution in closed form has been found for Eq. (10-22). A series solution is obtainable, however. The function ϕ is expanded in a Taylor series.

$$\phi = C_0 + C_1\eta + C_2\frac{\eta^2}{2!} + \frac{C_3\eta^3}{3!} + \cdots \tag{10-23}$$

Differentiating Eq. (10-23) with respect to η,

$$\phi' = C_1 + \frac{2C_2\eta}{2!} + \frac{3C_3\eta^2}{3!} + \frac{4C_4\eta^3}{4!} + \cdots \tag{10-24}$$

Since ϕ and ϕ' are zero at $\eta = 0$, C_0 and C_1 must be zero. The series values of ϕ, ϕ'', and ϕ''' are substituted in Eq. (10-22), and like powers of η are collected, giving

$$C_3 + C_4\eta + \left(\frac{C_2{}^2}{2!} + \frac{C_5}{2!}\right)\eta^2 + \cdots = 0 \tag{10-25}$$

Since η may have values from zero to infinity, the coefficients of η must be zero. So

$$C_3 = 0$$

$$C_4 = 0$$

$$\frac{C_2{}^2}{2!} + \frac{C_5}{2!} = 0$$

All constants that are not zero may be expressed in terms of C_2, which results in

$$\phi = \frac{C_2\eta^2}{2!} - \frac{C_2{}^2\eta^5}{5!} + \frac{11C_2{}^3\eta^8}{8!} - \frac{375C_2{}^4\eta^{11}}{11!} + \cdots \tag{10-26}$$

Equation (10-26) is a series solution of Eq. (10-22), and it satisfies the boundary condition at $\eta = 0$. The third boundary condition, i.e., at $\eta = \infty$, $\phi' = 2$, is used to evaluate C_2. Using numerical methods, Goldstein [19] obtained a value of 1.32824 for C_2. Table 10-1 shows the values of

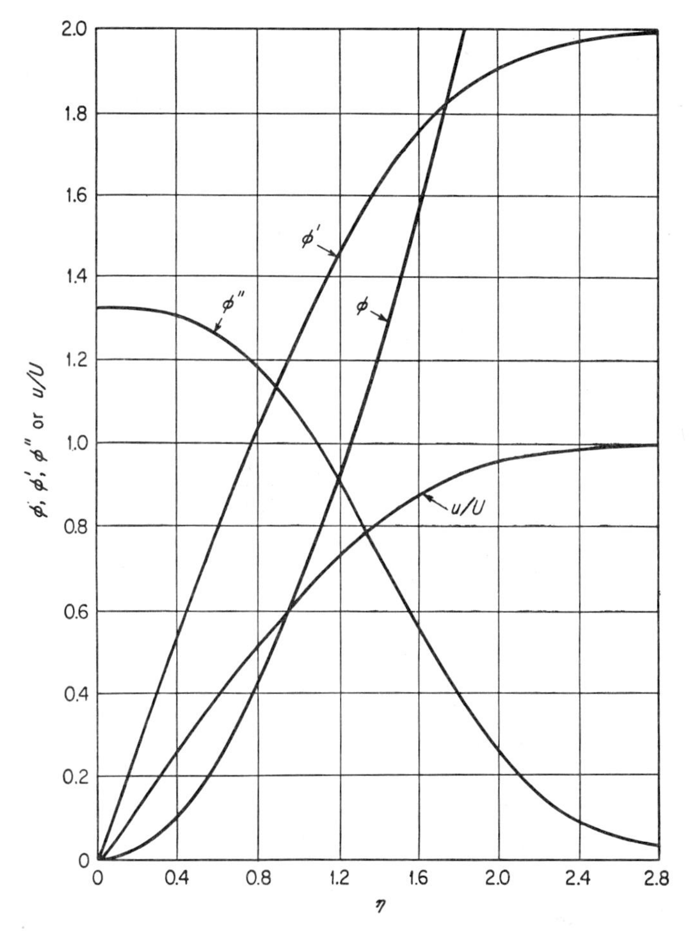

FIG. 10-3. Values of ϕ, ϕ', ϕ'', and u/U obtained from the solution of Eq. (10-22).

ϕ, ϕ', ϕ'' as functions of η determined by Howarth,[25] which are more accurate than the values originally given by Blasius.[1] Values of u/U from Eq. (10-17) are also included. From the quantities in Table 10-1, the flow conditions at any x and y may be obtained. In Fig. 10-3, ϕ, ϕ', and ϕ'' are plotted versus η to show graphically how these functions vary.

TABLE 10-1. VALUES OF ϕ, ϕ', ϕ'', AND u/U FOR FLOW PAST A THIN, FLAT PLATE †

η	ϕ	ϕ'	ϕ''	$\frac{u}{U}$
0	0	0	1.32824	0
0.2	0.0266	0.2655	1.3260	0.1328
0.4	0.1061	0.5294	1.3096	0.2647
0.6	0.2380	0.7876	1.2664	0.3938
0.8	0.4203	1.0336	1.1867	0.5168
1.0	0.6500	1.2596	1.0670	0.6298
1.2	0.9223	1.4580	0.9124	0.7290
1.4	1.2310	1.6230	0.7360	0.8115
1.6	1.5691	1.7522	0.5565	0.8761
1.8	1.9295	1.8466	0.3924	0.9233
2.0	2.3058	1.9110	0.2570	0.9555
2.2	2.6924	1.9518	0.1558	0.9759
2.4	3.0853	1.9756	0.0875	0.9878
2.6	3.4819	1.9885	0.0454	0.9943
2.8	3.8803	1.9950	0.0217	0.9915
3.0	4.2796	1.9980	0.0096	0.9990
3.2	4.6794	1.9992	0.0039	0.9996
3.4	5.0793	1.9998	0.0015	0.9999
3.6	5.4793	2.0000	0.0005	1.0000
3.8	5.8792	2.0000	0.0002	1.0000

† From L. Howarth, *Proc. Roy. Soc.* (*London*), **164A**:547 (1938).

Example 10-1

A thin, flat plate is immersed in a stream of air at atmospheric pressure flowing at a velocity of 20 ft/sec. Air temperature is 60°F. At a point 6 in. from the leading edge determine the distance from the plate at which the point velocity is half the main-stream velocity. At this point calculate v and $\partial u/\partial y$.

Solution

At 60°F

$$\text{Kinematic viscosity of air} = 1.58 \times 10^{-4}\ \text{ft}^2/\text{sec}$$

$$\text{Density of air} = 0.0765\ \text{lb}_m/\text{ft}^3$$

When $x = 6$ in., it is required to determine the value of y for $u/U = 0.5$. From either Table 10-1 or Fig. 10-3, when $u/U = 0.5$,

$$\eta = \frac{y}{2}\left(\frac{U}{x\nu}\right)^{1/2} = 0.78$$

and

$$\phi = 0.4$$

$$\phi' = 1.0$$

$$\phi'' = 1.20$$

Thus
$$\frac{y}{2}\left[\frac{(20)(12)}{(6)(1.58 \times 10^{-4})}\right]^{1/2} = 0.78$$

Giving
$$y = 3.1 \times 10^{-3} \text{ ft}$$
$$= 0.0372 \text{ in.}$$

From Eq. (10-19)
$$\frac{\partial u}{\partial y} = \frac{20}{4}\left[\frac{(20)(12)}{(6)(1.58 \times 10^{-4})}\right]^{1/2} 1.20$$
$$= 3.02 \times 10^{3} \text{ sec}^{-1}$$

From Eq. (10-21)
$$v = \frac{1}{2}\left[\frac{(20)(1.58 \times 10^{-4})(12)}{6}\right]^{1/2} [(0.78)(1.0) - 0.4] = 0.0151 \text{ ft/sec}$$

The conditions at the point $x = 6$ in., $y = 0.0372$ in. are shown in Fig. 10-4.

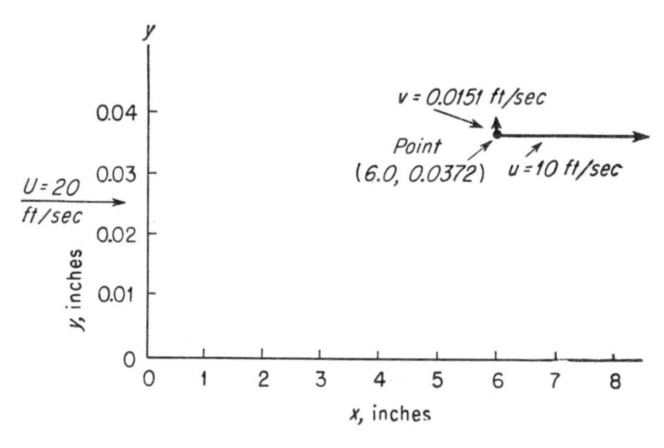

FIG. 10-4. Results of Example 10-1.

10-5. Von Kármán's Integral Momentum Equation

Von Kármán [29] analyzed the flow in the boundary layer, and using Newton's second law he derived an integral relationship for the velocity distribution in the boundary layer. He considered a two-dimensional region of fluid including the boundary layer and having a differential length dx. In this region the total rate of increase of momentum is equal in both magnitude and direction to the forces acting on the boundaries of the region. The relationship derived by von Kármán may also be obtained by integrating Eqs. (10-3) and (10-4) directly with respect to y between the limits $y = 0$ and $y = \delta$, where δ is large enough to include the boundary layer. Equations (10-3) and (10-4) become

$$\int_0^\delta u\frac{\partial u}{\partial x}\,dy + \int_0^\delta v\frac{\partial u}{\partial y}\,dy = \int_0^\delta -\frac{g_c}{\rho}\frac{\partial P}{\partial x}\,dy + \nu\int_0^\delta \frac{\partial^2 u}{\partial y^2}\,dy \qquad (10\text{-}27)$$

$$\int_0^\delta \frac{\partial u}{\partial x}\,dy + \int_0^\delta \frac{\partial v}{\partial y}\,dy = 0 \qquad (10\text{-}28)$$

The integration of Eqs. (10-27) and (10-28) is carried out holding x constant. The boundary conditions are as follows (see Fig. 10-5):

At $y = 0$ at the solid surface

$u = 0$

$v = 0$

$$\frac{\partial u}{\partial y} = \frac{\tau_w g_c}{\mu}$$

At $y = \delta$ at edge of boundary layer

$u = U$

$v = v_\delta$

$$\frac{\partial u}{\partial y} = 0$$

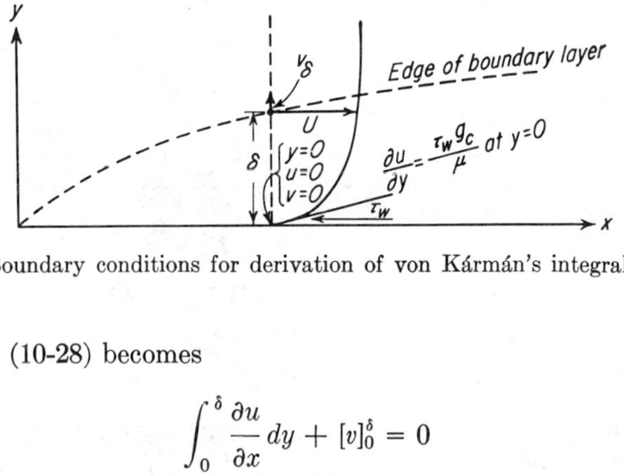

Fig. 10-5. Boundary conditions for derivation of von Kármán's integral momentum equation.

Equation (10-28) becomes

$$\int_0^\delta \frac{\partial u}{\partial x}\,dy + [v]_0^\delta = 0 \qquad (10\text{-}29)$$

giving

$$v_\delta = -\int_0^\delta \frac{\partial u}{\partial x}\,dy \qquad (10\text{-}30)$$

On integration, Eq. (10-27) becomes

$$\frac{\partial}{\partial x}\,\tfrac{1}{2}\int_0^\delta u^2\,dy + [uv]_0^\delta - \int_0^\delta u\frac{\partial v}{\partial y}\,dy = -\frac{\delta g_c}{\rho}\frac{\partial P}{\partial x} + \nu\left[\frac{\partial u}{\partial y}\right]_0^\delta \qquad (10\text{-}31)$$

Introducing the boundary conditions gives

$$\frac{\partial}{\partial x} \tfrac{1}{2} \int_0^\delta u^2\, dy + U v_\delta - \int_0^\delta u \frac{\partial v}{\partial y}\, dy = -\frac{\delta g_c}{\rho} \frac{\partial P}{\partial x} - \frac{\tau_w g_c}{\rho} \tag{10-32}$$

Substituting Eqs. (10-11), (10-28), and (10-30) into (10-32),

$$\frac{\partial}{\partial x} \tfrac{1}{2} \int_0^\delta u^2\, dy - U \int_0^\delta \frac{\partial u}{\partial x}\, dy + \int_0^\delta u \frac{\partial u}{\partial x}\, dy = \delta U \frac{\partial U}{\partial x} - \frac{\tau_w g_c}{\rho} \tag{10-33}$$

or

$$\frac{\partial}{\partial x} \tfrac{1}{2} \int_0^\delta u^2\, dy - U \frac{\partial}{\partial x} \int_0^\delta u\, dy + \frac{\partial}{\partial x} \tfrac{1}{2} \int_0^\delta u^2\, dy = \delta U \frac{\partial U}{\partial x} - \frac{\tau_w g_c}{\rho} \tag{10-34}$$

and collecting like terms,

$$\frac{\partial}{\partial x} \int_0^\delta u^2\, dy - U \frac{\partial}{\partial x} \int_0^\delta u\, dy = \delta U \frac{\partial U}{\partial x} - \frac{\tau_w g_c}{\rho} \tag{10-35}$$

Equation (10-35) may be rearranged to the following form on the basis that U is a function of x and not of y:

$$\frac{\tau_w g_c}{\rho} = \frac{\partial}{\partial x} \int_0^\delta u(U - u)\, dy + \frac{dU}{dx} \int_0^\delta (U - u)\, dy \tag{10-36}$$

which is the same as

$$\frac{\tau_w g_c}{\rho} = \frac{\partial}{\partial x} U^2 \int_0^\delta \frac{u}{U} \frac{U - u}{U}\, dy + \frac{dU}{dx} U \int_0^\delta \frac{U - u}{U}\, dy \tag{10-37}$$

Equations (10-36) and (10-37) are von Kármán's integral momentum equations for the boundary layer. They express the shear at the wall τ_w as a function of the point the velocity in the boundary layer u and the velocity at the edge of the boundary layer U. In order to solve Eq. (10-36) the point velocity must be known as a function of y.

For a flat plate in parallel flow, U is constant, and Eq. (10-36) becomes

$$\frac{\tau_w g_c}{\rho} = \frac{\partial}{\partial x} \int_0^\delta u(U - u)\, dy \tag{10-38}$$

which is von Kármán's integral momentum equation for the boundary layer on a flat plate.

10-6. The Boundary-layer Thickness from Blasius' Solution

As pointed out previously, the thickness of the boundary layer is theoretically infinite, and one of the boundary conditions for Blasius' solution is at an infinite distance from the wall. It is usual to define the boundary-layer thickness δ as the normal distance from the wall where the point velocity is within 1 per cent of the main-stream velocity. Thus δ is the value of y where $u/U = 0.99$. From Table 10-1, when $u/U = 0.99$, η has a value of 2.48. Thus

$$\frac{\delta}{2}\left(\frac{U}{\nu x}\right)^{1/2} = 2.48 \tag{10-39}$$

which is rearranged to

$$\frac{\delta}{x} = \frac{4.96}{\sqrt{\text{Re}_x}} \tag{10-40}$$

where Re_x is the local Reynolds number Ux/ν based on the distance from the leading edge of the plate.

10-7. Pohlhausen's Analysis of the Laminar Boundary Layer on a Flat Plate

Pohlhausen [43] used von Kármán's integral equation for a flat plate [Eq. (10-38)] to obtain expressions for the velocity profile and laminar boundary-layer thickness. He assumed three different forms of the velocity-profile curve,

$$u = C_1 y + C_2 y^3 \tag{10-41}$$

$$u = C_1 y + C_2 y^2 \tag{10-42}$$

$$u = C_1 y + C_2 y^2 + C_3 y^3 + C_4 y^4 \tag{10-43}$$

Equation (10-38) will be used with Eq. (10-41) to derive expressions for the velocity profile and boundary-layer thickness. For Eq. (10-41) the boundary conditions are

At $y = 0$

$$u = 0$$

At $y = \delta$

$$u = U$$

$$\frac{\partial u}{\partial y} = 0$$

From the conditions, C_1 and C_2 may be evaluated.

$$C_1 = 1.5\frac{U}{\delta}$$

$$C_2 = -\frac{U}{2\delta^3}$$

and Eq. (10-41) becomes

$$\frac{u}{U} = 1.5\frac{y}{\delta} - \frac{1}{2}\left(\frac{y}{\delta}\right)^3 \tag{10-44}$$

The shear stress at the wall (τ_w at $y = 0$) is determined from Eq. (10-44) by differentiating with respect to y and setting $y = 0$:

$$\frac{1}{U}\left(\frac{\partial u}{\partial y}\right)_{y=0} = \frac{1.5}{\delta} \tag{10-45}$$

thus

$$\frac{\tau_w g_c}{\rho} = 1.5\frac{U\nu}{\delta} \tag{10-46}$$

Substituting Eqs. (10-44) and (10-46) into Eq. (10-38),

$$\frac{1.5U\nu}{\delta} = \frac{\partial}{\partial x}\left[U^2\int_0^\delta\left(\frac{1.5y}{\delta} - \frac{y^3}{2\delta^3}\right)\left(1 - \frac{1.5y}{\delta} + \frac{y^3}{2\delta^3}\right)dy\right] \tag{10-47}$$

Integration of Eq. (10-47) with respect to y gives

$$\frac{1.5U\nu}{\delta} = \frac{\partial}{\partial x}\left(\tfrac{117}{840}U^2\delta\right) \tag{10-48}$$

Since U and ν are constant, the variables x and δ may be separated and the result integrated with respect to x.

$$\frac{(1.5)(840)\nu}{117U}\,dx = \delta\,d\delta \tag{10-49}$$

and

$$\frac{(1.5)(840)\nu x}{117U} = \frac{\delta^2}{2} + C \tag{10-50}$$

Introducing the boundary condition $\delta = 0$ at $x = 0$, the constant of integration becomes zero. Thus

$$\delta^2 = \frac{(2)(1.5)(840)}{117}\frac{\nu x}{U} \tag{10-51}$$

or

$$\delta = 4.64\left(\frac{\nu x}{U}\right)^{1/2} \tag{10-52}$$

By substituting Eq. (10-52) in Eq. (10-44) the velocity distribution in terms of y and x is obtained.

$$\frac{u}{U} = \frac{1.5}{4.64}\frac{y}{\sqrt{\nu x/U}} - \frac{1}{(2)(4.64)^3}\frac{y^3}{(\sqrt{\nu x/U})^3} \tag{10-53}$$

Equations (10-52) and (10-53) are, respectively, the boundary-layer-thickness and velocity-profile equations obtained from von Kármán's integral equations by assuming the velocity profile to be of the form given by Eq. (10-41). The laminar-boundary-layer thickness obtained by the Blasius

TABLE 10-2. PROPERTIES OF THE LAMINAR BOUNDARY LAYER ON A FLAT PLATE AS DETERMINED BY VARIOUS METHODS

Form of velocity-profile curve	Boundary-layer thickness	Displacement thickness	Drag coefficient	
			Local	Total
Blasius-Howarth solution	$\frac{\delta}{x} = \frac{4.96}{\sqrt{\text{Re}_x}}$	$\frac{\delta^*}{x} = \frac{1.721}{\sqrt{\text{Re}_x}}$	$f' = \frac{0.664}{\sqrt{\text{Re}_x}}$	$f = \frac{1.328}{\sqrt{\text{Re}_L}}$
Eq. (10-41)	$\frac{\delta}{x} = \frac{4.64}{\sqrt{\text{Re}_x}}$	$\frac{\delta^*}{x} = \frac{1.740}{\sqrt{\text{Re}_x}}$	$f' = \frac{0.648}{\sqrt{\text{Re}_x}}$	$f = \frac{1.296}{\sqrt{\text{Re}_L}}$
Eq. (10-42)	$\frac{\delta}{x} = \frac{5.5}{\sqrt{\text{Re}_x}}$	$\frac{\delta^*}{x} = \frac{1.833}{\sqrt{\text{Re}_x}}$	$f' = \frac{0.727}{\sqrt{\text{Re}_x}}$	$f = \frac{1.454}{\sqrt{\text{Re}_L}}$
Eq. (10-43)	$\frac{\delta}{x} = \frac{5.83}{\sqrt{\text{Re}_x}}$	$\frac{\delta^*}{x} = \frac{1.749}{\sqrt{\text{Re}_x}}$	$f' = \frac{0.686}{\sqrt{\text{Re}_x}}$	$f = \frac{1.372}{\sqrt{\text{Re}_L}}$

solution and by Pohlhausen [using Eqs. (10-41) to (10-43)] is shown in Table 10-2. In Fig. 10-6 the velocity profiles as given by the Blasius solution and by Eq. (10-53) are compared.

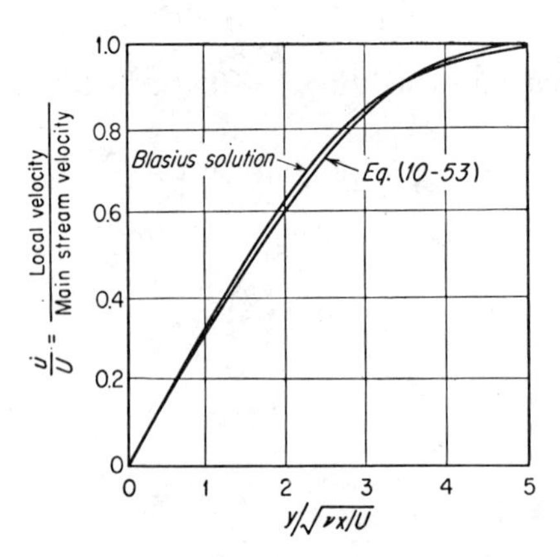

FIG. 10-6. Comparison of Blasius' solution with Eq. (10-53) for the laminar velocity distribution for flow over a flat plate in parallel flow.

10-8. The Displacement Thickness

A common term used in describing boundary layers is the displacement thickness δ^*, which is defined by the relation

$$\delta^* = \int_0^\delta \left(1 - \frac{u}{U}\right) dy \tag{10-54}$$

Referring to Fig. 10-7, δ^* is that thickness which makes area 1 equal to area 2 so that area 1 + 3 equals area 2 + 3; i.e., the displacement thick-

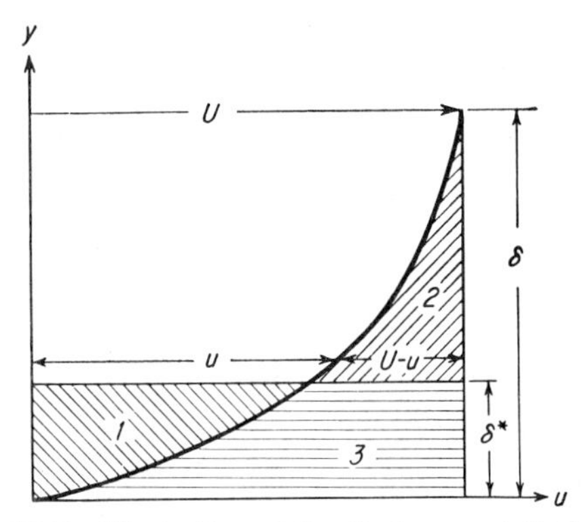

FIG. 10-7. Illustration of the displacement thickness.

ness is the thickness that a layer of fluid (area 1 + 3) would have if it had the same integrated velocity defect as the actual boundary layer (area 2 + 3). If the velocity profile is of the form given by Eq. (10-41),

$$\delta^* = \int_0^\delta \left(1 - 1.5\frac{y}{\delta} + \frac{1}{2}\frac{y^3}{\delta^3}\right) dy \tag{10-55}$$

$$\delta^* = \frac{3\delta}{8} \tag{10-56}$$

From Eq. (10-52)

$$\delta^* = 1.740\left(\frac{\nu x}{U}\right)^{1/2} \tag{10-57}$$

Table 10-2 gives values of the displacement thickness on a flat plate calculated from the various velocity-profile equations.

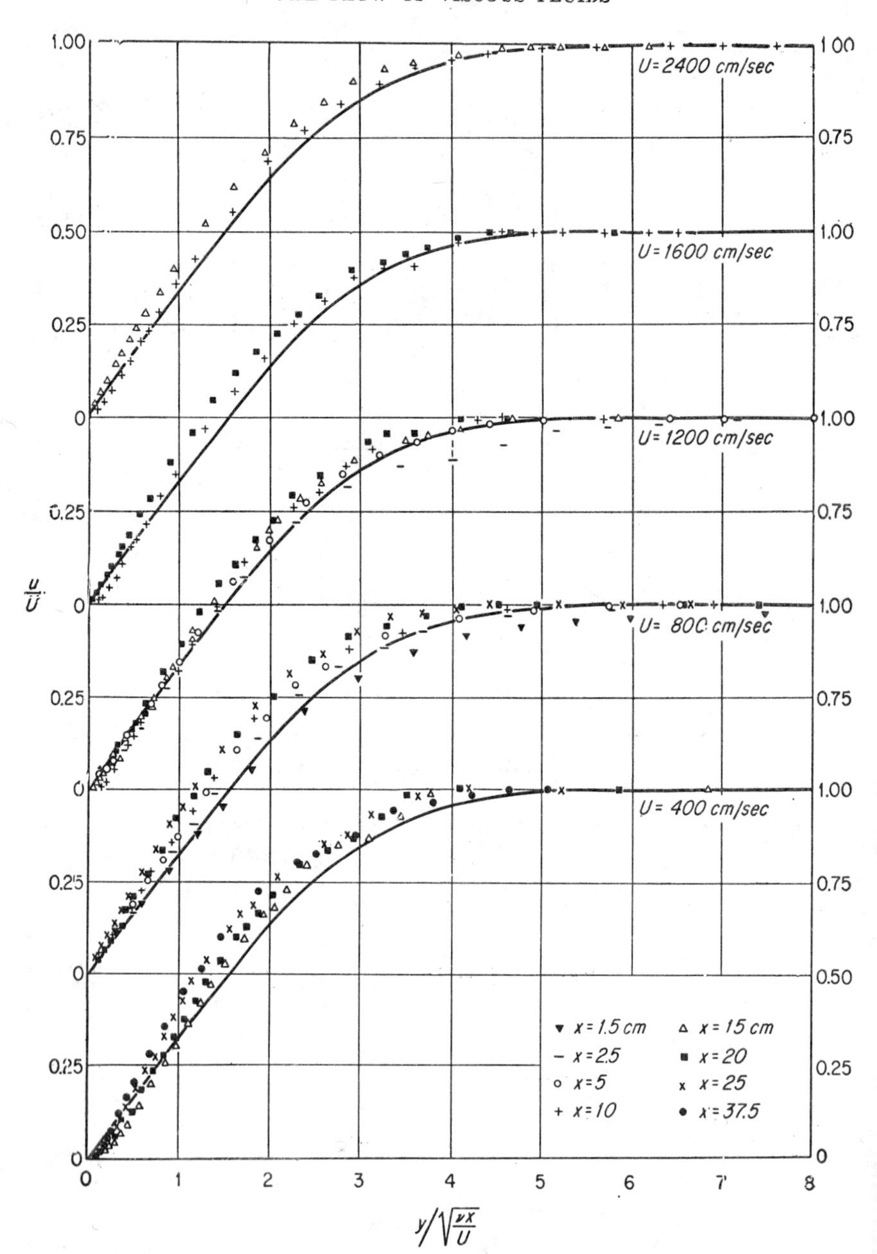

FIG. 10-8. Velocity distribution in the laminar boundary layer on a flat plate. (*From J. M. Burgers, Proc. Intern. Congr. Appl. Mech., 1st Congr., Delft, 1924, p.* 113.)

10-9. Experimental Laminar Velocity Profiles on Flat Plates

A considerable number of investigations of laminar velocity profiles on flat plates parallel to the flowing stream have been carried out, and the majority of the results agree well with the Blasius solution of the boundary layer. Burgers [2] determined velocity profiles for air flowing over smooth plates having lengths varying from 22 to 125 cm. The velocity distribution near the surface of these plates was measured at various values of x and

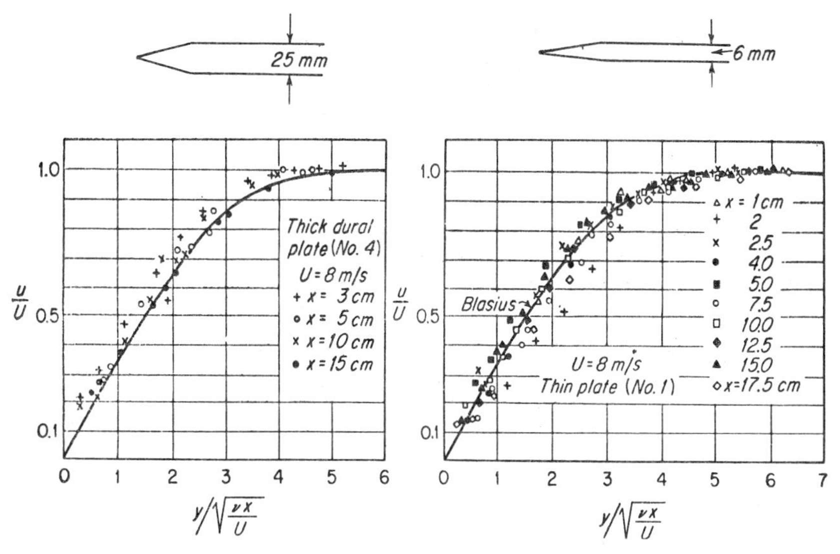

FIG. 10-9. Velocity distribution in the laminar boundary layer on a flat plate. (*From M. Hansen, NACA TM* 585, 1930.)

for various velocities. Burgers's data are shown in Fig. 10-8, where u/U is plotted versus $y/\sqrt{\nu x/U}$. The solid curve represents the velocity distribution given by Eq. (10-42). It appears that Blasius' solution or Eq. (10-53) would fit the data better. Figure 10-9 depicts some data obtained by Hansen [21] for both thin and thick plates. The reference curve is Blasius' solution. The data for the thick plate lie somewhat above the curve, while for the thin plate good agreement is obtained. Figure 10-10 shows some velocity-distribution data obtained by Dhawan.[6] The laminar-flow data agree with Blasius' predicted curve. Figure 10-10 also shows the velocity profile for a turbulent boundary layer at a value of $\mathrm{Re}_x = 2.19 \times 10^5$. This turbulent boundary layer was obtained by providing a disturbance at the leading edge of the plate.

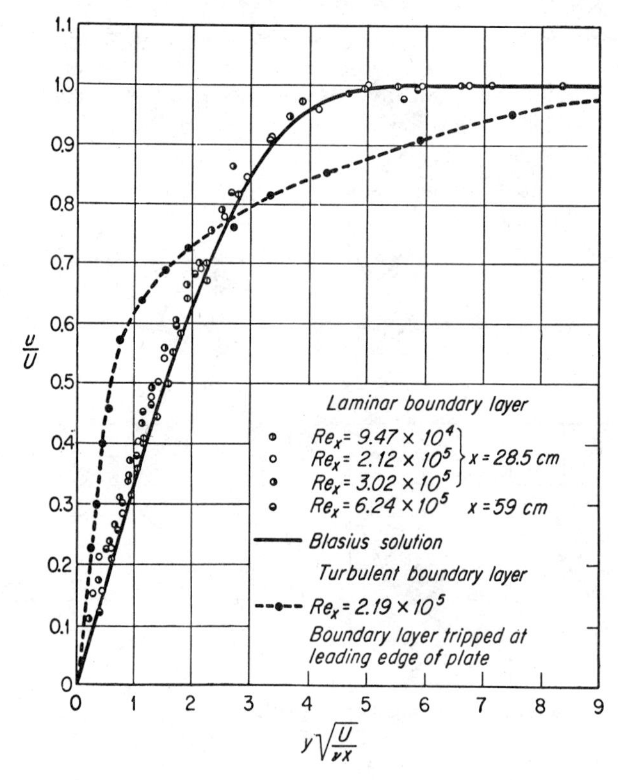

FIG. 10-10. Velocity profiles in the boundary layer on a flat plate. (*From S. Dhawan, NACA TN* 2567, 1952.)

10-10. Flow near the Leading Edge of a Flat Plate

The Blasius solution for the laminar boundary layer on a flat plate is restricted to positions relatively far from the leading edge. Since, in regions close to the leading edge, conditions (2) and (4) of Sec. 10-2 are not attained, the simplified differential equation (10-3) is not applicable. It is necessary to use the complete momentum equations (10-1) and (10-2) to analyze flow in the vicinity of the leading edge of the plate. Carrier and Lin [3] have made a study of such flow.

10-11. Drag Coefficients for Laminar Flow past a Flat Plate

The shear stress at the surface of the plate is

$$\tau_w g_c = \mu \left(\frac{\partial u}{\partial y}\right)_{y=0} \tag{10-58}$$

Combining Eq. (10-58) with Eq. (10-7), which defines the local coefficient of friction f',

$$f' = \frac{2\mu}{\rho U^2}\left(\frac{\partial u}{\partial y}\right)_{y=0} \tag{10-59}$$

The velocity gradient at the surface $\partial u/\partial y$ may be determined from Eq. (10-19) and the value of ϕ'' at $y = 0$. From Table 10-1, $\phi'' = 1.328$ at $y = 0$. Thus, from Eq. (10-19),

$$\left(\frac{\partial u}{\partial y}\right)_{y=0} = \frac{U}{4}\left(\frac{U\rho}{\mu x}\right)^{1/2} 1.328 \tag{10-60}$$

Substituting Eq. (10-60) into Eq. (10-59),

$$f' = \frac{0.664}{\sqrt{xU/\nu}} = \frac{0.664}{\sqrt{\mathrm{Re}_x}} \tag{10-61}$$

Using Eq. (10-9) to obtain the total drag coefficient for a plate of length L,

$$f = \frac{1}{L}\int_0^L \frac{0.664}{\sqrt{xU/\nu}}\,dx \tag{10-62}$$

from which

$$f = \frac{1.328}{\sqrt{\mathrm{Re}_L}} \qquad 2 \times 10^4 < \mathrm{Re}_L < 5 \times 10^5 \tag{10-63}$$

where $\mathrm{Re}_L = UL/\nu$ is the total Reynolds number for flow over the flat plate of length L. Equations (10-61) and (10-63) are the drag coefficients predicted by the Blasius solution.

Drag coefficients on flat plates may be determined experimentally by three methods. The total drag coefficient may be obtained by measuring the force exerted on a plate in parallel flow and calculating the total drag coefficient by Eq. (10-5). Errors are involved in this method, since the plates must be of finite width, and care must be taken to eliminate these edge effects. All relationships so far derived apply to flat plates of infinite span. Local drag coefficients may be obtained from the measured velocity distribution adjacent to the surface of the plate, or the local shear stress may be measured directly by determining the force exerted on a small movable element of surface. The latter method has proved satisfactory for determining local coefficients of friction.

Figure 10-11 shows some experimental values of the local drag coefficient determined by Dhawan,[6] who measured wall friction directly. The experimental data (including some obtained from velocity-profile measurements) agree with Eq. (10-61).

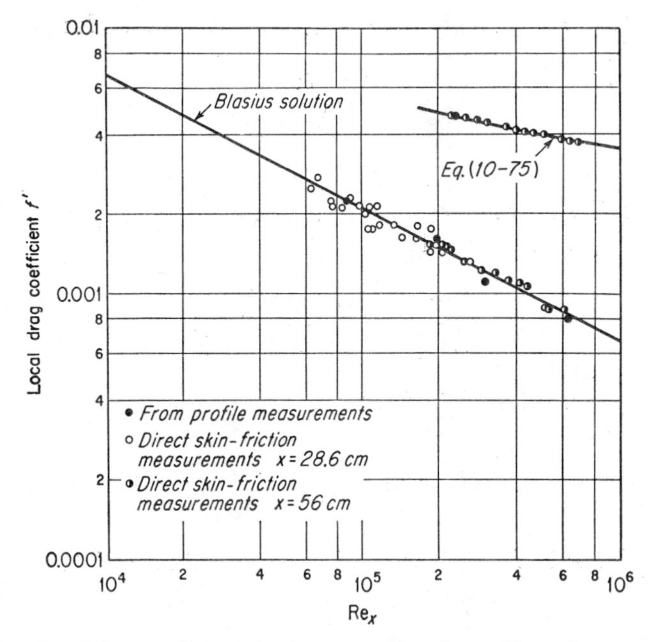

FIG. 10-11. Local drag coefficients for flow past a flat plate. (*From S. Dhawan, NACA TN* 2567, 1952.)

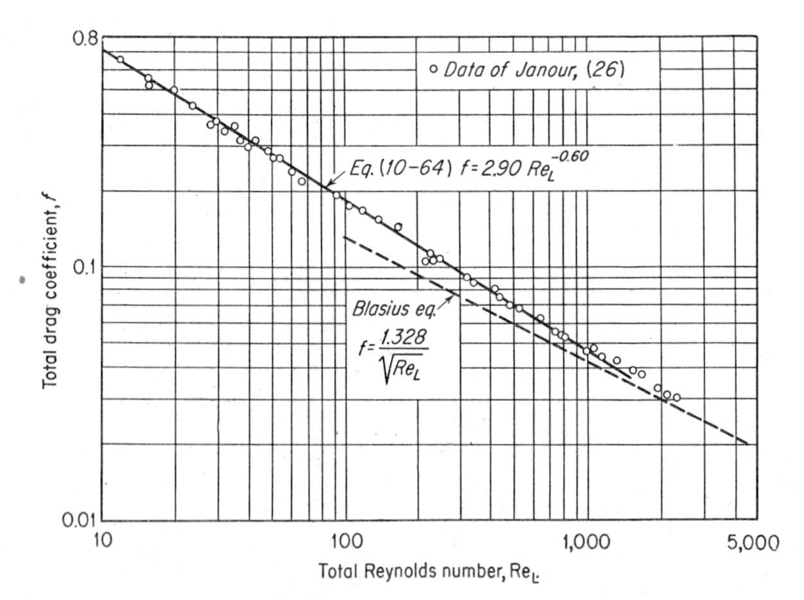

FIG. 10-12. Total drag coefficients for flow parallel to flat plates at low Reynolds numbers. (*From Z. Janour, NACA TM* 1316, 1951.)

Janour[26] obtained total drag coefficients for flat plates for very low Reynolds numbers and reported values of f higher than those given by Eq. (10-63) for the Reynolds-number range from 10 to 3,000. In this range

$$f = 2.90(\text{Re}_L)^{-0.60} \qquad 10 < \text{Re}_L < 3{,}000 \tag{10-64}$$

The Reynolds-number range for which Eq. (10-63) holds is from 2×10^4 to 5×10^5. At this upper value of the Reynolds number the boundary layer becomes turbulent. Janour suggests the following correction for edge effects:

$$F_{\text{edge}} = \frac{3.2\mu LU}{g_c} \qquad 10 < \text{Re}_L < 3{,}000 \tag{10-65}$$

where F_{edge} is the force of resistance of the two edges of the plate which are parallel to the direction of flow. Janour's experimental data are shown in Fig. 10-12, where the total drag coefficient (corrected for edge effects) is plotted versus the Reynolds number. Equations (10-63) and (10-64) are plotted for comparison.

Drag coefficients at very low Reynolds numbers have been determined by Janssen,[27] who solved Eqs. (10-1) and (10-4) by means of an analogue computer. The results are shown in Table 10-3, where values of the total drag coefficient are shown as a function of the Reynolds number.

TABLE 10-3. TOTAL DRAG COEFFICIENTS FOR FLOW PAST FLAT PLATES AT LOW REYNOLDS NUMBERS †

Re_L	f
0.1	22.22
1.0	2.80
10.0	0.57

† From E. Janssen, "Heat Transfer and Fluid Mechanics Institute, Preprints of Papers," p. 173, Stanford University Press, Stanford, Calif., 1956.

Example 10-2

The flat plate in Example 10-1 is 2 ft square. What force is exerted on it? (Neglect edge effects.)

Solution

$$\text{Re}_L = \frac{LU}{\nu} = \frac{(2)(20)}{1.58 \times 10^{-4}} = 25.3 \times 10^4$$

From Eq. (10-63)

$$f = \frac{1.328}{\sqrt{25.3 \times 10^4}} = 0.00264$$

The force exerted on the plate is calculated from Eq. (10-5) considering both sides in the area term.

$$F = \frac{(0.00264)(0.0765)(20)^2(2)(2)(2)}{(2)(32.2)} = 0.01 \text{ lb}_f$$

Example 10-3

For the flat plate in Example 10-1 determine the boundary-layer thickness and the local drag coefficient at a distance 1 ft from the leading edge.

Solution

At $x = 1$ ft

$$\mathrm{Re}_x = \frac{(1)(20)}{1.58 \times 10^{-4}} = 1.27 \times 10^5$$

From Eq. (10-61)

$$f' = \frac{0.664}{\sqrt{1.27 \times 10^5}} = 0.00186$$

This is the local drag coefficient 1 ft from the leading edge.

From Eq. (10-40)

$$\frac{\delta}{x} = \frac{4.96}{\sqrt{1.27 \times 10^5}} = 0.0139$$

When $x = 1$ ft,

$$\delta = 0.0139 \text{ ft} = 0.167 \text{ in.}$$

This is the boundary-layer thickness 1 ft from the leading edge.

10-12. The Transition from Laminar to Turbulent Flow on a Flat Plate

In the vicinity of the sharp leading edge of a flat plate, flow in the boundary layer is laminar, even though the fluid in the main stream is turbulent. As the laminar boundary layer increases in thickness, it becomes unstable, and flow becomes turbulent. Figure 10-13 is a schematic

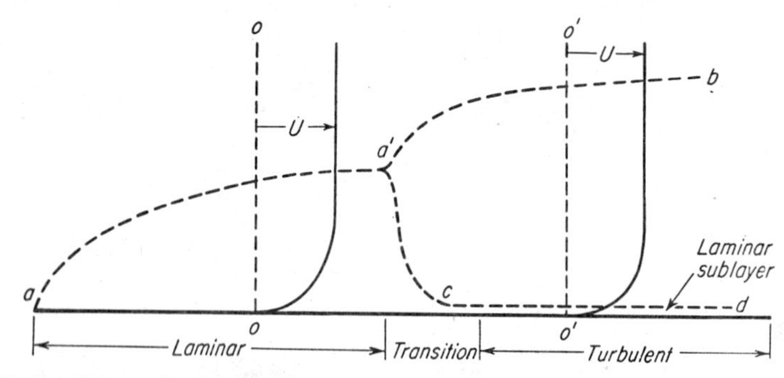

FIG. 10-13. Schematic sketch of laminar and turbulent boundary layers on a flat plate.

representation of the change from a laminar boundary layer to a turbulent boundary layer. For a certain distance along the plate a laminar zone exists, while beyond is a short transition region followed by the turbulent zone. The broken line aa' represents the outer edge of the laminar boundary layer, while $a'b$ is the outer edge of the turbulent boundary layer. It is assumed that a laminar sublayer exists adjacent to the wall for the turbulent boundary layer, so laminar flow always exists between the broken line $aa'cd$ and the surface of the plate.

The region of transition is marked by an increase in the wall shear stress and a change in the form of the velocity-distribution curve. The boundary-layer thickness increases much more rapidly in the turbulent zone than in the laminar zone. A turbulent and laminar velocity profile are shown in Fig. 10-10, and a series of velocity profiles in the transition region is shown in Fig. 10-14. These profiles, obtained by Schubauer and Klebanoff,[52] were

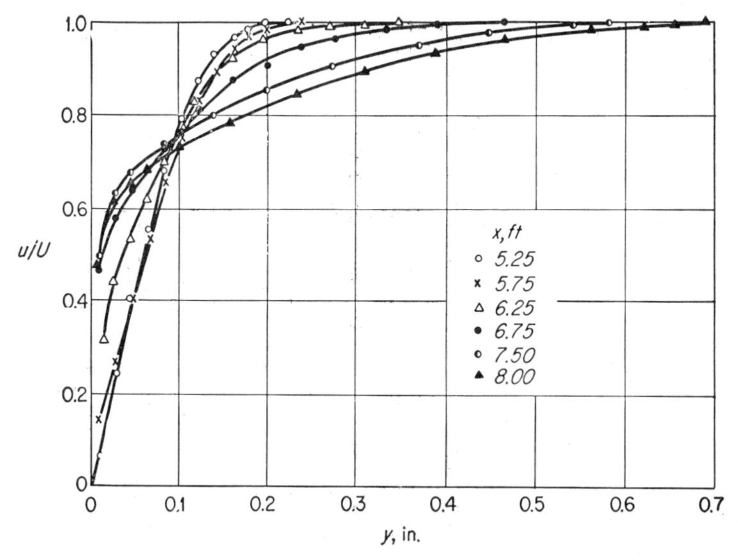

FIG. 10-14. Mean velocity profiles in the transition region on a flat plate. $U = 80$ ft/sec; free-stream turbulence = 0.03 per cent. (*From G. B. Schubauer and P. S. Klebanoff, NACA TN* 3489, 1955.)

measured on a flat plate at distances from 5.25 to 8.00 ft from the leading edge, which correspond to approximate values of Re_x from 2.6×10^6 to 4×10^6. The changing form of the velocity profile and rapid increase in boundary-layer thickness and wall shear stress are evident. The curve for $x = 5.25$ ft is for the laminar boundary layer, while at $x = 8.0$ ft the boundary layer is completely turbulent. Investigation with a hot-wire anemometer in the transition region showed that part of the time flow was laminar and part of the time it was turbulent. The fraction of the time when the flow was turbulent increased with increasing x and became unity at $x = 8$ ft.

The various changes that occur in boundary layers in the transition zone have been used to determine the point of transition. The increase in wall shear stress has been detected by hot-wire anemometers, pitot tubes, and direct wall-friction measurements.[6, 52, 56] Turbulence in the boundary layer has been detected by volatile chemicals on the wall or by injection of smoke or gas at the surface of the plate.[4, 37, 46] The transition zone may also be determined by measurement of the boundary-layer thickness. Hansen[21] determined the transition zone by plotting the quantity $\delta/\sqrt{\nu x/U}$ versus Re_x, as shown in Fig. 10-15. For the laminar boundary layer $\delta/\sqrt{\nu x/U}$ is constant. The break at a local Reynolds number of about 3×10^5 is the point of transition to a turbulent boundary layer.

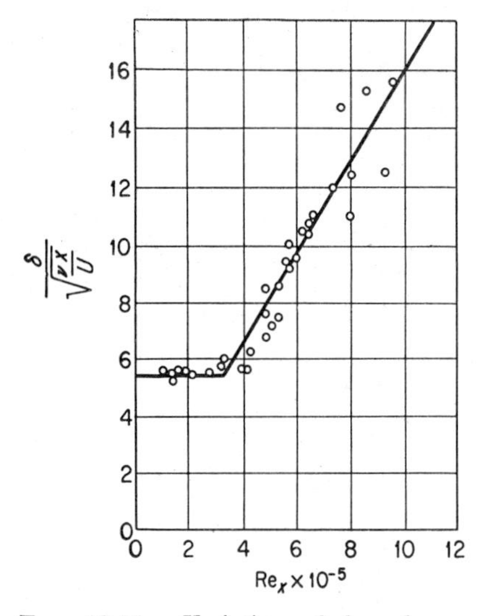

FIG. 10-15. Variation of boundary-layer thickness with local Reynolds number for flow over flat plates. (*From M. Hansen, NACA TM* 585, 1930.)

FIG. 10-16. Effect of main-stream turbulence on transition on a flat plate. (*From G. B. Schubauer and H. K. Skramstad, NACA Rept.* 909, 1948.)

The transition from a laminar to a turbulent boundary layer is affected by the roughness of the surface of the flat plate and the percentage turbulence of the main stream. For a smooth, flat plate with a sharp leading edge transition may occur at Reynolds numbers from 10^5 to 3×10^6. Dryden[7] reported transition Reynolds numbers of 1.1×10^6 for a turbulence of 0.5 per cent and 1×10^5 for a turbulence of 3 per cent. Hansen's result (transition at $\mathrm{Re}_x = 3 \times 10^5$) indicates that the turbulence was between 1 and 2 per cent for the plates he studied. The transition-region velocity profiles shown in Fig. 10-14 were measured for a free-stream turbulence of 0.03 per cent, and the transition zone extends from $\mathrm{Re}_x = 3 \times 10^6$ to 4×10^6. Comparable values of the transition Reynolds number are shown in Fig. 10-16, where Re_x is plotted versus the percentage turbulence. These curves were obtained by Schubauer and Skramstad.[53] The region below the curves corresponds to laminar flow, the transition zone is between the curves, and the turbulent zone lies above. The transition from laminar to turbulent flow is a sudden phenomenon, but the point of transition oscillates over the range of Reynolds numbers included in the transition region in Fig. 10-16. For flow of a turbulent stream past a flat plate transition may usually be considered to occur at a local Reynolds number of 5×10^5 if no information is available regarding the percentage turbulence of the free stream.

10-13. Means of Artificially Controlling the Laminar-Turbulent Transition on Flat Plates

The position of the laminar-turbulent transition on a flat plate may be moved either upstream or downstream in a number of ways. If the leading edge of the plate contains an irregularity to disturb the flow,[6] the boundary layer may be turbulent for the whole length of the plate. If the surface of the plate is rough, the transition will occur closer to the leading edge than for a smooth plate. Transition also occurs at a smaller value of Re_x if the plate is heated, and a similar effect is noted if flow occurs against a positive pressure gradient ($\partial P/\partial x$ positive).[35]

A common means of preventing transition to turbulent flow is to make the surface of the plate porous and to provide suction so that some of the fluid flows into the pores of the plate. This introduces new conditions in the solution of the boundary-layer equations, since at the surface the y component of velocity v is not zero but depends on the rate of flow into the porous surface.

10-14. Turbulent Boundary Layer on a Smooth, Flat Plate: Velocity Distribution

Velocity profiles in turbulent boundary layers have been obtained by van der Hegge-Zijnen,[24] Burgers,[2] and Hansen.[21] Figure 10-17 is a plot of log u versus log y for some data obtained by van der Hegge-Zijnen for air flowing over a plate at a point 59.1 in. from the leading edge of the plate. The laminar, transition, and turbulent zones in the boundary layer are indicated. Some plots obtained by Hansen are reproduced in Fig. 10-18, and experimental data obtained by Burgers are shown in Fig. 10-19. The insert in Fig. 10-18 gives the dimensions of the plate studied.

Using the same method as that used for circular tubes, von Kármán[28] determined the form of the velocity-profile equation for the turbulent portion of the boundary layer. At any point x on the plate the relationship between the point velocity u and the distance from the plate y is given by

$$\frac{u}{U} = \left(\frac{y}{\delta}\right)^{1/7} \tag{10-66}$$

This is the power law for the velocity distribution over flat plates. The data of van der Hegge-Zijnen [24] (Fig. 10-17) indicate that for the turbulent portion of the boundary layer the velocity is proportional to y to the 0.146

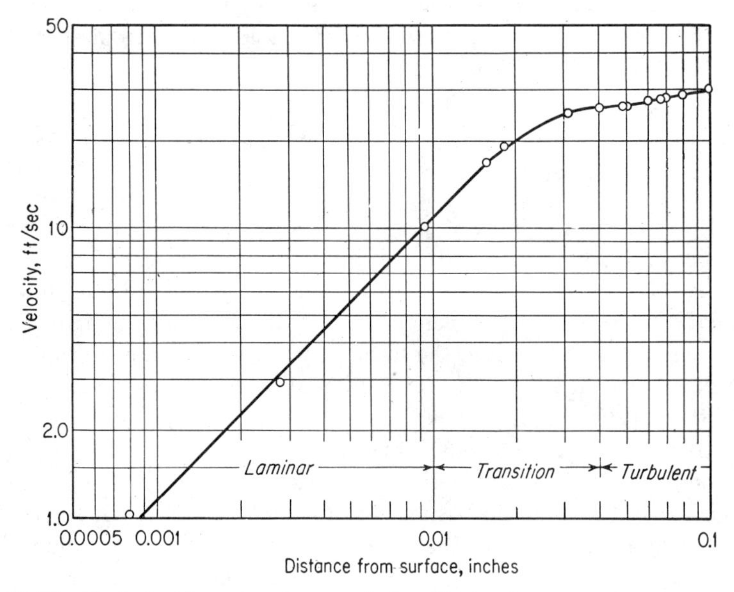

Fig. 10-17. Turbulent-boundary-layer velocity distribution on a flat plate. (*From B. G. van der Hegge-Zijnen, Thesis, Delft,* 1924.)

power, which is in good agreement with Eq. (10-66). Hansen's data (Fig. 10-18) indicate that the exponent on the term y/δ varies from 0.186 to 0.198, while the exponent given by Burgers's data (Fig. 10-19) varies from 0.15 to 0.18. Although velocity-distribution data for the turbulent bound-

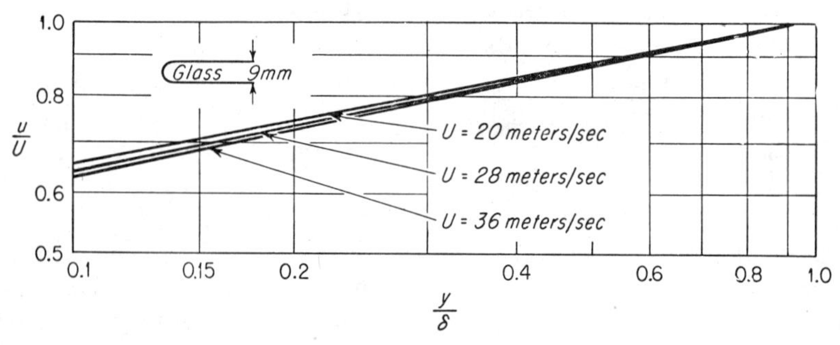

Fig. 10-18. Turbulent-boundary-layer velocity distribution on a flat plate. (*From M. Hansen, NACA TM* 585, 1930.)

ary layer obey some exponential law, it is not the one-seventh-power law of von Kármán. The exponent appears to lie in the range 0.15 to 0.20

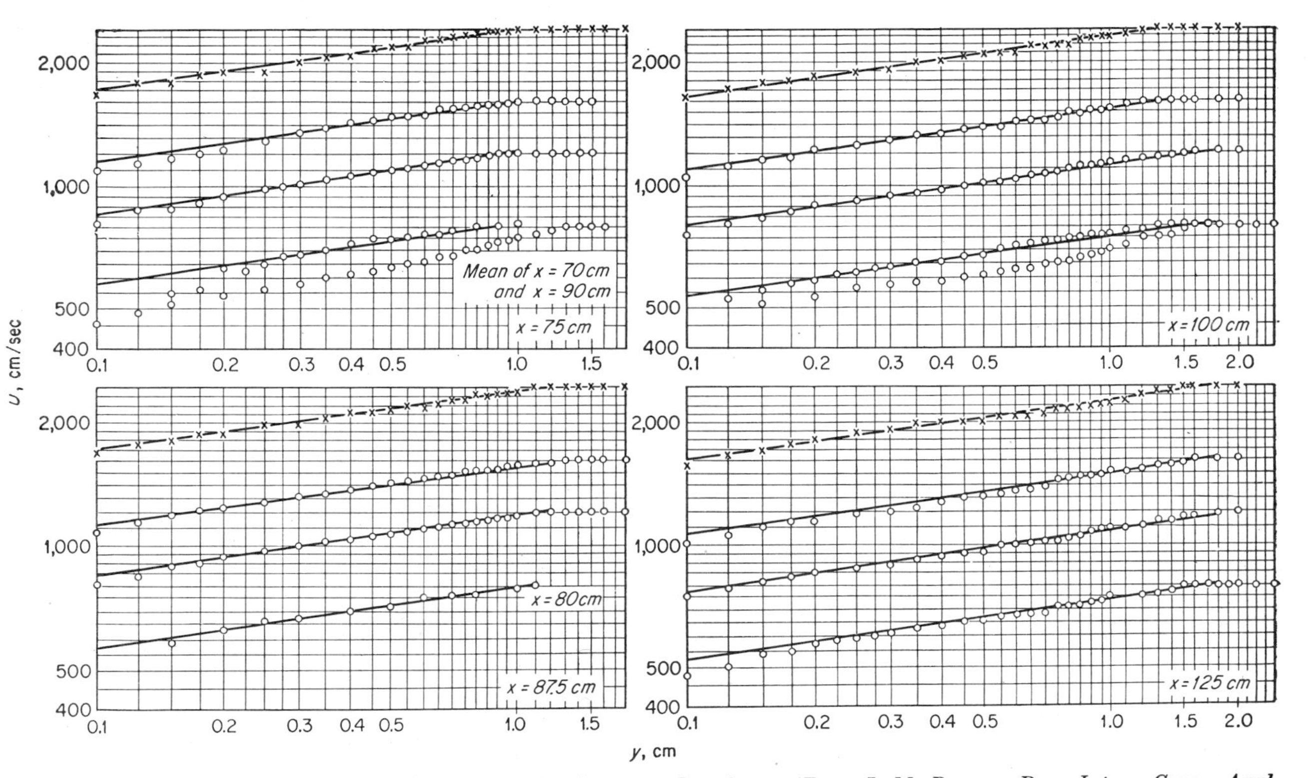

FIG. 10-19. Turbulent-boundary-layer velocity distribution on a flat plate. (*From J. M. Burgers, Proc. Intern. Congr. **Appl.** Mech., 1st Congr., Delft, 1924, p.* 113.)

and to depend on the velocity in the main stream and the type of plate over which the fluid is flowing.

The fact that the majority of the experimental data on the velocity profile in the turbulent boundary layer not do follow the one-seventh-power

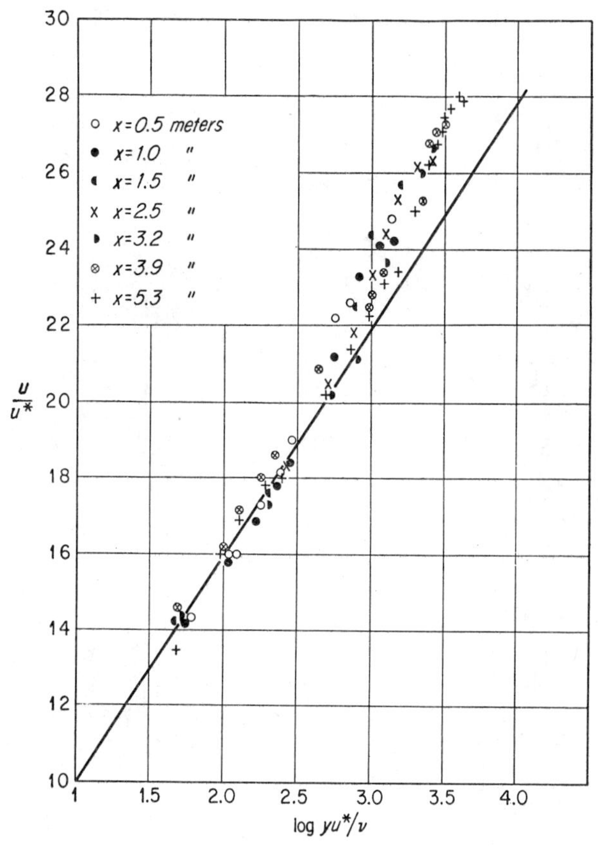

FIG. 10-20. Velocity distribution for the turbulent boundary layer on a flat plate plotted as u^+ versus y^+. (*From F. Schultz-Grunow, NACA TM* 986, 1941.)

law of von Kármán led Falkner [16] to propose a new formula, which agrees very well with most experimental data, especially at high Reynolds numbers.

$$\frac{u}{U} = \left(\frac{y}{\delta}\right)^{1/5} \tag{10-67}$$

It is seen that the exponent $1/7$ has been changed to $1/5$. Hansen's [21] data for the velocity profile on smooth plates indicate a better agreement with Eq. (10-67) than with Eq. (10-66).

Schultz-Grunow[55] correlated turbulent-boundary-layer data by plotting the dimensionless groups $u^+ = u/u^*$ versus $y^+ = yu^*/\nu$, where $u^* = \sqrt{\tau_w g_c/\rho}$. A plot of his results is shown in Fig. 10-20, where the straight line represents the equation

$$u^+ = 5.93 \log y^+ + 4.07 \tag{10-68}$$

At values of y^+ greater than 400 all the data deviate from Eq. (10-68). Schultz-Grunow recommends, on the basis of experimental work of Kempf,[30] that Eq. (7-35) be used in the range $10 < y^+ < 400$ for the velocity distribution in the turbulent boundary layer. This is the equation for the velocity distribution in the turbulent core for flow in a circular tube.

$$u^+ = 5.5 + 2.5 \ln y^+ \tag{7-35}$$

There appear to be three zones in the turbulent boundary layer of a flat plate. The laminar sublayer is adjacent to the surface and extends from $y^+ = 0$ to $y^+ = 10$. For the laminar sublayer

$$u^+ = y^+ \tag{7-37}$$

There is a zone from $y^+ = 10$ to $y^+ = 400$ where Eq. (7-35) represents the relationship between u^+ and y^+ and from a value of $y^+ = 400$ to the edge of the boundary layer Eq. (10-67) probably best represents the velocity distribution. Klebanoff[31] has described the portion of the boundary layer beyond $y^+ = 400$ as a region characterized by intermittency, which means that the thickness of the boundary layer varies with time because of turbulence fluctuations in the boundary layer and in the main stream.

10-15. Turbulent Boundary Layer on a Smooth, Flat Plate: Boundary-layer Thickness and Drag Coefficients

The thickness of the turbulent boundary layer is calculated from the Blasius friction equation (7-3) for the friction factors in circular tubes.

$$f = 0.079(\mathrm{Re})^{-1/4} \tag{7-3}$$

From Eq. (7-3) von Kármán[28] expressed the shear stress at the surface of the plate as

$$\tau_w = 0.0228 \frac{\rho U^2}{g_c} \left(\frac{\nu}{U\delta}\right)^{1/4} \tag{10-69}$$

Combining Eq. (10-69) with the integral equation (10-38), one obtains

$$\frac{\partial}{\partial x} \int_0^\delta u(U - u)\, dy = 0.0228 U^2 \left(\frac{\nu}{U\delta}\right)^{1/4} \tag{10-70}$$

Substituting for the value of u given by Eq. (10-66) and integrating gives an expression relating the boundary-layer thickness to the local Reynolds number and the distance from the leading edge of the plate.

$$\frac{\delta}{x} = 0.376(\text{Re}_x)^{-1/5} \qquad 5 \times 10^5 < \text{Re}_x < 10^7 \tag{10-71}$$

Equation (10-71) is valid in the local-Reynolds-number range $5 \times 10^5 < \text{Re}_x < 10^7$ since Eq. (10-66) is valid in this range. For local Reynolds numbers above 6.5×10^6 Falkner [16] recommends the relation

$$\frac{\delta}{x} = 0.1285(\text{Re}_x)^{-1/7} \tag{10-72}$$

A numerical comparison of Eqs. (10-71) and (10-72) is shown in Table 10-4.

TABLE 10-4. COMPARISON OF EQUATIONS FOR CALCULATING TURBULENT-BOUNDARY-LAYER THICKNESS ON FLAT PLATES

Reynolds number Re_x	$\frac{\delta}{x}$	
	By Eq. (10-72)	By Eq. (10-71)
10^4	0.0343	0.0598
10^5	0.0284	0.0376
10^6	0.0178	0.0238
10^7	0.0129	0.0149
10^8	0.0093	0.0102

Substituting Eq. (10-71) into Eq. (10-69),

$$\tau_w = 0.0228 \frac{\rho U^2}{g_c} \left[\frac{\nu (Ux/\nu)^{1/5}}{U(0.376x)} \right]^{1/4} \tag{10-73}$$

from which

$$\frac{2\tau_w g_c}{\rho U^2} = 0.0585(\text{Re}_x)^{-1/5} \tag{10-74}$$

and from Eq. (10-7)

$$f' = 0.0585(\text{Re}_x)^{-1/5} \tag{10-75}$$

Equation (10-75) gives the local drag coefficient for the turbulent boundary layer. Upon carrying out the integration indicated in Eq. (10-9), the total coefficient of friction is obtained.

$$f = 0.074(\text{Re}_L)^{-1/5} \qquad 5 \times 10^5 < \text{Re}_L < 10^7 \tag{10-76}$$

A number of other relationships for local and total drag coefficients have been derived for turbulent boundary layers on flat plates. In the high-Reynolds-number range Falkner recommends the relations

$$f' = 0.0262(\mathrm{Re}_x)^{-1/7} \tag{10-77}$$

and

$$f = 0.0306(\mathrm{Re}_L)^{-1/7} \qquad 6.5 \times 10^5 < \mathrm{Re}_L < 10^8 \tag{10-78}$$

Schlichting [48] derived an expression for the turbulent drag coefficient using von Kármán's momentum equation (10-38) and the logarithmic-distribution law [Eq. (7-35)] and obtained the semiempirical relation

$$f = \frac{0.455}{(\log \mathrm{Re}_L)^{2.58}} \tag{10-79}$$

Prandtl [45] suggested that an amount equivalent to $1{,}700/\mathrm{Re}_L$ be subtracted from the right side of Eqs. (10-76) and (10-79) to account for the laminar boundary layer on the front part of the plate. The Prandtl-Schlichting equation for the drag coefficient makes use of this correction; i.e.,

$$f = \frac{0.455}{(\log \mathrm{Re}_L)^{2.58}} - \frac{C}{\mathrm{Re}_L} \tag{10-80}$$

where the factor C is a function of the Reynolds number of transition as shown in Table 10-5.

TABLE 10-5. VALUES OF C TO BE USED IN EQ. (10-80)

Reynolds number of transition	C
3×10^5	1,050
5×10^5	1,700
10^6	3,300
3×10^6	8,700

Schultz-Grunow [55] derived the following relations for the local and total drag coefficients for flow on flat plates based on velocity distributions he measured:

$$f' = \frac{0.370}{(\log \mathrm{Re}_x)^{2.584}} \tag{10-81}$$

$$f = \frac{0.427}{(-0.407 + \log \mathrm{Re}_L)^{2.64}} \qquad 10^6 < \mathrm{Re}_L < 10^9 \tag{10-82}$$

Equations (10-76), (10-78), (10-79), and (10-82) are relationships giving the total drag coefficient for turbulent flow on a flat plate. These equations are compared in Table 10-6, where the drag coefficient predicted by each is shown for several Reynolds numbers. Equations (10-79) and (10-82) agree well over the whole Reynolds-number range and are recommended where precise values of the drag coefficient are required.

TABLE 10-6. COMPARISON OF DRAG-COEFFICIENT EQUATIONS FOR TURBULENT FLOW OVER FLAT PLATES

Re_L	Predicted f			
	Eq. (10-76)	Eq. (10-78)	Eq. (10-79)	Eq. (10-82)
10^5	0.0074	0.00592	0.00712	0.00762
10^6	0.00466	0.00426	0.00446	0.00454
10^7	0.00295	0.00306	0.00299	0.00291
10^8	0.00186	0.00220	0.00214	0.00204
10^9	0.00117	0.00159	0.00157	0.00146

Equation (10-63) (for the laminar boundary layer), Eq. (10-80) (with $C = 1{,}700$), and Eq. (10-79) are plotted in Fig. 10-21. Data of Kempf,[30] Wieselsberger,[62] and Falkner [16] are included.

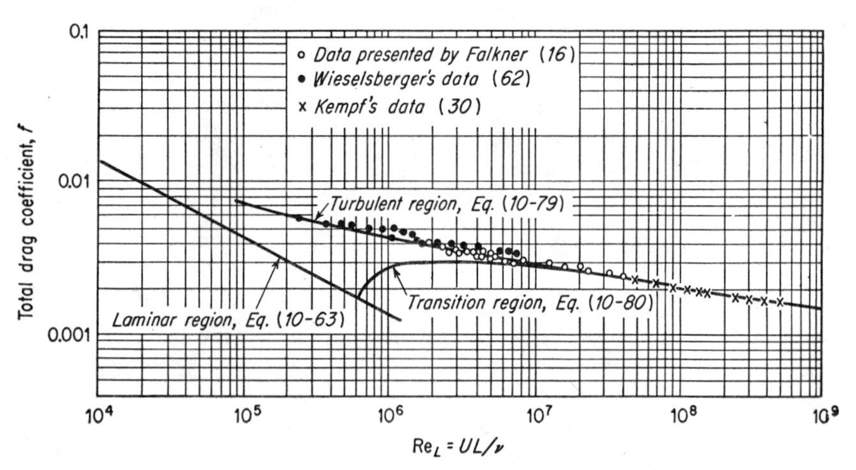

FIG. 10-21. Total drag coefficients for flow parallel to flat plates.

Example 10-4

Air at 60°F and 1 atm pressure is flowing parallel to a flat plate which is 4 ft square. The air velocity is 100 ft/sec. Calculate (*a*) the boundary-layer thickness 18 in. from the leading edge, (*b*) the total drag coefficient, and (*c*) the drag force on the plate.

Solution

For air at 60°F

$$\nu = 1.58 \times 10^{-4}\ \text{ft}^2/\text{sec}$$

$$\rho = 0.0765\ \text{lb}_m/\text{ft}^3$$

(*a*) At $x = 18$ in.

$$\text{Re}_x = \frac{(1.5)(100)}{1.58 \times 10^{-4}} = 9.49 \times 10^5$$

From Eq. (10-71)

$$\frac{\delta}{x} = (0.376)(9.49 \times 10^5)^{-1/5} = 0.0240$$

$$\delta = (1.5)(0.0240) = 0.036\ \text{ft}$$

$$= 0.432\ \text{in.}$$

This is the boundary-layer thickness at $x = 18$ in.

(*b*)

$$\text{Re}_L = \frac{(4)(100)}{1.58 \times 10^{-4}} = 2.53 \times 10$$

By Eq. (10-80), with $C = 1{,}700$,

$$f = \frac{0.455}{[\log (2.53 \times 10^6)]^{2.58}} - \frac{1{,}700}{2.53 \times 10^6} = 0.00306$$

(*c*) From Eq. (10-5) the total drag force exerted on the plate is

$$\frac{(0.00306)(0.0765)(100)^2(4)(4)(2)}{(2)(32.2)} = 1.16\ \text{lb}_f$$

10-16. The Admissible Roughness of Flat Plates

The admissible roughness of a flat plate is the roughness above which an increase in drag coefficient over that for a smooth plate would occur. A plate with a surface roughness below the admissible roughness may be considered hydraulically smooth. Schlichting [49] recommends the following relation for the determination of the admissible roughness:

$$\frac{U e_{admiss}}{\nu} = 10^2 \tag{10-83}$$

where e_{admiss} is the admissible height of the roughness projections.

10-17. Turbulent Flow past Thin, Rough Plates

A small amount of work has been done on the determination of velocity profiles in the boundary layer for air flowing over rough plates. Van der Hegge-Zijnen [23] studied velocity profiles on a plate having small quadrilateral pyramids on its surface and measuring 189.5 by 50 cm. The projections on the surface were described as quadrilateral pyramids arranged in vertical and horizontal rows. The mean height was 1.7 mm, and the spacing was 6.5 mm in the longitudinal direction and 6.33 mm in the transverse direction. Both the tops of the projections and the valleys were slightly rounded. Figure 10-22 shows a longitudinal cross section of this *waffle plate* investigated by van der Hegge-Zijnen. Velocity profiles were determined at values of x from 25 to 175 cm and at values of the average velocity from 811 to 3,200 cm/sec. A plot of the data obtained at values of $x = 175$ cm and $x = 25$ cm is shown in Fig. 10-23. At values of y greater than 0.50 cm the points lie approximately on a straight line. Van der Hegge-Zijnen investigated the possibility that the velocity in the boundary layer could be represented by the formula

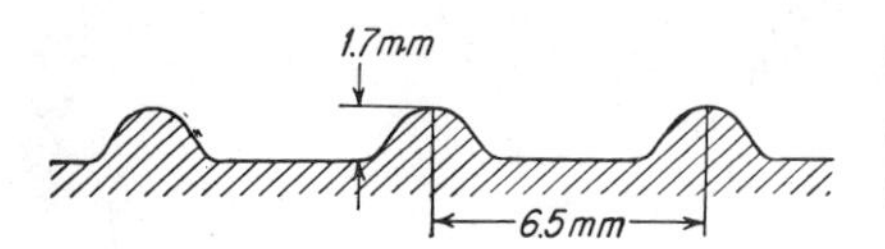

FIG. 10-22. Longitudinal cross section of a rough plate. [*From B. G. van der Hegge-Zijnen, Verhandel. Koninkl. Akad. Wetenschap. Amsterdam Afdeel. Natuurk.*, **31**: 499 (1928).]

$$u = y^n \tag{10-84}$$

The results of his experiments are summarized in Table 10-7, where mean values of the exponent n are tabulated as a function of x. At each value of x the value of n increases slightly with increasing velocity in the main

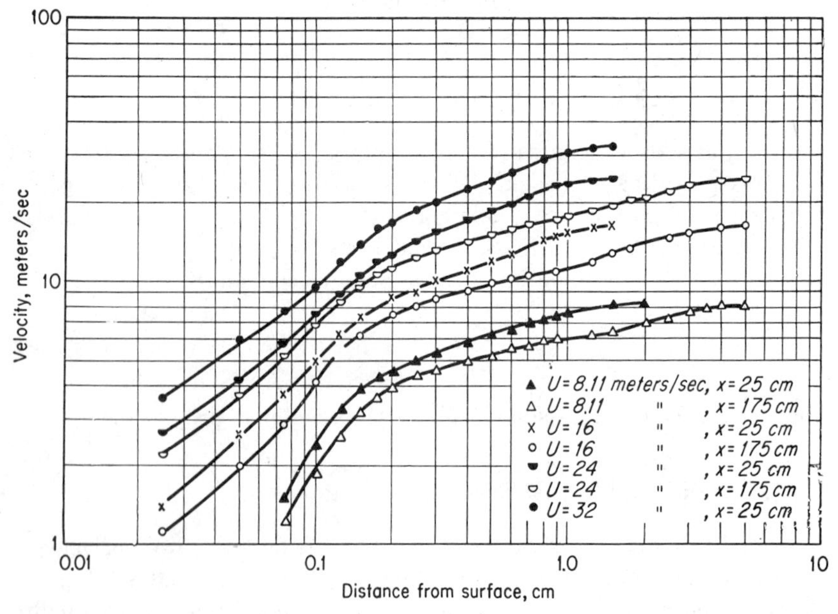

FIG. 10-23. Velocity distribution in the boundary layer on a rough plate. [*From B. G. van der Hegge-Zijnen, Verhandel. Koninkl. Akad. Wetenschap. Amsterdam Afdeel. Natuurk.*, **31**:499 (1928).]

stream. The average deviation given in Table 10-7 is the average deviation (from the average value) of the observed values of n at each velocity. It is seen that, as x increases, n decreases, indicating, according to van der Hegge-Zijnen, that perhaps the turbulent boundary layer was not fully developed at low values of x. Above $x = 100$ cm n becomes constant at about 0.25.

TABLE 10-7. VALUES OF n [IN EQ. (10-84)] OBTAINED BY VAN DER HEGGE-ZIJNEN †

Distance from leading edge of plate, cm	Average of n	Average deviation of observed values of n from average value
15	0.35	0.025
20	0.35	0.015
25	0.36	0.011
37.5	0.31	0.010
50	0.33	0.009
62.5	0.30	0.012
75	0.27	0.002
87.5	0.28	0.003
100	0.29	0.006
125	0.26	0.010
150	0.25	0.004
175	0.25	0.005

† B. G. van der Hegge-Zijnen, *Verhandel. Koninkl. Akad. Wetenschap. Amsterdam Afdeel. Natuurk.*, **31**:499 (1928).

Hansen [21] also made a study of the velocity profiles over rough plates and determined the value of the exponent n. No measurements are available on the roughness projections on the plates used by Hansen. He reported that for plates having a wavy surface the value of n varied from 0.189 to 0.194, increasing slightly as the velocity of the main stream increased. For a surface with a few shallow indentations the value varied from 0.215 to 0.233, and for a plate having well-defined projections on its surface the value of n was from 0.284 to 0.325. These results are somewhat in agreement with those of van der Hegge-Zijnen.

Van der Hegge-Zijnen derived the following formula for determining the thickness of the turbulent boundary layer on the surface of a rough plate:

$$\delta = 0.259 \left(\frac{x}{e}\right)^{2/3} e(1 - 0.00059x) \qquad x < 175 \text{ cm} \qquad (10\text{-}85)$$

The quantities x, e, and δ are measured in centimeters.

He also measured the thickness of the boundary layer over the plate and compared the measured value with that calculated from Eq. (10-85). This equation expresses the thickness of the boundary layer as a function of the distance from the leading edge x and the average height of the roughness projections e. Table 10-8 shows the calculated and experimental values of δ on a rough plate. The agreement between the experimental boundary-layer thickness and the calculated value is very good. Equation (10-85) indicates that the boundary-layer thickness is independent of the velocity in the main stream. Van der Hegge-Zijnen found experimentally that the boundary-layer thickness decreased somewhat as the velocity increased, the thickness decreasing about 20 per cent as the velocity increased from 800 to 3,200 cm/sec.

TABLE 10-8. VALUES OF BOUNDARY-LAYER THICKNESS AS CALCULATED AND AS OBSERVED EXPERIMENTALLY FOR FLOW PAST ROUGH PLATES

Distance from leading edge of plate, cm	Thickness of boundary layer	
	Obtained experimentally δ, cm	Obtained from Eq. (10-85) for $e = 1.7$ mm
25	1.11	1.21
50	1.84	1.89
75	2.63	2.44
100	3.01	2.92
125	3.69	3.32
150	3.80	3.70
175	3.72	4.03

10-18. Drag Coefficients for Turbulent Flow past Rough Plates

For a plate with a roughness below the admissible roughness any of the smooth-plate relationships may be used to calculate the drag coefficient. Schlichting [49] indicates that when the dimensionless group Ue/ν is above 2,000, the drag coefficient for rough plates is a function of the roughness only and not of the Reynolds number. This drag coefficient is given by the relation

$$f = \left(1.89 + 1.62 \log \frac{L}{e}\right)^{-2.5} \qquad 10^2 < \frac{L}{e} < 10^6 \qquad (10\text{-}86)$$

$$\frac{Ue}{\nu} > 2,000$$

II. FLOW NORMAL TO TWO-DIMENSIONAL BODIES

10-19. The Boundary Layer on Two-dimensional Bodies

Prandtl's momentum equations for the boundary layer are

$$u\frac{\partial u}{\partial x} + v\frac{\partial u}{\partial y} = -\frac{g_c}{\rho}\frac{\partial P}{\partial x} + \frac{\mu}{\rho}\frac{\partial^2 u}{\partial y^2} \tag{10-3}$$

$$\frac{\partial u}{\partial x} + \frac{\partial v}{\partial y} = 0 \tag{10-4}$$

For flow over a flat plate the term $-\frac{g_c}{\rho}\frac{\partial P}{\partial x}$ was neglected because the velocity at the edge of the boundary layer was independent of x. When flow takes place normal to thick bodies such as cylinders, air foils, and wedges, the velocity at the edge of the boundary layer is a function of x, and in solving Eqs. (10-3) and (10-4) the term $-\frac{g_c}{\rho}\frac{\partial P}{\partial x}$ must be considered.

When fluid flows past a two-dimensional body (say a cylinder), the fluid is accelerated as it passes over the forward portion of the body and is then decelerated after it passes the thickest part of the body. This is responsible for the phenomenon known as *separation* of the boundary layer.

Separation of the boundary layer occurs at the point on the body where the pressure gradient is zero. In Fig. 10-24 a section of a cylinder with fluid flowing past it is shown. The boundary-layer thickness increases

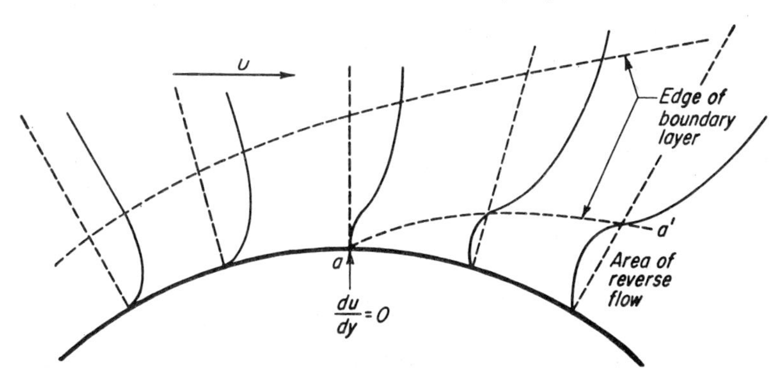

FIG. 10-24. Schematic sketch of a boundary layer on a circular cylinder.

with the distance in the direction of flow. The fluid in the main stream flowing past the cylinder is accelerated as a result of going around the cylinder, and the acceleration (increase in kinetic energy) is accompanied by a decrease in the pressure ($\partial P/\partial x$ is negative). However, as the fluid in the main stream goes past the cylinder, the expanding cross section of flow on the right side of Fig. 10-24 requires a deceleration of the fluid and a corresponding increase in the pressure; i.e., $\partial P/\partial x$ is positive. The boundary layer is thus flowing against an adverse pressure gradient as it moves around the cylinder, which results in a marked change in the velocity profile in the boundary layer. In order to maintain flow in the direction of the adverse pressure gradient, the boundary layer separates from the solid surface and continues in space.

Actually, the point of separation is a stagnation point. The velocity gradient $\partial u/\partial y$ at the surface is zero. At any point on the surface of the cylinder before separation $u = 0$ and $v = 0$, so, from Eq. (10-3),

$$\left(\frac{\partial P}{\partial x} = \frac{\mu}{g_c}\frac{\partial^2 u}{\partial y^2}\right)_{y=0} \tag{10-87}$$

Equation (10-87) indicates that for a change in sign of $\partial P/\partial x$ the term $\partial^2 u/\partial y^2$ changes in sign, so that the velocity-profile curve exhibits a point of inflection.

The above results are presented schematically in Fig. 10-24. The broken lines represent the limits of the boundary layer on a cylinder immersed in a flowing fluid. Various velocity profiles are drawn in this layer. Separation is shown to take place where $(\partial u/\partial y)_{y=0} = 0$. The velocity-profile curve has a point of inflection as it crosses the inner limit of the separated boundary layer.

Actual separation of fluid as it flows around a cylinder has been observed experimentally by studying flow patterns. In Fig. 10-25, consisting of a series of photographs obtained by Tietjens [59] for fluid flowing past a cylinder at various Reynolds numbers, the separation of the boundary layer can be easily seen, and the disturbance in the fluid downstream from the cylinder is indicated. At the point of separation of the boundary layer from the surface of the cylinder the velocity gradient at the surface is zero or nearly zero. Beyond the point of separation the fluid is flowing in the direction opposite to that in the main stream. The broken line aa' (Fig. 10-24) is the edge of the separated boundary layer and is the point where the reverse flow under the boundary layer changes to forward flow in the boundary layer. Thus the area behind the cylinder is an area of disturbed flow, being characterized by eddies which may be clearly seen in Fig. 10-25. The area of disturbance beyond the cylinder is the turbulent wake, which, as will be shown in a later section, dissipates a large amount of energy.

The Reynolds number for fluids flowing past immersed cylinders is expressed as d_0U/ν, where d_0 is the outside diameter of the cylinder, U is the velocity in the main stream, and ν is the kinematic viscosity of the fluid.

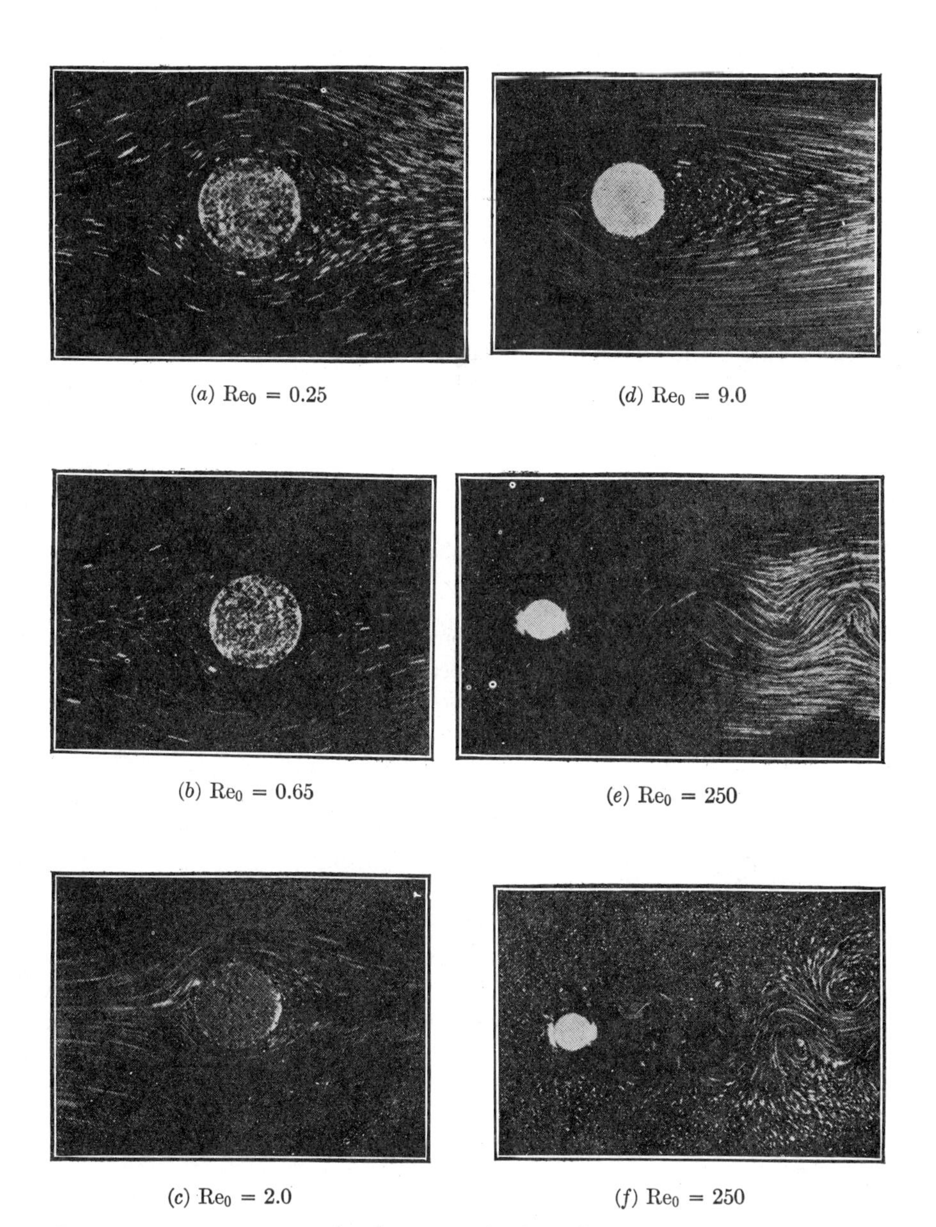

(*a*) $\text{Re}_0 = 0.25$

(*d*) $\text{Re}_0 = 9.0$

(*b*) $\text{Re}_0 = 0.65$

(*e*) $\text{Re}_0 = 250$

(*c*) $\text{Re}_0 = 2.0$

(*f*) $\text{Re}_0 = 250$

FIG. 10-25. Flow patterns for flow past circular cylinders. (*From O. J. Tietjens, Proc. Intern. Congr. Appl. Mech., 3rd Congr., Stockholm, 1930, vol.* 1, *p.* 331.)

10-20. Comments on the Solution of the Two-dimensional Boundary-layer Equation

The solution of Eqs. (10-3) and (10-4) requires a knowledge of the term $\partial P/\partial x$. Since Eq. (10-3) applies over the whole boundary layer, a relation similar to Eq. (10-11) may be obtained for the pressure gradient; i.e.,

$$-\frac{g_c}{\rho}\frac{\partial P}{\partial x_1} = u_{\max}\frac{\partial u_{\max}}{\partial x_1} \tag{10-88}$$

where $u_{\max}$ is the velocity at the edge of the boundary layer. Equations (10-3) and (10-4) are then rewritten for two-dimensional bodies as follows:

$$u\frac{\partial u}{\partial x_1} + v\frac{\partial u}{\partial y} = u_{\max}\frac{\partial u_{\max}}{\partial x_1} + \frac{\mu}{\rho}\frac{\partial^2 u}{\partial y^2} \tag{10-89}$$

$$\frac{\partial u}{\partial x_1} + \frac{\partial v}{\partial y} = 0 \tag{10-90}$$

where x_1 is the distance measured from the forward point of stagnation of the body and y is the distance measured normal to the surface of the body. These quantities are illustrated in Fig. 10-26. Only in the case of the flat plate are x and x_1 identical, the quantity x being the distance measured in the direction of the x axis in the rectangular coordinate system.

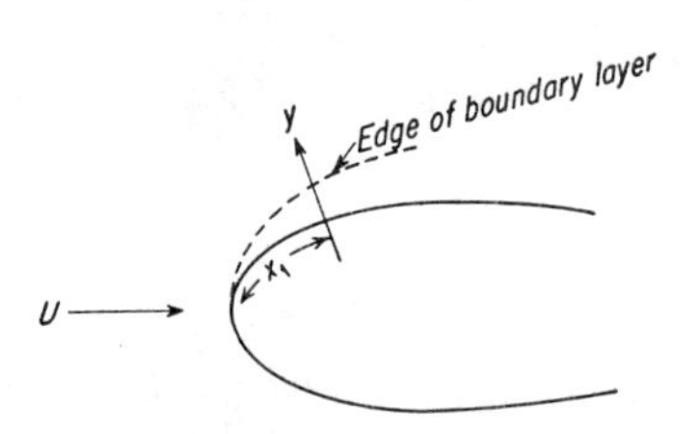

FIG. 10-26. Definition of distances x_1 and y for Eq. (10-89).

The following methods are employed in solving Eqs. (10-89) and (10-90):

1. Assume a velocity distribution in the boundary layer and use the von Kármán integral equation. This method was employed by Pohlhausen [43] and later refined by Dryden.[8] Their results agree well with experiment.

2. Assume a relation for the variation of $u_{\max}$ with x_1. It has been found that if the relation

$$u_{\max} = cx_1^{n'} \tag{10-91}$$

is assumed, Eq. (10-89) may be reduced to an ordinary nonlinear differential equation which may be solved. Falkner and Skan [17] used this method to determine the velocity distribution on the forward portion of a circular cylinder.

3. Assume that the velocity at the edge of the boundary layer varies the same as if the flow were irrotational. For a circular cylinder the veloc-

ity at the edge of the boundary layer would then be expressed by Eq. (3-57):

$$|\mathbf{V}| = 2U \sin \theta \tag{3-57}$$

where $|\mathbf{V}|$ is the absolute value of the velocity at the edge of the boundary layer. Since $|\mathbf{V}|$ has direction parallel to the surface of the cylinder, it corresponds to $u_{\max}$, the velocity at the edge of the boundary layer. Thus

$$u_{\max} = 2U \sin \theta \tag{10-92}$$

Meksyn [39] used this method to obtain an expression for the laminar boundary layer on cylindrical and elliptical cylinders and on spheres.

10-21. The Solution of Eq. (10-89) Assuming $u_{\max} = c{x_1}^{n'}$

If the velocity at the edge of the boundary layer is assumed to vary according to Eq. (10-91), it is possible to reduce Eq. (10-89) to an ordinary nonlinear differential equation. Defining the stream function ψ as in Eqs. (3-16) and (3-17), (also replacing x by x_1) satisfies the continuity equation. Defining the functions

$$\eta_1 = y \sqrt{\left(\frac{1 + n'}{2}\right)\left(\frac{u_{\max}\rho}{\mu x_1}\right)} \tag{10-93}$$

and

$$\psi = \left(\frac{2}{1 + n'} \frac{\mu x_1 u_{\max}}{\rho}\right)^{1/2} \phi \tag{10-94}$$

where ϕ is a function of η_1 only, and x_1 and y are measured according to Fig. 10-26. Equation (10-89) becomes

$$\phi''' + \phi''\phi + \lambda(1 - \phi'^2) = 0 \tag{10-95}$$

where

$$\lambda = \frac{2n'}{1 + n'} \tag{10-96}$$

Equation (10-95) is an ordinary nonlinear differential equation. This equation was solved numerically by Hartree,[22] and the solution is tabulated in Table 10-9, where values of $\phi''(0)$ and ϕ' are given for various values of η_1 and λ. This table may be used to calculate velocity profiles and local coefficients of friction for flow over any immersed body provided the value of n' is known. The velocity profile and local coefficient of friction are calculated from the relations [derived from Eqs. (10-93) and (10-94)]

$$\frac{u}{u_{\max}} = \phi' \tag{10-97}$$

$$\left(\frac{\partial u}{\partial y}\right)_{\eta_1=0} = u_{\max} \sqrt{\frac{1 + n'}{2} \frac{u_{\max}\rho}{\mu x_1}}\, \phi''(0) \tag{10-98}$$

TABLE 10-9. SOLUTION OF EQ. (10-95) †

λ	-0.198_8	-0.19	-0.18	-0.16	-0.14	-0.10	0	0.1	0.2	0.3
$\phi''(0)$	0.0000	0.086	0.128	0.190	0.239	0.319	0.4696	0.5870	0.686	0.774
η_1	ϕ'	ϕ'	ϕ'	ϕ'	ϕ'	ϕ'	ϕ'	ϕ'	ϕ'	ϕ'
0.0	0.0000	0.0000	0.0000	0.0000	0.0000	0.0000	0.0000	0.0000	0.0000	0.0000
0.1	0.0010	0.0095	0.0137_5	0.0198_5	0.0246_5	0.0324	0.0469_5	0.0582	0.0677	0.0760
0.2	0.0040	0.0209	0.0293	0.0413	0.0507	0.0659	0.0939	0.1154	0.1334	0.1490
0.3	0.0089	0.0343	0.0467	0.0643	0.0781	0.1003	0.1408	0.1715	0.1970	0.2189
0.4	0.0158	0.0495	0.0659	0.0889	0.1069	0.1356	0.1876	0.2265	0.2584	0.2858
0.5	0.0248	0.0665	0.0868	0.1151	0.1370	0.1718	0.2342	0.2803	0.3177	0.3495
0.6	0.0358	0.0855	0.1094	0.1427	0.1684	0.2088	0.2806	0.3328	0.3747	0.4100
0.7	0.0487	0.1063	0.1338	0.1719	0.2010	0.2466	0.3266	0.3839	0.4294	0.4672
0.8	0.0636	0.1289	0.1598	0.2023	0.2347	0.2849	0.3720	0.4335	0.4816	0.5212
0.9	0.0803	0.1533	0.1874	0.2341	0.2694	0.3237	0.4167	0.4815	0.5312	0.5718
1.0	0.0991	0.1794	0.2166	0.2671	0.3050	0.3628	0.4606	0.5274	0.5782	0.6190
1.2	0.1423	0.2364	0.2791	0.3362	0.3784	0.4415	0.5453	0.6135	0.6640	0.7033
1.4	0.1927	0.2991	0.3463	0.4083	0.4534	0.5194	0.6244	0.6907	0.7383	0.7743
1.6	0.2498	0.3665	0.4170	0.4820	0.5284	0.5948	0.6967	0.7583	0.8011	0.8326
1.8	0.3126	0.4372	0.4896	0.5555	0.6016	0.6660	0.7610	0.8160	0.8528	0.8791
2.0	0.3802	0.5095	0.5621	0.6269	0.6712	0.7314	0.8167	0.8637	0.8940	0.9151
2.2	0.4509	0.5814	0.6327	0.6944	0.7354	0.7896	0.8633	0.9019	0.9260	0.9421
2.4	0.5230	0.6509	0.6995	0.7561	0.7927	0.8398	0.9011	0.9315	0.9500	0.9617
2.6	0.5946	0.7162	0.7605	0.8107	0.8422	0.8817	0.9306	0.9537	0.9672	0.9754
2.8	0.6635	0.7754	0.8146	0.8574	0.8836	0.9153	0.9529	0.9697	0.9792	0.9847
3.0	0.7278	0.8273	0.8607	0.8959	0.9168	0.9413	0.9691	0.9808	0.9873	0.9908
3.2	0.7858	0.8713	0.8986	0.9265	0.9425	0.9607	0.9804	0.9883	0.9924	0.9943
3.4	0.8364	0.9071	0.9286	0.9499	0.9616	0.9746	0.9880	0.9931	0.9957	0.9970
3.6	0.8789	0.9352	0.9515	0.9669	0.9752	0.9841	0.9929	0.9961	0.9976	0.9984
3.8	0.9132	0.9563	0.9681	0.9789	0.9845	0.9904	0.9959	0.9978	0.9987	0.9991_5
4.0	0.9399	0.9716	0.9798	0.9871	0.9907	0.9944	0.9978	0.9988_5	0.9993	0.9995_5
4.2	0.9598	0.9822	0.9876	0.9924	0.9946	0.9969	0.9988	0.9994	0.9996_5	0.9997_5
4.4	0.9741	0.9893	0.9927	0.9957	0.9970	0.9983	0.9994	0.9997	0.9998_5	0.9999
4.6	0.9839	0.9938	0.9959	0.9977	0.9984	0.9991	0.9997	0.9998_5	0.9999_5	0.9999_5
4.8	0.9904	0.9965	0.9978	0.9988	0.9992	0.9996	0.9999	0.9999_5	1.0000	1.0000
5.0	0.9945	0.9981_5	0.9988_5	0.9994	0.9996	0.9998	0.9999_5	1.0000		
5.2	0.9969	0.9990	0.9994	0.9997	0.9998	0.9999	1.0000			
5.4	0.9984	0.9995	0.9997	0.9999	0.9999_5	1.0000				
5.6	0.9992	0.9997	0.9999	0.9999_5	1.0000					
5.8	0.9996_5	0.9999	0.9999_5	1.0000						
6.0	0.9998_5	0.9999_5	1.0000							
6.2	0.9999_5	1.0000								
6.4	1.0000									

† This table may be used for Eqs. (10-106), (10-115), and (10-135) by replacing λ with λ_1 and η_1 with η_2 and for Eqs. (10-109) and (10-112) by replacing λ with λ_2 and η_1 with η_3.

TO GIVE VALUES OF ϕ' ‡ AND $\phi''(0)$

λ	0.4	0.5	0.6	0.8	1.0	1.2	1.6	2.0	2.4
$\phi''(0)$	0.854	0.927	0.996	1.120	1.2326	1.336	1.521	1.687	1.837
η_1	ϕ'	ϕ'	ϕ'	ϕ'	ϕ'	ϕ'	ϕ'	ϕ'	ϕ'
0.0	0.0000	0.0000	0.0000	0.0000	0.0000	0.0000	0.0000	0.0000	0.0000
0.1	0.0834	0.0903	0.0966	0.1080	0.1183	0.1276	0.1441	0.1588	0.1720
0.2	0.1628	0.1756	0.1872	0.2081	0.2266	0.2433	0.2726	0.2980	0.3206
0.3	0.2382	0.2558	0.2719	0.3003	0.3252	0.3475	0.3859	0.4186	0.4472
0.4	0.3097	0.3311	0.3506	0.3848	0.4144	0.4405	0.4849	0.5219	0.5537
0.5	0.3771	0.4015	0.4235	0.4619	0.4946	0.5231	0.5708	0.6096	0.6424
0.6	0.4403	0.4670	0.4907	0.5317	0.5662	0.5959	0.6446	0.6834	0.7155
0.7	0.4994	0.5276	0.5524	0.5947	0.6298	0.6596	0.7076	0.7449	0.7752
0.8	0.5545	0.5834	0.6086	0.6512	0.6859	0.7150	0.7610	0.7858	0.8235
0.9	0.6055	0.6344	0.6596	0.7015	0.7350	0.7629	0.8058	0.8376	0.8624
1.0	0.6526	0.6811	0.7056	0.7460	0.7778	0.8037	0.8432	0.8717	0.8934
1.2	0.7351	0.7615	0.7837	0.8194	0.8467	0.8682	0.8997	0.9214	0.9373
1.4	0.8027	0.8258	0.8449	0.8748	0.8968	0.9137	0.9375	0.9530	0.9640
1.6	0.8568	0.8860	0.8917	0.9154	0.9324	0.9450	0.9620	0.9726	0.9799
1.8	0.8988	0.9141	0.9264	0.9443	0.9569	0.9658	0.9775	0.9845	0.9892
2.0	0.9305	0.9421	0.9514	0.9644	0.9732	0.9793	0.9871	0.9914	0.9944
2.2	0.9537	0.9621	0.9689	0.9779	0.9841	0.9879	0.9928	0.9954	0.9970
2.4	0.9700	0.9760	0.9807	0.9867	0.9905	0.9931	0.9961	0.9976	0.9985
2.6	0.9812	0.9852	0.9884	0.9922	0.9946	0.9962	0.9980	0.9989	0.9993
2.8	0.9886	0.9913	0.9933	0.9956	0.9971	0.9980	0.9990	0.9994_5	0.9996_5
3.0	0.9933	0.9952	0.9962	0.9976	0.9985	0.9989	0.9995	0.9997_5	0.9998_5
3.2	0.9962	0.9974	0.9979	0.9987	0.9992	0.9995	0.9998	0.9999	0.9999_5
3.4	0.9979	0.9986	0.9989	0.9993	0.9996	0.9997_5	0.9999	1.0000	1.0000
3.6	0.9989	0.9993	0.9995	0.9997	0.9998	0.9999	1.0000		
3.8	0.9994	0.9997	0.9997_5	0.9998_5	0.9999	1.0000			
4.0	0.9997	0.9999	0.9999	0.9999_5	1.0000				
4.2	0.9999	0.9999_5	0.9999_5	1.0000					
4.4	0.9999_5	1.0000	1.0000						
4.6	1.0000								
4.8									
5.0									
5.2									
5.4									
5.6									
5.8									
6.0									
6.2									
6.4									

‡ From D. R. Hartree, *Proc. Cambridge Phil. Soc.*, **33**:223 (1937).

The flow on the forward portion of sharp-edged wedges, circular cylinders, and parabolic cylinders closely follows the relation

$$u_{\max} = c x_1^{n'} \tag{10-91}$$

For circular and parabolic cylinders n' has a value of unity over a large part of the front. For sharp-edged wedges

$$n' = \frac{\alpha}{\pi - \alpha} \tag{10-99}$$

where 2α is the angle of the wedge (see Fig. 10-27).

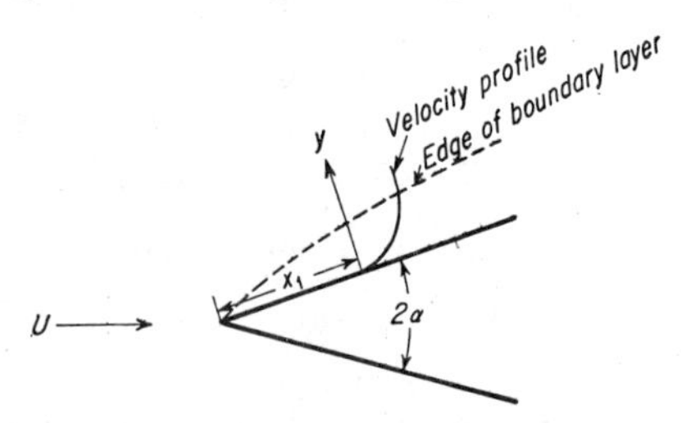

FIG. 10-27. Flow past a two-dimensional wedge.

10-22. The Pressure Distribution for Flow over Circular Cylinders

The sketch in Fig. 10-28 serves to define the various terms concerned with the boundary layer formed on cylinders. The broken line represents the edge of the boundary layer of thickness δ. The conditions in the undisturbed stream are given by the velocity U and the pressure P. The outer edge of the boundary layer is the point where $\partial u/\partial y$ becomes zero.

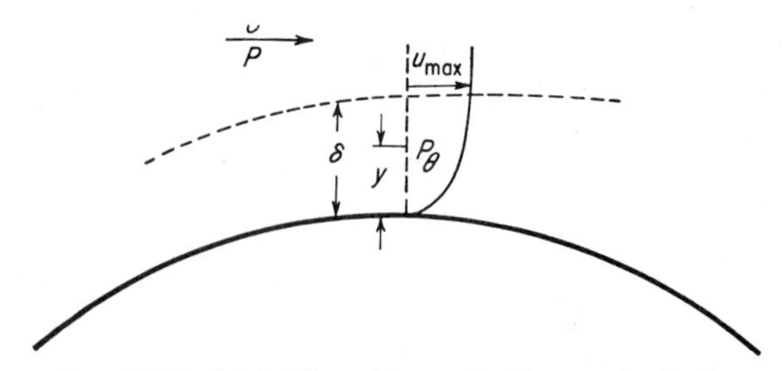

FIG. 10-28. Definition of terms for flow past cylinders.

The velocity at this point is $u_{\max}$; it differs from the main-stream velocity and varies with the angle θ, which indicates positions on the surface of the cylinder in relation to the leading edge. The static pressure P_θ in the boundary layer is constant across the thickness of the boundary layer Since flow in the main stream is considered irrotational, Bernoulli's steady-flow equation may be applied between the main stream and the edge of the boundary layer thus

$$P_\theta + \frac{\rho u_{\max}^2}{2g_c} = P + \frac{\rho U^2}{2g_c} \tag{10-100}$$

Solving for $u_{\max}$,

$$u_{\max} = U \sqrt{\frac{P - P_\theta}{\rho U^2/2g_c} + 1} \tag{10-101}$$

Therefore, from Eq. (10-101), the velocity at the edge of the boundary layer $u_{\max}$ may be calculated from the pressure distribution over the cylinder. When the effect of fluid viscosity can be neglected, the pressure distribution over the surface of the cylinder is given by Eq. (3-58), which is rearranged as

$$\frac{P_\theta - P}{\rho U^2/2g_c} = 1 - 4 \sin^2 \theta \tag{10-102}$$

Thom,[58] Green,[20] and Fage and Falkner [14] have obtained data on the distribution of the static pressure at the surface of a cylinder in a stream of

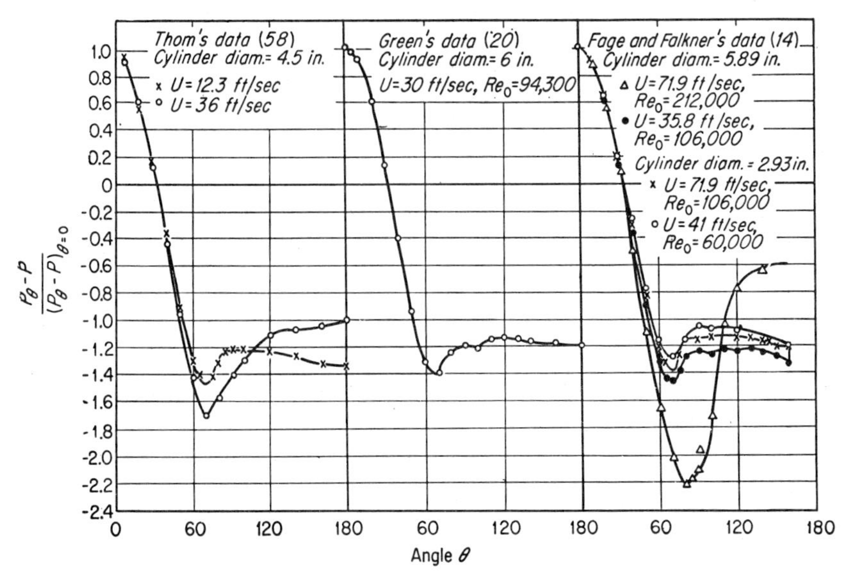

FIG. 10-29. Pressure distribution for air flow past circular cylinders.

air. Figure 10-29 shows the data of these workers plotted as $(P_\theta - P)/(P_\theta - P)_{\theta=0}$ † versus θ. Up to a value of $\theta = 60°$ all curves nearly coincide regardless of Reynolds number. Likewise, Eq. (10-102) is obeyed

† $(P_\theta - P)_{\theta=0} = \rho U^2/2g_c$.

well for values of θ up to 45°. Therefore Eq. (10-102) or Fig. 10-29 may be used to calculate the velocity at the edge of the boundary layer by means of Eq. (10-101). Green verified Eq. (10-101) as shown in Fig. 10-30, where measured values of u_{max} are plotted with values of u_{max} calculated

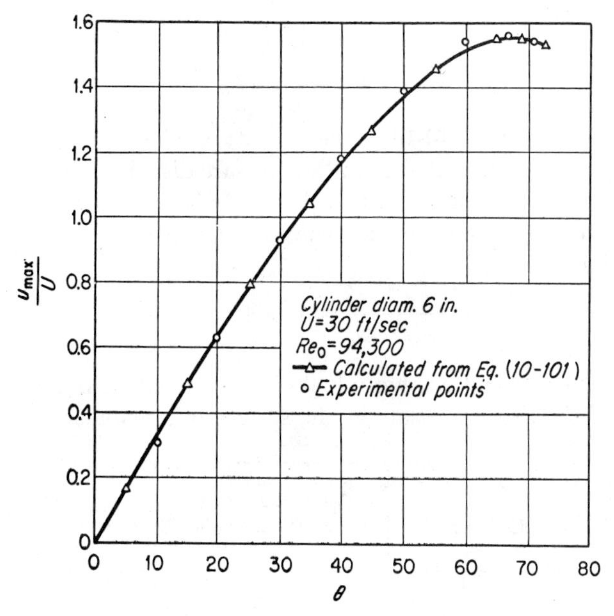

FIG. 10-30. Experimental and calculated values of u_{max} for the boundary layer on a circular cylinder. (*From J. J. Green, Brit. Aeronaut. Research Comm. R. & M.* 1313, 1930.)

from Eq. (10-101). The points were obtained experimentally, while the curve is a plot of Eq. (10-101) using the pressure distributions shown in Fig. 10-29.

10-23. Separation of the Boundary Layer on a Cylinder

The actual point of separation cannot be predicted from the conditions of flow in the main stream. It is influenced by the type of wind tunnel in which the cylinder is located and also by the history of the main stream before it comes in contact with the cylinder. The Reynolds number also affects the point of separation, as does the condition of the surface of the cylinder itself. Fage [13] studied visually the flow around a cylinder and determined the approximate values of θ where separation occurred. He also made measurements of the velocity gradient at the surface and determined the point where it was zero. These results are summarized in Table 10-10.

TABLE 10-10. POINTS OF BOUNDARY-LAYER SEPARATION FOR AIR FLOWING PAST A CIRCULAR CYLINDER †

Main-stream velocity, ft/sec	Value of θ	
	Where $\frac{du}{dy} = 0$	Where separation is observed visually
22.0	79.6	Near 84.6
26.9	84.5	94.5
39.2	84.8–89.8	Near 94.8
57.9	Near 94.8	Near 104.8
71.4	Near 94.8	Near 104.8

† From A. Fage, *Brit. Aeronaut. Research Comm. R. & M.* 1179, 1928.

10-24. The Laminar Boundary Layer on Single Circular Cylinders

A number of investigators have measured velocity distributions in the laminar boundary layer on a circular cylinder. The results of the investigations of Thom [58] and Green [20] are shown in Fig. 10-31. The circles show their experimental data, while the solid curves are predicted velocity profiles obtained from a numerical solution of Eqs. (10-89) and (10-90). In the curves shown in Fig. 10-31 it is seen that the boundary-layer thickness increases rapidly over the forward portion of the cylinder up to a value of $\theta = 50°$ and thence increases more slowly up to a value of $\theta = 90°$. As seen in Table 10-10, separation of the boundary layer takes place in the vicinity of $\theta = 90°$.

The velocity distribution in the boundary layer on the forward portion of a circular cylinder may be predicted by assuming that n' in Eq. (10-91) is unity. This is true over the forward part of the cylinder. Thus, from Eq. (10-96), λ is also unity. Therefore values of ϕ' may be obtained from Table 10-9 for $\lambda = 1$, and u/u_{max} may be calculated. The value of u_{max} for the particular point in question may be calculated from Eq. (10-101).

Meksyn [59] presented a more accurate solution for the laminar boundary layer on a circular cylinder, assuming that the pressure at the surface of the cylinder could be predicted from the relationships for irrotational flow. He also made correction for the region near the point of separation. The

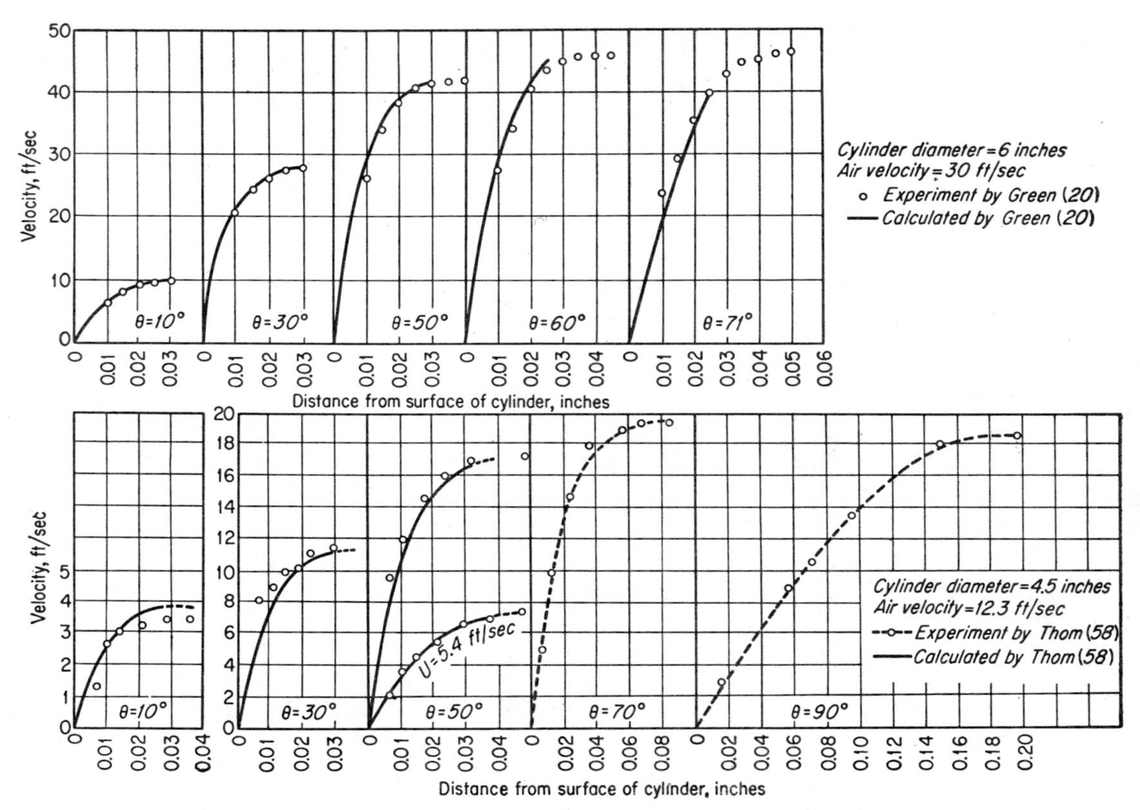

FIG. 10-31. Laminar-boundary-layer velocity profiles on circular cylinders. (*Data from J. J. Green, Brit. Aeronaut. Research Comm. R. & M.* 1313, 1930; *A. Thom, Brit. Aeronaut. Research Comm. R. & M.* 1176, 1928.)

relationships given below apply up to the point of separation. Meksyn defines the following quantities:

$$\lambda_1 = -\frac{\cos(180° - \theta)}{\sin^2[(180° - \theta)/2]} \tag{10-103}$$

$$\gamma = 2r_0 + \left(r_0 + y + \frac{r_0^2}{r_0 + y}\right)\cos(180° - \theta) \tag{10-104}$$

$$\eta_2 = \left[\frac{1}{2}\left(\frac{2U}{\nu\gamma}\right)^{1/2} 2\sin(180° - \theta)\right] y \tag{10-105}$$

The equation for the velocity profile at any angle θ is

$$\frac{u}{U} = \phi'[2\sin(180° - \theta)] \tag{10-106}$$

where ϕ' is the derivative of ϕ with respect to η_2.

Table 10-9 may also be used in calculating the velocity profile from Meksyn's solution. Given the angle where the point velocity is to be determined, λ_1 may be calculated. The value of η_2 may be calculated from the flow condition, the angle, and the value of y (measured normal to the cylinder). γ is calculated from y, r_0, and θ. When λ_1 and η_2 have been determined, ϕ' is obtained from Table 10-9. This value of ϕ' is substituted in Eq. (10-106) to determine u/U.

Example 10-5

Calculate the velocity at a radial distance of 0.0025 in. from the surface of a cylinder at a point 30° from the leading edge. The cylinder is 6 in. in diameter, the velocity of the air in the main stream is 30 ft/sec, and the Reynolds number is 94,300.

(*a*) Use Eqs. (10-91) with $n' = 1$, (10-93), (10-96), and (10-97).

(*b*) Use Eqs. (10-103) to (10-106).

Solution

(*a*) From Fig. 10-30, at $\theta = 30°$

$$\frac{u_{max}}{U} = 0.91$$

So

$$u_{max} = (30)(0.91) = 27.3 \text{ ft/sec}$$

$$\text{Kinematic viscosity of air} = \frac{(0.5)(30)}{94{,}300} = 1.59 \times 10^{-4} \text{ ft}^2\text{/sec}$$

At $\theta = 30°$

$$x_1 = 1.57 \text{ in}$$

$$= 0.131 \text{ ft}$$

By Eq. (10-93)

$$\eta_1 = \frac{0.0025}{12}\sqrt{\frac{27.3}{(1.59 \times 10^{-4})(0.131)}} = 0.238$$

By Eq. (10-96)

$$\lambda = \frac{(2)(1)}{1+1} = 1$$

From Table 10-9, at $\lambda = 1$ and $\eta_1 = 0.238$

$$\phi' = 0.266$$

Thus
$$u = (0.266)(27.3) = 7.27 \text{ ft/sec}$$

(*b*) From Eq. (10-103)

$$\lambda_1 = -\frac{\cos 150°}{\sin^2 75°} = 0.928$$

$$\gamma = (2)(3) + \left(3 + 0.0025 + \frac{9}{3.0025}\right)\cos 150° = 0.800 \text{ in.}$$

From Eq. (10-105)

$$\eta_2 = \frac{0.0025}{(2)(12)}\sqrt{\frac{(2)(30)(12)}{(1.59 \times 10^{-4})(0.800)}}(2 \sin 150°) = 0.247$$

From Table 10-9, at $\eta_2 = 0.247$ and $\lambda_1 = 0.928$

$$\phi' = 0.266$$

Thus
$$u = (30)(0.266)(2 \sin 150°) = 7.98 \text{ ft/sec}$$

The point velocities calculated by the two methods are within 10 per cent of each other.

10-25. The Laminar Boundary Layer on Single Elliptical Cylinders

When a fluid passes over an elliptical cylinder, the phenomena are similar to those observed when it passes over a circular cylinder. The leading edge of the ellipse is the point of stagnation, and from this point the boundary layer builds up and finally separates from the surface. Figure 10-32 is a sketch of an elliptical cylinder with a major axis of $2r_1$ and a minor axis of $2r_0$. The cylinder is oriented so that the major axis is parallel to the direction of net flow in the main stream, thus making one end of the major axis a point of stagnation. The broken line shows the limits of the boundary layer and its separation from the surface. Positions on the surface of the cylinder are measured along the surface from the stagnation point and are given the symbol x_1. A position on the surface may also be designated by the

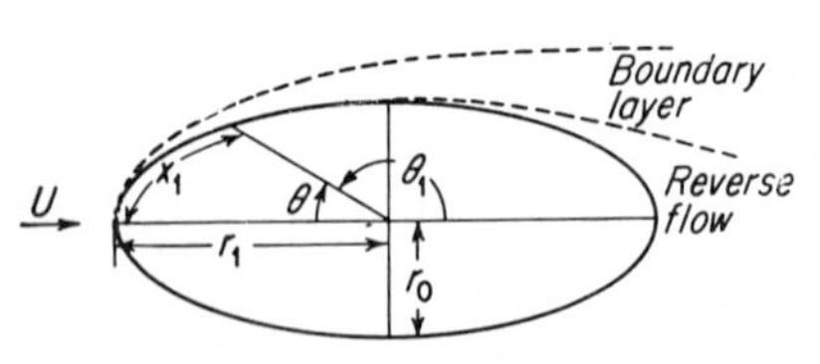

FIG. 10-32. Definition of terms for flow past elliptical cylinders.

angle which is made by a straight line through the point and the axis of the cylinder. Measured from the trailing edge of the ellipse, the angle is θ_1, and measured from the leading edge of the ellipse, it is θ. Thus $\theta + \theta_1 = 180°$.

Tietjens[59] obtained flow patterns of a fluid flowing across an immersed elliptical cylinder. Photographs of these flow patterns are reproduced in Fig. 10-33 for four different Reynolds numbers. At higher values of the

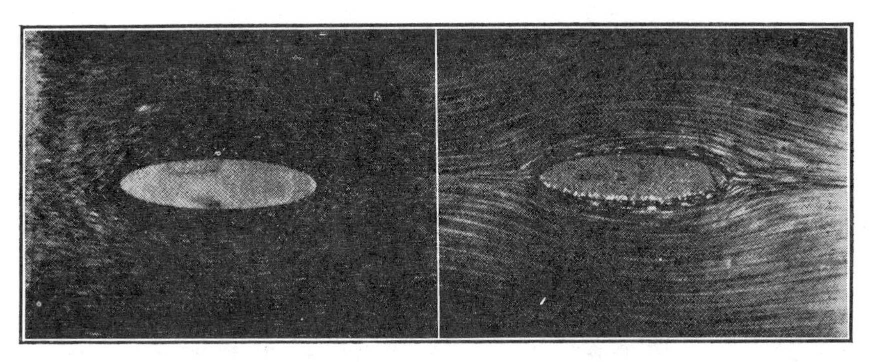

(*a*) $\text{Re}_0 = 0.3$ (*c*) $\text{Re}_0 = 3.3$

(*b*) $\text{Re}_0 = 1.3$ (*d*) $\text{Re}_0 = 150$

FIG. 10-33. Flow patterns for flow past elliptical cylinders. (*From O. J. Tietjens, Proc. Intern. Congr. Appl. Mech., 3rd Congr., Stockholm, 1930, vol.* 1., *p.* 331.)

Reynolds number the disturbance downstream from the ellipse increases, but it is not so great as in the case of circular cylinders. The Reynolds number for flow past an ellipse where the major axis is parallel to the flow is expressed as $2r_0U/\nu$.

Schubauer[51] made extensive measurements of the point velocities in the boundary layer formed on an elliptical cylinder with a major axis of 11.78 in. and minor axis of 3.98 in. Point velocities were also determined in the separated boundary layer. Figure 10-34 shows plots of Schubauer's data for values of x_1 of 0.180, 1.097, and 1.457 in. He found that the point of boundary-layer separation was at approximately $\theta_1 = 60°$ ($\theta = 120°$).

Meksyn[39] obtained a solution for the velocity distribution in the boundary layer on an elliptical cylinder, and the equations are similar to those he gave for circular cylinders. The solid line shown in Fig. 10-34 represents Meksyn's predicted velocity distribution.

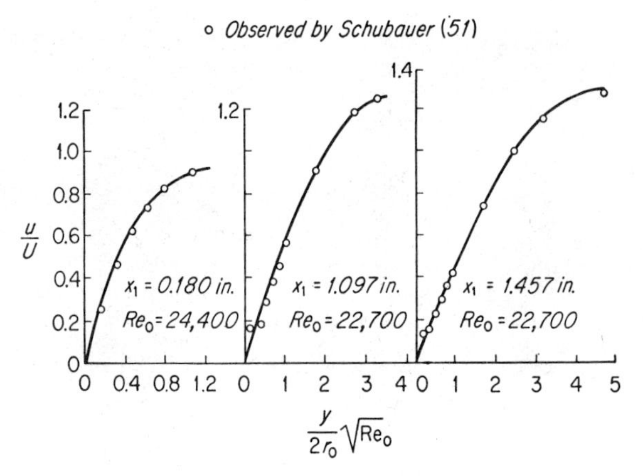

FIG. 10-34. Velocity profiles for flow past elliptical cylinders. [*From D. Meksyn, Proc. Roy. Soc. (London),* **192A**:545 (1948), *and G. B. Schubauer, NACA Rept.* 527, 1935.]

Meksyn introduces elliptic coordinates ξ and θ_1, which can be expressed in terms of the major and minor axes of the ellipse. The angle θ_1 is measured from the trailing edge of the ellipse, while ξ can be expressed in terms of r_0 and r_1 as follows:

$$\cosh \xi = \frac{r_1}{\sqrt{r_1^2 - r_0^2}} \qquad (10\text{-}107)$$

and

$$\sinh \xi = \frac{r_0}{\sqrt{r_1^2 - r_0^2}} \qquad (10\text{-}108)$$

The following quantities are defined in terms of the flow conditions, θ_1, and ξ:

$$\eta_3 = \left[\frac{1}{2}\left(\frac{2U}{\nu\gamma_1}\right)^{1/2}\left(\frac{\cosh 2\xi - \cos 2\theta_1}{2}\right)^{-1/2} e^{\xi} \sin \theta_1\right] y \qquad (10\text{-}109)$$

$$\gamma_1 = \sqrt{r_1^2 - r_0^2}\, e^{\xi}(1 + \cosh \xi \cos \theta_1) \qquad (10\text{-}110)$$

$$\lambda_2 = -\frac{(\cosh 2\xi - 1)\cos \theta_1}{(\cosh 2\xi - \cos 2\theta_1)\sin^2 \theta_1/2} \qquad (10\text{-}111)$$

The relationship for calculating the point velocity at the position being considered is

$$\frac{u}{U} = \left(\frac{\cosh 2\xi - \cos 2\theta_1}{2}\right)^{-1/2} \phi' e^{\xi} \sin \theta_1 \qquad (10\text{-}112)$$

where ϕ' is the derivative of ϕ with respect to η_3.

The above equation may be solved by the use of Table 10-9 in the same manner as Eq. (10-106).

10-26. The Drag of Cylinders Immersed in a Flowing Fluid

The drag of two-dimensional cylinders immersed in a flowing fluid is much greater than that of a flat plate. This fact is due to the separation of the boundary layers on two-dimensional bodies and the subsequent formation of a turbulent wake behind the body. The local and total drag coefficients are given by Eqs. (10-6) and (10-7), which are rearranged as

$$f_D = \frac{F}{A_P} \frac{2g_c}{\rho U^2} \qquad (10\text{-}113)$$

$$f' = \frac{2\tau_w g_c}{\rho U^2} \qquad (10\text{-}114)$$

where the area A_P is the projected area of the cylinder on a plane normal to the direction of flow. The local coefficient of friction is related to the shear stress at the surface, which, in turn, can be calculated from the velocity gradient at the surface. This quantity can be calculated from Eq. (10-98) if the value of n' in Eq. (10-91) is known. Meksyn [39] derived the following expression for the local coefficient of friction on circular cylinders:

$$f' = 16 \left(\frac{\nu}{2r_0 U}\right)^{1/2} \phi''(0) \sin^2 \frac{180° - \theta}{2} \cos \frac{180 - \theta}{2} \qquad (10\text{-}115)$$

Equation (10-115) predicts local friction coefficients on circular cylinders which agree well with measured values. The equation is valid up to the point of separation, which occurs theoretically when the velocity gradient (and hence f') is zero. By Eq. (10-115) separation occurs at $\theta = 95°$ (approximately). Fage and Falkner [14] measured values of the local coefficient of friction on a circular cylinder. Their results are shown in Fig. 10-35.

Total drag coefficients for flow past cylinders have been obtained by Wieselsberger,[61] who studied a range of Reynolds numbers from 5 to 10^6. The effect of the length-to-diameter ratio of the cylinder was also studied.

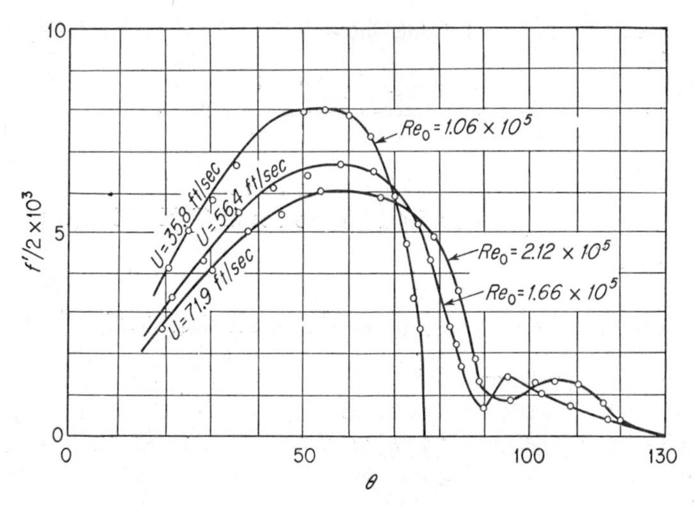

FIG. 10-35. Local coefficients of friction on circular cylinders. (*From A. Faye and V. M. Falkner, Brit. Aeronaut. Research Comm. R. & M.* 1369, 1931.)

The results of the work are shown in Fig. 10-36. The two curves shown are for different diameter-to-length ratios, the top curve being for a diameter-to-length ratio of 1:∞ and the lower curve being for a ratio of 1:5.

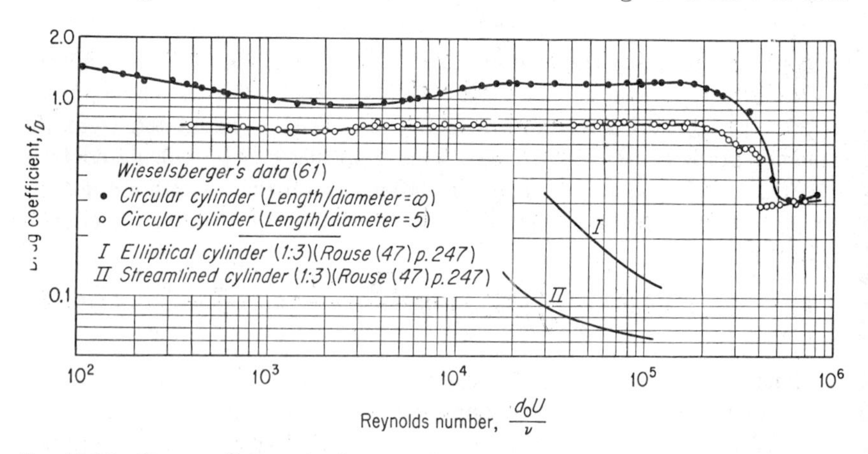

FIG. 10-36. Drag coefficients for flow past circular cylinders at high Reynolds numbers.

Wieselsberger investigated cylinders having diameters varying from 0.05 to 300 mm, and individual curves were obtained for each length-to-diameter ratio. The effect of this ratio is shown in Table 10-11.

Lamb [33] developed an expression for the coefficient of drag for flow past circular cylinders when no separation of the boundary layer occurs, in

TABLE 10-11. EFFECT OF THE LENGTH-TO-DIAMETER RATIO FOR FLOW PAST CIRCULAR CYLINDERS †

($\mathrm{Re}_0 = 88{,}000$)

Length / *Diameter*	*Drag coefficient* f_D
∞	1.2
40	0.98
20	0.92
10	0.82
5	0.74
3	0.74
2	0.68
1	0.63

† C. Wieselsberger, *Ergeb. Aerodyn. Versuchanstalt Göttingen* 2, p. 22, 1923.

which case the resistance is due only to viscous drag:

$$f_D = \frac{8\pi}{\mathrm{Re}_0(2.002 - \ln \mathrm{Re}_0)} \qquad \mathrm{Re}_0 < 0.5 \qquad (10\text{-}116)$$

This expression applies only for low Reynolds numbers and is plotted as a broken line in Fig. 10-37, where drag coefficients for low flow rates are plotted versus the Reynolds number. The circles on the figure represent Wieselsberger's experimental data.

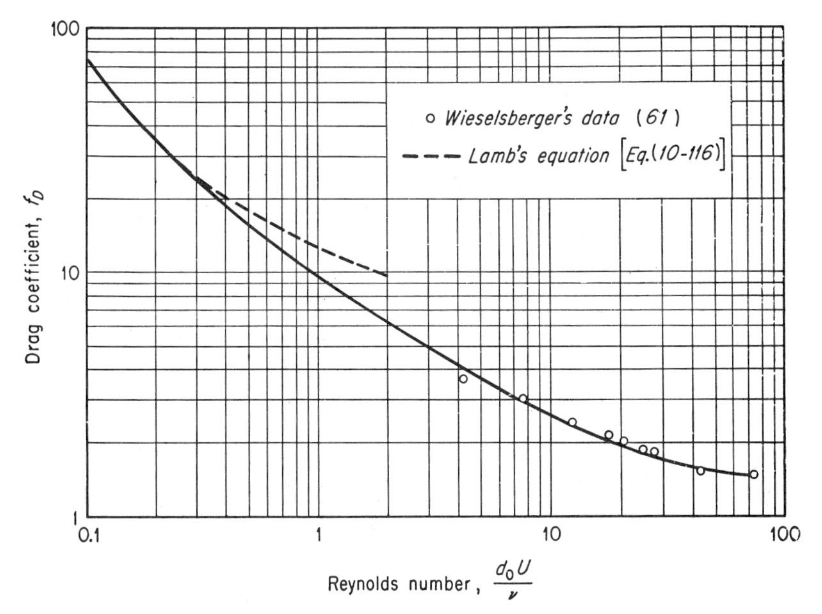

FIG. 10-37. Drag coefficients for flow past circular cylinders at low Reynolds numbers.

TABLE 10-12. DRAG COEFFICIENTS FOR FLOW NORMAL TO TWO-DIMENSIONAL CYLINDERS †

Direction of flow →	d_0, in.	d_1, in.	$\frac{r}{d_0}$	f_D	Range of Re_0	Critical Re_0
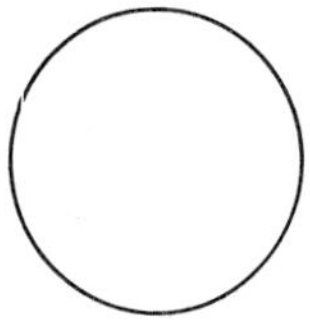	12.00 4.00 1.00	12.00 4.00 1.00		1.00	10^4–2×10^5	4×10^5
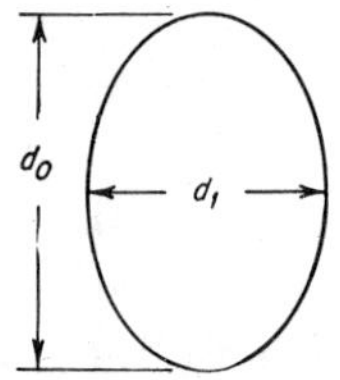	12.00 4.00 1.00	6.00 2.00 0.50		1.6	2×10^5– 8×10^5	1.4×10^6
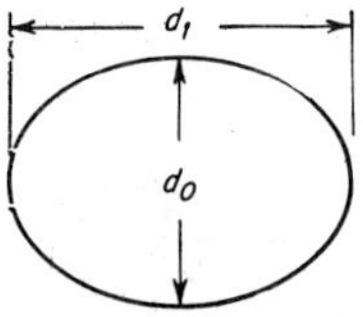	6.00 2.00 0.50	12.00 4.00 1.00		0.6	10^4–5×10^4	10^5
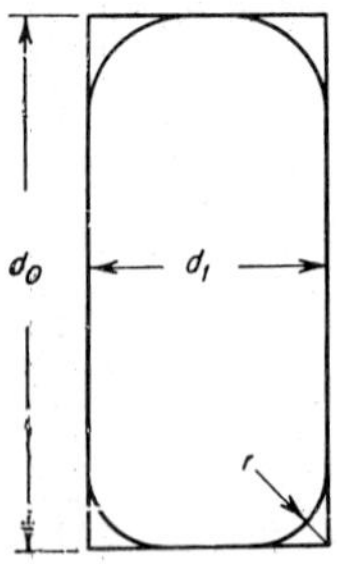	12.00 4.00 1.00	6.00 2.00 0.50	0.021 0.021 0.021	2.2	2×10^4–10^6	
	12.00	6.00	0.083	1.9	1.6×10^5– 6×10^5	1.2×10^6
	12.00 4.00	6.00 2.00	0.250 0.250	1.6	3×10^4– 2×10^5	6×10^5

† From K. Delany and N. E. Sorenson, *NACA TN* 3038, 1953.

TABLE 10-12. DRAG COEFFICIENTS FOR FLOW NORMAL TO TWO-DIMENSIONAL CYLINDERS (*Continued*)

Direction of flow →	d_0, in.	d_1, in.	$\frac{r}{d_0}$	f_D	Range of Re_0	Critical Re_0
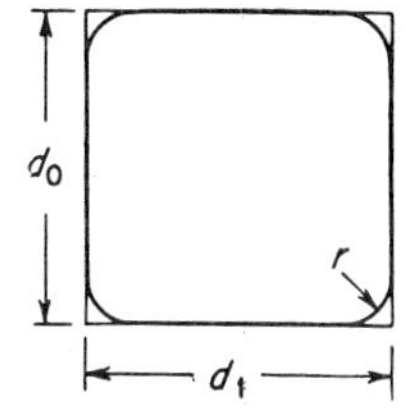	12.00	12.00	0.021	2.0	2×10^4–2.8×10^6	
	4.00	4.00	0.021			
	1.00	1.00	0.021			
	12.00	12.00	0.167	1.2	3×10^5–10^6	1.4×10^6
	12.00	12.00	0.333	1.0	5×10^4–4×10^5	5.6×10^5
	1.00	1.00	0.333			
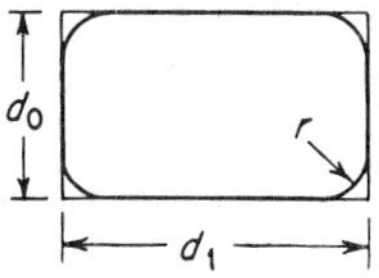	6.00	12.00	0.042	1.4	1.8×10^4–7×10^5	8×10^5
	2.00	4.00	0.042			
	0.50	1.00	0.042			
	6.00	12.00	0.167	0.7	10^5–5×10^5	5×10^5
	6.00	12.00	0.500	0.4	10^5–10^6	
	2.00	4.00	0.500			
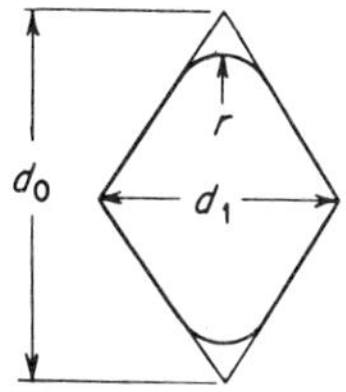	12.00	6.00	0.021	1.8	2×10^4–4×10^6	
	4.00	2.00	0.021			
	1.00	0.50	0.021			
	12.00	6.00	0.083	1.7	2×10^5	
	12.00	6.00	0.167	1.7	10^5–4×10^5	6×10^5
	4.00	2.00	0.167			
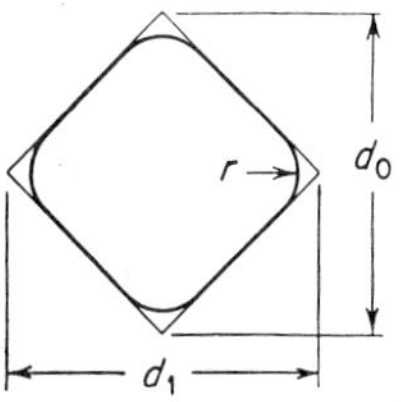	16.97	16.97	0.015	1.5	2×10^4–2×10^6	
	5.66	5.66	0.015			
	1.41	1.41	0.015			
	16.97	16.97	0.118	1.5	2×10^5–10^6	
	16.97	16.97	0.235	1.5	4×10^4–2×10^5	6×10^5
	5.66	5.66	0.235			

TABLE 10-12. DRAG COEFFICIENTS FOR FLOW NORMAL TO TWO-DIMENSIONAL CYLINDERS (*Continued*)

Direction of flow →	d_0, in.	d_1, in.	$\frac{r}{d_0}$	f_D	Range of Re_0	Critical Re_0
	6.00 2.00 0.50	12.00 4.00 1.00	0.042 0.042 0.042	1.1	10^4–10^6	
	6.00	12.00	0.167	1.1	8×10^4–5×10^5	
	6.00 2.00	12.00 4.00	0.333 0.333	1.1	2×10^4–2×10^5	4×10^5
	12.00 4.00 1.00	12.00 4.00 1.00	0.021 0.021 0.021	1.2	5×10^4–10^6	
	12.00	12.00	0.083	1.3	1.8×10^5–10^6	
	12.00 4.00	12.00 4.00	0.250 0.250	1.1	4×10^4–2×10^5	6×10^5
	12.00 4.00 1.00	12.00 4.00 1.00	0.021 0.021 0.021	2.0	10^4–10^6	
	12.00	12.00	0.083	1.9	2×10^5–10^6	
	12.00 4.00	12.00 4.00	0.250 0.250	1.3	2×10^4–3×10^6	4×10^5

Finn [18] studied the drag of a circular cylinder at Reynolds numbers from 0.05 to 5.0 and obtained agreement with Eq. (10-116) up to a Reynolds number of 1.0.

The resistance provided by an elliptical cylinder and a streamlined body, such as a strut or an airfoil, is considerably less than that provided by a circular cylinder. Streamlining the rear portion of a solid body results in reducing the turbulent wake behind the body, thus leading to a reduction of the drag coefficient for these solid bodies. Curve I in Fig. 10-36 is a plot given by Rouse [47] for the drag coefficient f_D over an elliptical cylinder.

Curve II is a plot of f_D versus Re_0 for a streamlined strut. The range of Reynolds numbers covered by curves I and II in Fig. 10-36 is small, but the curves serve to indicate the relative values of the drag coefficients for bodies of various shapes.

Delany and Sorenson [5] conducted an extensive investigation on the resistance of cylinders of various shapes placed with axis normal to a flowing stream of air. They studied cylindrical, elliptical, square, rectangular, and triangular cylinders and determined the effect of rounding the sharp edges of the square, rectangular, and triangular cylinders. Their results are tabulated in Table 10-12, where the drag coefficient is given for a range of Reynolds numbers. In this range the drag coefficient is within about 10 per cent of the value given in Table 10-12. The critical Reynolds numbers (see Sec. 10-27) are also tabulated to show when the laminar boundary layer becomes turbulent.

10-27. The Transition from Laminar to Turbulent Flow on Cylinders

The flow of fluids over two-dimensional circular cylinders can conveniently be divided into three regimes. At low Reynolds numbers (less than $Re_0 = 0.5$) flow is laminar over almost the whole cylinder, and the only resistance to flow is due to viscous surface friction, little energy being dissipated in a turbulent wake. As the Reynolds numbers increase from 0.5, the drag coefficient becomes relatively constant and remains so over a fairly wide range of Reynolds numbers. This regime of flow, in which the drag coefficient is constant, is characterized by laminar flow over the forward portion of the cylinder and a fairly wide turbulent wake behind the cylinder. The boundary layer is laminar up to the point of separation. For circular cylinders the drag coefficient is constant over an approximate range of Reynolds numbers from 10^3 to 10^5.

The third regime of flow begins at the point where the drag coefficient decreases rapidly to a value less than one-half of its previous constant value. At this point flow in the boundary layer changes from laminar to turbulent. Beyond this point of rapid decrease the drag coefficient is again almost constant with respect to the Reynolds number. The critical Reynolds number is the Reynolds number at which transition from laminar to turbulent flow in the boundary layer occurs.

The critical Reynolds number is greatly influenced by the intensity of turbulence of the main stream. This fact was evident in the case of flow over flat plates where transition took place at Reynolds numbers between 10^5 and 3×10^6. For flow over circular cylinders, Fage [12] reported a critical-Reynolds-number range of 1.05×10^5 to 2.60×10^5 for a stream with 2.2 per cent turbulence and 4.8×10^4 to 1.48×10^5 with 5.3 per cent turbulence. The effect of intensity and scale of turbulence on the position of laminar-turbulent transition on an elliptical cylinder is shown by the curve in Fig. 10-38, which was obtained by Schubauer.[50] In this figure, the quantity $\frac{\sqrt{\overline{u'^2}}}{U}\left(\frac{r_0}{L_x}\right)^{1/5}$ is plotted versus x_1/r_0 where $\sqrt{\overline{u'^2}}/U$ is the turbulence intensity, L_x is the scale [defined by Eq. (5-34)], r_0 is one-half the minor axis of the elliptical cylinder, and x_1 is the distance measured along the surface from the leading edge to the point where transition occurs.

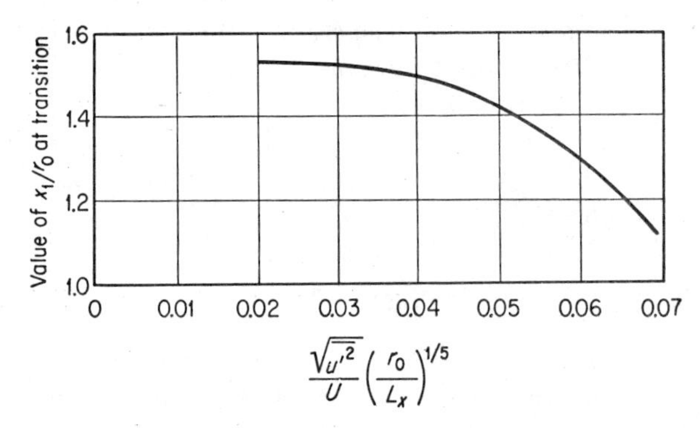

FIG. 10-38. Effect of main-stream turbulence intensity and scale on boundary-layer transition on an elliptical cylinder. (*From G. B. Schubauer, NACA Rept.* 652, 1939.)

The sharp decrease in the drag coefficient that occurs at the transition point is due to the displacement of the point of separation toward the trailing edge of the cylinder. This has the effect of reducing the size of the turbulent wake and thus reducing the drag coefficient. The shear stress at the wall for the turbulent boundary layer is much greater than for the laminar boundary layer; however, the contribution of surface friction to the total drag coefficient is small compared to the contribution of the turbulent wake. Therefore, any reduction in the size of the wake has the effect of materially reducing the drag coefficient.

III. FLOW PAST BODIES OF REVOLUTION: SPHERES, ELLIPSOIDS, DISKS

10-28. The Three-dimensional Boundary-layer Equation for Bodies of Revolution

When flow takes place past an immersed body of revolution such as a sphere, the motion of the fluid near the surface of the body is three-dimensional with respect to a rectangular coordinate system. However, such flow is symmetrical about the axis of the body. Millikan[40] derived the boundary-layer equations for such axially symmetrical flow over blunt-nosed bodies of revolution. They are as follows:

The momentum equation:

$$u \frac{\partial u}{\partial x_1} + v \frac{\partial u}{\partial y} = -\frac{g_c}{\rho}\frac{\partial P}{\partial x_1} + \nu \left(\frac{\partial^2 u}{\partial y^2} + \frac{1}{r}\frac{\partial r}{\partial y}\frac{\partial u}{\partial y}\right) \tag{10-117}$$

The continuity equation:

$$\frac{\partial(ru)}{\partial x_1} + \frac{\partial(rv)}{\partial y} = 0 \tag{10-118}$$

where x_1 = distance measured from the leading edge of body
y = distance measured normal to surface of body
r = radius of transverse cross section at point (x_1,y)

At the surface ($y = 0$) the transverse cross section of the body has a radius r_t. The radius of curvature at any point is designated as r_c. These quantities are illustrated in Fig. 10-39. Equations (10-117) and (10-118) are based on the assumption that y is small compared to r_c and δ is small compared to r. This is true except in the vicinity of the leading edge, where r approaches zero. For the two-dimensional case, r is infinite, so Eqs. (10-117) and (10-118) become identical with Eqs. (10-89) and (10-90).

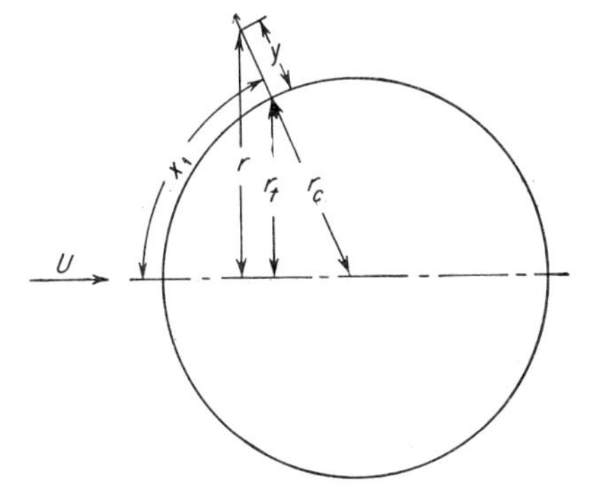

FIG. 10-39. Definition of terms for flow past spheres.

Solution of Eqs. (10-117) and (10-118) will yield expressions for the velocity profile in the boundary layer, the boundary thickness, and the local drag coefficients. Unfortunately, however, Eqs. (10-117) and (10-118) can be solved only in special cases. The alternative method of determining boundary-layer thickness and local drag coefficient has been to use an integral momentum equation similar to the one for two-dimensional flow [Eq. (10-38)]. Millikan [40] derived the following three-dimensional integral momentum equation from Eqs. (10-117) and (10-118):

$$\frac{\partial}{\partial x_1}\int_0^\delta u^2\,dy - u_{\max}\frac{\partial}{\partial x_1}\int_0^\delta u\,dy + \frac{1}{r_t}\frac{\partial r_t}{\partial x_1}\left(\int_0^\delta u^2\,dy - u_{\max}\int_0^\delta u\,dy\right) = \delta u_{\max}\frac{\partial u_{\max}}{\partial x_1} - \frac{\tau_w g_c}{\rho} \tag{10-119}$$

To solve Eq. (10-119) the following must be known or assumed:

1. The form of the velocity profile in the boundary layer. Usually the form of the profile is assumed.

2. The variation of the static pressure along the surface of the body. This will give $\partial u_{max}/\partial x_1$ by the use of Eq. (10-88). In solving Eq. (10-119) either the experimental pressure distribution or the pressure distribution for irrotational flow past the body is used.

3. The variation of r_t with respect to x_1. This is known when the shape of the body is specified.

10-29. The Pressure Distribution for Flow past a Sphere

As pointed out above, it is necessary to know the pressure distribution over an immersed body of revolution if Eq. (10-119) is to be solved. For the flow of a nonviscous fluid past a sphere it may be shown by methods presented in Chap. 3 that the pressure distribution is given by the relation

$$\frac{P_\theta - P}{\rho U^2/2g_c} = 1 - 9/4 \sin^2 \theta \qquad (10\text{-}120)$$

where θ = angle measured from forward stagnation point of sphere
P_θ = static pressure at angle θ
P = static pressure of undisturbed stream

From Bernoulli's equation (3-27)

$$\frac{u_{max}}{U} = 1.50 \sin \theta \qquad (10\text{-}121)$$

The pressure distribution for the flow of air past spheres was determined experimentally by Fage.[11] His data for three Reynolds numbers are shown in Fig. 10-40, where $(P_\theta - P)/(\rho U^2/2g_c)$ is plotted versus θ. Up to a value of $\theta = 60°$ the three curves are close together. Separation takes place in the range $80° < \theta < 100°$.

Tomotika[60] reported an empirical equation for u_{max}, the velocity at the edge of the boundary layer, using measured pressure distributions.

$$\frac{u_{max}}{U} = 1.50\theta - 0.364020\theta^3 - 0.0246680\theta^5 \qquad \theta < 80° \qquad (10\text{-}122)$$

where θ is in radians. Equations (10-120) and (10-122) are also plotted in Fig. 10-40. At Reynolds numbers below the critical Eq. (10-122) predicts the pressure distribution, while at Reynolds numbers above the critical Eq. (10-121) predicts the pressure distribution. Both equations apply up to $\theta = 80°$.

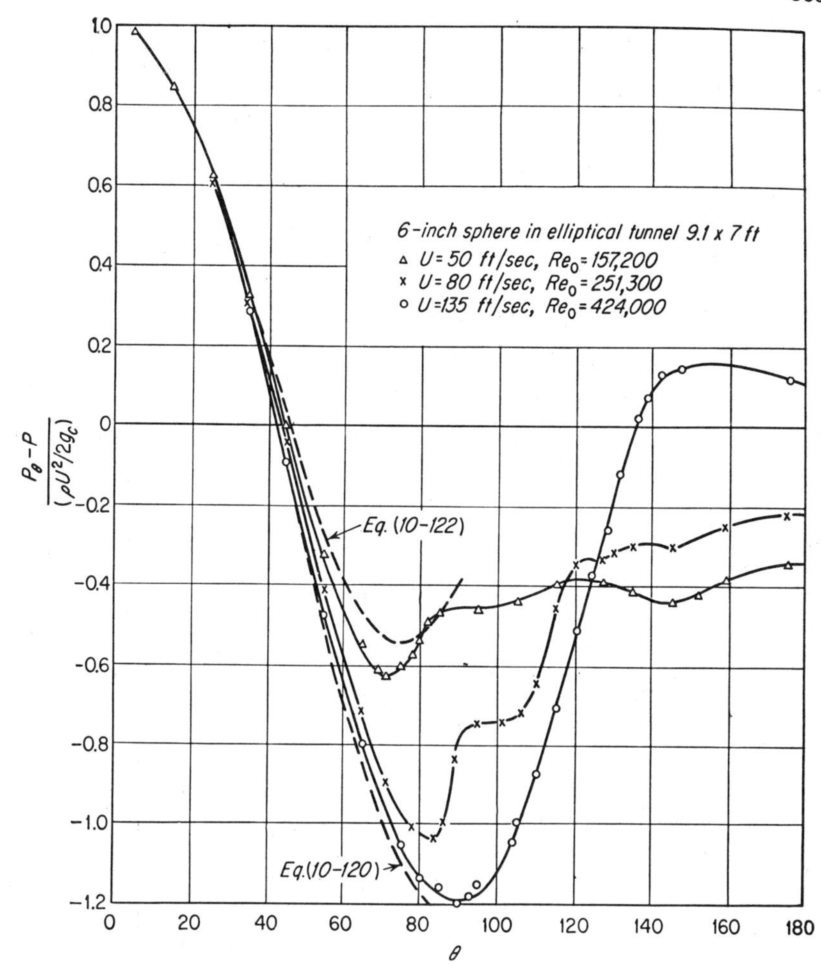

FIG. 10-40. Pressure distribution for flow past spheres. (*From A. Fage, Brit. Aeronaut. Research Comm. R. & M.* 1766, 1937.)

10-30. The Laminar Boundary Layer on a Sphere

Tomotika [60] solved Eq. (10-119) by assuming that the velocity distribution in the laminar boundary layer could be represented by the relation

$$u = c_1 y + c_2 y^2 + c_3 y^3 + c_4 y^4 \tag{10-123}$$

He obtained the following equation for the laminar-boundary-layer thickness:

$$\frac{U}{\nu r_0}\frac{\partial(\delta^2)}{\partial\theta} = \phi_1(D)\frac{U}{u_{\max}} - \phi_2(D)\cot\theta\frac{U}{\partial u_{\max}/\partial\theta} + \phi_3(D)\frac{\delta^4}{U\nu^2}\frac{\partial^2 u_{\max}}{\partial\theta^2} \tag{10-124}$$

$$\phi_1(D) = \frac{7{,}257.6 - 1{,}336.2D + 37.92D^2 + 0.8D^3}{213.12 - 5.76D - D^2} \tag{10-125}$$

$$\phi_2(D) = \frac{426.24D - 3.84D^2 - 0.4D^3}{213.12 - 5.76D - D^2} \tag{10-126}$$

$$\phi_3(D) = \frac{3.84 + 0.8D}{213.12 - 5.76D - D^2} \tag{10-127}$$

and

$$D = \frac{\delta^2}{r_0\nu}\frac{\partial u_{\max}}{\partial\theta} \tag{10-128}$$

A graphical solution of Eq. (10-124) is given in Fig. 10-41, in which the "theoretical" curve is based on the values of $u_{\max}$ given by Eq. (10-121)

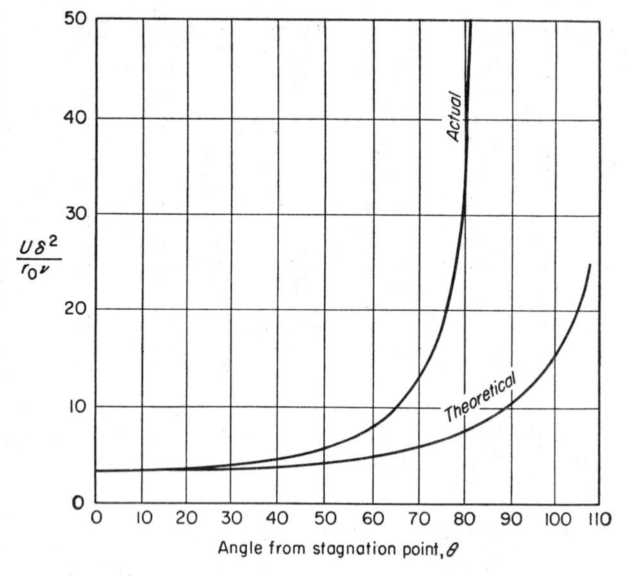

FIG. 10-41. Boundary-layer thickness on a sphere. (*From S. Tomotika, Brit. Aeronaut. Research Comm. R. & M.* 1678, 1935.)

and the "actual" curve is based on values of $u_{\max}$ given by the empirical relation [Eq. (10-122)]. The actual curve may be used to predict the laminar-boundary-layer thickness on the surface of a sphere up to $\theta = 80°$

Millikan [40] solved Eq. (10-119) assuming a parabolic velocity profile in the laminar boundary layer, such as given by Eq. (10-42). He gives the following general equation for the thickness of the laminar boundary on the forward portion of a body of revolution:

$$\delta^2 = \frac{30\nu}{u_{\max}^9 r_t^2} \int_0^{x_1} u_{\max}^8 r_t^2 \, dx_1 \tag{10-129}$$

Example 10-6

Using the equation derived by Millikan for the thickness of the laminar boundary layer on a body of revolution, calculate the laminar-boundary-layer thickness on a

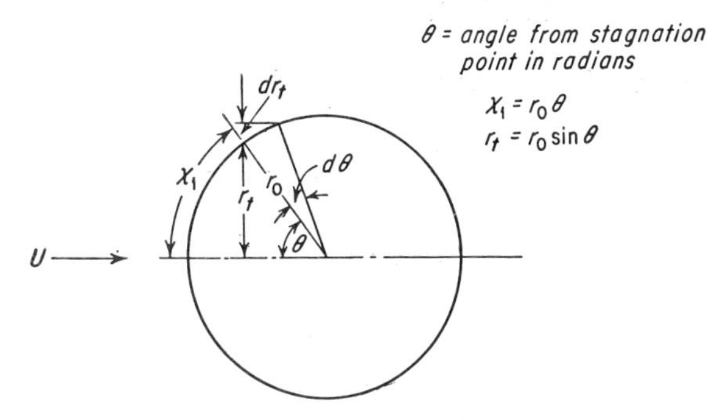

FIG. 10-42. Quantities used in Examples 10-6 and 10-7.

sphere of radius r_0 in a fluid flowing with a velocity of U. Use the value of $u_{\max}$ given by Eq. (10-121), and determine $\delta^2 U/\nu r_0$ as a function of θ.

Solution

From Fig. 10-42, when θ is in radians,

$$x_1 = r_0\theta$$

and

$$r_t = r_0 \sin\theta$$

From Eq. (10-121)

$$u_{\max} = 1.5U \sin\theta$$

Substituting the above into Eq. (10-129) results in

$$\frac{\delta^2 U}{\nu r_0} = \frac{20}{\sin^{11}\theta} \int_0^{\theta} \sin^{10}\theta \, d\theta$$

which may be integrated graphically or analytically to give $\delta^2 U/\nu r_0$ as a function of θ.

10-31. Drag Coefficients for Bodies of Revolution

Various bodies of revolution are shown in Fig. 10-43. The radius in the plane normal to the direction of flow is r_0, and the Reynolds number for the flow is d_0U/ν, where $d_0 = 2r_0$. The projected area of the body is πr_0^2.

The over-all length of the body in the direction of flow is L, and the ratio $2r_0/L$ is the diameter-to-length ratio.

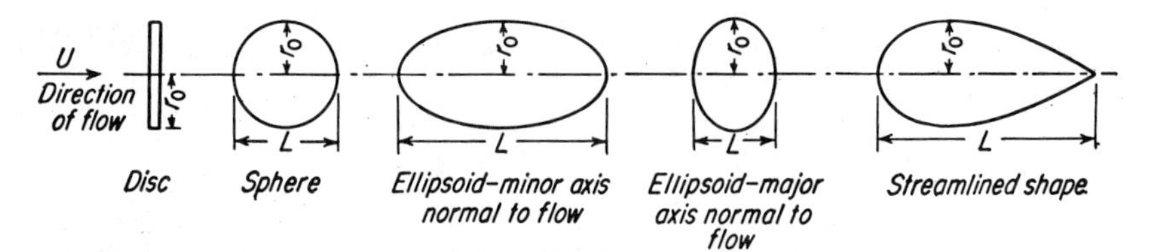

FIG. 10-43. Bodies of revolution.

The drag on a sphere moving very slowly in a viscous fluid was determined by Stokes. In this case, the boundary layer does not separate from the sphere, and the force of resistance by Stokes' equation is

$$F = \frac{3\pi d_0 \mu U}{g_c} \tag{10-130}$$

Oseen [42] has proposed the following relation for the flow of a viscous fluid past a sphere:

$$F = \frac{3\pi d_0 \mu U}{g_c}(1 + \tfrac{3}{16}\mathrm{Re}_0) \tag{10-131}$$

From Eq. (10-6) the total drag coefficient given by Stokes' equation is

$$f_D = \frac{24}{\mathrm{Re}_0} \qquad \mathrm{Re}_0 < 1.0 \tag{10-132}$$

while Oseen's equation gives

$$f_D = \frac{24}{\mathrm{Re}_0}(1 + \tfrac{3}{16}\mathrm{Re}_0) \qquad \mathrm{Re}_0 < 2.0 \tag{10-133}$$

Equations (10-132) and (10-133) apply for the motion of spheres in an infinite fluid, the effects of neighboring walls being negligible. Ladenburg [32] proposed the following relation for the drag coefficient for flow past a sphere in a cylinder of diameter d_w:

$$\frac{f_D}{1 + 2.4(d_0/d_w)} = \frac{24}{\mathrm{Re}_0} \tag{10-134}$$

In Fig. 10-44 some experimental data of Liebster [34] giving drag coefficients for spheres at low Reynolds numbers are plotted. Equations (10-132) and (10-133) are plotted as broken curves. Oseen's equation agrees with the experimental data up to a Reynolds number of 2.0, whereas Stokes' equation is good only up to a Reynolds number of 1.0. The solid circles in Fig. 10-44 represent Liebster's [34] data for the movement of spheres in cylinders. These points are values of $f_D/[1 + 2.4(d_0/d_w)]$ plotted versus the Reynolds number, and they agree fairly well with Stokes' equation.

Drag coefficients for bodies of revolution at Reynolds numbers higher than those covered in Fig. 10-44 can be obtained from Fig. 10-45, where

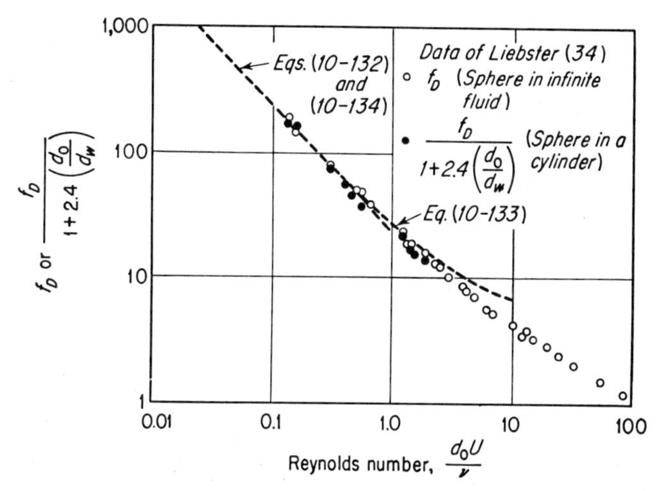

FIG. 10-44. Drag coefficients for flow past spheres at low Reynolds numbers.

Wieselsberger's data [61] on disks, spheres, and ellipsoids are plotted. The two ellipsoids Wieselsberger studied had diameter-to-length ratios $2r_0/L$ of 1.33 and 0.67, respectively.

The actual effect of the diameter-to-length ratio is indicated in a plot of f_D versus Re_0 presented by Müller [41] and shown in Fig. 10-46. Only a small range of Reynolds numbers is covered, but the plot serves to indicate that for ellipsoids the coefficient of drag decreases with decreasing values of $2r_0/L$. As the value of $2r_0/L$ decreases, the body becomes more streamlined, and consequently the resistance to the flow of fluids becomes less. The two lower curves of Fig. 10-46 compare the drag coefficient of an ellipsoid having $2r_0/L = 0.333$ with an airship hull having a value of $2r_0/L = 0.166$. The drag coefficients for these two bodies are very nearly the same. The upper curve of Fig. 10-46 shows the coefficient of drag for disks for which $2r_0/L$ is infinite.

Local coefficients of friction on spheres may be predicted by a relationship derived by Meksyn [38] that is valid up to $\theta = 80°$.

$$f' = 6\sqrt{3}\left(\frac{\nu}{2r_0U}\right)^{1/2} \phi''(0) \sin^2 \frac{180° - \theta}{2} \cos \frac{180° - \theta}{2} \qquad (10\text{-}135)$$

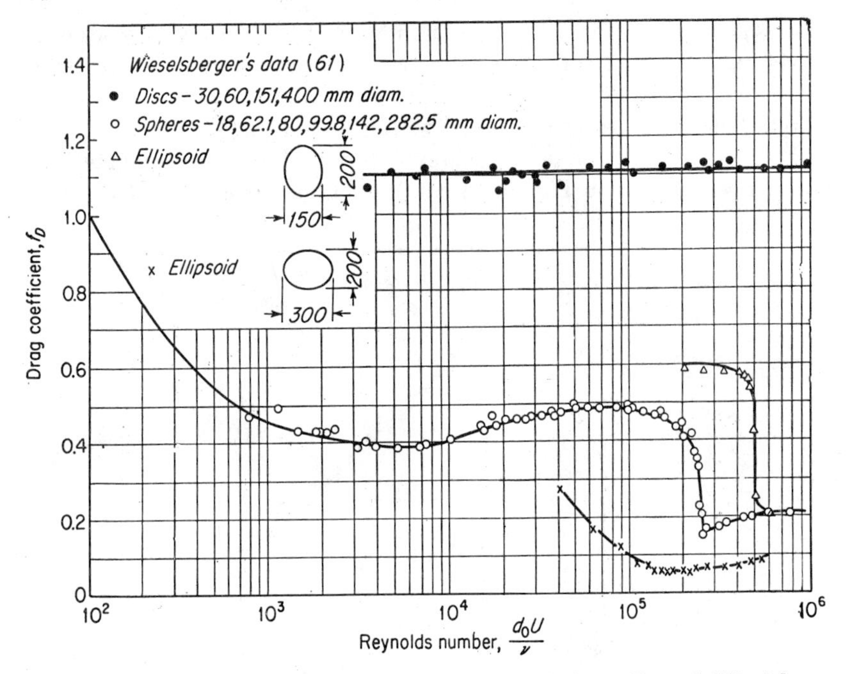

FIG. 10-45. Drag coefficients for flow past bodies of revolution. (*From C. Wieselsberger, Ergeb. Aerodyn. Versuchanstalt Göttingen* 2, *p.* 22, 1923.)

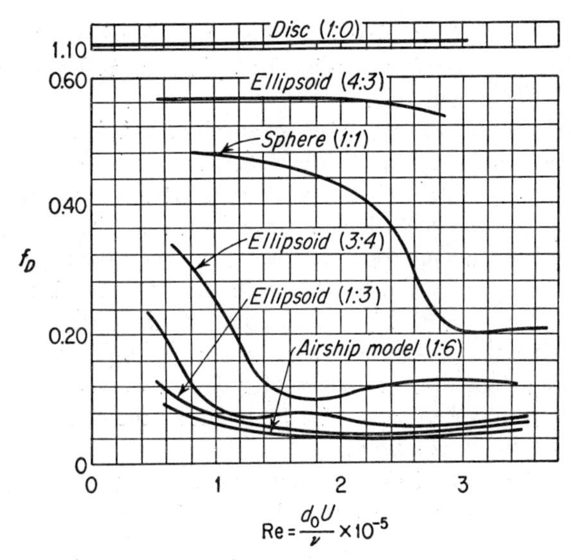

FIG. 10-46. Drag coefficients for flow past bodies of revolution. (*From W. Müller, "Mathematische Strömungslehre," Springer-Verlag OHG, Berlin,* 1928.)

Equation (10-135) may be solved using Table 10-9, determining $\phi''(0)$ for any specified value of λ_1 [defined in Eq. (10-103)]. Fage's [11] experimental values of f' for spheres are shown in Fig. 10-47.

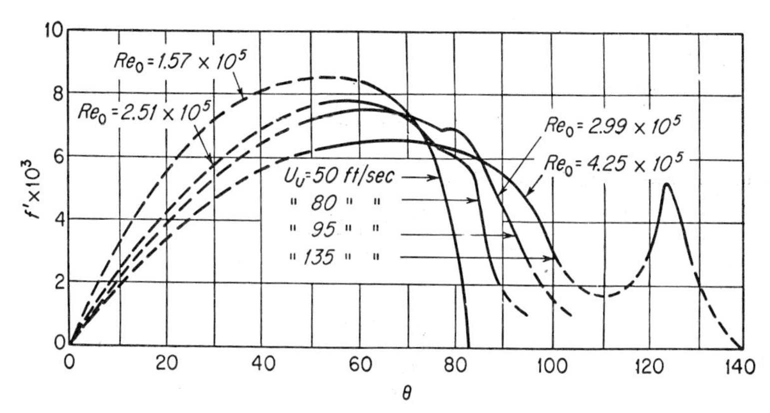

FIG. 10-47. Local coefficients of friction for flow past spheres. (*From A. Fage, Brit. Aeronaut. Research Comm. R. & M.* 1766, 1937.)

Example 10-7

For a Reynolds number of 157,000 calculate the drag coefficient for a sphere from the pressure distribution given in Fig. 10-40 and the local coefficients of friction given in Fig. 10-47.

Solution

It is necessary to consider separately the forces due to pressure and the forces due to viscous shear stresses.

Force due to pressure. Considering a differential element of area subtended by an angle $d\theta$ (see Fig. 10-42), the force exerted in the direction of flow is given by

$$dF_{\text{pressure}} = (P_\theta - P)2\pi r_t \, dr_t = (P_\theta - P)\pi \, d(r_t)^2$$

where P_θ is the pressure at the angle θ and $2\pi r_t \, dr_t$ is the projected area of the differential element of surface.

$$F_{\text{pressure}} = \int_0^\pi (P_\theta - P)\pi r_0^2 \, d(\sin^2 \theta)$$

But the drag coefficient due to pressure is defined by

$$F_{\text{pressure}} = \frac{f_{\text{pressure}} \rho U^2 \pi r_0^2}{2g_c}$$

Thus

$$f_{\text{pressure}} = \int_0^\pi \frac{P_\theta - P}{\rho U^2/2g_c} \, d(\sin^2 \theta)$$

$$= 0.464$$

(evaluated by graphical integration from values obtained from Fig. 10-40).

Force due to shear stress at the surface. Considering the same differential element, the force exerted in the direction of flow due to surface shear stress is

$$dF_{\text{shear}} = (\tau_w 2\pi r_t r_0 \, d\theta) \sin\theta$$

where $2\pi r_t r_0 \, d\theta$ is the area of the differential element of spherical surface and τ_w is the shear stress at the surface.

Since $r_t = r_0 \sin\theta$,

$$F_{\text{shear}} = \int_0^\pi 2\pi r_0^2 \tau_w \sin^2\theta \, d\theta$$

The drag coefficient due to shear at the surface is defined as

$$F_{\text{shear}} = \frac{f_{\text{shear}} \rho U^2 \pi r_0^2}{2g_c}$$

So

$$f_{\text{shear}} = 2\int_0^\pi \frac{2\tau_w g_c}{\rho U^2} \sin^2\theta \, d\theta$$

$$= 2\int_0^\pi f' \sin^2\theta \, d\theta$$

$$= 0.009$$

(evaluated from Fig. 10-47).

The total drag coefficient is

$$f_{\text{pressure}} + f_{\text{shear}} = 0.464 + 0.009$$

$$= 0.473$$

This value agrees well with Fig. 10-45.

10-32. Transition from Laminar to Turbulent Flow on Bodies of Revolution

Phenomena similar to those on two-dimensional cylinders are observed on bodies of revolution when transition from laminar to turbulent flow occurs. The point of boundary-layer separation moves downstream, and a sharp decrease in the drag coefficient is observed. For spheres, the drag coefficient remains between 0.4 and 0.5 for Reynolds numbers from 10^3 to 10^5. When transition occurs, the drag coefficient drops rapidly to a value in the vicinity of 0.2 and remains relatively constant thereafter.

The critical Reynolds number, i.e., the Reynolds number where the drag coefficient decreases, is influenced by the turbulence of the main stream flowing past the body. In fact, a drag-coefficient–Reynolds-number curve in the transition region is obtained for each percentage turbulence of the main stream. Of the numerous investigations of the effect of turbulence on the critical Reynolds numbers of spheres, the work of Dryden, Schubauer, Mock, and Skramstad [10] results in good correlation of the factors

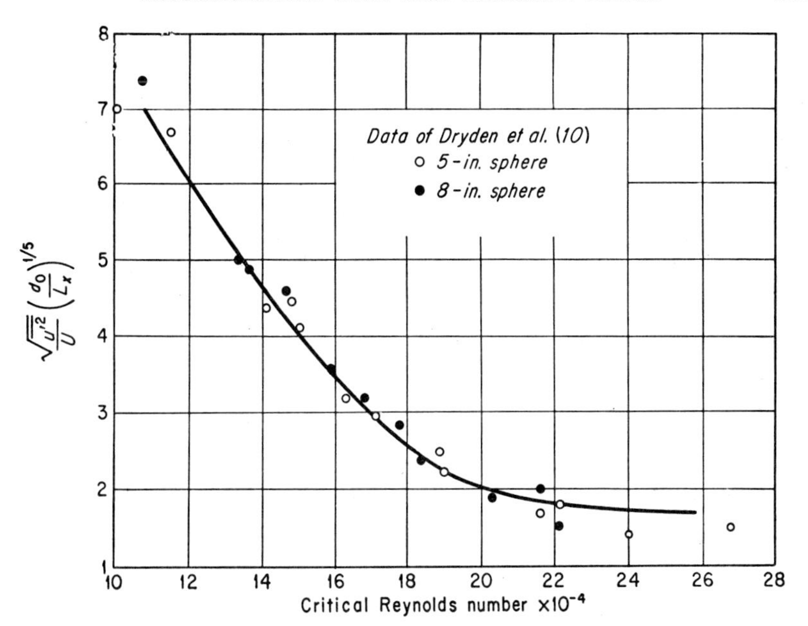

FIG. 10-48. Effect of turbulence on the critical Reynolds number for flow past spheres. (*From H. L. Dryden et al., NACA Rept.* 581, 1937.)

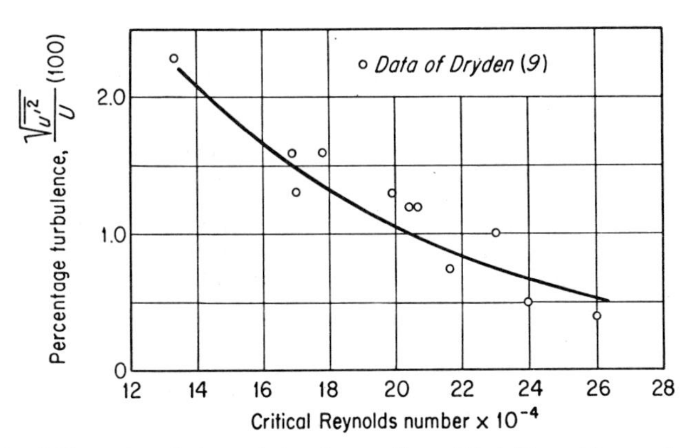

FIG. 10-49. Effect of turbulence intensity on the critical Reynolds number for flow past spheres. (*From H. L. Dryden, NACA Rept.* 392, 1931.)

involved. These workers showed that the critical Reynolds number is related to the dimensionless term $\frac{\sqrt{\overline{u'^2}}}{U}\left(\frac{d_0}{L_x}\right)^{1/5}$, where $\sqrt{\overline{u'^2}}/U$ is the intensity of turbulence and L_x is the scale of turbulence. Figure 10-48 shows the effect of turbulence on critical Reynolds numbers for spheres. In this plot the experimental data were obtained for spheres of two diameters. All empirical data, however, may be represented by a single curve. The critical Reynolds number plotted in Fig. 10-48 is the Reynolds number where the drag coefficient has a value of 0.3.

Some useful results for spheres obtained by Dryden [9] are shown in Fig. 10-49, where the critical Reynolds number (where $f_D = 0.3$) is plotted versus the percentage turbulence of the main stream. These results may be used to predict approximate critical Reynolds numbers for spheres or, conversely, to predict the percentage turbulence of the stream flowing past the sphere.

10-33. Two- and Three-dimensional Turbulent Boundary Layers

Turbulent boundary layers on circular cylinders and bodies of revolution are much more difficult to analyze than the laminar boundary layer. In solving the integral momentum equation (10-119) for laminar flow it is necessary to assume the form of the laminar velocity profile. This assumption allows the calculation of the shear stress at the wall. When solving Eq. (10-119) for the turbulent boundary layer, an assumption must also be made regarding the shear stress at the surface of the body. Another difficulty arises in that the actual position of the turbulent boundary layer is unknown. Transition from laminar to turbulent flow on cylinders and bodies of revolution is accompanied by the sharp decrease in the drag coefficient as indicated in Figs. 10-36 and 10-45. However, this does not mean that the boundary layer is completely turbulent. Presumably it will be laminar over the forward portion of the body, the point of transition from laminar to turbulent boundary layer moving toward the leading edge as the Reynolds numbers increase above the critical value.

Millikan,[40] Mager,[36] and Tetervin and Lin [57] have solved the integral momentum equation for the turbulent boundary layer on cylinders and bodies of revolution. Millikan assumes that the turbulent velocity profile follows the one-seventh-power law and that the shear stress at the wall varies in the same manner as on a flat plate. Tetervin and Lin assume that the turbulent velocity profile follows some power law and employ experimental values of the wall shear stress to obtain a general solution to the integral momentum equation for turbulent two- and three-dimensional boundary layers.

IV. FLOW PAST FLAT PLATES—NORMAL AND INCLINED TO THE STREAM

10-34. Manner of Flow of a Fluid Impinging on a Flat Plate

As fluid flows past a flat plate placed normal to the direction of flow, a boundary layer forms on the plate and separates at the edges. The flow is characterized by a turbulent wake behind the plate which is always

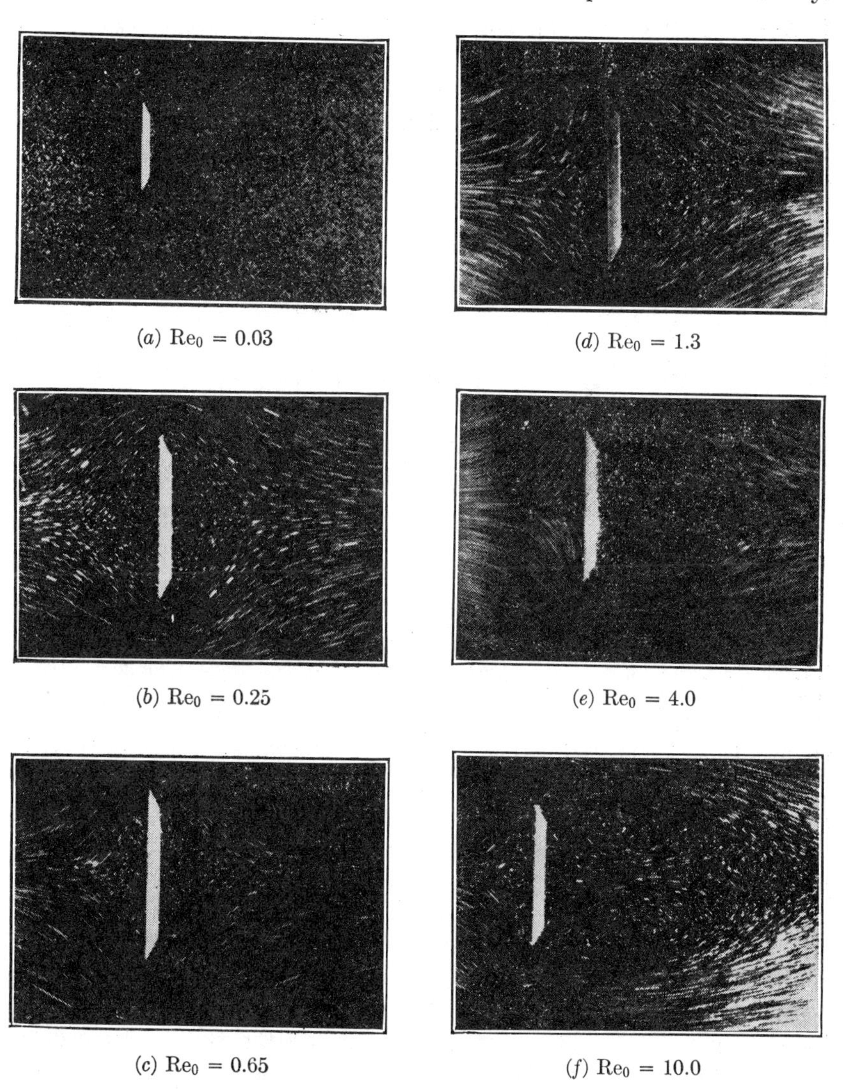

(a) $\mathrm{Re}_0 = 0.03$ (d) $\mathrm{Re}_0 = 1.3$

(b) $\mathrm{Re}_0 = 0.25$ (e) $\mathrm{Re}_0 = 4.0$

(c) $\mathrm{Re}_0 = 0.65$ (f) $\mathrm{Re}_0 = 10.0$

FIG. 10-50. Flow patterns for a fluid impinging on a flat plate. (*From O. J. Tietjens, Proc. Intern. Congr. Appl. Mech., 3rd Congr., Stockholm, 1930, vol.* 1, *p.* 331.)

about as wide as the plate itself. Flow patterns for fluids flowing past flat plates placed normal to the direction of flow obtained by Tietjens[59] are shown in Fig. 10-50. The photographs in this figure are for various Reynolds numbers.

10-35. The Drag of Flat Plates Normal and Inclined to the Stream

For flow at Reynolds numbers above 1,000 the drag coefficient of a flat plate normal or inclined to the stream is constant and independent of the Reynolds number. Schubauer and Dryden[54] reported drag coefficients for a flat plate placed normal to a flowing stream. They also showed the effect

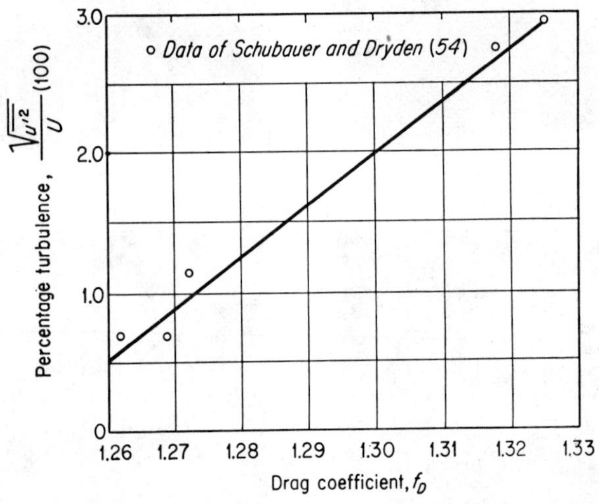

FIG. 10-51. Effect of percentage turbulence on the drag coefficient for flow normal to a flat plate. (*From G. B. Schubauer and H. L. Dryden, NACA Rept.* 546, 1936.)

of turbulence in the main stream on the drag coefficient. Figure 10-51 is a plot of f_D versus percentage turbulence for flow past a plate placed normal to flow. The circles represent the data of Schubauer and Dryden.

Fage and Johansen[15] determined drag coefficients for a flat plate inclined at various angles to the flowing stream. The results are shown in Fig. 10-52, where the drag coefficient f_D is plotted versus the angle of incidence, which is the angle the flat plate makes with the flowing stream. Except for a small decrease near an angle of incidence of 15°, the drag coefficient increases as the angle increases up to 90°. The drag coefficient for an angle of 90° is about 20 per cent below the values obtained by Schubauer and Dryden. Figures 10-51 and 10-52 may be used to predict drag coefficients for flat plates normal or inclined to the direction of flow of the fluid.

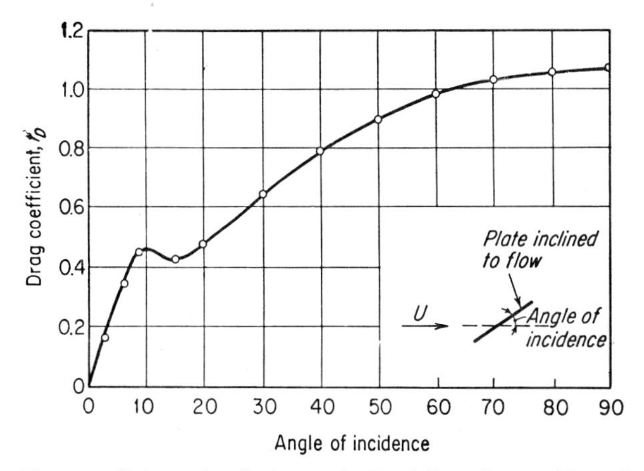

FIG. 10-52. Drag coefficients for flow past inclined flat plates. [*Data from A. Fage and F. C. Johansen, Proc. Roy. Soc. (London),* **116A**:170 (1927).]

BIBLIOGRAPHY

1. Blasius, H.: *Z. Math. u. Phys.*, **56**:1 (1908).
2. Burgers, J. M.: *Proc. Intern. Congr. Appl. Mech., 1st Congr., Delft, 1924,* p. 113.
3. Carrier, G. F., and C. C. Lin: *Quart. Appl. Math.*, **6**:63 (1948).
4. Charters, A. C., Jr.: *NACA TN* 891, 1943.
5. Delany, K., and N. E. Sorenson: *NACA TN* 3038, 1953.
6. Dhawan, S.: *NACA TN* 2567, 1952.
7. Dryden, H. L.: *NACA Rept.* 562, 1936.
8. Dryden, H. L.: *NACA Rept.* 497, 1934.
9. Dryden, H. L.: *NACA Rept.* 392, 1931.
10. Dryden, H. L., G. B. Schubauer, W. C. Mock, and H. K. Skramstad: *NACA Rept.* 581, 1937.
11. Fage, A.: *Brit. Aeronaut. Research Comm. R. & M.* 1766, 1937.
12. Fage, A.: *Brit. Aeronaut. Research Comm. R. & M.* 1765, 1937.
13. Fage, A.: *Brit. Aeronaut. Research Comm. R. & M.* 1179, 1928.
14. Fage, A., and V. M. Falkner: *Brit. Aeronaut. Research Comm. R. & M.* 1369, 1931.
15. Fage, A., and F. C. Johansen: *Proc. Roy. Soc. (London)*, **116A**:170 (1927).
16. Falkner, V. M.: *Aircraft Eng.*, **15**:65 (1943).
17. Falkner, V. M., and S. W. Skan: *Brit. Aeronaut. Research Comm. R. & M.* 1314, 1930.
18. Finn, R. K.: *J. Appl. Phys.*, **24**:771 (1953).
19. Goldstein, S. (ed.): "Modern Developments in Fluid Dynamics," Oxford University Press, London, 1938.
20. Green, J. J.: *Brit. Aeronaut. Research Comm. R. & M.* 1313, 1930.
21. Hansen, M.: *NACA TM* 585, 1930.
22. Hartree, D. R.: *Proc. Cambridge Phil. Soc.*, **33**:223 (1937).
23. Hegge-Zijnen, B. G. van der: *Verhandel. Koninkl. Akad. Wetenschap. Amsterdam Afdeel. Natuurk.*, **31**:499 (1928).
24. Hegge-Zijnen, B. G. van der: Measurements of the Velocity Distribution in the Boundary Layer along a Plane Surface, Thesis, Delft, 1924.
25. Howarth, L.: *Proc. Roy. Soc. (London)*, **164A**:547 (1938).
26. Janour, Z.: *NACA TM* 1316, 1951.

27. Janssen, E.: "Heat Transfer and Fluid Mechanics Institute, Preprints of Papers," p. 173, Stanford University Press, Stanford, Calif., 1956.
28. Kármán, T. von: *NACA TM* 1092, 1946.
29. Kármán, T. von: *Z. angew. Math. u. Mech.*, **1**:233 (1921).
30. Kempf, G.: "Neue Ergebnisse der Widerstandsforschung," Werft, Reederei, Hafen, p. 234, 1929.
31. Klebanoff, P. S.: *NACA TN* 3178, 1954.
32. Ladenburg, R.: *Ann. Physik*, **23**:447, 1907.
33. Lamb, H.: *Phil. Mag.*, **21**:112 (1911).
34. Liebster, H.: *Ann. Physik*, **82**:541 (1927).
35. Liepmann, H. W., and G. H. Fila: *NACA TN* 1196, 1947.
36. Mager, A.: *NACA Rept.* 1067, 1952.
37. Main-Smith, J. D.: *Brit. Aeronaut. Research Comm. R. & M.* 2755, 1954.
38. Meksyn, D.: *Proc. Roy. Soc. (London)*, **194A**:218 (1948).
39. Meksyn, D.: *Proc. Roy. Soc. (London)*, **192A**:545 (1948).
40. Millikan, C. B.: *Trans. ASME*, **54**:APM-3, 1932.
41. Müller, W.: "Mathematische Strömungslehre," Springer-Verlag OHG, Berlin, 1928.
42. Oseen, C. W.: *Arkiv. Mat. Astron. Fysik*, **6**:75 (1910); **9**:1 (1913).
43. Pohlhausen, K.: *Z. angew. Math. u. Mech.*, **1**:252 (1921).
44. Prandtl, L.: *NACA TM* 452, 1928.
45. Prandtl, L.: *Ergeb. Aerodyn. Versuchanstalt Göttingen*, 3, p. 1, 1927.
46. Preston, J. H., and N. E. Sweeting: *Brit. Aeronaut. Research Comm. R. & M.* 2014, 1947
47. Rouse, H.: "Elementary Mechanics of Fluids," John Wiley & Sons, Inc., New York, 1948.
48. Schlichting, H.: "Boundary Layer Theory," McGraw-Hill Book Company, Inc., New York, 1955.
49. Schlichting, H.: *NACA TM* 1218, 1949.
50. Schubauer, G. B.: *NACA Rept.* 652, 1939.
51. Schubauer, G. B.: *NACA Rept.* 527, 1935.
52. Schubauer, G. B., and P. S. Klebanoff: *NACA TN* 3489, 1955.
53. Schubauer, G. B., and H. K. Skramstad: *NACA Rept.* 909, 1948.
54. Schubauer, G. B., and H. L. Dryden: *NACA Rept.* 546, 1936.
55. Schultz-Grunow, F.: *NACA TM* 986, 1941.
56. Simmons, L. F. G., and A. F. C. Brown: *Brit. Aeronaut. Research Comm. R. & M.* 1547, 1934.
57. Tetervin, N., and C. C. Lin: *NACA Rept.* 1046, 1951.
58. Thom, A.: *Brit. Aeronaut. Research Comm. R. & M.* 1176, 1928.
59. Tietjens, O. J.: *Proc. Intern. Congr. Appl. Mech., 3rd Congr. Stockholm, 1930*, vol. 1, p. 331.
60. Tomotika, S.: *Brit. Aeronaut. Research Comm. R. & M.* 1678, 1935.
61. Wieselsberger, C.: *Ergeb. Aerodyn. Versuchanstalt Göttingen* 2, p. 22, 1923.
62. Wieselsberger, C.: *Ergeb. Aerodyn. Versuchanstalt Göttingen* 1, p. 120, 1923.

CHAPTER 11

FLOW IN THE SHELL SIDE OF MULTITUBE HEAT EXCHANGERS

11-1. Introduction

The flow of fluids in the shell side of multitube heat exchangers is a combination of immersed flow and conduit flow. As far as the heat-exchanger tubes are concerned, they are immersed in the fluid flowing in the shell. The heat-exchanger shell itself is actually a closed conduit with fluid flowing in it. The effect of the shell on the dynamics of flow in it is probably very small, since a large portion of the free space in the shell is occupied by the tubes. Such dimensions as tube shape, tube size, tube spacing, and tube configuration are the main variables affecting the mechanism of flow in heat-exchanger shells.

In this chapter flow in heat-exchanger shells is discussed from two standpoints. First, the flow of fluid normal to banks of tubes is considered. In most gas and air heaters, such as automobile radiators and room heaters, the gas flows perpendicular to a bank of tubes carrying the hot fluid. In this case, the cross section of the heat exchanger perpendicular to the gas flow is usually rectangular. The second type to be considered is flow in multitube, baffled heat exchangers. In these the shell-side fluid is caused, by means of a baffle arrangement, to flow both perpendicular and parallel to the tubes. The purpose of the baffles is to prevent the formation of stagnant areas in the heat exchanger.

11-2. Tube Arrangements

It is usual, in the construction of multitube heat exchangers, to arrange the tubes in rows. The rows of tubes perpendicular to the flowing stream are called transverse rows; the rows parallel to the flow are called longitudinal rows.

There are two ways of arranging tubes in tube banks:

1. In the *in-line arrangement*, shown in Fig. 11-1*a*, the tubes in adjacent transverse rows are in line with each other in the direction of flow of the fluid. The center-to-center distance between tubes in adjacent transverse rows is the longitudinal pitch S_L. The center-to-center distance between the longitudinal rows is the transverse pitch S_T. Both S_L and S_T

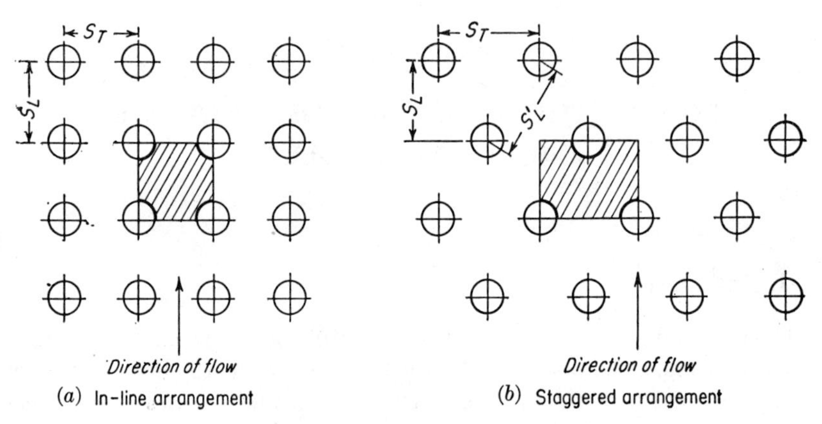

FIG. 11-1. Tube arrangements.

are shown in Fig. 11-1*a* for the in-line arrangement. There are two types of in-line arrangements:

a. Square in-line arrangement, in which $S_L = S_T$.

b. Rectangular in-line arrangement, in which $S_L < S_T$ or $S_L > S_T$.

For the in-line arrangements the minimum cross section presented to the flowing stream is the distance between the longitudinal rows $S_T - d_0$, where d_0 is the outside diameter of the tubes.

2. The *staggered arrangement* is shown in Fig. 11-1*b*. The tubes in adjacent transverse rows are not in line with the direction of flow of the fluid. The distance S_T is the center-to-center spacing of tubes in the transverse rows, while S_L is the center-to-center spacing of transverse rows. The center-to-center distance between the tubes in adjacent transverse rows is S_L', which is related to S_L and S_T by

$$S_L' = \sqrt{S_L{}^2 + \left(\frac{S_T}{2}\right)^2} \tag{11-1}$$

There are two types of staggered tube arrangements:

a. Staggered triangular arrangement, in which $S_T < S_L'$ or $S_T > S_L'$. When $S_T < S_L'$, the minimum cross section of flow is $S_T - d_0$, while if $S_T > S_L'$, the minimum cross section of flow is $S_L' - d_0$. The staggered square arrangement is a special case of the staggered triangular arrangement in which $S'_L = S_T/\sqrt{2}$.

b. The staggered equilateral-triangular arrangement, in which $S'_L = S_T$.

11-3. Flow Patterns for Flow Perpendicular to Banks of Tubes

As might be expected, flow perpendicular to a cylindrical tube in a bank of tubes is similar to flow past a single cylinder in an infinite fluid. A boundary layer forms on the forward portion of the cylinder and ultimately separates from the surface, producing a turbulent wake behind the cylinder.

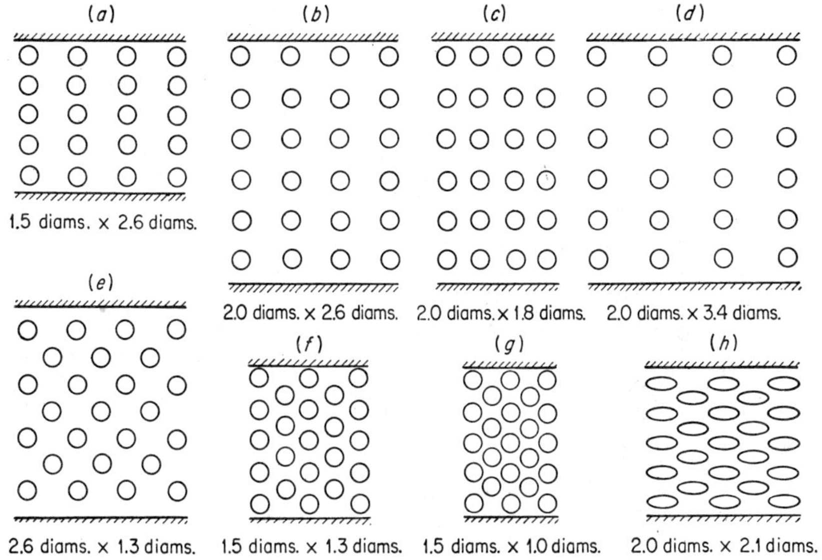

(a) 1.5 diams. x 2.6 diams.
(b) 2.0 diams. x 2.6 diams.
(c) 2.0 diams. x 1.8 diams.
(d) 2.0 diams. x 3.4 diams.
(e) 2.6 diams. x 1.3 diams.
(f) 1.5 diams. x 1.3 diams.
(g) 1.5 diams. x 1.0 diams.
(h) 2.0 diams. x 2.1 diams.

FIG. 11-2. Banks of tubes studied by Wallis. [*From R. P. Wallis, Engineering*, **148:** 423 (1934).]

The presence of adjacent cylinders, however, affects the thickness and velocity distribution in the boundary layer. The nature of the turbulent wake is determined to a large extent by the tube arrangement. Drag coefficients for individual tubes in a tube bank will be affected by surrounding tubes and will thus be different from those for a single tube with no tubes surrounding it. It is usual in measuring the resistance to flow across tube banks to determine the drag of rows of tubes rather than to determine drag coefficients for individual tubes.

Wallis [15] studied visually the flow of fluids perpendicular to tube banks by observing the motion of fine aluminum powder placed on the surface of water flowing perpendicular to the tubes. The axes of the tubes were

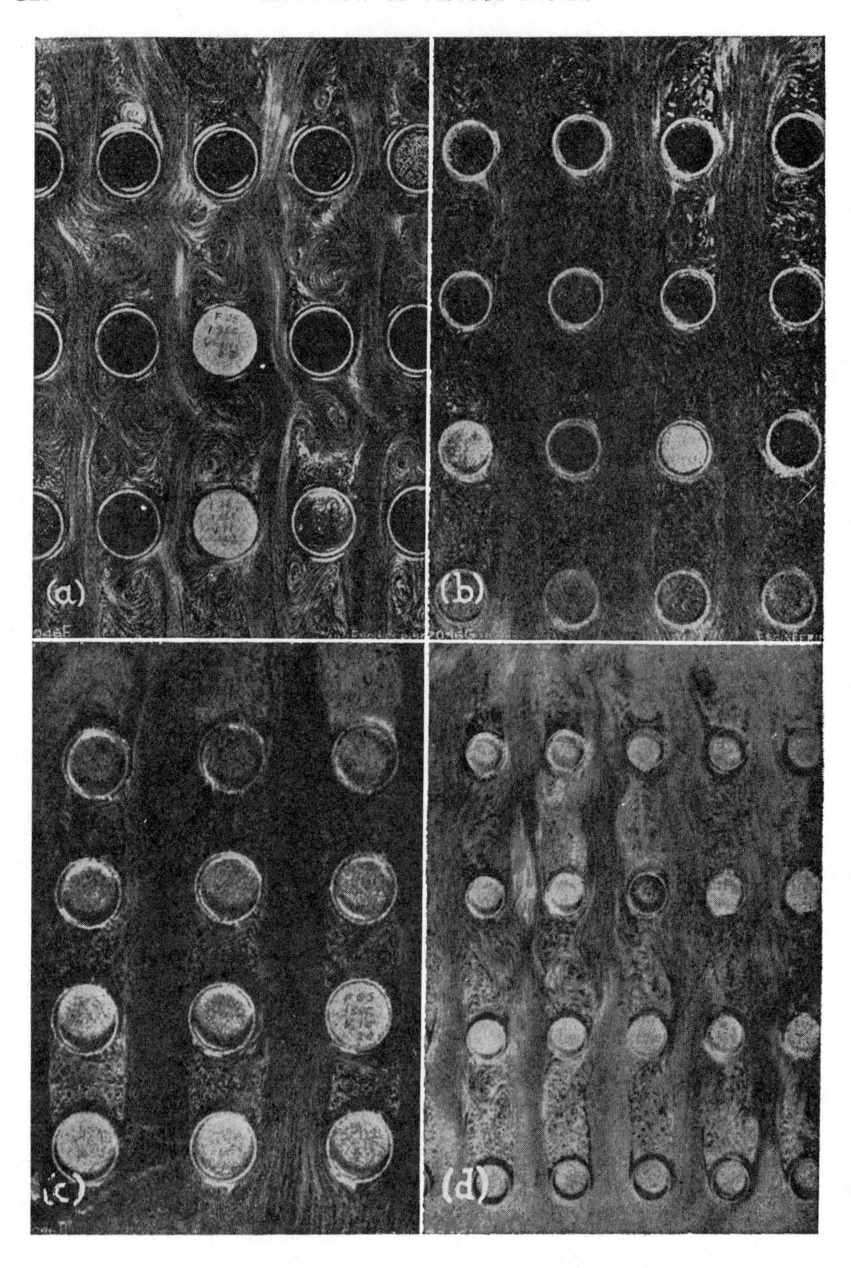

FIG. 11-3. Flow patterns for flow past tube banks.

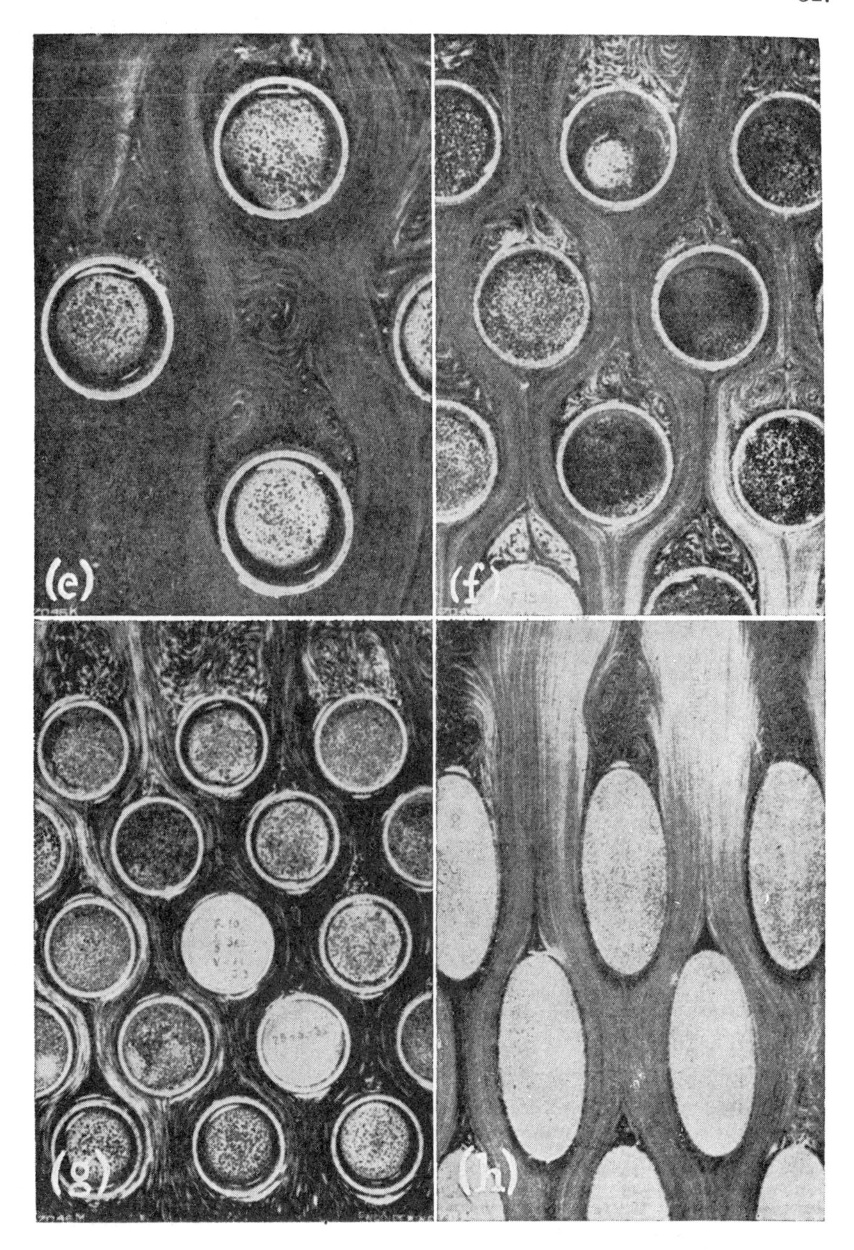

[*From R. P. Wallis, Engineering*, **148**:423 (1934).]

vertical. After a number of trials, Wallis was able to obtain some good photographs of the flow patterns which occurred as the water flowed through the tube bank. He studied the tube groupings depicted in Fig. 11-2, i.e., four different in-line arrangements, three different staggered arrangements, and one staggered arrangement in which the tubes are elliptical in section. The designation given for the distances between the tubes in Fig. 11-2 is based on the tube diameter. For example, in Fig. 11-2, the designation "1.5 diams. × 2.6 diams." means that S_T is $1.5d_0$ and S_L is $2.6d_0$, where d_0 is the diameter of the tubes. For the arrangement of the elliptical tubes in Fig. 11-2 the designation "2.0 diams. × 2.1 diams." means that S_T is 2.0 times the minor axis of the ellipse and S_L is 2.1 times the minor axis of the ellipse, where S_T and S_L are again center-to-center distances.

The flow patterns for the in-line arrangements in Fig. 11-2 are shown in Fig. 11-3. They are all similar. The patterns around the tubes in the first transverse row are similar to the patterns shown in Fig. 10-25 for flow around a single circular cylinder. The separation of the boundary layer and the turbulent wake behind the tubes are evident. However, it appears that the turbulent wake continues to the next tube in the next transverse row, and only a very thin boundary layer forms on that tube. The spaces between the tubes in all transverse rows contain a turbulent wake, while in the unobstructed space between the longitudinal rows there is no evidence of excessive eddying or turbulence. Figure 11-3 gives the flow patterns observed by Wallis for the staggered tube arrangements shown in Fig. 11-2. When the tubes are widely spaced, a turbulent wake occurs behind each tube and extends nearly up to the next tube, which is two transverse rows away. However, a boundary layer is formed on the forward part of each tube in the bundle, and separation of this boundary layer takes place. For the closely spaced staggered arrangements (Fig. 11-3) the turbulent wake behind each tube is considerably reduced. With these spacings the tubes are placed so that they are not in the turbulent wake of the tubes immediately upstream, with the result that energy dissipation is likely to be reduced. The only place where there is a large turbulent wake is behind the last transverse row of tubes. The flow pattern for the elliptical tubes is shown in Fig. 11-3. With this arrangement there is still less evidence of turbulent-wake formation than in the case of the circular tubes, and thus the energy loss due to flow past a bank of elliptical tubes should be less than with circular tubes.

11-4. Flow in Baffled Heat Exchangers

Commercial multitube heat exchangers contain baffles to guide the fluid through the equipment and to prevent stagnant regions from forming. The baffles are of various types, each type producing a different kind of flow pattern in the shell of the heat exchanger. The common forms of baffles installed in heat exchangers are as follows.

1. *Orifice baffles* (Fig. 11-4) extend through the cross section of the shell. The holes for the tubes are considerably larger than the tube, and the fluid flows through this annular orifice at each point where the tube passes through the baffle. The arrows in Fig. 11-4 indicate the general direction of fluid flow; it is mainly parallel to the tubes, but a small amount of crossflow occurs.

2. *Disk-and-doughnut baffles* are shown in Fig. 11-5. This arrangement consists of disks and doughnuts, or annular rings, placed alternately along

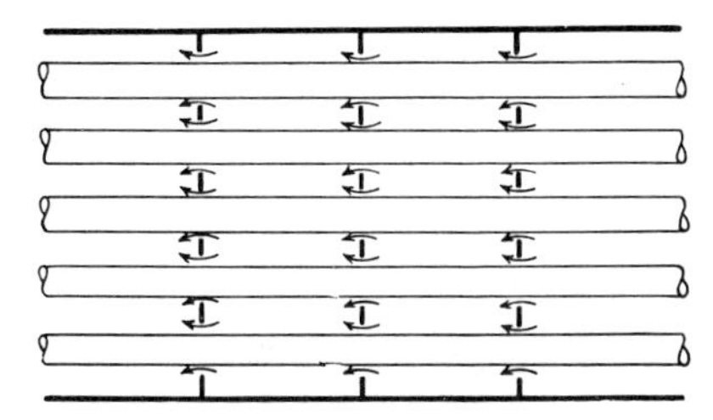

FIG. 11-4. Orifice baffles.

FIG. 11-5. Disk-and-doughnut baffles.

the length of the tube bundle. Since the clearance between the baffle and the tubes where they go through the baffle is small, most of the flow takes place parallel to the tubes, as indicated by the arrows. A greater portion of the tube bank is subjected to crossflow with this baffle arrangement than with orifice baffles.

3. *Segmental baffles* are the type most frequently found in commercial heat exchangers (see Fig. 11-6). The segmental baffle is a segment of a

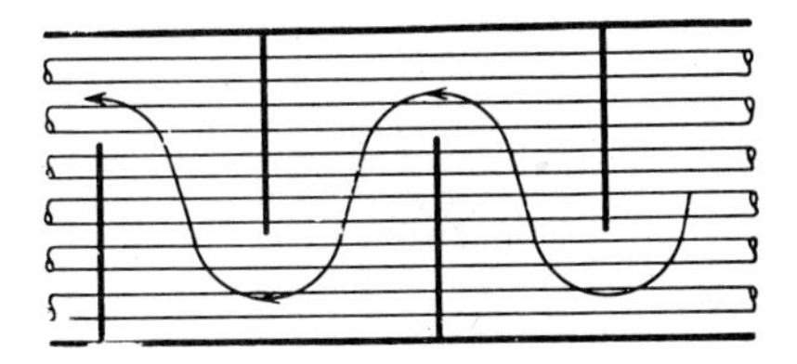

FIG. 11-6. Segmental baffles.

circle and is installed in the tube bundle in such a way that the fluid flows back and forth across the tube bank. The arrows in Fig. 11-6 indicate the general direction of fluid flow through a tube bundle containing segmental baffles. In general, the segmental baffles cause more crossflow than any other type of baffle. In the typical heat exchanger containing segmental baffles a large portion of the tubes have fluid flowing perpendicular to them.

The shaded areas in Fig. 11-7 indicate the approximate regions where flow is parallel to the tube in a tube bundle containing segmental baffles.

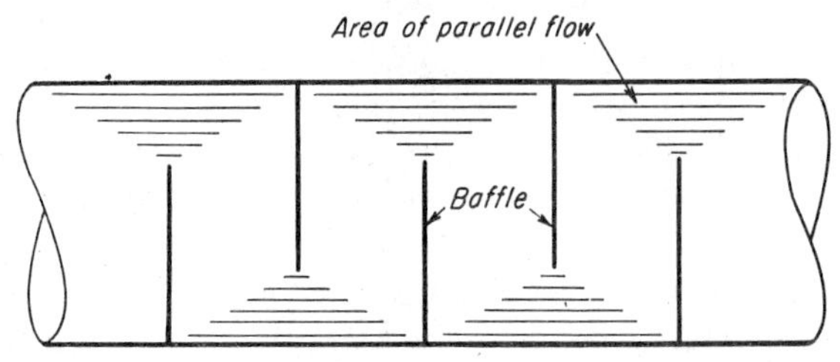

Fig. 11-7. Areas of parallel flow in a heat exchanger with segmental baffles.

The amount of crossflow and extent of stagnant regions in a heat exchanger containing segmental baffles depends on the baffle spacing, the size of opening between baffle and shell, and the amount of baffle overlap.

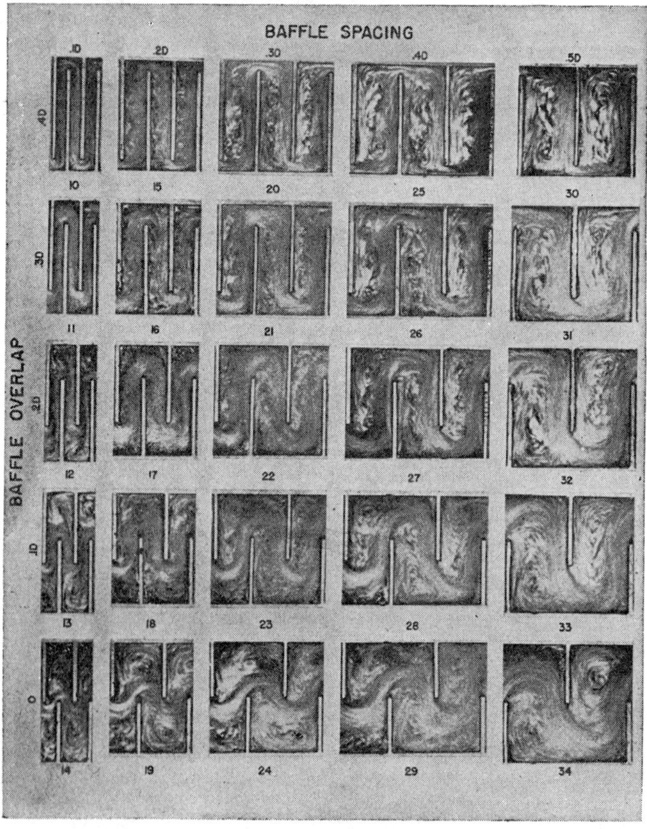

Fig. 11-8. Flow patterns for flow past two-dimensional baffles. (*From A. Y. Gunter, H. R. Sennstrom, and S. Kopp, paper* 47-A-103, *presented at the annual meeting of the ASME, Atlantic City, N.J.*, 1947.)

Gunter, Sennstrom, and Kopp [8] obtained photographs of the flow pattern of a fluid flowing past segmental baffles of different heights and spacing. They observed the motion of fine aluminum powder on the surface of a fluid flowing through a two-dimensional system of baffles. The series of photographs in Fig. 11-8 shows the flow patterns obtained for five different baffle spacings and five positions of the baffles relative to each other (baffle overlap), giving twenty-five different systems investigated. Flow of the fluid is from the left.

The edge of each baffle is a point of separation of the boundary layer from the baffle. A turbulent wake extends behind each baffle, increasing in size as the baffle spacing increases. However, the main path of the fluid, as indicated by smooth streamlines, takes up a considerable portion of the space between the baffles. The dark areas in the photographs, particularly in the corners at the base of the baffle, are evidence of regions of stagnation, where the heat-transfer surface is probably very ineffective. In general, areas of stagnation are reduced by close baffle spacing and large baffle overlap. The photographs also indicate that the major portion of the flow is across the tubes.

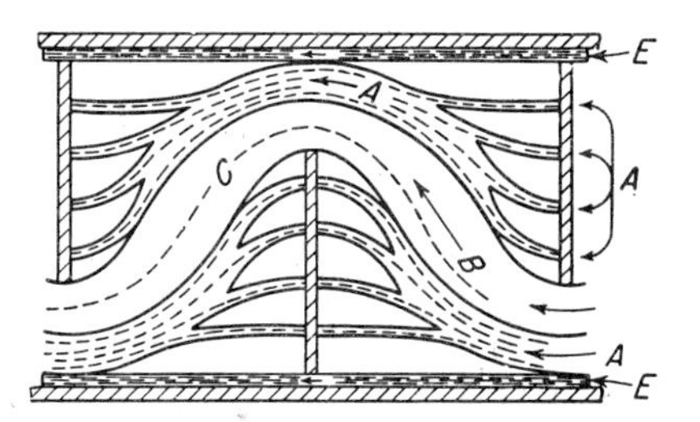

Fig. 11-9. Fluid streams through heat-exchanger shell. (*From T. Tinker, "Proceedings of the General Discussion on Heat Transfer," Institution of Mechanical Engineers, London, and American Society of Mechanical Engineers, New York,* 1951, *p.* 89.)

For convenience in construction it is necessary that commercial baffled heat exchangers have clearance both between the tubes and the baffles and between the baffles and the shell. These clearances should be as small as possible, but often they are of considerable magnitude, and their effect on the flow in the heat exchanger must be taken into account. Since the clearances allow leakage of fluid past the baffles, the effectiveness of the baffles is reduced. Tinker's [14] representation of the flow of the fluid stream through the shell side of a commercial heat exchanger containing segmental baffles is shown in Fig. 11-9. It is seen that the stream approaching a particular baffle can be divided into four parts:

1. Stream A, which flows through the tube holes in the baffles
2. Stream B, which flows across the bank, through the baffle opening or window, across the tube bank, and so on through the exchanger
3. Stream E, which flows between the baffles and the shell and which is an annular stream of thickness equal to the clearance between the baffles and the shell
4. Stream C, which is a bypass or recirculating stream between streams B and E

It is apparent that stream B should constitute as large a fraction as possible of the total flow. If all clearances were zero, stream B could conceivably be 100 per cent of the total flow. Evidently, as much as possible should be done to reduce streams C and E, since neither one is effective in transferring heat to or from the tubes. The effect of stream A is unknown. The fact that it goes through the baffle holes and thereby contacts the tubes at a relatively high velocity indicates that it may contribute to the effectiveness of heat transfer in the heat exchanger.

11-5. The Equivalent Diameter of Banks of Tubes

A number of diameter terms are used in friction-factor and Reynolds-number calculations for flow across banks of tubes. The dimensions by which a bank of tubes is described are as follows:

1. Tube diameter d_0
2. Spacing of longitudinal rows S_T
3. Spacing of transverse rows S_L
4. Distance between tubes in adjacent transverse rows S_L'
5. Number of transverse rows of tubes N_T
6. Dimensions of shell

Many investigators have used either the tube diameter d_0 or the clearance between tubes in transverse rows $d_c = S_T - d_0$ as the length term to be used in the Reynolds number. A correlation of friction data must then involve dimensionless groups containing the other variables, S_L, S_L', N_T, and shell dimensions. Other investigators have defined equivalent diameters which include all of the dimensional variables of the tube bank in an effort to obtain a correlation of data involving a minimum number of dimensionless groups.

A common equivalent diameter for tube banks is that based on the free cross-sectional area and the wetted perimeter, i.e.,

$$d_e = \frac{(4)(\text{free cross-sectional area})}{\text{wetted perimeter}} \tag{11-2}$$

Considering the shaded portions in Fig. 11-10, d_e may be given as

$$d_e = 4\,\frac{S_T S_L - \pi d_0^2/4}{\pi d_0} \tag{11-3}$$

The volumetric equivalent diameter is defined as

$$d_v = 4\,\frac{\text{free volume in tube bank}}{\text{exposed surface area of tubes}} \tag{11-4}$$

For large banks of bare tubes d_e and d_v are essentially equal, but for small

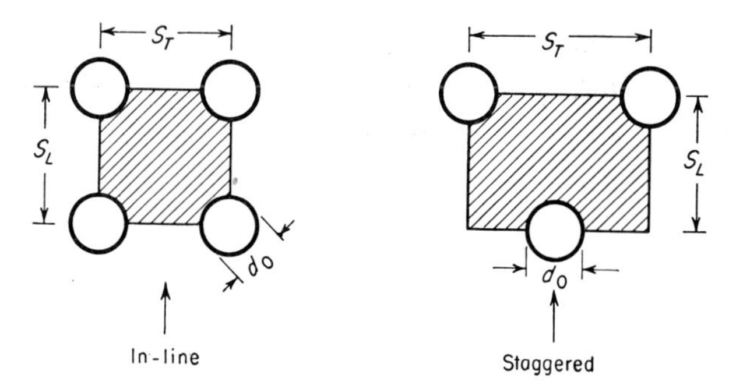

FIG. 11-10. Method of calculating d_e for tube banks.

banks of bare tubes and banks of finned tubes they are not. Kays, London, and Lo[13] defined an equivalent diameter based on the minimum free area of flow perpendicular to the flowing stream and on the length of the tube bank. This equivalent diameter is similar to the volumetric diameter and is expressed as

$$d_v' = \frac{(4)(\text{minimum area of flow})(L)}{\text{heat-transfer area in exchanger}} \tag{11-5}$$

where L is the effective flow length of the tube bank and is defined as

$$L = N_T S_L \tag{11-6}$$

For the staggered arrangement the minimum area of flow is determined from S_T if S_T is less than S_L' and from S_L' if S_T is greater than S_L'.

The velocity used in the calculation of the Reynolds number is that through the minimum area of flow. The mass velocity through the minimum area is $G_{\max}(m/L^2t)$. In Table 11-1 the various diameter terms described above are defined and the corresponding Reynolds number shown.

11-6. Definition of Friction Factors for Tube Banks

Various friction factors are defined for flow across banks of tubes. The definitions more or less consider the various dimensions of the system. The usual friction factor for tube banks is defined as

$$f_{TB} = \frac{2(-\Delta P_f)g_c\rho}{4G_{\max}^2}\frac{1}{N'} \tag{11-7}$$

where N' is the number of major restrictions encountered in flow through the tube bank. For all in-line arrangements and for staggered arrangements where S_T is less than S_L'

$$N' = N_T \tag{11-8}$$

TABLE 11-1. DIAMETER TERMS USED FOR TUBE BANKS

Term	Symbol and definition	Reynolds number
Tube diameter	d_0	$\mathrm{Re}_0 = \dfrac{d_0 G_{\max}}{\mu}$
Tube clearance	$d_c = S_T - d_0$	$\mathrm{Re}_c = \dfrac{d_c G_{\max}}{\mu}$
Equivalent diameter	$d_e = \dfrac{(4)(\text{free cross-sectional area})}{\text{wetted perimeter}} = 4\dfrac{S_T S_L - \pi d_0^2/4}{\pi d_0}$	$\mathrm{Re} = \dfrac{d_e G_{\max}}{\mu}$
Volumetric equivalent diameter	$d_v = 4\dfrac{\text{free volume in tube bank}}{\text{exposed surface area of tubes}}$	$\mathrm{Re}_v = \dfrac{d_v G_{\max}}{\mu}$
Modified volumetric equivalent diameter	$d_v' = \dfrac{(4)(\text{minimum area of flow})(N_T S_L)}{\text{heat-transfer area in exchanger}}$	$\mathrm{Re}_v' = \dfrac{d_v' G_{\max}}{\mu}$

where N_T is the number of transverse rows. For staggered arrangements where S_T is greater than S_L'

$$N' = N_T - 1 \tag{11-9}$$

The friction factor defined by Eq. (11-7) has been used in recent correlations of Bergelin et al.[3] Chilton and Genereaux[4] defined the following friction factor:

$$f_{CG} = \frac{2(-\Delta P_f)g_c\rho}{4G_{\max}^2}\frac{d_v}{L} \tag{11-10}$$

From Eq. (11-6)

$$f_{CG} = \frac{2(-\Delta P_f)g_c\rho}{4G_{\max}^2}\frac{1}{N_T}\frac{d_v}{S_L} \tag{11-11}$$

Thus, when $N' = N_T$,

$$f_{CG} = f_{TB}\frac{d_v}{S_L} \tag{11-12}$$

Gunter and Shaw[7] defined the following friction factors for tube banks:

For in-line arrangements:

$$f_{GS} = 4\frac{2(-\Delta P_f)g_c\rho}{4G_{\max}^2}\frac{d_v}{L}\left(\frac{S_T}{d_v}\right)^{0.4}\left(\frac{S_T}{S_L}\right)^{0.6} \tag{11-13}$$

For staggered arrangements:

$$f'_{GS} = 4\,\frac{2(-\Delta P_f)g_c\rho}{4G_{\max}^2}\,\frac{d_v}{L}\left(\frac{S_T}{d_v}\right)^{0.4}\left(\frac{S_T}{S'_L}\right)^{0.6} \tag{11-14}$$

As before, it may be shown that, when $N' = N_T$,

$$f_{GS} = 4f_{TB}\,\frac{d_v}{S_L}\left(\frac{S_T}{d_v}\right)^{0.4}\left(\frac{S_T}{S_L}\right)^{0.6} \tag{11-15}$$

and

$$f'_{GS} = 4f_{TB}\,\frac{d_v}{S_L}\left(\frac{S_T}{d_v}\right)^{0.4}\left(\frac{S_T}{S'_L}\right)^{0.6} \tag{11-16}$$

Kays, London, and Lo [13] defined the friction factor

$$f_{KL} = \frac{2(-\Delta P_f)g_c\rho}{4G_{\max}^2}\,\frac{d'_v}{L} \tag{11-17}$$

and when $N' = N_T$

$$f_{KL} = f_{TB}\,\frac{d'_v}{S_L} \tag{11-18}$$

11-7. Friction Factors for Laminar Flow across Banks of Tubes

When a fluid flows across a tube bank in such a way that little or no separation of the boundary layer occurs on the individual tubes, the flow is considered laminar in nature. All frictional resistance associated with such flow is viscous resistance. The friction factor is inversely proportional to the Reynolds number in this regime of flow. Laminar flow exists up to a value of $\text{Re}_v = 100$.

There has been extensive investigation of flow across banks of tubes for both laminar and turbulent flow. This work is important since friction factors for flow across banks of tubes have been used to predict the friction losses in commercial heat exchangers.[14] A study of laminar flow across tube banks has been conducted by Bergelin and coworkers,[1,2] who studied laminar crossflow for seven different tube arrangements and obtained the following expression for the friction factor:

$$f_{TB}\left(\frac{\mu_b}{\mu_w}\right)^n = \frac{70}{\text{Re}_v}\left(\frac{d_0}{S_m}\right)^{1.6} \qquad \text{Re}_v < 100 \qquad S_m/d_0 = 1.25 \text{ or } 1.50 \tag{11-19}$$

where $S_m = S_T$ for in-line arrangements and $S_m = S_T$ or S'_L, whichever is smaller, for staggered arrangements. The viscosity correction is used when flow is nonisothermal, where μ_b is viscosity of the fluid at the average bulk temperature and μ_w is the fluid viscosity at the tube-wall temperature.

The exponent n is given by the relation

$$n = 0.57(\text{Re}_v)^{-0.25} \qquad \text{Re}_v < 300 \tag{11-20}$$

The empirical relationship of Eq. (11-19) is restricted to two values of S_m/d_0, 1.25 and 1.50.

Chilton and Genereaux [4] recommend the following equation for laminar flow across banks of tubes:

$$f_{CG} = \frac{26.5}{\text{Re}_v} \tag{11-21}$$

Gunter and Shaw [7] reviewed a large amount of data on crossflow in tube banks and reported the relations

$$f_{GS} \text{ or } f'_{GS} = \frac{180}{\text{Re}_v} \tag{11-22}$$

For predicting friction factors for laminar flow across tube banks Eq. (11-19) is recommended if $1.25 < S_m/d_0 < 1.50$. Outside this range of S_m/d_0 Eq. (11-22) is recommended. Equation (11-21) gives friction factors which are low.

Example 11-1

A tube bank has the following dimensions:

$d_0 = 0.375$ in.
$N_T = 10$ transverse rows
$N' = 10$
$S_T = 15/32$ in.
$S'_L = 15/32$ in. $= S_m$
$S_L = 13/32$ in.
$d_v = 0.0225$ ft
$L = 0.338$ ft
Tube length = 6 in.
Minimum free area of flow = 0.0254 ft^2
Number of tubes in transverse row = 7
Tube arrangement: staggered, equilateral triangle

Oil is flowing across this tube bank at a mass velocity (based on minimum flow area) of 241,000 lb$_m$/(hr)(ft^2), giving a value of $\text{Re}_v = 45.3$. Calculate the value of $-\Delta P_f$ from Eqs. (11-19), (11-21), and (11-22). The density of the oil is 53.29 lb$_m$/ft^3.

Solution

By Eq. (11-19)

$$f_{TB} = \frac{70}{45.3}\left[\frac{(0.375)(32)}{15}\right]^{1.6} = 1.08$$

(The experimental value of f_{TB} is 1.06.)

From Eq. (11-7)

$$-\Delta P_f = \frac{(1.08)(2)(241{,}000)^2(10)}{(32.2)(3{,}600)^2(53.29)} = 56.4 \text{ lb}_f/\text{ft}^2$$

By Eq. (11-21)

$$f_{CG} = \frac{26.5}{45.3} = 0.586$$

By Eq. (11-10)

$$-\Delta P_f = \frac{(0.586)(2)(241{,}000)^2(0.338)}{(32.2)(3{,}600)^2(53.29)(0.0225)} = 46.0 \text{ lb}_f/\text{ft}$$

By Eq. (11-22)

$$f'_{GS} = \frac{180}{45.3} = 3.97$$

By Eq. (11-14)

$$-\Delta P_f = \frac{(3.97)(241{,}000)^2}{(2)(32.2)(3{,}600)^2(53.29)} \frac{0.338}{0.0225} \left(\frac{0.0225}{0.0391}\right)^{0.4} = 62.6 \text{ lb}_f/\text{ft}^2$$

Equation (11-19) gives results intermediate between those of Eqs. (11-21) and (11-22).

11-8. Transition Flow across Tube Banks

Extensive investigation of the friction loss for flow across tube banks indicates that transition to turbulent flow occurs at a value of Re_v in the neighborhood of 100. The transition region extends over a fairly wide Reynolds-number range. Bergelin, Brown, and Doberstein [1] determined friction factors in the transition region of flow for five different tube arrangements. Their results are reproduced in Fig. 11-11, where $f_{TB}\left(\frac{\mu_b}{\mu_w}\right)^{0.14}$ is plotted versus Re_0, the Reynolds number based on the tube diameter d_0. These authors specify the transition region to be in the range $200 < \text{Re}_0 < 5{,}000$. For staggered tube arrangements transition is characterized by a smooth friction-factor curve, while for in-line arrangements an irregularity is noted in the curve much the same as for closed conduits. This fact indicates that the transition process for staggered arrangements differs from that for in-line arrangements.

Figure 11-12 is a plot of f_{KL} [defined by Eq. (11-17)] versus Re'_v showing data obtained by Kays, London, and Lo [13] for seven different banks of tubes. The banks contained tubes 9.75 in. long and had 13 to 26 transverse rows and 10 to 20 tubes per row. Six staggered arrangements and one in-line arrangement are represented in Fig. 11-12. Again the transition curve for the in-line arrangement is different from that for the staggered arrangement. Figures 11-11 and 11-12 may be used to predict friction factors for transition flow across banks of tubes.

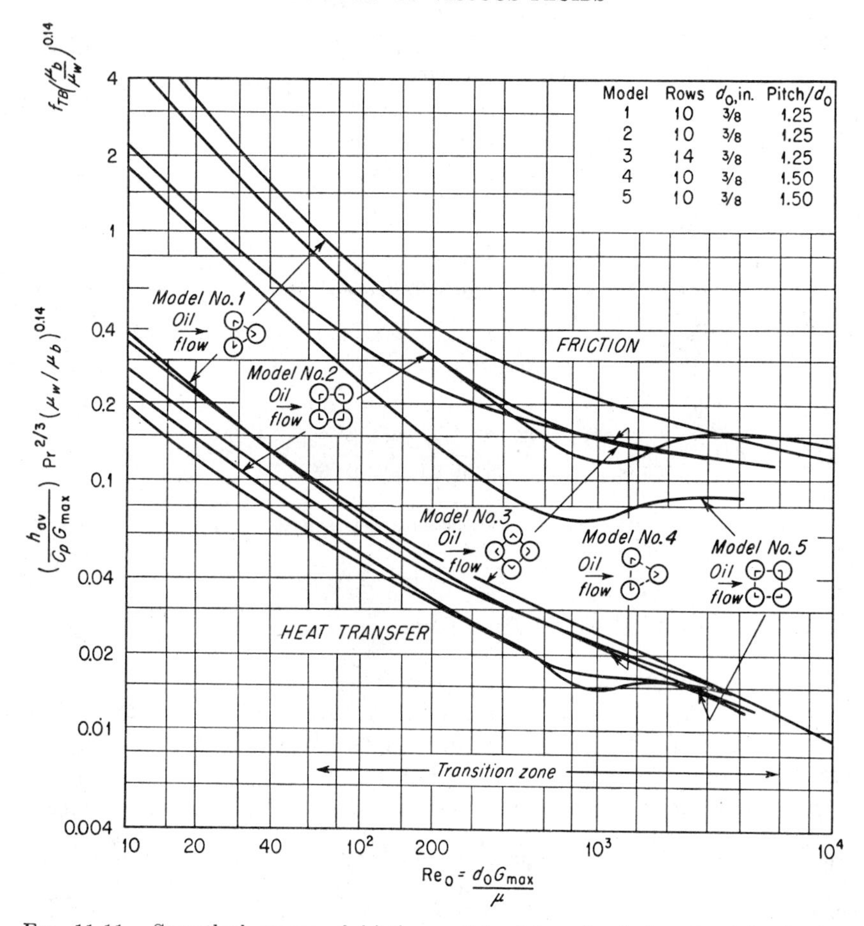

FIG. 11-11. Smoothed curves of friction and heat-transfer factors for tube banks. [*From O. P. Bergelin, G. A. Brown, and S. C. Doberstein, Trans. ASME*, **74**:953 (1952).]

11-9. Turbulent Flow across Tube Banks

Several relationships for predicting friction factors for flow across banks of tubes have been developed from a large number of experimental data on tube banks of various dimensions. Chilton and Genereaux [4] propose the following relationships:

For staggered arrangements:

$$f_{TB} = 0.75(\mathrm{Re}_c)^{-0.2} \tag{11-23}$$

For in-line arrangements:

$$f_{TB} = 0.33(\mathrm{Re}_c)^{-0.2} \tag{11-24}$$

(For both arrangements $100 < \mathrm{Re}_c < 20{,}000$.)

Arrangement	Symbol	Curve	d_0, in.	S_T, in.	S_L, in.	d_v', ft.
Staggered	△	A	0.375	0.4688	0.4688	0.0125
"	□	B	0.375	0.5625	0.375	0.0196
"	⌓	C	0.375	0.5625	0.5625	0.0298
"	▽	D	0.375	0.750	0.375	0.0327
"	●	E	0.375	0.9375	0.2813	0.0271
"	○	F	0.375	0.4688	0.5625	0.0256
In-line	x	G	0.250	0.3125	0.375	0.0166

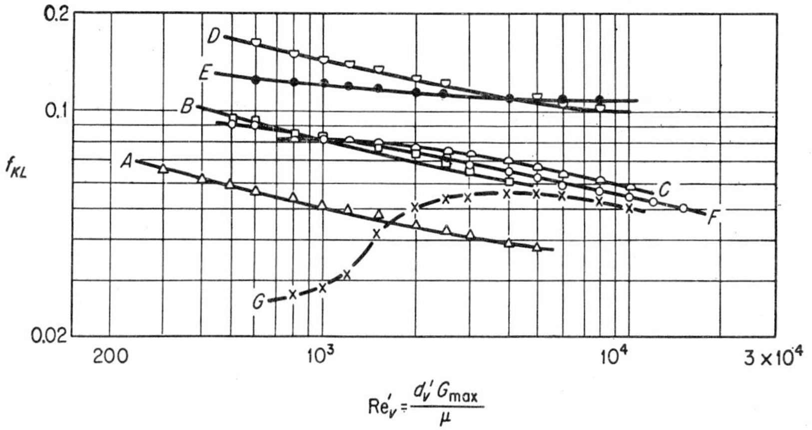

f_{KL}

$$Re_v' = \frac{d_v' G_{max}}{\mu}$$

FIG. 11-12. Friction factors for flow across tube banks. [*Data from W. M. Kays, A. L. London, and R. K. Lo, Trans. ASME,* **76**:387 (1954).]

In Eqs. (11-23) and (11-24) Re_c is based on the clearance between tubes in transverse rows (see Table 11-1).

Grimison[6] gave a complete friction-factor plot for flow across tube banks based on the friction-factor data from numerous tube banks. Jakob[9] studied the same data as Grimison and obtained the following equations for the friction factor:

For the staggered arrangement:

$$f_{TB} = (Re_0)^{-0.16}\left[0.25 + \frac{0.1175}{(S_T/d_0 - 1)^{1.08}}\right] \tag{11-25}$$

For in-line arrangements:

$$f_{TB} = (Re_0)^{-0.15}\left[0.044 + \frac{0.08(S_L/d_0)}{(S_T/d_0 - 1)^{0.43+(1.13d_0/S_L)}}\right] \tag{11-26}$$

(For both arrangements $5{,}000 < Re_0 < 40{,}000$.)

Gunter and Shaw[7] reviewed the available friction-factor data for both plain and extended-surface tube banks and recommended the following relationship for both types of surface:

For both arrangements:

$$f_{GS} \text{ or } f'_{GS} = 1.92(\text{Re}_v)^{-0.145} \qquad (11\text{-}27)$$

(For bare tubes, $500 < \text{Re}_v < 200{,}000$)
(For extended surfaces, $100 < \text{Re}_v < 20{,}000$)

Example 11-2

Given the two tube banks described in Table 11-2, determine the pressure drop due to friction at values of $\text{Re}_v = 2{,}000$ and $40{,}000$ by means of Eqs. (11-23) to (11-27).

TABLE 11-2

Dimensions, ft	Tube bank	
	No. 1 staggered	No. 2 in-line
d_0	0.0313	0.0313
S_T	0.0391	0.0393
S_L	0.0338	0.0391
S'_L	0.0391	
d_v	0.0225	0.0308
d_c	0.0078	0.0078
L	0.338	0.391
N'	10	10

Solution

For tube bank 1, at $\text{Re}_v = 2{,}000$

$$\text{Re}_c = 2{,}000 \frac{d_c}{d_v} = \frac{(2{,}000)(0.0078)}{0.0225} = 693$$

$$\text{Re}_0 = 2{,}000 \frac{d_0}{d_v} = \frac{(2{,}000)(0.0313)}{0.0225} = 2{,}780$$

By Eq. (11-23)

$$f_{TB} = (0.75)(693)^{-0.2} = 0.202$$

From Eq. (11-7)

$$\frac{-\Delta P_f \, g_c \rho}{G_{\max}^2} = \frac{(0.202)(4)(10)}{2} = 4.04$$

Similar calculations may be made for each bank and each Reynolds number. The results are shown in Table 11-3.

TABLE 11-3

Re_v			$\frac{-\Delta P_f\, g_c \rho}{G^2_{max}}$
Tube Bank No. 1			
2,000	Re_c = 693 Re_0 = 2,780	f_{TB} [Eq. (11-23)] = 0.202 f_{TB} [Eq. (11-25)] = 0.212 f'_{GS} [Eq. (11-27)] = 0.638	4.04 4.24 3.84
40,000	Re_c = 13,900 Re_0 = 55,600	f_{TB} [Eq. (11-23)] = 0.111 f_{TB} [Eq. (11-25)] = 0.131 f'_{GS} [Eq. (11-27)] = 0.413	2.22 2.62 2.49
Tube Bank No. 2			
2,000	Re_c = 507 Re_0 = 2,030	f_{TB} [Eq. (11-24)] = 0.0948 f_{TB} [Eq. (11-26)] = 0.215 f_{GS} [Eq. (11-27)] = 0.638	1.90 4.30 3.67
40,000	Re_c = 10,100 Re_0 = 40,600	f_{TB} [Eq. (11-24)] = 0.0515 f_{TB} [Eq. (11-26)] = 0.137 f_{GS} [Eq. (11-27)] = 0.413	1.03 2.74 2.37

For the staggered arrangements comparable values of $(-\Delta P_f\, g_c \rho)/G^2_{max}$ are obtained by all friction-factor equations. For in-line arrangements the equations of Jakob and Gunter and Shaw agree fairly well in predicting friction loss. For design purposes use of either Jakob's equations (11-25) and (11-26) or Gunter and Shaw's equation is recommended to predict friction loss for turbulent flow past banks of tubes. When flow is non-isothermal, an additional term is included to account for the variation in the viscosity of the fluid. The right sides of Eqs. (11-7), (11-10), (11-13), and (11-14) are multiplied by the term $(\mu_b/\mu_w)^{0.14}$.

11-10. Turbulent Flow across Banks of Tubes Having Noncircular Cross Section

There has been little investigation of flow past banks of tubes of non-circular cross section. Streamlined, rather than circular, tubes in a tube bank would offer much less resistance to flow and therefore should be more effective in transferring heat from the standpoint of power required per unit of heat transferred. Joyner and Palmer [10] determined friction factors for tube banks containing streamlined tubes and showed that certain shapes of tube were superior to round tubes from the standpoint of power consumption. Since obvious difficulties arise in the construction of heat exchangers containing noncircular tubes, little has been done commercially along this line. Joyner and Palmer found that the friction loss for flow across banks of noncircular tubes could be expressed by the relation

$$\frac{2(-\Delta P_f) g_c \rho}{4G_{max}^2} \frac{d_e}{L} = c \left(\frac{d_e G_{max}}{\mu} \right)^{-0.2} \tag{11-28}$$

where d_e is the equivalent diameter defined in Table 11-1, i.e., four times

TABLE 11-4. TUBE BANKS OF STREAMLINED TUBES STUDIED BY JOYNER AND PALMER †

(All staggered arrangements)

Tube shape and size ⟶ Direction of flow	$\frac{d_c}{d_0}$	Value of c ‡ to be used in Eq. (11-28)	Range of $\frac{d_e G_{max}}{\mu}$
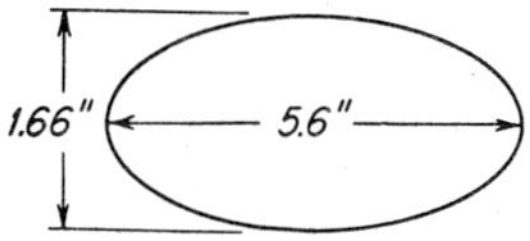 1.66″ 5.6″	0.5, 1.0	0.049	1.5×10^4–6×10^4
	1.5, 2.4	0.063	2.5×10^4–1.5×10^5
1.66″ 5.6″	0.5	0.033	1.5×10^4–4.5×10^4
	1.0	0.045	2.9×10^4–10^5
	1.5	0.055	1.5×10^4–9×10^4
	d_c, in.		
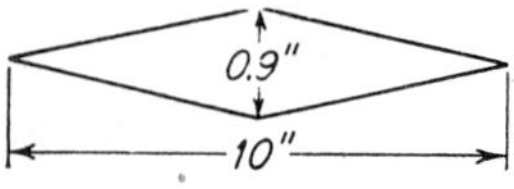 0.9″ 10″	0.83 1.66 2.49	0.039	1.5×10^4–7×10^4
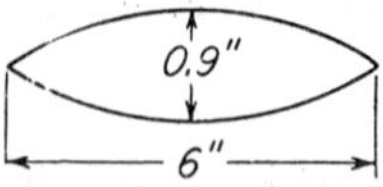 0.9″ 6″	0.83 1.66 2.49	0.047	10^4–7×10^4

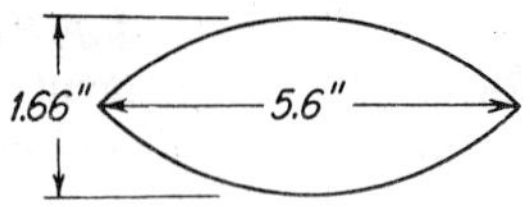

† From U. T. Joyner and C. B. Palmer, *NACA WR* L-609 (formerly *ARR*, January, 1943).

‡ For all runs the average value of c is 0.049.

the free cross-sectional area divided by the wetted perimeter, and c is a coefficient dependent on tube shape and tube spacing. The tubes studied by Joyner and Palmer and the values of c to be used in Eq. (11-28) are given in Table 11-4. Only staggered arrangements were studied, and the clearance between tubes in adjacent rows was one-half the clearance d_c between tubes in transverse rows. These authors report that tubes shaped like the third diagram in Table 11-4 are best suited for use in tube banks on the basis of heat transfer and consistently low friction factor.

11-11. Flow across Banks of Fin Tubes and Tubes with Extended Surfaces

Tube banks made up of fin tubes are even more complicated geometrically than banks of smooth tubes. A type of tube bank often encountered is the radiator-core type, made up of a bank of tubes held together by continuous fins placed close to each other, as shown in Fig. 11-13. Fluid flow

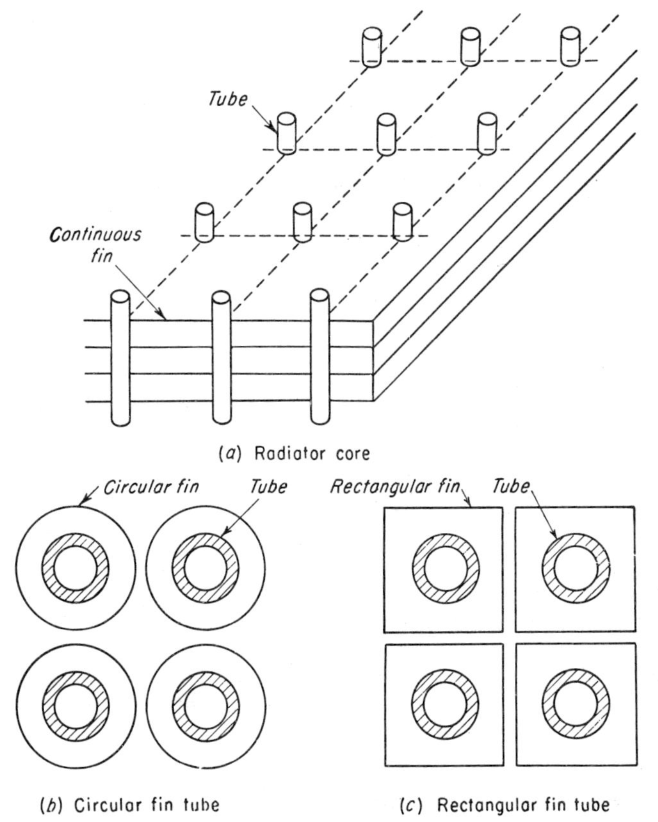

FIG. 11-13. Types of tube banks with extended surfaces.

through these tube banks takes place perpendicular to the tubes and parallel to the fins. Other tube banks are made up of fin tubes each carrying its own fins. The fins may be either circular or rectangular, as shown in the cross section in Fig. 11-13.

Gunter and Shaw [7] studied the existing pressure-drop data for crossflow of fluids across banks of fin tubes in an attempt to obtain a correlation for

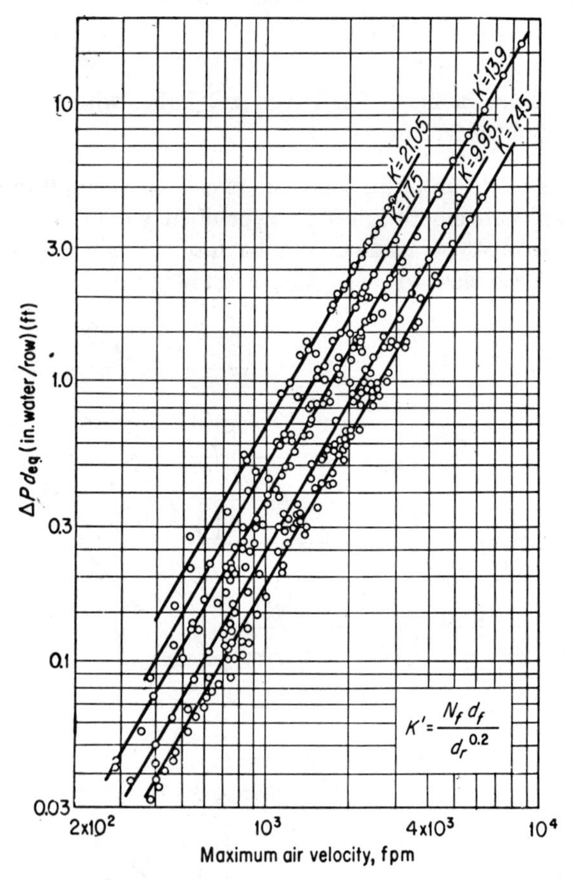

FIG. 11-14. Friction losses for air flow across banks of transverse-fin tubes. (*From D. L. Katz et al., Univ. Mich. Eng. Research Inst. Rept. on Project* M592, *August*, 1952.)

all types of surfaces. They recommend Eq. (11-22) for laminar flow across banks of fin tubes and Eq. (11-27) for turbulent flow. Turbulent flow exists at Reynolds numbers Re_v greater than 100.

Katz et al.[11] correlated a considerable amount of data for air flowing across banks of transverse-fin tubes. The data were from some thirty different units, and twenty-seven different types of fin tubes were encountered. Most tube-bank units consisted of tubes on equilateral-triangular pitch. The frontal-face area varied from 0.605 to 480 ft², and the number of transverse rows of tubes ranged from 1 to 8. The number of tubes per transverse row varied from 3 to 23.

The correlation these investigators obtained for air is shown in Fig. 11-14, where the product of the pressure drop in inches of water per row and the

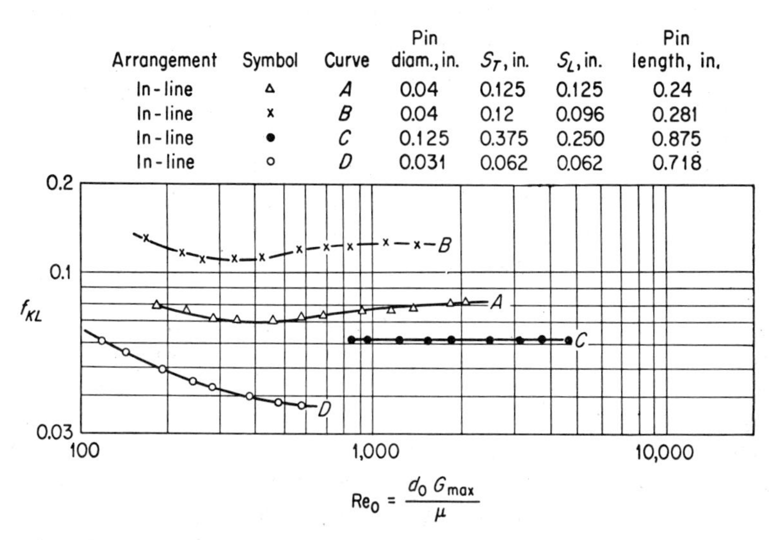

FIG. 11-15. Friction factors for flow in pin-fin heat exchangers. (*From W. M. Kays, Stanford Univ. Dept. Mech. Eng. Tech. Rept.* 19, *Navy Contract* N6-onr-251 *Task Order* 6, *August,* 1953.)

equivalent diameter of the tube bank is plotted versus the maximum lineal velocity of air through the unit. The equivalent diameter for a fin-tube unit is defined as follows:

$$d_{eq} = \frac{N_f d_f}{12} \left(\frac{S_T}{d_r}\right)^2 \tag{11-29}$$

The various curves shown in Fig. 11-14 represent constant values of the parameter $K' = N_f d_f / d_r^{0.2}$, which is a function of the dimensions of the fin tube. In calculating K' and d_{eq} (in feet), d_f, d_r, and S_T are all in inches.

Kays [12] reported friction factors for the flow of fluids through pin-fin heat exchangers, which consist of parallel, flat heating (or cooling) surfaces connected by cylindrical pins arranged in an orderly pattern. Friction factors obtained by Kays for four pin-fin heat exchangers are plotted in Fig. 11-15, where f_{KL} is plotted versus Re_0. The pins are arranged in line with the flow and range from 0.031 to 0.125 in. in diameter and 0.24 to 0.88 in. in length. Dimensional data for each heat exchanger are given on Fig. 11-15. In calculating the value of d'_v [for Eq. (11-17)] the area to be used is all the wetted area in the heat exchanger.

11-12. Friction Losses in Commercial Baffled Tubular Heat Exchangers

It is most important from a practical standpoint to be able to predict friction losses for flow through commercial heat exchangers. However, because of the complicated geometry of such systems, few good relationships exist for the accurate calculation of friction losses. One of the main difficulties involved in such a calculation is in the determination of the amount of fluid which recirculates or flows through clearances between baffles and tubes (see Fig. 11-9). If all baffles could be installed without appreciable clearance, consideration could be confined to friction losses for crossflow and friction losses through baffle openings or baffle windows.

The most complete relation for predicting friction losses in commercial heat exchangers was devised by Tinker,[14] who took into consideration all the various clearances involved in a heat exchanger. The relation depends on an accurate knowledge of friction factors for flow across banks of tubes as well as the friction loss for flow through the baffle windows. Tinker compared his predicted-friction-loss equation with a large number of experimental data and obtained fair agreement, although considerable scattering of the data occurred. For a detailed consideration of Tinker's equation, the reader is referred to the original article.

Donohue [5] proposed equations for predicting friction loss for flow in heat exchangers. He considered only crossflow and flow through the baffle window and neglected any flow which might occur through baffle clearances. For the baffle-window loss the suggested equation is

$$(-\Delta P_f)_w = \frac{2.9 G_w{}^2}{\text{sp.gr.} \times 10^{13}} \tag{11-30}$$

where $(-\Delta P_f)_w$ = friction loss through baffle window, $\text{lb}_f/\text{in.}^2$
G_w = mass velocity through window, $\text{lb}_m/(\text{hr})(\text{ft}^2)$
sp.gr. = specific gravity of fluid

Donohue proposes the following equations for the crossflow friction loss:

Laminar flow, $\text{Re}_0 < 100$:

$$f'_{TB} = \frac{2(-\Delta P_f) g_c \rho}{4 G_{\text{max}}^2} \frac{1}{N_1} = \frac{15}{(S_T/d_0 - 1)(d_0 G_{\text{max}}/\mu)} \tag{11-31}$$

Turbulent flow, $500 < \text{Re}_0 < 30{,}000$:

$$f'_{TB} = \frac{2(-\Delta P_f) g_c \rho}{4 G_{\text{max}}^2} \frac{1}{N_1} = \frac{0.75}{(S_T/d_0 - 1)^{0.2}(d_0 G_{\text{max}}/\mu)^{0.2}} \tag{11-32}$$

where N_1 is the number of tube rows traversed by the fluid in crossflow in going from one baffle window to the next (see Fig. 11-16 for illustration of the baffle window and the definition of N_1). The term $-\Delta P_f$ is the friction loss for the crossflow in one baffle space. The total friction loss for crossflow in a heat exchanger is the value of $-\Delta P_f$ as determined by Eqs. (11-31) and (11-32) multiplied by the number of baffle spaces.

Williams and Katz [16] conducted a study of heat transfer and friction losses in small baffled heat exchangers containing both plain and fin tubes.

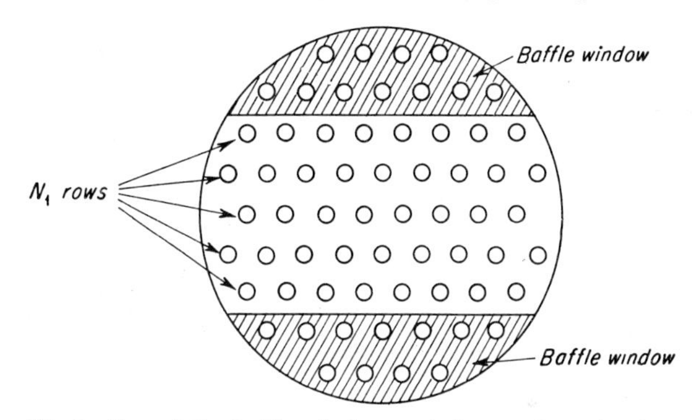

FIG. 11-16. Illustration of the baffle window and the number of tube rows between the baffle windows.

They obtained relationships of the same form as given by Eqs. (11-31) and (11-32) for the crossflow of fluid in the heat-exchanger tube bank. The results of these workers and dimensional data on the heat exchangers studied are summarized in Table 11-5. Fluids used in the heat exchangers

TABLE 11-5. FRICTION FACTORS FOR CROSSFLOW IN BAFFLED HEAT EXCHANGERS †

Tube O.D., in.	Root diameter, in.	Fins per in.	Fin height, in.	Tube spacing,‡ in.	Shell I.D., in.	Number of segmental baffles	f'_{TB} Laminar flow $Re_0 < 100$	f'_{TB} Turbulent flow $Re_0 > 7500$	Value of c to be used in Eq. (17-90)
Plain tubes									
0.751	....	....	....	0.94	7.97	9	$100 (Re_0)^{-1}$	$1.51 (Re_0)^{-0.2}$	0.356
0.504	....	....	....	0.625	7.97	9	$230 (Re_0)^{-1}$	$3.0 (Re_0)^{-0.2}$	0.302
0.621	....	....	....	0.750	6.01	11	$430 (Re_0)^{-1}$	$3.6 (Re_0)^{-0.2}$	0.400
Finned tubes §									
0.735	0.639	18.42	0.048	0.94	7.97	9	$100 (Re_0)^{-1}$	$1.51 (Re_0)^{-0.2}$	0.255
0.486	0.378	19.33	0.054	0.625	7.97	9	$130 (Re_0)^{-1}$	$2.0 (Re_0)^{-0.2}$	0.205
0.620	0.520	18.10	0.050	0.750	6.01	11	$160 (Re_0)^{-1}$	$1.75 (Re_0)^{-0.2}$	0.324

† From R. B. Williams and D. L. Katz, *Trans. ASME*, **74**:1307 (1952).

‡ All tube arrangements are equilateral triangular. Tube lengths are 46.64 in.

§ For finned tubes the effective diameter is taken as that of a plain tube having the same inside diameter and the same weight of metal as the finned tube. This effective diameter is used in the calculation of Re_0.

were water, glycerin, and oil. The friction factors shown in Table 11-5 are somewhat higher than those predicted by Eqs. (11-31) and (11-32).

In a commercial heat exchanger the measured friction loss for flow across the tubes is generally lower than that predicted by any of the crossflow-friction-factor equations. This phenomenon, which was observed by Donohue,[5] is due to the fact that part of the total flow goes through the baffle clearances.

BIBLIOGRAPHY

1. Bergelin, O. P., G. A. Brown, and S. C. Doberstein: *Trans. ASME*, **74**:953 (1952).
2. Bergelin, O. P., A. P. Colburn, and H. L. Hull: *Univ. Delaware Eng. Expt. Sta. Bull.* 2, 1950.
3. Bergelin, O. P., G. A. Brown, H. L. Hull, and F. W. Sullivan: *Trans. ASME*, **72**:881 (1950).
4. Chilton, T. H., and R. P. Genereaux: *Trans. AIChE*, **29**:161 (1933).
5. Donohue, D. A.: *Ind. Eng. Chem.*, **41**:2499 (1949).
6. Grimison, E. D.: *Trans. ASME*, **59**:583 (1937).
7. Gunter, A. Y., and W. A. Shaw: *Trans. ASME*, **67**:643 (1945).
8. Gunter, A. Y., H. R. Sennstrom, and S. Kopp: A Study of Flow Patterns in Baffled Heat Exchangers, paper 47-A-103, presented at the annual meeting of the ASME, Atlantic City, N.J., 1947.
9. Jakob, M.: *Trans. ASME*, **60**:381 (1938).
10. Joyner, U. T., and C. B. Palmer: *NACA WR* L-609 (formerly *ARR*, January, 1943).
11. Katz, D. L., E. H. Young, R. B. Williams, G. Balekjian, and R. P. Williamson: Correlation of Heat Transfer and Pressure Drop for Air Flowing across Banks of Finned Tubes, *Univ. Mich. Eng. Research Inst. Rept. on Project* M592, August, 1952.
12. Kays, W. M.: *Stanford Univ. Dept. Mech. Eng. Tech. Rept.* 19, Navy Contract N6-onr-251 Task Order 6, August, 1953.
13. Kays, W. M., A. L. London, and R. K. Lo: *Trans. ASME*, **76**:387 (1954).
14. Tinker, T.: "Proceedings of the General Discussion on Heat Transfer," Institution of Mechanical Engineers, London, and American Society of Mechanical Engineers, New York, 1951, pp. 89, 97, 110.
15. Wallis, R. P.: *Engineering*, **148**:423 (1934).
16. Williams, R. B., and D. L. Katz: *Trans. ASME*, **74**:1307 (1952).

PART III

CONVECTION HEAT TRANSFER

The transfer of energy in the form of heat is an operation frequently encountered in all phases of engineering work. In virtually every process, whether it is electrical, mechanical, chemical, or atomic, heat transfer is involved either by one or all of the three modes of transfer: conduction, convection, or radiation. In many cases the engineer must design equipment which will economically remove or add as much heat as possible from a given process, while in other cases the converse is desirable, i.e., economically preventing heat from being transferred. All three modes of heat transfer are important industrially, and each finds application under certain operating conditions.

Conduction of heat in a solid occurs under the influence of a temperature gradient, and the transfer is affected by an exchange of vibrational kinetic energy between the individual molecules. Pure conduction in liquids and gases seldom occurs because of the difficulty of preventing the molecules from moving under the influence of density differences. The transfer of heat through a fluid flowing in laminar motion is considered to be largely by conduction. The differential equation giving the temperature as a function of distance and time for three-dimensional heat conduction in a solid or stationary fluid is obtainable from the energy equation (2-48) by letting all velocities be zero.

Heat transfer by radiation is significant at high temperatures. Energy is emitted by matter at high temperatures, and this energy, upon striking another body, may be absorbed, reflected, or transmitted. The portion of the energy that is absorbed is transformed, in most cases, into heat. If a hot and a cold body are so arranged that some of the energy emitted from the hot body strikes the cold body, there will be a net interchange of energy between the two. The energy will manifest itself by increasing the temperature of the cold body. An equilibrium will be reached, in which the energy lost by each body will equal the energy gained.

The transfer of heat between a solid surface and a fluid flowing past the surface is one of the most common means of heating and cooling fluids.

Convection heat transfer is the transfer of heat from one point in a fluid to another point by actual movement of fluid particles. If the movement is brought about by density differences, the heat is transferred by natural convection; if the movement is created by actually pumping the fluid, the heat is then transferred by forced convection.

Convection heat transfer is significantly affected by the mechanics of fluid flow occurring adjacent to the solid surface. The following chapters are devoted to a consideration of forced-convection heat transfer between fluids and closed conduits, between fluids and immersed bodies, and in multitube heat exchangers. The mechanism of forced-convection heat transfer is described, and theoretical and empirical relationships are presented which predict heat-transfer coefficients for the various systems considered. The analogy between momentum transfer and heat transfer is discussed, and consideration is given to the relation between heat transfer and friction loss. Also treated is heat transfer with liquid metals, which has attained considerable importance, particularly in connection with nuclear power plants.

CHAPTER 12

THE CONVECTION-HEAT-TRANSFER COEFFICIENT. DIMENSIONAL ANALYSIS IN CONVECTION HEAT TRANSFER

12-1. Heat Transfer between a Solid Wall and a Turbulent-flowing Fluid

Since forced-convection heat transfer is brought about by the movement of fluids and the mixing of the fluid particles, the mechanism of fluid flow must be known in order to understand the mechanism of heat transfer and to explain phenomena occurring during the process. For heat transfer to take place, the temperature of the conduit wall must be different from the temperature of the fluid. The flow patterns of the fluid particles flowing past a solid wall are sketched in Fig. 12-1. Immediately adjacent to

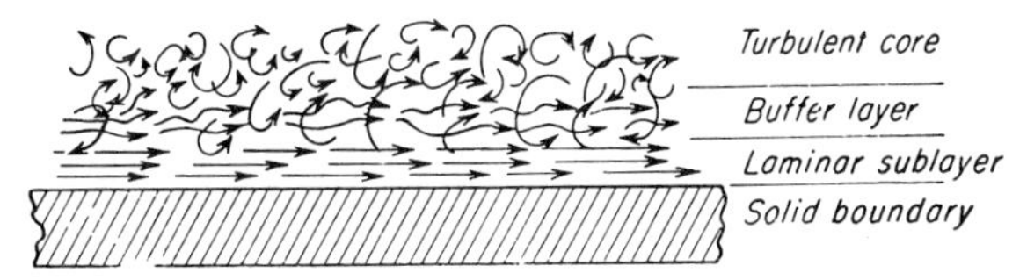

FIG. 12-1. Motion of a fluid flowing past a solid boundary.

the wall is a layer of fluid which is either in laminar flow or approaches it. In this layer there is no mixing of the fluid. The edge of the laminar layer is indicated in the sketch by wavy arrows, and beyond them is the turbulent core, in which circular arrows represent the eddies of turbulent flow. The layer of fluid immediately adjacent to the wall will be at the same temperature as the wall. Since there is no mixing of the fluid in the laminar layer, heat must be transferred by conduction through this layer. If heat is being transferred from the wall to the fluid, the temperature at the edge of the laminar layer will be less than the wall temperature but greater than the temperature in the turbulent fluid. In the turbulent core the particles have a circular, eddying motion, and the eddies sweep the edge of the laminar layer and probably penetrate it, taking with them into the turbulent core fluid which is at a higher temperature than the fluid in the core. The hotter fluid is rapidly mixed with the colder fluid in the turbulent core, the result being that heat is rapidly transferred from the edge of the laminar layer to the center of the stream. In his visual studies of turbulent flow Reynolds [2] found that the colored band mixed very quickly with the fluid as soon as it was injected into the stream, and it was not possible to detect an uncolored laminar region adjacent to the wall. When heat transfer occurs during turbulent flow, the temperature drop across the turbulent core is small, while the temperature drop across the laminar layer is quite large. Since the rate of heat transfer is proportional to the temperature drop and inversely proportional to the thermal resistance of the material through which heat is flowing, it is apparent that the laminar layer provides the greatest resistance to heat flow, while the turbulent core provides very little resistance to heat flow.

12-2. The Convection-heat-transfer Coefficient

The rate of heat transfer from a solid wall to a fluid flowing past it is proportional to the area of the surface and the temperature difference between the solid and the fluid; i.e.,

$$dq_w \propto \Delta T \, dA_w \tag{12-1}$$

where dq_w is the heat-transfer rate from a small element of surface dA_w and ΔT is the temperature difference between the surface and the fluid. The definition of ΔT differs according to whether the heat transfer takes place during flow in a conduit or flow past an immersed body.

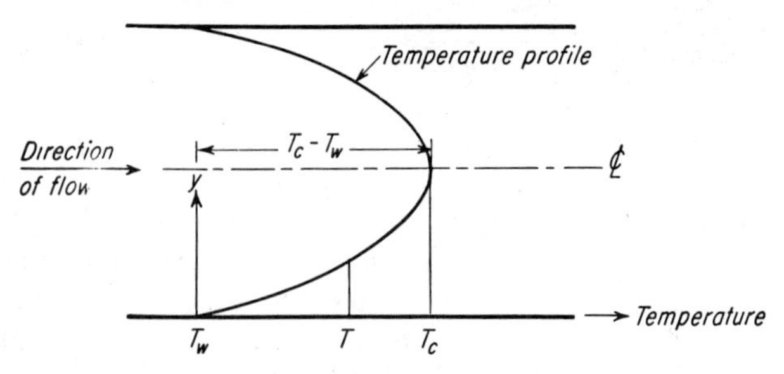

FIG. 12-2. Schematic temperature profile for cooling a fluid flowing in a closed conduit.

For fully developed conduit flow the temperature profile in the fluid for cooling is shown schematically in Fig. 12-2. The proportionality shown in Eq. (12-1) may be expressed as

$$dq_w = h \, dA_w \, (T_w - T_b) \tag{12-2}$$

where dA_w is the wetted area of a differential element dx of conduit length and T_b is the bulk temperature of the fluid, i.e., the temperature obtained by thorough mixing of the fluid.† The term h is the proportionality factor and is defined as the local convection-heat-transfer coefficient. It is also called the film coefficient of heat transfer, since in turbulent flow the laminar film adjacent to the wall provides most of the resistance to heat transfer. In many cases of heat transfer in ducts the coefficient varies along the length, and a mean value must be used. The mean value of the coefficient h_m is

$$h_m = \frac{1}{L} \int_0^L h \, dx \tag{12-3}$$

and the mean rate of heat transfer per unit area over the length L is

$$\frac{q_w}{A_w} = h_m(T_w - T_b)_{\text{l.m.}} \tag{12-4}$$

where q_w/A_w is the mean rate of heat transfer per unit area over the length L and $(T_w - T_b)_{\text{l.m.}}$ is the logarithmic mean temperature difference between the wall and the fluid.

Another mean heat-transfer coefficient frequently used is based on the arithmetic mean temperature difference between the wall and the fluid; i.e.,

$$\frac{q_w}{A_w} = h_{\text{a.m.}}(T_w - T_b)_{\text{a.m.}} \tag{12-5}$$

When heat transfer takes place during flow over immersed bodies, the temperature profile has the form shown schematically in Fig. 12-3. At

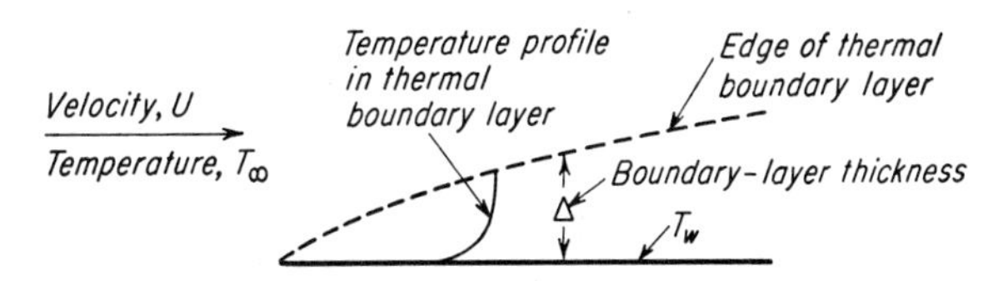

FIG. 12-3. Temperature profile for flow over an immersed body.

the surface of the solid the fluid is at the temperature of the surface. Some distance from the surface the fluid is at the temperature T_∞ of the undisturbed stream. Thus the local convection-heat-transfer coefficient is defined by the relation

$$dq_w = h \, dA_w \, (T_w - T_\infty) \tag{12-6}$$

The region over which the fluid temperature changes from that of the wall

† In the experimental determination of heat-transfer coefficients the bulk temperature of the fluid leaving the heated section of a conduit is obtained by passing the fluid through mixing chambers before measuring its temperature.

to that of the undisturbed main stream is the *thermal boundary layer*, the thickness of which is designated by Δ.

Equation (12-2) [or Eq. (12-6)] may be rearranged into an Ohm's-law type of relation

$$\text{Resistance} = \frac{\text{potential difference}}{\text{flux}}$$

$$\frac{1}{h} = \frac{T_w - T_b}{dq_w/dA_w} \tag{12-7}$$

and the reciprocal of the heat-transfer coefficient becomes equivalent to a resistance to heat transfer. For turbulent flow past solid surfaces this resistance to heat transfer is made up of the resistance of the laminar sublayer, the buffer layer, and the turbulent core.

12-3. Relation of the Heat-transfer Coefficient to the Temperature Gradient at the Surface

The local convection-heat-transfer coefficient may be related to the temperature gradient in the fluid at the surface of the solid. Considering Fig. 12-2, the rate of heat transfer dq_w to the fluid through the area dA_w is given by Eq. (12-2). This amount of heat must pass through the surface of area dA_w into the fluid by pure conduction, assuming that laminar flow exists adjacent to the wall. Therefore, considering y as defined in Fig. 12-2, from Eq. (1-9)

$$dq_w = -k\,dA_w\left(\frac{\partial T}{\partial y}\right)_{y=0} \tag{12-8}$$

Combining Eqs. (12-2) and (12-8),

$$h = -\frac{k}{T_w - T_b}\left(\frac{\partial T}{\partial y}\right)_{y=0} \tag{12-9}$$

Equation (12-9) relates the heat-transfer coefficient to the thermal conductivity of the fluid, the temperature gradient at the wall, and the temperature difference between wall and fluid for heat transfer in closed conduits. For flow past immersed bodies

$$h = -\frac{k}{T_w - T_\infty}\left(\frac{\partial T}{\partial y}\right)_{y=0} \tag{12-10}$$

12-4. Application of Dimensional Analysis to Heat Transfer

The local heat-transfer coefficient is a function of the physical properties of the fluid which is flowing past the surface. In addition, the mechanism of fluid flow also has a significant effect on the coefficient. Such physical properties as fluid density, viscosity, heat capacity, and thermal conductivity are considered to have an effect on the heat-transfer coefficient, as do the dimensional variables required for geometrical description of the system and the point where the local heat-transfer coefficient is being considered. The dimensional variables include diameter (or equivalent diameter) of closed conduits and distance from entrance or leading edge. The flow property to be considered is the fluid velocity (average velocity in a closed conduit or undisturbed main-stream velocity for flow past immersed bodies). Heat-transfer data and convection-heat-transfer coefficients have been correlated by means of dimensionless groups derived from the above-mentioned variables.

Example 12-1

The convection-heat-transfer coefficient for turbulent flow in a circular tube is a function of the following variables: tube diameter d_w, distance from tube entrance x, fluid velocity U, viscosity μ, density ρ, heat capacity at constant pressure C_p, and thermal conductivity k. Using Buckingham's method as described in Chap. 6, determine the dimensionless groups by which the variables of the system may be arranged.

Solution

The quantities to be considered and their dimensions are tabulated below [x is not included since it will appear in x/d_w as one of the dimensionless groups (see Sec. 6-2)].

Quantity	*Dimensions*
k	mL/t^3T
d_w	L
U	L/t
μ	m/Lt
h	m/t^3T
C_p	L^2/t^2T
ρ	m/L^3

$$n = 7 \qquad j = 4$$

$$i = n - j = 3$$

Three independent dimensionless groups are obtainable. Since $j = 4$, four variables are selected which between them contain the four fundamental quantities, mass, length, time, and temperature. Selecting d_w, U, μ, and k, the three groups are

$$\Pi_1 = d_w^{a1} U^{b1} \mu^{c1} k^{d1} \rho$$

$$\Pi_2 = d_w^{a2} U^{b2} \mu^{c2} k^{d2} h$$

$$\Pi_3 = d_w^{a3} U^{b3} \mu^{c3} k^{d3} C_p$$

and since Π_1, Π_2, and Π_3 are dimensionless,

$$\begin{array}{lll} a_1 = 1 & a_2 = 1 & a_3 = 0 \\ b_1 = 1 & b_2 = 0 & b_3 = 0 \\ c_1 = -1 & c_2 = 0 & c_3 = 1 \\ d_1 = 0 & d_2 = -1 & d_3 = -1 \end{array}$$

from which

$$\Pi_1 = \frac{d_w U \rho}{\mu} \qquad \text{(Reynolds number)}$$

$$\Pi_2 = \frac{h d_w}{k} \qquad \text{(Nusselt number)}$$

$$\Pi_3 = \frac{C_p \mu}{k} \qquad \text{(Prandtl number)}$$

For fully developed turbulent flow in a circular tube the data may be correlated by the three dimensionless groups obtained in the above example if the variables considered are the only ones affecting the heat-transfer coefficient. Most heat-transfer data have been correlated empirically using these groups. However, the fact that the fluid temperature varies across the section of the conduit requires other refinements to be included in the correlation. One that is commonly used is the dimensionless group μ_b/μ_w, where μ_b is the viscosity of the fluid at the bulk temperature and μ_w is the viscosity of the fluid at the wall temperature. Other improvements in empirical correlations involve evaluating the fluid properties at the so-called *film temperature*, which is the arithmetic average between the bulk temperature and the wall temperature.

12-5. Use of the Energy Equation to Obtain Dimensionless Groups in Heat Transfer

Using the methods of Klinkenberg and Mooy [1] described in Sec. 6-8, it is possible to obtain the dimensionless groups in heat transfer from the energy equation. Neglecting heat generation q' and dissipation Φ, the energy equation for three-dimensional motion becomes

$$\frac{\partial T}{\partial t} + u\frac{\partial T}{\partial x} + v\frac{\partial T}{\partial y} + w\frac{\partial T}{\partial z} = \frac{k}{C_p \rho}\left(\frac{\partial^2 T}{\partial x^2} + \frac{\partial^2 T}{\partial y^2} + \frac{\partial^2 T}{\partial z^2}\right) \qquad (12\text{-}11)$$

Equation (12-11) is dimensionally homogeneous, and division by one of the terms will yield dimensionless groups. Table 12-1, which is similar to Table 6-3, shows the various dimensionless groups obtainable from Eq. (12-11).

12-6. The Physical Significance of Dimensionless Groups in Heat Transfer

The physical significance of the dimensionless groups appearing in forced-convection heat transfer may be determined from Table 12-1. Of particular interest, however, are the Nusselt group (hd_w/k for tubes) and the Prandtl group $C_p\mu/k$, the significance of which is not evident from Table 12-1.

Table 12-1. Dimensionless Groups Obtainable from the Energy Equation

	Unsteady-state term	Convection terms	Conduction terms	Boundary condition
	$\frac{\partial T}{\partial t}$	$u\frac{\partial T}{\partial x} + v\frac{\partial T}{\partial y} + w\frac{\partial T}{\partial z}$	$\frac{k}{C_p\rho}\left(\frac{\partial^2 T}{\partial x^2} + \frac{\partial^2 T}{\partial y^2} + \frac{\partial^2 T}{\partial z^2}\right)$	Heat-transfer coefficient
	$\frac{T}{t}$	$\frac{UT}{L}$	$\frac{k}{C_p\rho}\frac{T}{L^2}$	$\frac{hT}{C_p\rho L}$
Conduction in solids or stationary fluids	← (unsteady-state ↔ conduction) →		$\frac{k}{C_p\rho}\frac{t}{L^2}$ Fourier number	
Heat transfer during fluid flow in closed conduits or past immersed bodies		$\frac{C_p\rho UL}{k}$ Peclet number = (Reynolds)(Prandtl) (convection ↔ conduction)		
		(convection ↔ boundary condition)		$\frac{h}{C_p\rho U}$ Stanton number = $\frac{\text{Nusselt}}{\text{Peclet}}$
			(conduction ↔ boundary condition)	$\frac{hL}{k}$ Nusselt number

Considering heat transfer in tubes, and multiplying each side of Eq. (12-9) by d_w/k,

$$\text{Nu} = \frac{hd_w}{k} = -\frac{d_w}{T_w - T_b}\left(\frac{\partial T}{\partial y}\right)_{y=0} \tag{12-12}$$

also

$$\text{Nu} = \frac{d_w}{k}\frac{dq_w}{dA_w}\frac{1}{T_w - T_b} \tag{12-13}$$

The term on the right of Eq. (12-12) may be considered as the ratio of the temperature gradient at the wall to the temperature gradient $(T_w - T_b)/d_w$ across the fluid in the pipe. This is equal to the Nusselt number.

TABLE 12-2. SIGNIFICANCE OF SOME DIMENSIONLESS GROUPS IN CONVECTION HEAT TRANSFER

Group	Name of group	Significance
$\frac{hL}{k}$	Nusselt (Nu)	Ratio of temperature gradients [see Eq. (12-12)]
$\frac{C_p\mu}{k}$	Prandtl (Pr)	$\frac{\text{Molecular diffusivity of momentum}}{\text{Molecular diffusivity of heat}}$
$\frac{C_p\rho UL}{k}$	Peclet (Pe)	$\frac{\text{Heat transfer by convection}}{\text{Heat transfer by conduction}}$
$\frac{h}{C_p\rho U}$	Stanton (St)	$\frac{\text{Wall heat-transfer rate}}{\text{Heat transfer by convection}}$
$\frac{\mu_b}{\mu_w}$	Viscosity ratio	$\frac{\text{Viscosity of fluid at bulk temperature}}{\text{Viscosity of fluid at wall temperature}}$
$\text{Pe}\frac{d_w}{L}\frac{\pi}{4}$ or $\text{Pe}\frac{d_w}{L}$	Graetz (Gz or Gz′)	Same as Peclet number except effect of d_w/L considered (entrance regions)
$\frac{gL^3}{\nu^2}\beta(T_w - T_\infty)$ Natural convection	Grashof (Gr)	$\frac{\text{(Buoyancy forces)(inertia force)}}{\text{(Viscous forces)}^2}$

The Prandtl number may be obtained as follows:

$$\text{Pr} = \frac{C_p\mu}{k} = \frac{\mu/\rho}{k/C_p\rho} = \frac{\nu}{\alpha} \tag{12-14}$$

where ν is the kinematic viscosity of the fluid or the molecular diffusivity of momentum and α is the molecular diffusivity of heat. The Prandtl number is the ratio two molecular diffusivities, that of momentum to that of heat.

Table 12-2 shows various dimensionless groups which are of importance in heat transfer by convection and their significance.

12-7. Comment on the Grashof Group in Natural Convection

When heat transfer takes place by natural convection, movement of the fluid occurs because of density differences brought about by temperature gradients. Therefore, the variation of density with temperature as represented by the coefficient of thermal expansion β is an important factor in natural convection. The rate of natural-convection heat transfer is also determined by the magnitude of the temperature difference $T_w - T_\infty$ between the wall and the fluid. The Grashof group contains these variables.

Example 12-2

The natural-convection heat-transfer coefficient on a hot vertical plate immersed in a cold fluid is a function of the following variables:

Quantity	*Dimensions*
β	$1/T$
$T_w - T_\infty$	T
x	L
k	mL/t^3T
C_p	L^2/t^2T
ρ	m/L^3
μ	m/Lt
g	L/t^2

Determine the dimensionless groups into which the variables of the system may be arranged.

Solution

By Buckingham's method five dimensionless groups are obtainable from the variables:

$$\Pi_1 = \frac{C_p\mu}{k} \qquad \text{(Prandtl number)}$$

$$\Pi_2 = \frac{hx}{k} \qquad \text{(Nusselt number)}$$

$$\Pi_3 = \beta(T_w - T_\infty)$$

$$\Pi_4 = \frac{x\mu\beta g}{k}$$

$$\Pi_5 = \frac{x^2 k \rho^2}{\mu^3 \beta}$$

The Grashof number used in correlating natural-convection data is the product of Π_3, Π_4, and Π_5.

BIBLIOGRAPHY

1. Klinkenberg, A., and H. H. Mooy: *Chem. Eng. Progr.*, **44**:17 (1948).
2. Reynolds, O.: *Trans. Roy. Soc.* (*London*), **174A**:935 (1883).

CHAPTER 13

HEAT TRANSFER DURING LAMINAR FLOW IN CLOSED CONDUITS

13-1. Heat Transfer during Laminar Flow

When heat transfer occurs during laminar flow of a fluid, the transfer through the fluid is by conduction alone. No mixing of the fluid, like that occurring during turbulent flow, takes place. In practice it is difficult to obtain truly laminar flow during heat transfer except in very small passages. Natural convection currents are usually present, and under these conditions conduction alone is not the only mode of heat transfer to be considered.

In the present chapter heat transfer during laminar flow in closed conduits is analyzed theoretically, and the analytical results are compared with some of the experimental data available. It is possible, since turbulence is absent, to set up the continuity, momentum, and energy differential equations for a given system. With the application of the appropriate boundary conditions, the solution of these equations will give the temperature and velocity at any point in the conduit. The results may be expressed in terms of the heat-transfer coefficient or, if preferable, the Nusselt number.

It should be emphasized that the boundary conditions have a significant effect on the analytical results. The two extreme cases ordinarily considered in studies of this sort are that of uniform wall temperature along the whole length of the duct and that of uniform rate of heat transfer along the length of the duct. Other cases frequently considered are linear variation of the wall temperature along the tube and linear increase of the axial temperature of the fluid along the tube. These four conditions are shown schematically in Fig. 13-1. The determination of the arbitrary constants appearing in the solution of the differential equation involves the application of one of these four conditions. Analytical investigation of laminar heat transfer also involves a consideration of the hydrodynamical condition existing in the conduit. The heat transfer may be taking place at the entrance to the conduit, where the velocity profile is also developing, or it may take place far from the entrance, where the laminar velocity profile is fully developed. It is also possible to consider laminar heat transfer during *slug* flow, in which a uniform velocity profile is assumed. The

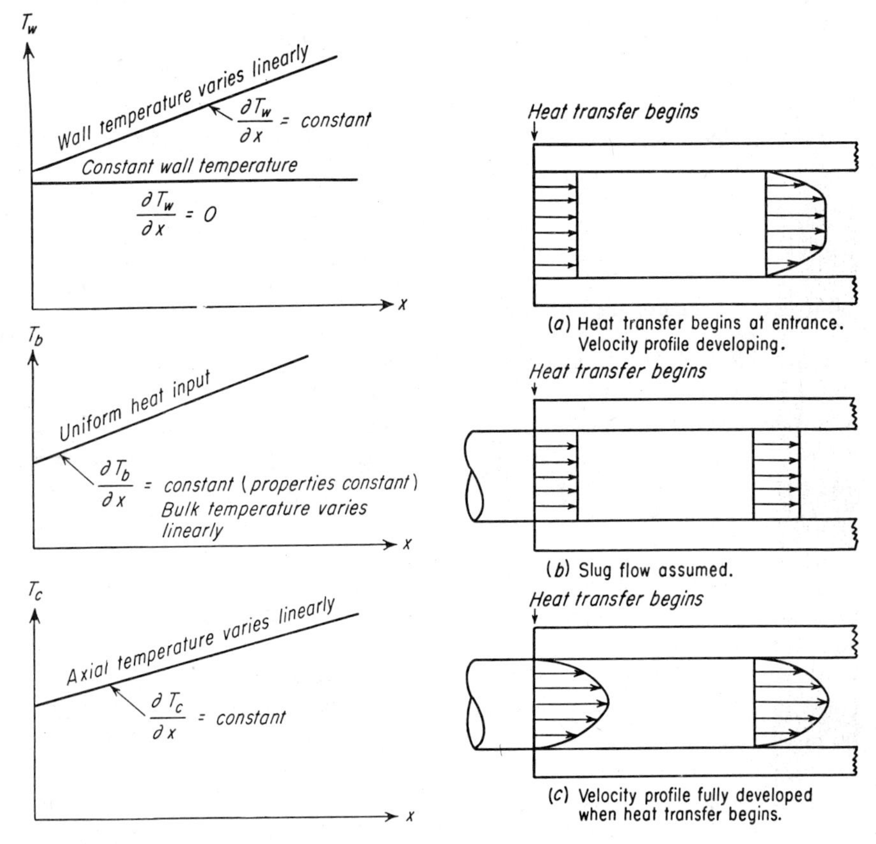

FIG. 13-1. Various boundary conditions used in analytical heat-transfer studies.

FIG. 13-2. Various hydrodynamical conditions existing in a duct when heat transfer begins.

various types of hydrodynamical conditions to be considered during laminar-flow heat transfer are shown in Fig. 13-2.

Combining the conditions of Fig. 13-2 with those of Fig. 13-1, it is seen that there are twelve possible combinations of boundary conditions and hydrodynamical conditions. Not all of these have been investigated, and in this chapter only a few will be discussed to indicate the limits for each set of conditions.

13-2. The Differential Equations for Laminar-flow Heat Transfer in a Closed Conduit

Considering a horizontal conduit with flow taking place in the x direction, the momentum and continuity differential equations become

$$u\frac{\partial u}{\partial x}+v\frac{\partial u}{\partial y}+w\frac{\partial u}{\partial z}=-\frac{g_c}{\rho}\frac{\partial P}{\partial x}+\nu\left(\frac{\partial^2 u}{\partial x^2}+\frac{\partial^2 u}{\partial y^2}+\frac{\partial^2 u}{\partial z^2}\right) \tag{13-1}$$

$$\frac{\partial u}{\partial x}+\frac{\partial v}{\partial y}+\frac{\partial w}{\partial z}=0 \qquad \text{for incompressible fluids} \tag{2-10}$$

Neglecting temperature effects due to viscous friction,

$$u\frac{\partial T}{\partial x}+v\frac{\partial T}{\partial y}+w\frac{\partial T}{\partial z}=\frac{k}{C_p\rho}\left(\frac{\partial^2 T}{\partial x^2}+\frac{\partial^2 T}{\partial y^2}+\frac{\partial^2 T}{\partial z^2}\right) \tag{13-2}$$

Eqs. (13-1) and (13-2) apply at any cross section of the conduit; however, when fully developed laminar flow exists in a uniform conduit, Eqs. (13-1) and (13-2) reduce to Eqs. (4-32) and (2-59) respectively.

I. LAMINAR-FLOW HEAT TRANSFER IN CIRCULAR TUBES

13-3. The Leveque Solution

One of the simplest solutions for the laminar-flow heat-transfer coefficient in circular tubes is that of Leveque.[19] The analysis applies directly to laminar-flow heat transfer on a flat plate, but the results may be easily applied

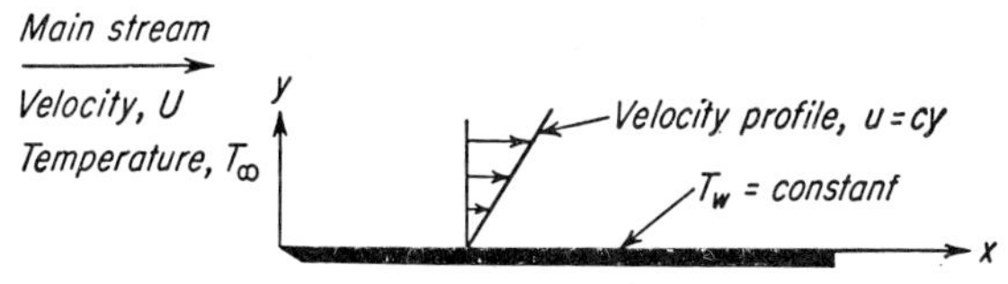

FIG. 13-3. Conditions for the Leveque solution.

to circular tubes (for this reason it is presented here, rather than in Chap. 17). The Leveque solution sets up the problem mathematically and yields a solution in the region near the wall. The temperature distribution along the tube and the heat-transfer coefficient may be readily calculated from the solution.

Consider a fluid flowing over a surface under the following conditions (see Fig. 13-3):

(1) The fluid properties are constant.
(2) The surface temperature is uniform at T_w.

(3) The undisturbed-fluid temperature is T_∞.
(4) Heat transfer is due to conduction alone.
(5) The velocity of the fluid is

$$u = cy, \; v = 0, \; w = 0$$

where y = direction outward normal to surface
u = velocity in x direction
c = constant
v = y-direction velocity
w = z-direction velocity

The fluid temperature T is a function of x and y. For small values of y, $\partial^2 T/\partial x^2 <<< \partial^2 T/\partial y^2$. Also $\dfrac{\partial^2 T}{\partial z^2}$ may be considered negligible. Eq. (13-2) becomes

$$cy \frac{\partial T}{\partial x} = \alpha \frac{\partial^2 T}{\partial y^2} \tag{13-3}$$

where
$$\alpha = \frac{k}{C_p \rho}$$

Conditions:

At $x = 0$
$y > 0$
$T = T_\infty$

At $x > 0$
$y = 0$
$T = T_w$

Equation (13-3) may be transformed to an ordinary differential equation by introducing a new variable X, where

$$X = y\left(\frac{c}{9\alpha x}\right)^{1/3} \tag{13-4}$$

Then Eq. (13-3) becomes

$$\frac{d^2T}{dX^2} + 3X^2\frac{dT}{dX} = 0 \tag{13-5}$$

Conditions:

At $X = 0$
$T = T_w$

At $X = \infty$
$T = T_\infty$

Equation (13-5) has the following solution, in which the boundary conditions have been introduced:

$$\frac{T - T_w}{T_\infty - T_w} = \frac{1}{0.893} \int_0^X e^{-X^3}\, dX \tag{13-6}$$

The temperature T is now expressed as a function of x and y (since X is a function of x and y). The integral shown on the right-hand side of Eq. (13-6) has been evaluated [2] through the use of a series. Several values of it are shown in Table 13-1.

TABLE 13-1. VALUES OF THE INTEGRAL $\int_0^X e^{-X^3}\, dX$ †

X	$\int_0^X e^{-X^3}\, dX$	X	$\int_0^X e^{-X^3}\, dX$
0.0	0	1.05	0.8246
0.05	0.0500	1.10	0.8390
0.10	0.1000	1.15	0.8510
0.15	0.1500	1.20	0.8609
0.20	0.1996	1.25	0.8689
0.25	0.2490	1.30	0.8752
0.30	0.2980	1.35	0.8801
0.35	0.3463	1.40	0.8838
0.40	0.3937	1.45	0.8866
0.45	0.4400	1.50	0.8886
0.50	0.4849	1.55	0.8901
0.55	0.5282	1.60	0.8911
0.60	0.5695	1.65	0.8918
0.65	0.6087	1.70	0.8922
0.70	0.6454	1.75	0.8925
0.75	0.6796	1.80	0.8927
0.80	0.7110	1.85	0.8928
0.85	0.7395	1.90	0.8929
0.90	0.7651	1.95	0.8929
0.95	0.7877	2.00	0.8930
1.0	0.8075		

† For a more complete table refer to M. Abramowitz, *J. Math. and Phys.*, **30**:162 (1951).

Equation (13-6) may be used to determine the heat-transfer coefficient as a function of x. From Eq. (12-10)

$$h = -\frac{k}{T_w - T_\infty}\left(\frac{\partial T}{\partial y}\right)_{y=0} \tag{12-10}$$

Also at constant x

$$\frac{\partial T}{\partial y} = \frac{\partial T}{\partial X}\frac{\partial X}{\partial y} \tag{13-7}$$

From Eq. (13-4)

$$\frac{\partial X}{\partial y} = \left(\frac{c}{9\alpha x}\right)^{1/3} \tag{13-8}$$

From Eq. (13-6)

$$\frac{\partial T}{\partial X} = \frac{T_\infty - T_w}{0.893} e^{-X} \tag{13-9}$$

(The derivative of an integral with respect to the upper limit is the integrand.) Thus

$$\frac{\partial T}{\partial y} = \frac{T_\infty - T_w}{0.893} e^{-X^3} \left(\frac{c}{9\alpha x}\right)^{1/3} \tag{13-10}$$

Therefore, since $X = 0$ when $y = 0$,

$$\left(\frac{\partial T}{\partial y}\right)_{y=0} = \frac{T_\infty - T_w}{0.893} \left(\frac{c}{9\alpha x}\right)^{1/3} \tag{13-11}$$

Substituting Eq. (13-11) into (12-10) gives

$$h = \frac{k}{0.893} \left(\frac{c}{9\alpha x}\right)^{1/3} \tag{13-12}$$

or

$$\mathrm{Nu}_x = \frac{hx}{k} = \frac{x}{0.893} \left(\frac{c}{9\alpha x}\right)^{1/3} \tag{13-13}$$

where Nu_x is the local Nusselt number on the surface a distance x from the leading edge. Equation (13-13) expresses the local Nusselt number as a function of x, the thermal diffusivity of the fluid, and c, the slope of the velocity profile at the surface.

Equation (13-12) may be modified to give the Nusselt number in the entrance region of a circular tube. If it is assumed that the velocity distribution in the laminar boundary layer in the entrance of a circular tube may be expressed by Eq. (4-44), i.e.,

$$u = 2U\left[1 - \left(\frac{r}{r_w}\right)^2\right] \tag{4-44}$$

and

$$\left(\frac{\partial u}{\partial r}\right)_{r=r_w} = \frac{-4U}{r_w} \tag{4-47}$$

Since

$$r = r_w - y \tag{13-14}$$

then

$$\frac{\partial u}{\partial r} = -\frac{\partial u}{\partial y}\dagger \tag{13-15}$$

† The partial derivatives are used here. Total derivatives would be correct also since u is a function only of y in this case.

so

$$\left(\frac{\partial u}{\partial y}\right)_{y=0} = \frac{4U}{r_w} \tag{13-16}$$

This is the value of c which may be used in Eq. (13-12).

Substituting Eq. (13-16) in (13-12),

$$h = \frac{k}{0.893}\left(\frac{4U}{9r_w\alpha x}\right)^{1/3} \tag{13-17}$$

and

$$\frac{hd_w}{k} = \frac{d_w}{0.893}\left(\frac{4U}{9r_w\alpha x}\right)^{1/3} \tag{13-18}$$

which simplifies to

$$\text{Nu} = 1.077(\text{Pe})^{1/3}\left(\frac{d_w}{x}\right)^{1/3} \qquad 100 < \text{Pe}\frac{d_w}{x} < 5{,}000 \tag{13-19}$$

Equation (13-19) is the Leveque solution for the Nusselt number for laminar flow in the entrance region of a circular tube. It is based on the assumption that the velocity gradient at the wall has a value given by Eq. (13-16). This assumption is not true at the immediate entrance to the tube, where the velocity gradient is infinite; however, the velocity profile at the wall develops rapidly, and the gradient soon approaches that given by Eq. (13-16). In general, Eq. (13-19) is applicable in the range $100 < \text{Pe}(d_w/x) < 5{,}000$, and in this region it agrees with empirical data.

Example 13-1

Determine the range of length-to-diameter ratios over which Eq. (13-19) may be used to predict the laminar-flow Nusselt number for a fluid with a Prandtl number of 1.0 and at a Reynolds number of 1,000.

Solution

Since Pe = RePr,

$$\text{Pe} = (1{,}000)(1.0) = 1{,}000$$

Thus, from the range given with Eq. (13-19),

$$100 < 1{,}000\frac{d_w}{x} < 5{,}000$$

or, solving for x,

$$10d_w > x > 0.2d_w$$

Therefore, when Pe = 1,000, Eq. (13-19) may be applied between a point $\frac{1}{5}$ diameter to a point 10 diameters downstream from the entrance.

13-4. The Classical Graetz Problem; Velocity Profile Fully Developed

One of the earliest analyses of laminar-flow heat transfer in tubes was made by Graetz [9] in 1885; it has been thoroughly described by Drew [7] and Jakob.[11] The analysis has been extended to include a variety of boundary conditions.

The conditions of the classical Graetz problem for laminar-flow heat transfer in circular tubes (see Fig. 13-4) are as follows:

(1) The fluid properties are constant.
(2) The laminar parabolic velocity profile is assumed to be established before heating or cooling of the fluid.
(3) At $x = 0$ the temperature of the tube wall changes from T_∞ to T_w and is uniform at this value for $x > 0$. T_∞ is the fluid temperature as it enters the heating or cooling section.

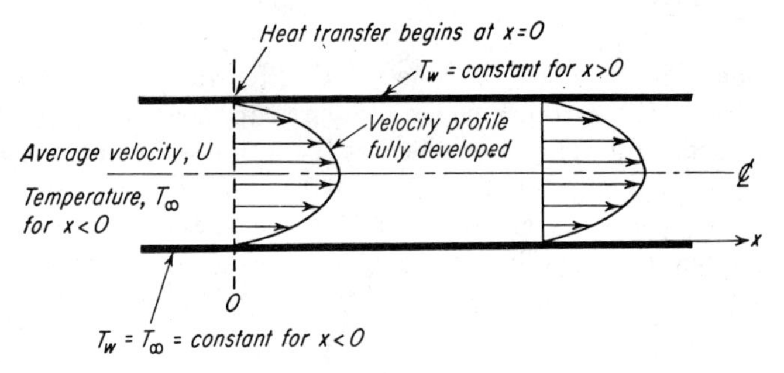

Fig. 13-4. Boundary conditions for the classical Graetz solution.

Since the parabolic velocity profile is fully developed, the momentum and continuity equations are satisfied by

$$u = 2U\left[1 - \left(\frac{r}{r_w}\right)^2\right] \tag{4-44}$$

Rewriting the energy equation (13-2) in terms of cylindrical coordinates, assuming radial symmetry and neglecting $\partial^2 T/\partial x^2$,

$$u\,\frac{\partial T}{\partial x} = \frac{k}{C_p \rho}\left[\frac{1}{r}\frac{\partial}{\partial r}\left(r\,\frac{\partial T}{\partial r}\right)\right] \tag{13-20}$$

Combining Eqs. (4-44) and (13-20),

$$2U\left[1 - \left(\frac{r}{r_w}\right)^2\right]\frac{\partial T}{\partial x} = \frac{k}{C_p \rho}\left[\frac{1}{r}\frac{\partial}{\partial r}\left(r\,\frac{\partial T}{\partial r}\right)\right] \tag{13-21}$$

Boundary conditions:

At $x = 0$ at any r
$T = T_\infty$

At $x > 0$
$r = r_w$
$T = T_w$

Solution of Eq. (13-21) gives T as a function of x and r. Introduction of the above boundary conditions will give a particular solution of Eq. (13-21) and permit the evaluation of the arbitrary constants appearing in the solution. This partial differential equation is solved by assuming that $T - T_w$ is the product of two functions, one which is a function of x and the other a function of r. The solution takes the form of an infinite series as follows:

$$\frac{T - T_\infty}{T_w - T_\infty} = \sum_{n=0}^{n=\infty} c_n \phi_n \left(\frac{r}{r_w}\right) \exp \frac{-\beta_n^2 (x/r_w)}{\text{Pe}} \tag{13-22}$$

where c_n are coefficients, $\phi_n(r/r_w)$ are functions of r/r_w determined by the boundary conditions, and β_n^2 are exponents determined by the boundary conditions.

Detailed solutions of Eq. (13-22) are given by Jakob [11] and Sellars, Tribus, and Klein.[25] Graetz originally evaluated the first three values of c_n and β_n and the first three functions $\phi_0(r/r_w)$, $\phi_1(r/r_w)$, and $\phi_2(r/r_w)$. Abramowitz [1] obtained very accurate values of the first five c_n coefficients, while Schenk and Dumore [24] reported the first five values of the coefficients and exponents of the terms in the series solution. Sellars, Tribus, and Klein [25] made a valuable contribution to the classical Graetz solution by reporting the first ten values of c_n and β_n^2.

The solution to Eq. (13-21) will not be considered in detail, but it is summarized in Table 13-2, and the practical results of the solution are expressed in equation form. The local Nusselt number for laminar flow in circular tubes predicted by the Graetz solution is

$$\text{Nu} = \frac{\displaystyle\sum_{n=0}^{n=\infty} \frac{c_n \phi_n'(1)}{2} \exp \frac{-\beta_n^2 (x/r_w)}{\text{Pe}}}{\displaystyle 2 \sum_{n=0}^{n=\infty} \frac{c_n \phi_n'(1)}{2\beta_n^2} \exp \frac{-\beta_n^2 (x/r_w)}{\text{Pe}}} \tag{13-23}$$

where the individual terms in the series may be obtained from Table 13-2.

TABLE 13-2. SUMMARY OF SOLUTION OF EQ. (13-21) †

Equation	$2U\left[1-\left(\frac{r}{r_w}\right)^2\right]\frac{\partial T}{\partial x}=\frac{k}{C_p\rho}\left[\frac{1}{r}\frac{\partial}{\partial r}\left(r\frac{\partial T}{\partial r}\right)\right]$
Boundary conditions	$x = 0$ all r $T = T_\infty$ $x > 0$ $r = r_w$ $T = T_w$
Solution	$\frac{T-T_\infty}{T_w-T_\infty}=\sum_{n=0}^{n=\infty} c_n\phi_n\left(\frac{r}{r_w}\right)\exp\frac{-\beta_n^2(x/r_w)}{\text{Pe}}$
where	$\beta_n = 4n + 8/3$ $c_n = (-1)^n(2.84606)\beta_n^{-2/3}$
for r small	$\phi_n\left(\frac{r}{r_w}\right)=J_0\left(\frac{\beta_n r}{r_w}\right)$
for r medium	$\phi_n\left(\frac{r}{r_w}\right)=\sqrt{\frac{2r_w}{\pi\beta_n r}}$ $\times\frac{\cos\left[(\beta_n/2)(r/r_w)\sqrt{1-(r/r_w)^2}+(\beta_n/2)\sin^{-1}(r/r_w)-\pi/4\right]}{[1-(r/r_w)^2]^{1/4}}$
for r close to r_w	$\phi_n\left(\frac{r}{r_w}\right)=\sqrt{\frac{2(1-r/r_w)}{3}}\left\{(-1)^n J_{1/3}\left[\frac{\beta_n\sqrt{8}}{3}\left(1-\frac{r}{r_w}\right)^{3/2}\right]\right\}$
	$\text{Nu}=\frac{\sum_{n=0}^{n=\infty}\frac{c_n\phi_n'(1)}{2}\exp\frac{-\beta_n^2(x/r_w)}{\text{Pe}}}{2\sum_{n=0}^{n=\infty}\frac{c_n\phi_n'(1)}{2\beta_n^2}\exp\frac{-\beta_n^2(x/r_w)}{\text{Pe}}}$ $-\frac{c_n\phi_n'(1)}{2}=1.0128\beta_n^{-1/3}$

† From J. R. Sellars, M. Tribus, and J. S. Klein, *Trans. ASME*, **78**:441 (1956); M. Jakob, "Heat Transfer," vol. 1, John Wiley & Sons, Inc., New York, 1949.

TABLE 13-2. SUMMARY OF SOLUTION OF EQ. (13-21) (*Continued*)

n	β_n	β_n^2	c_n	$\frac{-c_n\phi_n'(1)}{2}$
0	2.667	7.113	+1.480	0.7303
1	6.667	44.49	−0.8035	0.5381
2	10.667	113.8	+0.5873	0.4601
3	14.667	215.1	−0.4750	0.4137
4	18.667	348.5	+0.4044	0.3818
5	22.667	513.8	−0.3553	0.3579
6	26.667	711.1	+0.3189	0.3390
7	30.667	940.5	−0.2905	0.3236
8	34.667	1202	+0.2677	0.3106
9	38.667	1495	−0.2489	0.2995

$\frac{r}{r_w}$	$\phi_0\left(\frac{r}{r_w}\right)$	$\phi_1\left(\frac{r}{r_w}\right)$	$\phi_2\left(\frac{r}{r_w}\right)$
0	1	1	1
0.1	0.9818	0.8923	0.753
0.2	0.9290	0.6067	0.206
0.3	0.8456	0.2367	−0.290
0.4	0.7382	−0.1062	−0.407
0.5	0.6147	−0.3399	−0.204
0.6	0.4833	−0.4317	0.104
0.7	0.3506	−0.3985	0.278
0.8	0.2244	−0.3051	0.278
0.9	0.1069	−0.1637	0.144
1.0	0	0	0

Hausen [10] proposed the following equation as representing the Graetz solution for constant wall temperature and parabolic velocity distribution. This equation is for the mean Nusselt number over a length of pipe x.

$$\mathrm{Nu}_m = 3.66 + \frac{0.0668[(x/d_w)/\mathrm{Pe}]^{-1}}{1 + 0.04[(x/d_w)/\mathrm{Pe}]^{-2/3}} \tag{13-24}$$

where $$\mathrm{Nu}_m = \frac{h_m d_w}{k} \quad \text{for circular tubes} \tag{13-25}$$

The mean Nusselt number is useful in calculating the total heat transferred over a given length of pipe by Eq. (12-4).

Equation (13-23) may be used to evaluate the Nusselt number for fully developed laminar flow in a circular tube as a function of the distance from the beginning of the heat-transfer section. Application of Eq. (13-23) requires the use of a large number of terms when $(x/r_w)/\text{Pe}$ is small. For this case Sellars, Tribus, and Klein [25] propose the relation

$$\text{Nu} = 1.357 \left(\frac{x/r_w}{\text{Pe}}\right)^{-1/3} \qquad \frac{x/r_w}{\text{Pe}} \leq 0.01 \tag{13-26}$$

which is essentially the same as that obtained by the Leveque solution shown in Eq. (13-19).

From Eq. (13-23) the value of the Nusselt number as $(x/r_w)/\text{Pe}$ becomes large is

$$\text{Nu}_\infty = 3.656 \qquad \frac{x/r_w}{\text{Pe}} > 0.25 \tag{13-27}$$

where Nu_∞ is the asymptotic value of the Nusselt number when $(x/r_w)/\text{Pe}$

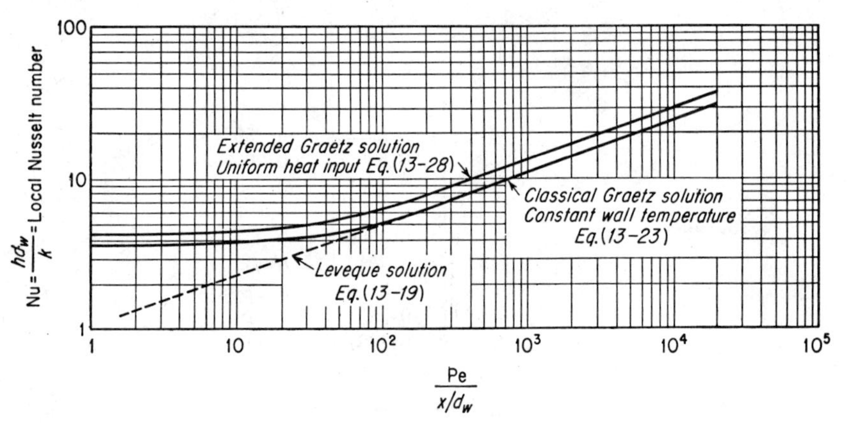

FIG. 13-5. Local Nusselt number for laminar flow in tubes. Velocity profile fully developed.

becomes greater than 0.25. Equation (13-23) is plotted in Fig. 13-5 in the range $0.0001 < (x/d_w)/\text{Pe} < 1.0$.

13-5. Extension of the Graetz Problem; Velocity Profile Fully Developed

Sellars, Tribus, and Klein [25] provided a useful extension of the classical Graetz solution by considering boundary conditions other than constant wall temperature. Their results are given below, and the reader is referred to the original article for details of the mathematical solution.

Case 1. *Constant-wall-heat Flux.* The differential equation (13-21) is applicable. The boundary conditions are (see Fig. 13-1)

At $x = 0$ at any r

$$T = T_\infty$$

At all x

$$\frac{q_w}{A_w} = \text{const}$$

$$\frac{\partial T_b}{\partial x} = \text{const}$$

The Nusselt number is

$$\text{Nu} = \frac{1}{{}^{11}\!/_{48} + {}^{1}\!/_{2} \sum_{m=0}^{m=\infty} \frac{\exp\{[-\beta_m{}^2(x/r_w)]/\text{Pe}\}}{\beta_m{}^4 \phi'_m(-\beta_m{}^2)}} \tag{13-28}$$

where:

m	$\beta_m{}^2$	$-\phi'_m(-\beta_m{}^2)$
0	25.64	8.854×10^{-3}
1	84.62	2.062×10^{-3}
2	176.40	9.435×10^{-4}

For small values of $(x/r_w)/\text{Pe}$

$$\text{Nu} = 1.639 \left(\frac{x/r_w}{\text{Pe}}\right)^{-1/3} \qquad \frac{x/r_w}{\text{Pe}} < 0.01 \tag{13-29}$$

For large values of $(x/r_w)/\text{Pe}$

$$\text{Nu}_\infty = 4.364 \qquad \frac{x/r_w}{\text{Pe}} \geq 0.25 \tag{13-30}$$

Equation (13-28) is plotted in Fig. 13-5.

Case 2. *Linear Wall Temperature.* The differential equation (13-21) is applicable. The boundary conditions are

At $x = 0$ for any r

$$T = T_\infty$$

At $x > 0$

$$r = r_w$$

$$T = T_w$$

$$T_w - T_\infty = cx \qquad \text{where } c = \text{const}$$

The Nusselt number becomes

$$\mathrm{Nu} = \frac{\tfrac{1}{2} + 4 \sum_{n=0}^{n=\infty} \dfrac{c_n \phi'(1)}{2\beta_n{}^2} \exp \dfrac{-\beta_n{}^2 (x/r_w)}{\mathrm{Pe}}}{\tfrac{88}{768} + 8 \sum_{n=0}^{n=\infty} \dfrac{c_n \phi'(1)}{2\beta_n{}^4} \exp \dfrac{-\beta_n{}^2 (x/r_w)}{\mathrm{Pe}}} \tag{13-31}$$

For small values of $(x/r_w)/\mathrm{Pe}$

$$\mathrm{Nu} = 2.035 \left(\frac{x/r_w}{\mathrm{Pe}} \right)^{-1/3} \qquad \frac{x/r_w}{\mathrm{Pe}} < 0.01 \tag{13-32}$$

For large values of $(x/r_w)/\mathrm{Pe}$

$$\mathrm{Nu}_\infty = 4.364 \qquad \frac{x/r_w}{\mathrm{Pe}} \geq 0.5 \tag{13-33}$$

The limiting Nusselt number for case 2 is the same as for case 1; however, for the former the limiting Nusselt number is reached when $(x/r_w)/\mathrm{Pe} = 0.5$, while in the latter it is reached when $(x/r_w)/\mathrm{Pe} = 0.25$. This means that the thermal entrance length (at constant Peclet number) for constant wall temperature (case 1) is one-half the thermal entrance length for linear wall temperature (case 2). The thermal entrance length is that distance from the beginning of heat transfer at which the Nusselt number becomes independent of length.

13-6. Laminar-flow Heat Transfer; Velocity Profile Developing

Since heat transfer often occurs near the actual entrance of a tube, the velocity profile is not parabolic but is developing. The Graetz solutions based on the parabolic velocity distribution are not valid in these circumstances. This problem was considered by Kays,[13] who obtained a numerical solution of Eq. (13-20) in which the laminar velocity profile was assumed to develop according to the relation derived by Langhaar [18] [see Eq. (9-4)]. Kays showed that his numerical solution was valid by solving Eq. (13-20) for constant wall temperature and assuming the velocity profile to be parabolic. His results agree with the classical Graetz solution.

Kays considered three boundary conditions:

(1) Constant wall temperature (Pr = 0.7)
(2) Constant heat input (Pr = 0.7)
(3) Constant temperature difference (Pr = 0.7)

The results of the numerical solution are shown in Figs. 13-6 and 13-7. In Fig. 13-6 values of the local Nusselt number hd_w/k are plotted versus $(x/d_w)/\mathrm{Pe}$ for the three boundary conditions. Experimental data of Kroll [17]

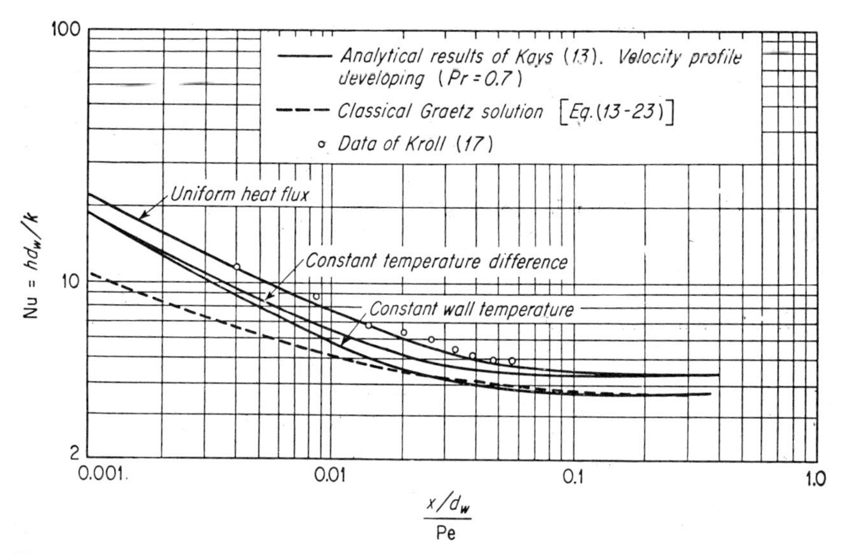

FIG. 13-6. Local Nusselt numbers for laminar flow in the entrance of a tube (Pr = 0.7).

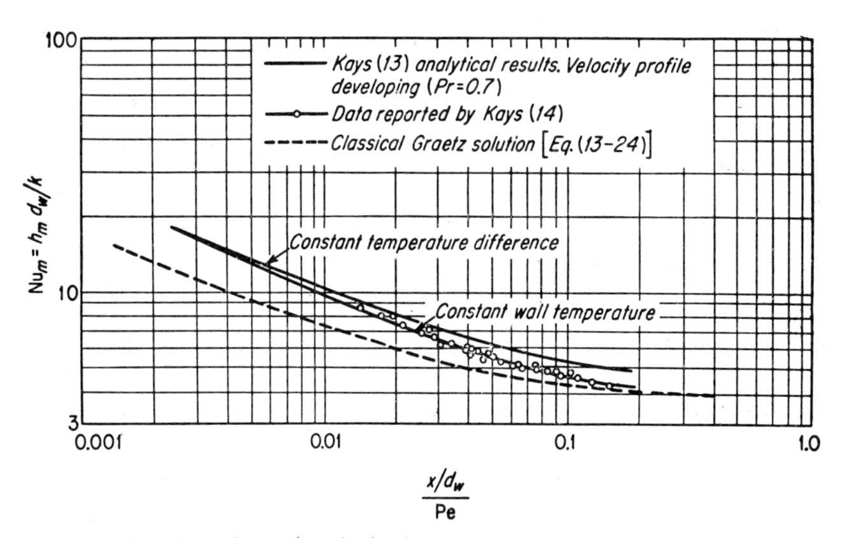

FIG. 13-7. Mean Nusselt numbers for laminar flow in the entrance of a tube (Pr = 0.7).

are included, indicating good agreement with the numerical solution. Equation (13-23), the classical Graetz solution, is shown in Fig. 13-6 for comparison. Figure 13-7 is a plot of the mean Nusselt number over a length of tube x versus $(x/d_w)/\text{Pe}$ for two of the boundary conditions considered by Kays. Experimental data reported by Kays [14] are shown in Fig. 13-7. Good agreement with the numerical solution is indicated.

Kays also reported relationships for the mean and local Nusselt numbers for the boundary conditions he studied. These relationships are summarized below.

1. Constant wall temperature ($\text{Pr} = 0.7$):

$$\text{Nu}_m = 3.66 + \frac{0.104[(x/d_w)/\text{Pe}]^{-1}}{1 + 0.016[(x/d_w)/\text{Pe}]^{-0.8}} \tag{13-34}$$

2. Constant temperature difference ($\text{Pr} = 0.7$):

$$\text{Nu}_m = 4.36 + \frac{0.10[(x/d_w)/\text{Pe}]^{-1}}{1 + 0.016[(x/d_w)/\text{Pe}]^{-0.8}} \tag{13-35}$$

3. Constant heat input ($\text{Pr} = 0.7$):

$$\text{Nu} = 4.36 + \frac{0.036[(x/d_w)/\text{Pe}]^{-1}}{1 + 0.0011[(x/d_w)/\text{Pe}]^{-1}} \tag{13-36}$$

Equations (13-34) to (13-36) may be used to predict Nusselt numbers for laminar flow in the entrance section of circular tubes. They are strictly applicable to the flow of air or the common gases since they were derived for a fluid with a Prandtl number of 0.7.

13-7. Empirical Correlation of Laminar-flow Heat-transfer Data

Experimental laminar-flow heat-transfer data are not plentiful; nor have all the available data been obtained under conditions for which analytical solutions are available, i.e., constant wall temperature, constant heat input, etc. Another difficulty encountered in comparing analytical and empirical results is that of variable fluid properties. Most theoretical work assumes constant fluid properties, but in practice the assumption is valid only in the limit where temperature differences approach zero. In dealing with experimental data the further question arises of whether natural convection plays a significant role in the heat transfer. Some experimental results are shown in Figs. 13-6 and 13-7, and relatively good agreement with theoretical work is obtained.

Boehm [3] studied the laminar-flow heat transfer for the cooling of oil in a tube. The heat transfer was studied after the velocity profile was established, and Nusselt numbers based on the arithmetic mean temperature difference were determined over a section having a length-to-diameter ratio x/d_w of 124. Since the data were obtained at essentially constant wall temperature, they may be compared with classical Graetz solution. Figure 13-8 is a plot of $\mathrm{Nu}_{a.m.}$ versus $(x/d_w)/\mathrm{Pe}$, where $\mathrm{Nu}_{a.m.} = h_{a.m.}d_w/k$, in which $h_{a.m.}$ is defined by Eq. (12-5). Boehm's data fall much below the curve for Eq. (13-24) and agree quite well with a curve for cooling obtained by Kraussold.[16] Kraussold's curve for heating lies above that for cooling. Nusselt's [22] laminar-flow data for air are plotted on Fig. 13-8. They were

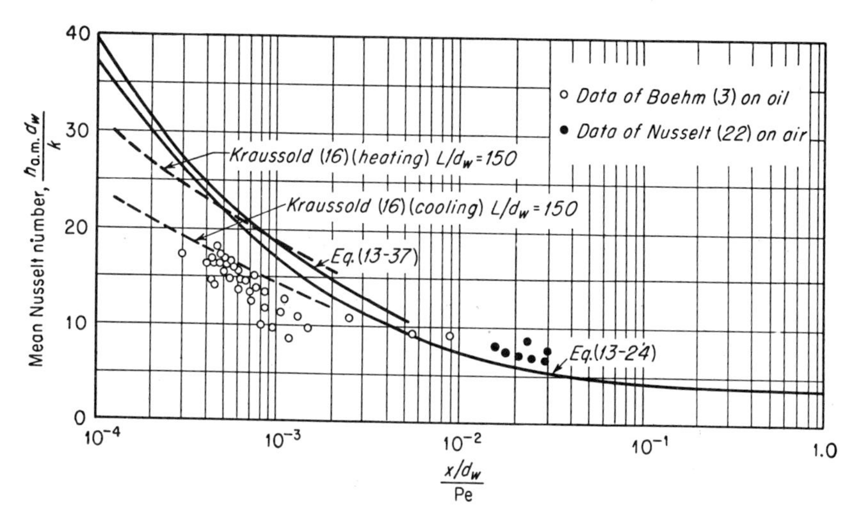

FIG. 13-8. Mean Nusselt numbers for fully developed laminar flow in tubes.

obtained at fairly high values of $(x/d_w)/\mathrm{Pe}$ and fall close to Eq. (13-24). The following equation, proposed by Sieder and Tate,[26] is also plotted on Fig. 13-8:

$$\mathrm{Nu}_{a.m.} = 1.86 \left(\frac{x/d_w}{\mathrm{Pe}}\right)^{-1/3} \left(\frac{\mu_b}{\mu_w}\right)^{0.14} \tag{13-37}$$

(Properties evaluated at bulk temperature)

where μ_b is the viscosity of the fluid at its arithmetic mean bulk temperature. Equation (13-37) gives Nusselt numbers somewhat higher than the analytical Graetz solution.

The relatively poor agreement of experimental data with theoretical results is probably due to natural-convection effects, which are difficult to eliminate.

13-8. The Effect of Variable Fluid Properties and Natural Convection

The effect of variable fluid properties is manifested in two ways. The velocity profile is no longer parabolic but becomes either more pointed or more blunt, depending on how the viscosity changes with temperature. The work of Deissler [6] in analyzing the effect on laminar-flow velocity profiles has already been mentioned (Sec. 4-11 and Figs. 4-12 and 4-13). Natural convection also occurs, mainly because of variation in the fluid density. The natural-convection effect is most noticeable during laminar flow in vertical tubes.

Martinelli and Boelter [20] studied laminar-flow heat transfer in vertical tubes and proposed the relation

$$\mathrm{Nu}_{\mathrm{a.m.}} = 1.75F_1\sqrt[3]{\mathrm{Gz}_{\mathrm{a.m.}} + 0.0722F_2\left(\frac{\mathrm{GrPr}}{x/d_w}\right)_w^{0.84}} \tag{13-38}$$

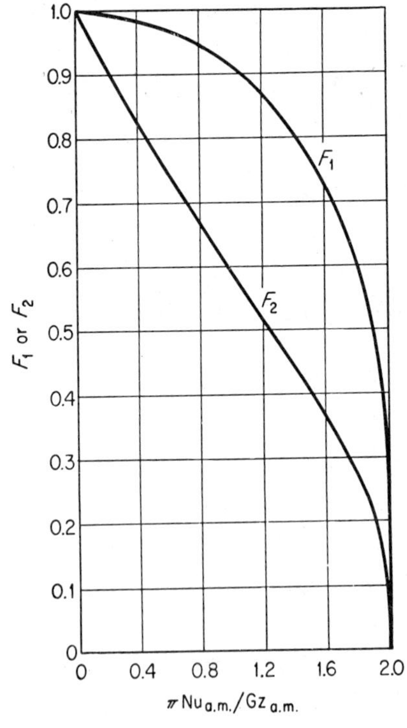

Fig. 13-9. Factors F_1 and F_2 to be used in Eq. (13-38). [*From R. C. Martinelli and L. M. K. Boelter, Univ. Calif. (Berkeley) Publs. Eng.*, **5**:23 (1942).]

Conditions:

(1) $\mathrm{Gz}_{\mathrm{a.m.}}$ determined using fluid properties at arithmetic mean bulk temperature.
(2) Gr and Pr evaluated at tube-wall temperature.
(3) Temperature difference in Gr is $T_w - T_\infty$, the initial temperature difference.
(4) Applies to heating with upward flow and cooling with downward flow.
(5) For cooling with upward flow and heating with downward flow change $+0.0722$ to -0.0722.

Equation (13-38) may be used to predict Nusselt numbers for laminar flow in vertical tubes. Although it is an empirical equation, it has the same form as an analytical relationship derived by Martinelli and Boelter. The functions F_1 and F_2 are given in Fig. 13-9.

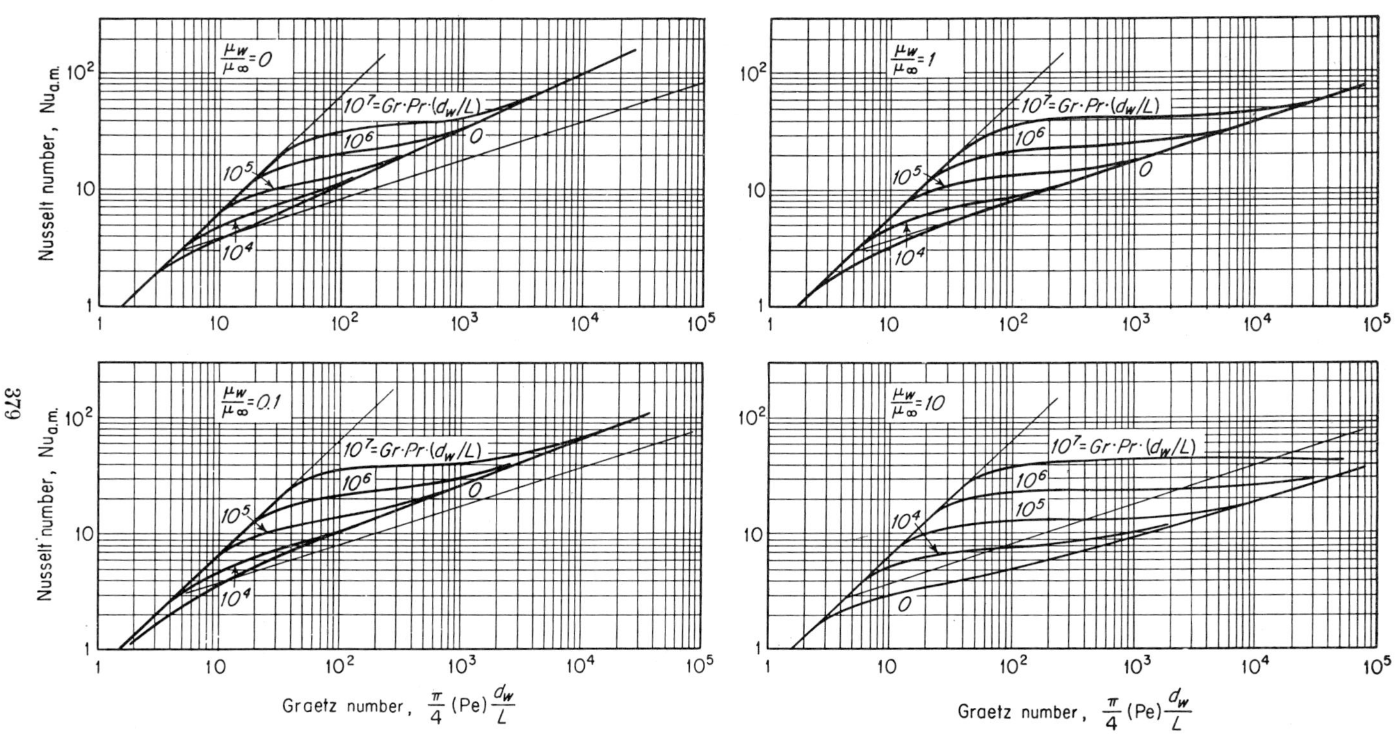

FIG. 13-10. Predicted Nusselt numbers for laminar flow in tubes showing effect of free convection and variable viscosity. [*From R. L. Pigford, Chem. Eng. Progr. Symposium Ser.*, [17] **51**:79 (1955).]

Pigford[23] has given charts (Fig. 13-10) for prediction of laminar-flow Nusselt numbers based on a theoretical analysis. The charts are a plot of $Nu_{a.m.}$ versus $Gz_{a.m.}$ for various values of $GrPr/(L/d_w)$. Each chart is at a constant value of u_w/u_∞ so that the effect of viscosity variation of the fluid is considered.

A tentative correlation for laminar flow in horizontal tubes is given by Eubank and Proctor.[8]

$$Nu_{a.m.}\left(\frac{\mu_w}{\mu_b}\right)^{0.14} = 1.75\sqrt[3]{Gz_{a.m.} + 0.04\left(\frac{GrPr}{x/d_w}\right)_b^{0.75}} \tag{13-39}$$

Conditions:

(1) Applies for laminar-flow heat transfer in circular tubes.
(2) $Gz_{a.m.}$, Gr, and Pr evaluated at arithmetic mean bulk temperature.
(3) Temperature difference in Gr is arithmetic mean temperature difference.

Example 13-2

Oil at 70°F is flowing upward through a vertical ½-in.-OD 16 BWG copper tube. After a sufficient calming section the temperature of the tube wall is maintained at 212°F. What length of tube (from the start of heating) is required for a 20°F rise in the oil temperature? The oil flow rate is 300 lb/hr.

Properties of oil:

Thermal conductivity: 0.085 Btu/(hr)(ft²)(°F)/ft
Specific gravity: 1.0
Heat capacity: 0.5 Btu/(lb_m)(°F)
Viscosity at

70°F: 25 centipoises
80°F: 20 centipoises
90°F: 15 centipoises
212°F: 3 centipoises

Solution

Average bulk temperature of oil = 80°F

Tube ID = 0.370 in.

$$Re = \frac{(4)(300)(12)}{(\pi)(0.370)(20)(2.42)} = 259$$

$$Pr = \frac{(0.5)(20)(2.42)}{0.085} = 285$$

$$Pe = 7.30 \times 10^4$$

$$(Pe)^{1/3} = 41.7$$

$$\left(\frac{\mu_b}{\mu_w}\right)^{0.14} = \left(\frac{20}{3}\right)^{0.14} = 1.304$$

From Eq. (13-37)

$$\mathrm{Nu}_{a.m} = (1.86)(41.7)\left(\frac{d_w}{x}\right)^{1/3}(1.304) = 101\left(\frac{d_w}{x}\right)^{1/3}$$

$$h_{a.m.} = \frac{(0.085)(101)}{d_w}\left(\frac{d_w}{x}\right)^{1/3} = \frac{8.59}{d_w}\left(\frac{d_w}{x}\right)^{1/3}$$

$$\Delta T_{a.m} = \frac{(212 - 70) + (212 - 90)}{2} = 132°\mathrm{F}$$

$$q = h_{a.m.}\pi d_w x \,\Delta T_{a.m.} = (300)(0.5)(20)$$

$$h_{a.m.} = \frac{(300)(0.5)(20)}{\pi d_w x(132)} = \frac{7.24}{d_w x}$$

Therefore

$$\frac{7.24}{d_w x} = \frac{8.59}{d_w}\left(\frac{d_w}{x}\right)^{1/3}$$

from which

$$\left(\frac{d_w}{x}\right)^{2/3} = 0.0364$$

and

$$\frac{d_w}{x} = 0.0069$$

giving

$$x = 54 \text{ in.}$$
$$= 4.5 \text{ ft}$$

The length may also be calculated using Eq. (13-24). Assuming the average bulk temperature to be 80°F,

$$\mathrm{Pe} = 7.30 \times 10^4$$

Assuming a length of 54 in.,

$$\frac{\mathrm{Pe}}{x/d_w} = \frac{7.30 \times 10^4}{54/0.370} = 500$$

From Eq. (13-24)

$$\mathrm{Nu}_m = 3.66 + \frac{(0.0668)(500)}{1 + 0.04(500)^{2/3}} = 13.16$$

giving

$$h_m = \frac{(13.16)(0.085)(12)}{0.370} = 36.3 \text{ Btu/(hr)(ft}^2\text{)(°F)}$$

$$\Delta T_{l.m.} = \frac{(212 - 70) - (212 - 90)}{\ln[(212 - 70)/(212 - 90)]} = 132$$

Thus

$$q = \frac{(36.3)(\pi)(0.370)(x)(132)}{12} = 3{,}000$$

giving

$$x = 6.45 \text{ ft}$$
$$= 77 \text{ in.}$$

This does not check with original assumption of x. Further trials give $x = 98$ in. Equation (13-24) predicts a much longer length than Eq. (13-37) largely because the viscosity ratio is neglected in Eq. (13-24). If the viscosity ratio is considered, Eq. (13-24) predicts a length of 62 in.

II. LAMINAR-FLOW HEAT TRANSFER IN ANNULI

13-9. Theoretical and Empirical Relationships for Laminar-flow Heat Transfer in Annuli

Little theoretical work has been done on laminar-flow heat transfer in annuli compared with that done on circular tubes. The relatively complicated form of the laminar velocity profile [Eq. (4-62)] makes the solution of Eq. (13-20) difficult.

Jakob and Rees [12] derived theoretical relationships for the temperature difference between the two walls of the annulus. They also gave relationships for the heat-transfer coefficient based on this temperature difference.

There have been analytical investigations of annular laminar-flow heat transfer assuming a uniform velocity profile (slug flow). Trefethen [28] obtained the following relation for the limiting Nusselt number for such slug flow:

$$(\mathrm{Nu}_s)_\infty = \frac{h(d_2 - d_1)}{k}$$

$$= \frac{8(d_2/d_1 - 1)[(d_2/d_1)^2 - 1]^2}{4(d_2/d_1)^4 \ln (d_2/d_1) - 3(d_2/d_1)^4 + 4(d_2/d_1)^2 - 1} \qquad (13\text{-}40)$$

where the heat is transferred through the inner tube of the annulus. If heat is transferred through the outer tube, the positions of d_2 and d_1 should be reversed in Eq. (13-40). Values of $(\mathrm{Nu}_s)_\infty$ as a function of d_2/d_1 are given in Table 13-3.

TABLE 13-3. VALUES OF THE LIMITING NUSSELT NUMBER FOR SLUG FLOW IN ANNULI

(Constant rate of heat input; inside tube heated)

$\frac{d_2}{d_1}$	$(\mathrm{Nu}_s)_\infty$
1.00	6.00
2.00	6.36
3.00	6.93
3.59	7.32
4.00	7.57
5.00	8.21
11.00	11.91
21.00	17.50
101.00	50.80

The empirical equation usually recommended is that obtained by Chen, Hawkins, and Solberg,[4] who studied laminar-flow heat transfer from the inner wall in four annuli with diameter ratios d_2/d_1 ranging from 1.09 to 2.0. The Nusselt number based on the arithmetic mean heat transfer coefficient and the equivalent diameter may be predicted from the following relation:

$$\mathrm{Nu}_{\mathrm{a.m.}} = 1.02(\mathrm{Re})^{0.45}(\mathrm{Pr})^{0.5}\left(\frac{\mu_b}{\mu_1}\right)_{\mathrm{a.m.}}^{0.14}\left(\frac{d_e}{L}\right)^{0.4}\left(\frac{d_2}{d_1}\right)^{0.8}(\mathrm{Gr})^{0.05} \qquad 200 < \mathrm{Re} < 2{,}000 \qquad (13\text{-}41)$$

(Re based on d_e)

where μ_b is the viscosity at the arithmetic mean bulk temperature of the fluid and μ_1 is the viscosity at the temperature T_1 of the inner wall of the annulus. Other properties are evaluated at the bulk temperature of the fluid.

III. LAMINAR-FLOW HEAT TRANSFER IN NONCIRCULAR DUCTS

13-10. The Graetz Problem for Infinite Parallel Planes; Velocity Profile Developed

When heat transfer takes place during fully developed laminar flow between infinite parallel planes with both planes heated (or cooled), the differential equation for the temperature of the fluid is

$$1.5U\left[1 - \left(\frac{y_c}{b/2}\right)^2\right]\frac{\partial T}{\partial x} = \frac{k}{C_p\rho}\frac{\partial^2 T}{\partial y_c{}^2} \qquad (13\text{-}42)$$

where b is the spacing of the planes and y_c is the distance measured normal to the mid-plane. Solution of Eq. (13-42) with appropriate boundary conditions will give T as a function of x and y_c. Sellars, Tribus, and Klein [25] solved Eq. (13-42) for two boundary conditions:

(1) Constant wall temperature:

At $x = 0$

$T = T_\infty$ for all y_c

At $x > 0$

$$y_c = \pm\frac{b}{2}$$

$$T = T_w = \text{const}$$

(2) Uniform heat flux:

At $x = 0$

$$T = T_\infty \qquad \text{for all } y_c.$$

At $x > 0$

$$y_c = \pm \frac{b}{2}$$

$$\frac{q_w}{A_w} = \text{const}$$

(A uniform heat flux means that dT_b/dx is constant for constant fluid properties.)

These workers reported series solutions for the temperature and the local rate of heat transfer, and the results are applicable to laminar-flow heat transfer between parallel planes and in flat rectangular ducts.

Norris and Streid [21] proposed the following relations for laminar-flow heat transfer in flat ducts for the case of constant wall temperatures:

$$\text{Nu}_{\text{a.m.}} = 1.85 \left(\frac{x/d_e}{\text{Pe}}\right)^{-1/3} \qquad \frac{x/d_e}{\text{Pe}} < 0.0005 \tag{13-43}$$

$$\text{Nu}_m = 1.85 \left(\frac{x/d_e}{\text{Pe}}\right)^{-1/3} \qquad \frac{x/d_e}{\text{Pe}} < 0.014 \tag{13-44}$$

$$\text{Nu}_\infty = 7.60 \qquad \frac{x/d_e}{\text{Pe}} > 0.014 \tag{13-45}$$

These relations are the results of a theoretical study of laminar-flow heat transfer between parallel planes and may be applied to predict Nusselt numbers for laminar flow in flat rectangular ducts with heat being transferred through all sides which are all at the same uniform temperature.

13-11. The Graetz Problem for Rectangular, Square, and Triangular Ducts; Fully Developed Velocity Profile; All Sides Heated

The differential equation for laminar-flow heat transfer in a rectangular duct is

$$u \frac{\partial T}{\partial x} = \frac{k}{C_p \rho} \left(\frac{\partial^2 T}{\partial y^2} + \frac{\partial^2 T}{\partial z^2}\right) \tag{13-46}$$

In this equation flow is in the x direction, and the cross section of the duct is parallel to the yz plane. The solution of Eq. (13-46) to give T as a function of x, y, and z requires a knowledge of the fully developed laminar velocity profile in the rectangular duct. This profile is given by Eq. (4-87), which, because of its complicated nature, prevents an analytical solution of

TABLE 13-4. LIMITING NUSSELT NUMBERS FOR LAMINAR-FLOW HEAT TRANSFER IN NONCIRCULAR DUCTS †

	Nu_∞	
	Constant wall temperature	Uniform heat flux
Equilateral triangle.........	....	3.00
Values of a/b for rectangles:		
1 (square)................	2.89	3.63
0.713....................	....	3.78
0.500....................	3.39	4.11
0.333....................	....	4.77
0.25.....................	....	5.35
0 (parallel planes)........	7.60	8.24

† From S. H. Clark and W. M. Kays, *Trans. ASME*, **75**:859 (1953).

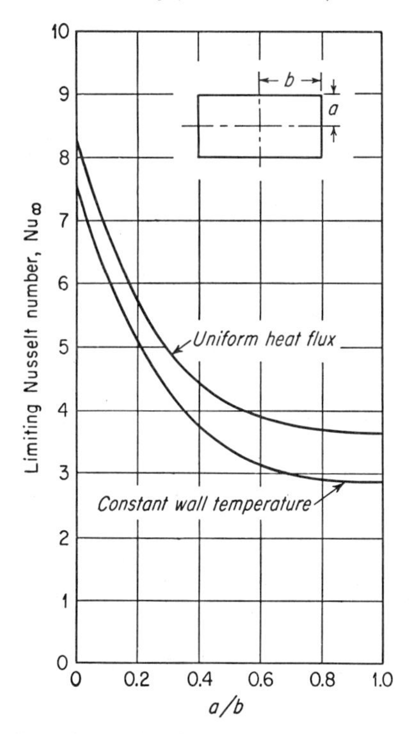

FIG. 13-11. Limiting Nusselt number for laminar flow in rectangular ducts. [*From S. H. Clark and W. M. Kays, Trans. ASME,* **75**:859 (1953).]

Eq. (13-46). Clark and Kays [5] solved Eq. (13-46) numerically and determined values of the limiting Nusselt number for two boundary conditions, uniform wall temperature and uniform heat flux. Their results are shown in Table 13-4. Figure 13-11 is a plot of Nu_∞ versus a/b, where a and b are defined by Fig. 4-19. The Nusselt numbers given in Table 13-4 apply for very long ducts. Clark and Kays indicate fair agreement of these predicted Nusselt numbers with measured values.

13-12. Laminar-flow Heat Transfer in Flat Ducts; Velocity Profile Developing

The problem of laminar heat transfer in the entrance section of flat ducts has been analyzed by Sparrow.[27] The flat ducts considered were approximately two parallel planes with heat being transferred through each plane. Sparrow solved the equation

$$u \frac{\partial T}{\partial x} + v \frac{\partial T}{\partial y} = \frac{k}{C_p \rho} \frac{\partial^2 T}{\partial y^2} \tag{13-47}$$

by getting it in integral form [see Eq. 17-6)] and assuming that the velocity profile develops according to Eq. (9-13) (see Sec. 9-8). Sparrow's analytical results may be represented by the following equation:

$$\mathrm{Nu}_m = \frac{0.664(\mathrm{Gz}')^{1/2}}{(\mathrm{Pr})^{1/6} F_1} \sqrt{1 + 7.3 \left(\frac{\mathrm{Pr}}{\mathrm{Gz}'}\right)^{1/2}} \qquad 0.01 \le \mathrm{Pr} \le 2 \tag{13-48}$$

The function F_1 is plotted in Fig. 13-12. For Prandtl numbers greater than 2.0 the following relationship holds:

$$\mathrm{Nu}_m = \frac{0.664(\mathrm{Gz}')^{1/2}}{(\mathrm{Pr})^{1/6}} \sqrt{1 + 6.27 \left(\frac{\mathrm{Pr}}{\mathrm{Gz}'}\right)^{4/6}} \tag{13-49}$$

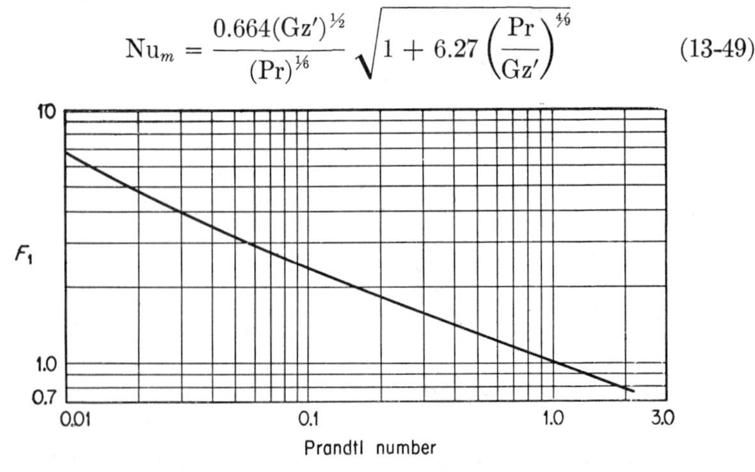

FIG. 13-12. Value of factor F_1 to be used in Eq. (13-48).

In Eqs. (13-48) and (13-49) $Gz' = Pe(d_e/x)$, and Pe and Nu_m are based on the equivalent diameter of infinite parallel planes, which is twice the spacing of the planes. Equations (13-48) and (13-49) are plotted in Fig. 13-13 for $100 < Gz' < 100{,}000$ and for Prandtl numbers of 0.01, 1, and 10. The curve labeled "parabolic velocity profile" is plotted from results of Norris and Streid,[21] whose investigation assumed the velocity profile to be fully developed. The upper curve labeled "slug flow" gives the Nusselt number as a function of Gz′ when the velocity is assumed uniform across the duct.

In Fig. 13-13 the effect of the Prandtl number on heat transfer in the entrance region of a duct is clearly shown. As the Prandtl number approaches zero, the Nusselt number approaches that predicted by assuming

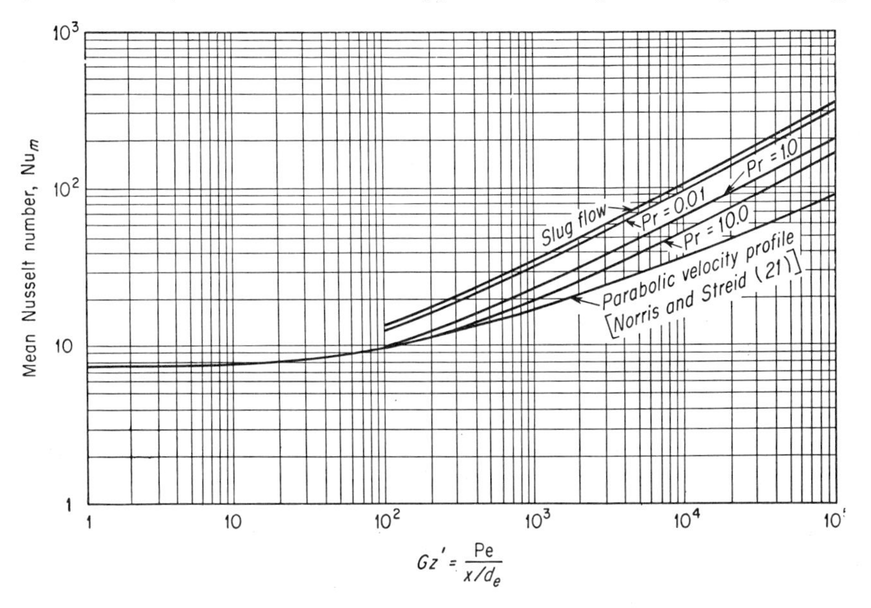

FIG. 13-13. Predicted mean Nusselt numbers for laminar flow between parallel planes. (*From E. M. Sparrow, NACA TN* 3331, 1955.)

slug flow, and, conversely, when the Prandtl number becomes very large, the Nusselt number approaches that obtained when the velocity profile is fully developed (parabolic). This fact is utilized in predicting rates of heat transfer with liquid metals, which have a very low Prandtl number. A reasonable value for the Nusselt number for such fluids may be obtained by assuming slug flow in the duct. This will be considered more fully in Chap. 16.

13-13. Empirical Relationship for Laminar-flow Heat Transfer in Rectangular Ducts

As with circular tubes, laminar-flow heat-transfer data for rectangular ducts are scarce. Kays and Clark [15] have published data obtained in the study of radiator cores in which the air passages were rectangular ducts. The following tentative empirical relationships obtained from the data should be applicable to rectangular ducts that are geometrically similar to the ones studied. The relations apply for air only.

1. For uniform wall temperature:

$$\frac{\text{Nu}_{\text{a.m.}}}{\text{Nu}_\infty} = 1 + \left(0.003 + 0.039\,\frac{a}{b}\right)\frac{\text{Pe}}{x/d_e} \qquad 500 < \text{Re} < 2{,}000 \quad \text{Pr} \approx 0.7 \qquad (13\text{-}50)$$

2. For uniform heat flux:

$$\frac{\text{Nu}_{\text{a.m.}}}{\text{Nu}_\infty} = 1 + \left(0.003 + 0.019\,\frac{a}{b}\right)\frac{\text{Pe}}{x/d_e} \qquad 500 < \text{Re} < 2{,}000 \quad \text{Pr} \approx 0.7 \qquad (13\text{-}51)$$

Equations (13-50) and (13-51) are based on data obtained on a square duct 0.18 by 0.18 in. with $x/d_e = 100$ and a rectangular duct 0.151 by 0.395 in. with $x/d_e = 55$. All tests were made on air. The value of Nu_∞ may be obtained from Fig. 13-11 for various values of a/b.

Example 13-3

The cross section of an air heater consists of a large number of small rectangular ducts having inside measurements of 0.2 by 0.4 in. The length of each duct is 4 in. The walls of the ducts are maintained at 150°F. Air at 70°F flows through the heater at a superficial mass velocity of 2,600 $\text{lb}_m/(\text{hr})(\text{ft}^2)$. The cross-sectional area of the rectangular ducts is 85 per cent of the total cross-sectional area of the radiator. Determine the temperature rise of the air flowing through the radiator.

Solution

Assume that the average bulk temperature of air is 85°F. At this temperature

$$\mu = 0.0180 \text{ centipoise}$$

$$\text{Pr} = 0.71$$

$$k = 0.0155 \text{ Btu/(hr)(ft}^2\text{)(°F)/ft}$$

$$C_p = 0.24 \text{ Btu/(lb}_m\text{)(°F)}$$

$$\text{Equivalent diameter of duct} = \frac{(4)(0.2)(0.4)}{(2)(0.2 + 0.4)} = 0.267 \text{ in.}$$

$$\text{Re} = \frac{0.267}{12}\,\Big|\,\frac{2{,}600}{0.85}\,\Big|\,\frac{}{0.0180}\,\Big|\,\frac{}{2.42} = 1{,}560$$

$$\frac{\text{Pe}}{x/d_e} = \frac{(1{,}560)(0.71)}{4/0.267} = 74$$

From Eq. (13-50)

$$\mathrm{Nu}_{\mathrm{a.m.}}/\mathrm{Nu}_{\infty} = 1 + \left[0.003 + \frac{(0.039)(0.2)}{0.4}\right](74) = 2.67$$

From Fig. 13-11

$$\mathrm{Nu}_{\infty} = 3.4$$

Thus $\mathrm{Nu}_{\mathrm{a.m.}} = (3.4)(2.67) = 9.08$

$$h_{\mathrm{a.m.}} = \frac{(9.08)(0.0155)(12)}{0.267} = 6.33 \text{ Btu/(hr)(ft}^2\text{)(°F)}$$

In each duct (letting T_2 be the exit temperature),

$$q = \frac{(2{,}600)(0.2)(0.4)(0.24)(T_2 - 70)}{(0.85)(144)}$$

$$= \frac{(6.33)(1.2)(4)}{144} \frac{(150 - 70) + (150 - T_2)}{2}$$

from which

$$(0.408)(T_2 - 70) = (0.211)\left(115 - \frac{T_2}{2}\right)$$

giving

$$T_2 = 103°\mathrm{F}$$

The air leaves the heater at 103°F.

BIBLIOGRAPHY

1. Abramowitz, M.: *J. Math. and Phys.*, **32**:184 (1953).
2. Abramowitz, M.: *J. Math. and Phys.*, **30**:162 (1951).
3. Boehm, J.: *Wärme*, **66**:143 (1943).
4. Chen, C. Y., G. A. Hawkins, and H. L. Solberg: *Trans. ASME*, **68**:99 (1946).
5. Clark, S. H., and W. M. Kays: *Trans. ASME*, **75**:859 (1953).
6. Deissler, R. G.: *NACA TN* 2410, 1951.
7. Drew, T. B.: *Trans. AIChE*, **26**:26 (1931).
8. Eubank, O. C., and W. S. Proctor: S.M. Thesis, Department of Chemical Engineering, Massachusetts Institute of Technology, 1951.
9. Graetz, L.: *Ann. Phys. u. Chem.*, **25**:337 (1885).
10. Hausen, H.: *Verfahrenstechnik Beih. Z. Ver. deut. Ing.* 4, p. 91, 1943.
11. Jakob, M.: "Heat Transfer," vol. 1, John Wiley & Sons, Inc., New York, 1949.
12. Jakob, M., and K. A. Rees: *Trans. AIChE*, **37**:619 (1941).
13. Kays, W. M.: *Trans. ASME*, **77**:1265 (1955).
14. Kays, W. M.: *Stanford Univ. Dept. Mech. Eng. Tech. Rept.* 14, Navy Contract N6-onr-251 Task Order 6, June 15, 1951.
15. Kays, W. M., and S. H. Clark: *Stanford Univ. Dept. Mech. Eng. Tech. Rept.* 17, Navy Contract N6-onr-251 Task Order 6, Aug. 15, 1953.
16. Kraussold, H.: *VDI-Forschungsheft* 351, 1931.
17. Kroll, C. L.: Heat Transfer and Pressure Drop for Air Flowing in Small Tubes, Sc.D. Thesis, Department of Chemical Engineering, Massachusetts Institute of Technology, 1951.
18. Langhaar, H. L.: *Trans. ASME*, **64**:A-55 (1942).
19. Leveque, J.: *Ann. mines*, [12] **13**:201, 305, 381 (1928).
20. Martinelli, R. C., and L. M. K. Boelter: *Univ. Calif.* (*Berkeley*) *Publs. Eng.*, **5**:23 (1942).

21. Norris, R. H., and D. D. Streid: *Trans. ASME*, **62**:525 (1940).
22. Nusselt, H.: Habilitationsschrift, Dresden, 1909.
23. Pigford, R. L.: *Chem. Eng. Progr. Symposium* Ser., [17] **51**:79 (1955).
24. Schenk, J., and J. M. Dumore: *Appl. Sci. Research*, **4A**:39 (1953).
25. Sellars, J. R., M. Tribus, and J. S. Klein: *Trans. ASME*, **78**:441 (1956).
26. Sieder, E. N., and G. E. Tate: *Ind. Eng. Chem.*, **28**:1429 (1936).
27. Sparrow, E. M.: *NACA TN* 3331, 1955.
28. Trefethen, L. M.: "Proceedings of the General Discussion on Heat Transfer," Institution of Mechanical Engineers, London, and American Society of Mechanical Engineers, New York, 1951, p. 436.

CHAPTER 14

TURBULENT-FLOW HEAT TRANSFER IN CLOSED CONDUITS. EMPIRICAL CORRELATIONS FOR HIGH-PRANDTL-NUMBER FLUIDS

14-1. Introduction

The transfer of heat to or from fluids flowing turbulently in closed conduits is one of the most important modes of industrial heat transfer. Virtually all industrial heat-transfer equipment contains closed conduits, usually circular tubes, through which either the hot or cold fluid flows. It is not surprising that empirical correlations for turbulent-flow heat transfer in tubes were developed quite early or that they have constantly been modified and improved as experimental methods have become refined and industrial needs have grown.

The complicated nature of turbulent flow prevents an analytical approach to the problem like that which can be made for laminar flow. The energy equation is applicable both for turbulent flow and laminar flow. In turbulent flow, however, heat is transferred by convection as well as conduction, and a knowledge of the turbulent velocity fluctuations is required to obtain a solution of the energy equation. Nevertheless, various studies on the basic mechanism of turbulent heat transfer have continued throughout the years, so that in addition to empirical correlations (which say nothing about mechanism) there are theoretical and semiempirical relationships based on fundamental knowledge of the processes occurring. The latter are considered in detail in Chap. 15. In the present chapter the more important empirical correlations of turbulent heat-transfer data in closed conduits are presented. The relationships given apply for fluids with a Prandtl number greater than 0.7.

Early experimental work on turbulent heat transfer in tubes was mainly on air and water, covering a Prandtl-number range of 0.7 to 10.0. Later studies were made on various high-viscosity oils having Prandtl numbers extending up to 1,000. Since most of the work was conducted using moderate temperature differences between the conduit wall and the fluid, the empirical correlations which were developed were applicable to these conditions. In recent years new heat-transfer media and heat transfer at extremely high temperature differences have required modification of existing relationships and the development of new correlations.

Among these new media the liquid metals are of importance. This group of fluids is characterized by very low Prandtl numbers (of the order of 0.01). Heat transfer with liquid metals is considered separately in Chap. 16.

14-2. Methods of Correlating Heat-transfer Data

The empirical correlations of turbulent heat-transfer data for closed conduits have been obtained using the dimensionless groups described in Chap. 12. Of these, the Nusselt, Reynolds, and Prandtl numbers have been used most frequently. However, other groups, namely, the Peclet and Stanton numbers, which are derived from the above-mentioned three, have also been used. It was seen in Chap. 13 that the Nusselt number and Peclet number RePr and the diameter-to-length ratio are sufficient for correlating laminar-flow heat-transfer data in the absence of natural-convection effects. For turbulent-flow heat transfer, however, these two groups are not sufficient.

The two common forms of equations relating these dimensionless groups for turbulent-flow heat transfer are

$$\mathrm{Nu} = c_1(\mathrm{Re})^{n_1}(\mathrm{Pr})^{n_2}\left(\frac{d_w}{L}\right)^{n_3} \tag{14-1}$$

$$\mathrm{St} = c_2(\mathrm{Re})^{n_4}(\mathrm{Pr})^{n_5}\left(\frac{d_w}{L}\right)^{n_6} \tag{14-2}$$

where c_1 and c_2 are constants and $n_1, \ldots, n_6$ are exponents.

Both equations are equivalent, but Eq. (14-2) has certain advantages for correlating experimental data. The mean heat-transfer coefficient for a circular tube of length L may be calculated from the relation

$$h_m(\pi d_w L)(T_w - T_b)_{\mathrm{l.m.}} = \frac{d_w{}^2\pi}{4} GC_p(T_{b_2} - T_{b_1}) \tag{14-3}$$

where T_{b_1} and T_{b_2} are, respectively, the inlet and outlet bulk temperature of the fluid. From Eq. (14-3) the Nusselt and Stanton numbers become

$$\mathrm{Nu}_m = \frac{d_w{}^2}{4L}\frac{GC_p}{k}\frac{T_{b_2} - T_{b_1}}{(T_w - T_b)_{\mathrm{l.m.}}} \tag{14-4}$$

$$\mathrm{St}_m = \frac{d_w}{4L}\frac{T_{b_2} - T_{b_1}}{(T_w - T_b)_{\mathrm{l.m}}} \tag{14-5}$$

The Stanton number may be determined without knowing or taking into account the physical properties of the fluid. In calculating the Nusselt number, one must include the mass velocity, the heat capacity, and the thermal conductivity. An error in any one of these is therefore included in the Nusselt number. Another advantage of correlating data by Eq. (14-2) is the fact that, at constant Prandtl number, the Stanton number varies as $(\mathrm{Re})^{-0.2}$, while the Nusselt number varies as $(\mathrm{Re})^{0.8}$. Equation (14-1) requires a greater range of ordinates.

14-3. The Effect of Temperature Difference on Turbulent Heat Transfer

If fluid properties were constant, the use of Eqs. (14-1) and (14-2) to correlate data would be quite simple. However, the temperature of the fluid not only varies across the section of the conduit but also along the length of the conduit. Since physical properties change with temperature, there is always the problem of which temperatures to use for evaluating the properties. In early work, where temperature differences were low and only air and water were studied, the bulk temperature of the fluid was suitable for evaluation of all fluid properties. With heat transfer with oils, in which the viscosity varies greatly with temperature, it was necessary to use an additional dimensionless group μ_b/μ_w to obtain satisfactory correlation of data. Recently it has become common practice to evaluate all fluid properties at a so-called *film temperature* rather than using a viscosity-ratio correction. The usual film temperature for evaluating properties is

$$T_{0.5} = \frac{T_w + T_b}{2} \tag{14-6}$$

i.e., it is the arithmetic average of the wall and bulk temperatures. Among the various other film temperatures which have been defined are

$$T_{0.4} = T_b + 0.4(T_w - T_b) \tag{14-7}$$

$$T_{0.6} = T_b + 0.6(T_w - T_b) \tag{14-8}$$

These have also been used frequently in correlating data.

I. TURBULENT-FLOW HEAT TRANSFER IN CIRCULAR TUBES

14-4. Fully Developed Turbulent Flow; Moderate Temperature Difference

A large number of turbulent heat-transfer data had been obtained by various investigators up to 1936. Most of these data were correlated by three equations. Dittus and Boelter [9] proposed the equation

$$\left(\frac{h_m d_w}{k}\right)_b = 0.023 \left(\frac{d_w G}{\mu}\right)_b^{0.8} \left(\frac{C_p \mu}{k}\right)_b^n \tag{14-9}$$

Conditions:

(1) Fluid properties evaluated at arithmetic mean bulk temperature
(2) Re > 10,000
(3) 0.7 < Pr < 100
(4) $n = 0.4$ for heating, 0.3 for cooling
(5) $L/d_w > 60$

Colburn [5] obtained the following equation, in which the Stanton group is used instead of the Nusselt group:

$$\left(\frac{h_m}{GC_p}\right)_b \left(\frac{C_p \mu}{k}\right)_{T0.5}^{2/3} = 0.023 \left(\frac{d_w G}{\mu}\right)_{T0.5}^{-0.2} \tag{14-10}$$

Conditions:

(1) Fluid properties, except C_p in Stanton group, evaluated at film temperature $T_{0.5}$ [Eq. (14-6)]
(2) Re > 10,000
(3) 0.7 < Pr < 160
(4) $L/d_w > 60$

Equations (14-9) and (14-10) are not applicable for fluids with very high Prandtl numbers. Sieder and Tate [27] proposed an equation which is applicable to fluids whose viscosity changes greatly for small temperature changes.

$$\left(\frac{h_m}{GC_p}\right)_b \left(\frac{C_p \mu}{k}\right)_b^{2/3} \left(\frac{\mu_w}{\mu_b}\right)^{0.14} = 0.023 \left(\frac{d_w G}{\mu}\right)_b^{-0.2} \dagger \tag{14-11}$$

Conditions:

(1) Fluid properties evaluated at bulk temperature (except μ_w)
(2) Re > 10,000
(3) 0.7 < Pr < 16,700
(4) $L/d_w > 60$

† Sieder and Tate employed a constant of 0.027. Drexel and McAdams [10] obtained a constant of 0.021 for air. The constant of 0.023 is considered satisfactory to correlate most available data.

The curves shown in Fig. 14-1 are recommended for determining heat-transfer coefficients in circular tubes for all regions of flow and for fluids with Prandtl numbers greater than 0.7. For values of the Reynolds num-

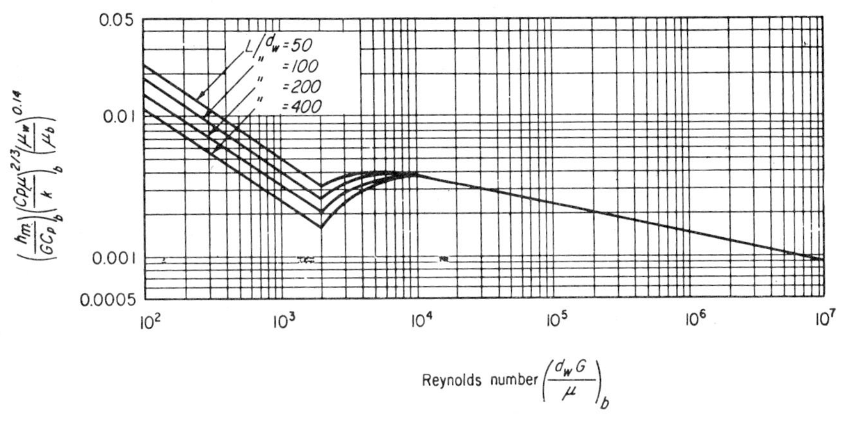

FIG. 14-1. Recommended curves for determining heat-transfer coefficients in circular tubes.

ber less than 2,000 Eq. (13-37) is plotted at various values of the diameter-to-length ratio. For values of the Reynolds number greater than 10,000, the single curve represents Eq. (14-11).

14-5. Additional Correlations of Turbulent-flow Heat-transfer Data

1. *Air at High Temperatures and High* ΔT. Humble, Lowdermilk, and Desmon [14] studied the transfer of heat to air flowing in a smooth circular tube by investigating various entrance configurations, tube lengths, and tube-wall temperatures. They obtained the empirical relationship

$$\left(\frac{h_m d_w}{k}\right)_{T_{0.5}} = 0.034 \left(\frac{d_w G}{\mu}\right)_{T_{0.5}}^{0.8} \left(\frac{C_p \mu}{k}\right)_{T_{0.5}}^{0.4} \left(\frac{L}{d_w}\right)^{-0.1} \tag{14-12}$$

Conditions:

(1) Fluid properties evaluated at film temperature [Eq. (14-6)]
(2) $10{,}000 < \text{Re} < 500{,}000$
(3) $30 < L/d_w < 120$
(4) $600 < T_w < 3050°\text{R}$
(5) $0.8 < T_w/T_b < 3.5$

Equation (14-12) was substantiated by Weiland and Lowdermilk.[29]

2. *Water at High Pressures and Temperatures.* Kaufman and Henderson [16] studied heat transfer to water at high pressures (200 and 2,000 psig) and high temperatures (up to 560°F) and found that Eq. (14-9) satisfac-

torily represented the data up to Reynolds numbers of 10^6. Kaufman and Isley [15] studied heat transfer to water at gauge pressures up to 200 in. Hg flowing in a tube with $L/d_w = 50$. For horizontal flow and flow up or down in a vertical tube these authors obtained the equation

$$\left(\frac{h_m d_w}{k}\right)_b = 0.0168 \left(\frac{d_w G}{\mu}\right)_b^{0.84} \left(\frac{C_p \mu}{k}\right)_b^{0.4} \tag{14-13}$$

Conditions:

(1) Fluid properties evaluated at bulk temperature
(2) $10{,}000 < \mathrm{Re} < 50{,}000$
(3) $L/d_w = 50$

Equation (14-13) was checked by Grele and Gedeon.[11]

3. *Molten Sodium Hydroxide.* Heat-transfer coefficients for heating molten sodium hydroxide flowing in a tube with $L/d_w = 100$ were found by Grele and Gedeon [11] to be about 20 per cent above those predicted by Eq.(14-9).

14-6. Heat Transfer in Tubes Containing Turbulence Promoters

The major resistance to heat transfer between solid boundaries and turbulent fluids is the laminar sublayer adjacent to the wall. The resistance of this laminar layer is proportional to its thickness, and any reduction of the thickness will result in a comparable increase in the rate of heat transfer between the wall and the fluid.

One means of reducing the thickness of the laminar layer is by increasing the intensity and scale of the turbulence of the flowing stream. This is done by placing turbulence promoters, usually spiral wires or strips, in the conduit. They have the effect of increasing not only the rate of heat transfer but also the resistance to flow; so actually the heat transferred per unit of power expended may not be changed.

Studies of the effect of turbulence promoters on the rate of heat transfer have been made by Royds,[24] Seigel,[26] and Colburn and King,[6] who showed that turbulence promoters materially increase the rate of heat transfer. A variety of turbulence promoters were investigated, and Fig. 14-2 shows Seigel's results. Here the heat-transfer coefficient for water flowing in a ⅝-in.-OD copper tube is plotted against the flow rate in gallons per minute. All tubes containing turbulence promoters have considerably higher heat-transfer coefficients than the empty tube. These results are in no way general, but they serve to indicate at least semiquantitatively the effect of turbulence promoters on heat transfer.

Another means of increasing turbulence of the flowing stream is by vibrations or pulsations. Martinelli and Boelter [20] have studied the effect of

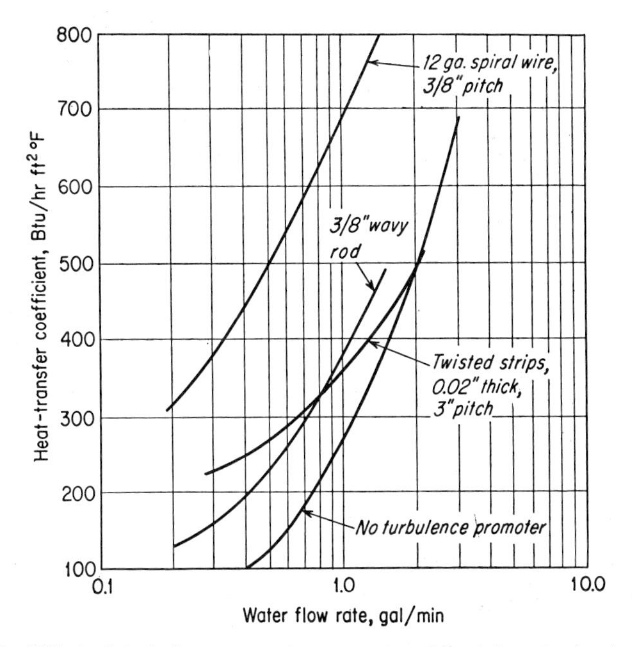

FIG. 14-2. Effect of turbulence promoters on rates of heat transfer in circular tubes. [*From L. G. Seigel, Heating, Piping, Air Conditioning,* **18**:111 (June, 1946).]

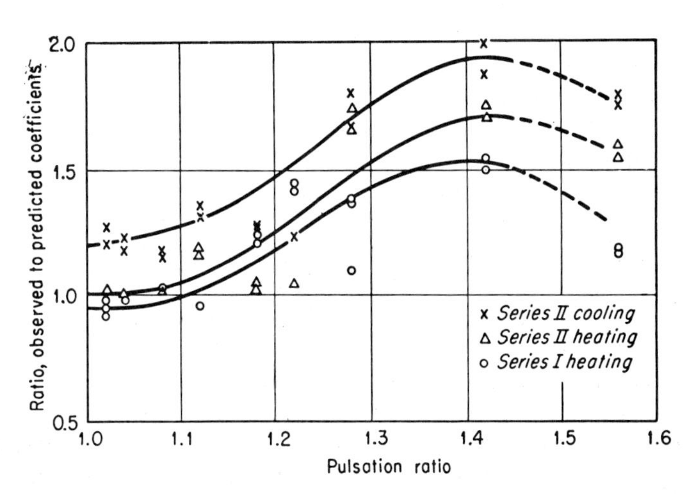

FIG. 14-3. Effect of pulsations on heat-transfer coefficients in circular tubes. [*From F. B. West and A. T. Taylor, Chem. Eng. Progr.,* **48**:39 (1952).]

vibrations on heat transfer from a cylinder. West and Taylor [30] determined heat-transfer coefficients for turbulent flow of water inside tubes to which water was supplied by a reciprocating pump. The pulsations from the pump brought about a considerable increase in the heat-transfer coefficient over that predicted by Eq. (14-9). The results of West and Taylor are shown in Fig. 14-3, where the ratio of observed to predicted [from Eq. (14-9)] heat-transfer coefficients is plotted versus the pulsation ratio. The pulses from the reciprocating pump were partially damped by an air chamber on the discharge side of the pump. The pulsation ratio is defined as the ratio of the maximum volume to the minimum volume of air in the air chamber over the cycle of one pulsation. Figure 14-3 clearly shows the increase in heat-transfer coefficients due to pulsation, the maximum increase occurring at a pulsation ratio of about 1.4.

14-7. Turbulent Heat Transfer in Rough Tubes

The condition of the surface in contact with the fluid affects the heat-transfer coefficient for turbulent flow. The heat-transfer coefficient for a rough surface is higher than for a smooth surface because the roughness projections on the surface disturb the laminar layer. If the roughness elements are of a sufficient height to project beyond the laminar layer and into the turbulent core, there will be a turbulent wake behind each element. This wake will disturb the laminar film, and the turbulent eddies from the wake will penetrate it. The result is a reduction of the resistance of the laminar film to heat transfer and likewise an increase in the heat-transfer coefficient. No general correlation showing the effect of wall roughness on the heat-transfer coefficient exists. There have been investigations in which one type of wall roughness was studied, but the results cannot be used to predict heat-transfer coefficients for other types of wall roughness.

Cope [7] made a study of the relation between heat transfer and friction loss in rough circular pipes. The pipes investigated were roughened by cutting both left-hand and right-hand threads in them, so that the surface was actually covered by small pyramids. Heat-transfer coefficients and friction factors in these roughened tubes are shown in Fig. 14-4, where the friction factor and the dimensionless term $\mathrm{Nu}\sqrt{\mathrm{Pr}}$ are both plotted versus the Reynolds number. The data indicate that the rate of heat transfer is proportional to the roughness of the tubes. Cope reported, however, that smooth pipes are more efficient than rough pipes when compared on the basis of the amount of heat transferred per unit of power used to pump the fluid through the pipe. The turbulent wake which formed behind each pyramid on the surface contributed less to the increase in heat transfer than to the increase in energy dissipation.

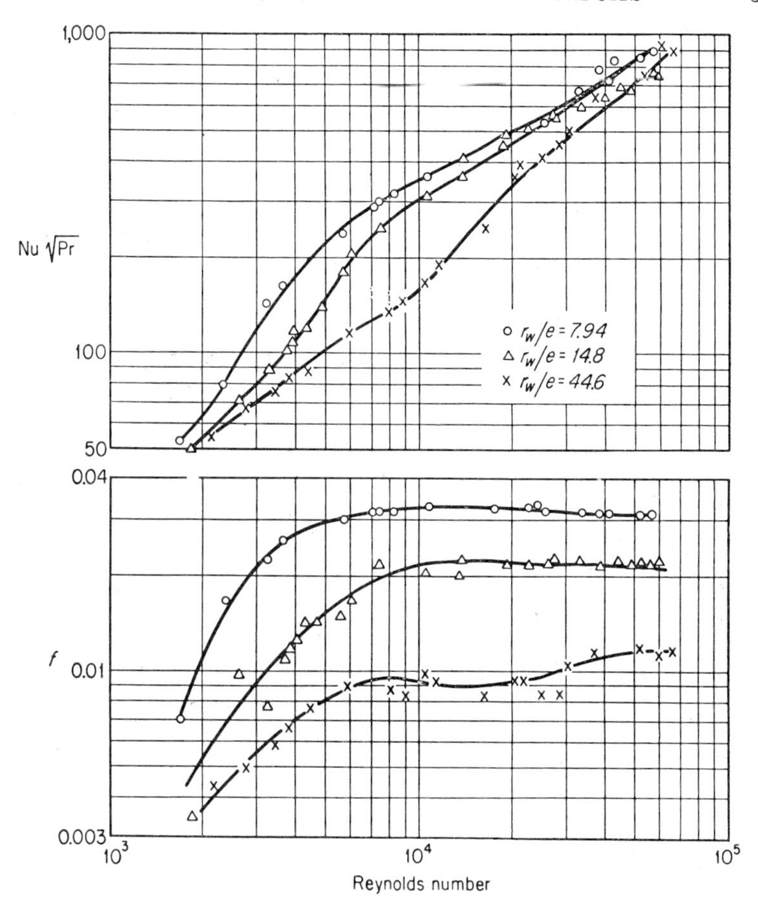

FIG. 14-4. Heat-transfer coefficients and friction losses in rough pipes. [*From W. F. Cope, Proc. Inst. Mech. Engrs. (London)*, **145**:99 (1941).]

A more general correlation dealing with wall roughness is given by Sams,[25] who investigated heat transfer from tubes with square threads on their inner walls. The experimental data could be represented within ±15 per cent by the relation

$$\left(\frac{h_m d_w}{k}\right)_{T0.5} = 0.040 \left(\frac{d_w u_f^*}{\nu}\right)_{T0.5} \left(\frac{C_p \mu}{k}\right)_{T0.5} \qquad (14\text{-}14)$$

Conditions:

(1) d_w = inside diameter of pipe
(2) Properties evaluated at $T_{0.5}$ [Eq. (14-6)]

(3) $u_f^* = U\sqrt{f_f/2}$

where $f_f/2 = 0.0036\left(\frac{s}{W}\right)^{0.8}\left(\frac{e}{W}\right)^{1.70}$

s = thread spacing
W = thread thickness
e = thread height

(4) $500 < d_w u_f^*/\nu < 2 \times 10^4$

14-8. Turbulent Heat Transfer at the Entrance to a Circular Tube

The empirical equations previously given in this chapter apply [with the exception of Eq. (14-12)] where both the velocity and temperature

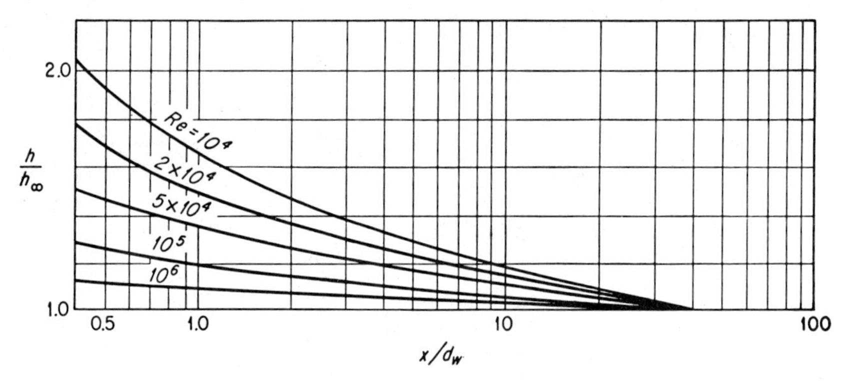

Fig. 14-5. Local heat-transfer coefficients in the entrance section of a tube for turbulent flow. (*From I. T. Aladyev, NACA TM* 1356, 1954.)

profiles are fully developed. They can generally be used to predict the mean heat-transfer coefficient in a pipe in which L/d_w is greater than 60. They cannot be used to predict coefficients in the entrance section of tubes.

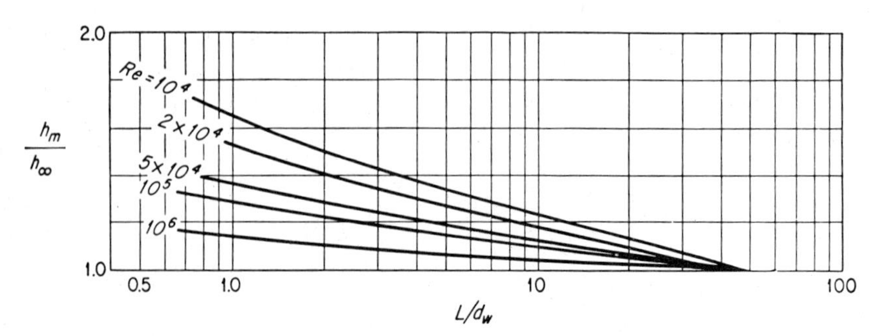

Fig. 14-6. Mean heat-transfer coefficients in the entrance section of a tube for turbulent flow. (*From I. T. Aladyev, NACA TM* 1356, 1954.)

There has been a considerable amount of investigation of mean and local turbulent heat-transfer coefficients in entrance regions of tubes; however, few general correlations showing the effect of entrance conditions on the heat-transfer coefficient exist. Stanton [28] showed that mean heat-transfer coefficients in tubes were the same between values of L/d_w from 30 to 60. Nusselt [23] studied the effect of entrance length and recommended introducing the factor $(L/d_w)^{-0.054}$, so that the resultant equation for the heat-transfer coefficient becomes

$$\left(\frac{h_m d_w}{k}\right)_b = 0.036 \left(\frac{d_w G}{\mu}\right)_b^{0.8} \left(\frac{C_p \mu}{k}\right)_b^{1/3} \left(\frac{L}{d_w}\right)^{-0.054} \qquad 10 < L/d_w < 400 \qquad (14\text{-}15)$$

Equation (14-15) has been used extensively for predicting heat-transfer coefficients in entrance sections.

Aladyev [1] measured local and mean heat-transfer coefficients for water flowing in the entrance region of a circular tube in which both the velocity and temperature profiles were developing. The tube wall was essentially at a constant temperature. Aladyev's results are shown in Fig. 14-5, where the ratio h/h_∞ is plotted versus x/d_w at constant values of the Reynolds number, and in Fig. 14-6, where the ratio of h_m/h_∞ is plotted versus L/d_w.

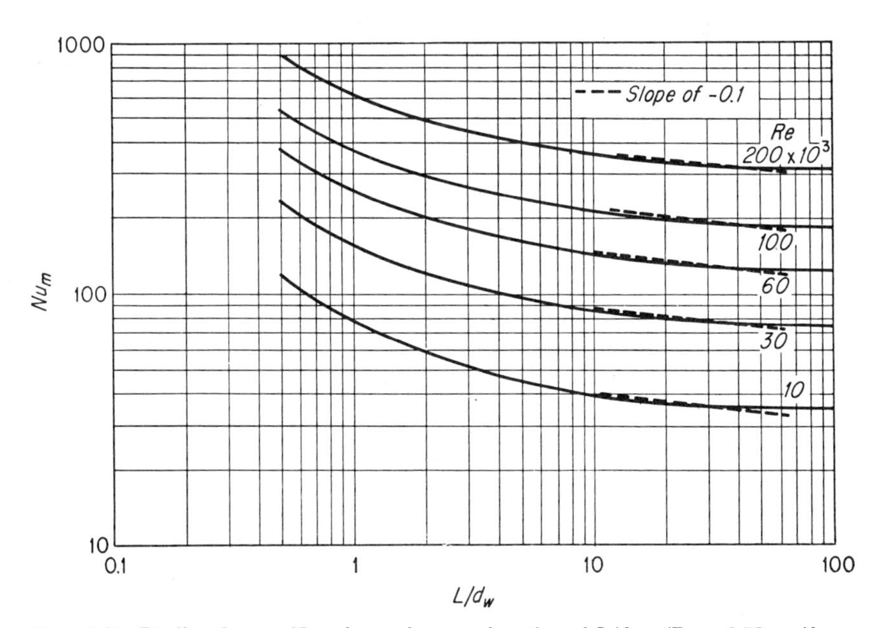

FIG. 14-7. Predicted mean Nusselt number as a function of L/d_w. (Pr = 0.73; uniform heat flux; uniform initial velocity and temperature profiles.) (*From R. G. Deissler, NACA TN* 3016, 1953.)

Aladyev's work covers a Reynolds-number range from 10^4 to 10^5, the curves for Re $= 10^6$ being extrapolated. The curves in Figs. 14-5 and 14-6 indicate that the local coefficient h is constant 40 diameters from the entrance and the mean coefficient h_m is constant 50 diameters from the entrance. Deissler [8] has presented analytical results for predicting heat-transfer coefficients in circular tubes. Figure 14-7 shows the predicted curve Deissler obtained, indicating the variation of the mean Nusselt number Nu_m with L/d_w at various Reynolds numbers. Conditions for Fig. 14-7 are constant

TABLE 14-1. VALUES OF F_1 TO USE IN EQ. (14-16) †

$$\left(\frac{L}{d_w} > 5\right)$$

Entrance	Description	F_1
	Bellmouth	0.7
	Bellmouth with screen	1.2
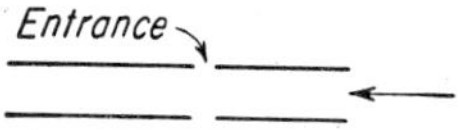 Entrance	Short calming section with sharp-edged entrance	3 (approx.)
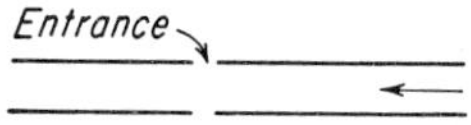 Entrance	Long calming section with sharp-edged entrance	1.4
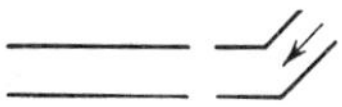	45°-angle-bend entrance	5 (approx.)
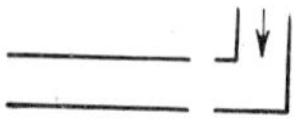	90°-angle-bend entrance	7 (approx.)
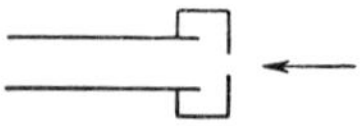	1-in. square-edged orifice entrance	16 (approx.)
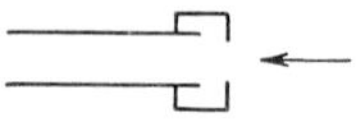	1.5-in. square-edged orifice entrance	7 (approx.)

† From L. M. K. Boelter, G. Young, and H. W. Iverson, *NACA TN* 1451, 1948.

heat input, velocity and temperature profile developing, and Prandtl number = 0.73; i.e., the curves apply for air. The broken lines in Fig. 14-7 are drawn at a slope of −0.1, indicating a basis for the empirical equation obtained by Humble et al.[14] [Eq. (14-12)]. Deissler's results show that the mean Nusselt number becomes essentially constant at values of $L/d_w > 30$.

Boelter, Young, and Iverson [2] studied the effect of entrance configuration on heat-transfer coefficients in circular tubes. They investigated the heat transfer to air in the entrance section of a tube in which the wall temperature was essentially constant and recommend the following equation:

$$\frac{h_m}{h_\infty} = 1 + F_1 \frac{d_w}{L} \qquad \frac{L}{d_w} > 5 \tag{14-16}$$

where experimental values of F_1 are given in Table 14-1 for the various types of entrance section illustrated. The results for the bellmouth entrance are substantially in agreement with the results obtained by Deissler.

II. TURBULENT-FLOW HEAT TRANSFER IN ANNULI

14-9. Turbulent Heat Transfer in Smooth Annuli

The heat-transfer coefficient for a fluid flowing in a smooth annulus with heat flowing through the inner tube wall is expressed by equations similar to those used for circular tubes. An additional term is included, however, to account for the geometry of the system, and it is usually some function of the diameter of the tubes making up the annulus. Wiegand [31] studied a large number of annular-heat-transfer data and recommended the following relationship for predicting heat-transfer coefficients for flow in annuli:

$$\left(\frac{h_m d_e}{k}\right)_b = 0.023 \left(\frac{d_e G}{\mu}\right)_b^{0.8} \left(\frac{C_p \mu}{k}\right)_b^{0.4} \left(\frac{d_2}{d_1}\right)^{0.45} \tag{14-17}$$

Conditions:

(1) Properties evaluated at bulk temperature
(2) $d_e = d_2 - d_1$
(3) $\mathrm{Re} > 10^4$

Monrad and Pelton [22] recommend the following equation, which gives results very close to Eq. (14-17) and is subject to the same conditions:

$$\left(\frac{h_m d_e}{k}\right)_b = 0.020 \left(\frac{d_e G}{\mu}\right)_b^{0.8} \left(\frac{C_p \mu}{k}\right)_b^{1/3} \left(\frac{d_2}{d_1}\right)^{0.53} \tag{14-18}$$

Subsequent experimental results of Miller, Byrnes, and Benforado [21] are substantially in agreement with Eq. (14-17). These workers found that the velocity profile was established in 20 equivalent diameters and the Nusselt number was constant 4 equivalent diameters from the annulus entrance.

14-10. Turbulent-flow Heat Transfer in Annuli Containing Finned Tubes

1. *Transverse-finned Tubes.* Knudsen and Katz [17] determined heat-transfer coefficients for water flowing in annuli containing transverse-helical-finned tubes. They obtained a relation similar to Eq. (14-9), but additional terms defining the dimensions of the finned tubes are included. The following equation applies for fin-height–fin-spacing ratios e/s from 1 to 2 and was determined for the finned tubes described in Table 7-2.

$$\left(\frac{h_m d_e}{k}\right)_b = 0.039 \left(\frac{d_e G_{\max}}{\mu}\right)_b^{0.87} \left(\frac{C_p \mu}{k}\right)_b^{0.4} \left(\frac{s}{d_e}\right)^{0.4} \left(\frac{e}{d_e}\right)^{-0.19} \qquad 1 < \frac{e}{s} < 2 \qquad (14\text{-}19)$$

The equivalent diameter used in Eq. (14-19) is $d_2 - d_f$. A finned tube with a value of $e/s = 3$ gave heat-transfer coefficients somewhat higher than those predicted by Eq. (14-19), probably as a result of extensive turbulence which was observed visually in the fin space of this particular tube.

The finned tubes studied by Knudsen and Katz were compared with each other on the basis of heat transferred per unit of power required to pump the fluid through the annulus. The results indicated that there is an optimum fin height and fin spacing. Some finned tubes investigated were more economical than others from the standpoint of heat transferred per unit of power.

2. *Spined Tubes.* Hobson and Weber [13] investigated the rate of heat transfer between air and three different spined tubes in annuli. In all, eight different annuli were studied, and these workers obtained the relation

$$\left(\frac{h_m d_e}{k}\right)_b = 0.00168 \left(\frac{d_e G_{\max}}{\mu}\right)_b^{1.09} \left(\frac{C_p \mu}{k}\right)_b^{1/3} \qquad 4{,}000 < \mathrm{Re} < 40{,}000 \qquad (14\text{-}20)$$

$$d_e = d_2 - d_f$$

3. *Longitudinal-finned Tubes.* Gunter and Shaw [12] and de Lorenzo and Anderson [18] conducted studies of heat transfer in annuli containing longitudinal-finned tubes and presented their results in graphical form. The outstanding feature of their work is the comparison of continuous and noncontinuous longitudinal fins. The noncontinuous fins have higher heat-transfer coefficients since a thick boundary layer cannot form on them, and they promote turbulence of the annular stream.

Clark and Winston[4] give the curve shown in Fig. 14-8 to predict heat-transfer coefficients in annuli containing longitudinal-finned tubes. In this figure the term $\mathrm{St}(\mathrm{Pr})^{2/3}(\mu_w/\mu_b)^{0.14}$ is plotted versus $\mathrm{Re}\sqrt{\pi L/p}$, where the Reynolds number is based on the equivalent diameter d_e, defined in Table 11-1, L is the heated length of the finned tube, and p is the wetted

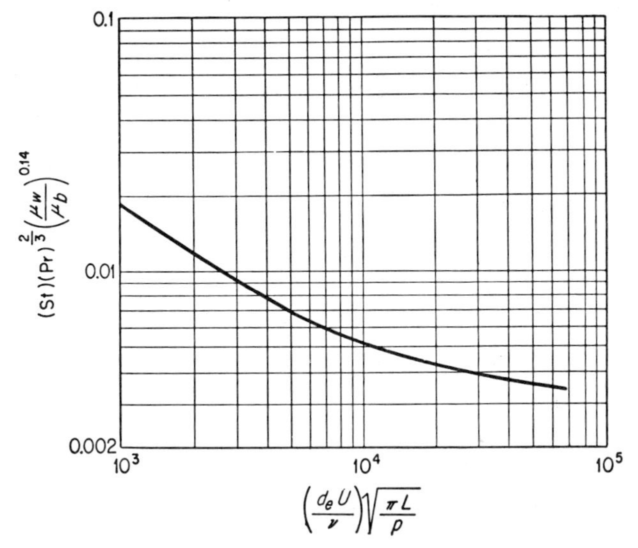

FIG. 14-8. Rates of heat transfer in annuli containing longitudinal-finned tubes. [*From L. Clark and R. E. Winston, Chem. Eng. Progr.,* **51**:147 (1955).]

perimeter of the channel between two longitudinal fins. Beyond a value of $\mathrm{Re}\sqrt{\pi L/p}$ of 60,000, Eq. (14-11) is recommended with d_w replaced by d_e.

III. TURBULENT HEAT TRANSFER IN NONCIRCULAR DUCTS

14-11. Turbulent Flow in Smooth Noncircular Ducts

The work of Drexel and McAdams[10] and Boelter, Sanders, and Romie[3] indicates that for moderate temperature differences Eq. (14-9) or (14-10) with d_w replaced by d_e may be used to predict turbulent heat-transfer coefficients in noncircular ducts. Boelter et al. found that local heat-transfer coefficients became constant about 30 equivalent diameters from the entrance.

For heat transfer at high temperature differences Lowdermilk, Weiland, and Livingood[19] obtained the following relation for air flowing in square, rectangular, and triangular ducts:

$$\left(\frac{h_m d_e}{k}\right)_{T0.5} = 0.023\left(\frac{d_e G}{\mu}\right)^{0.8}_{T0.5}\left(\frac{C_p\mu}{k}\right)^{0.4}_{T0.5} \qquad (14\text{-}21)$$

Conditions:

(1) Properties evaluated at film temperature [Eq. (14-6)]
(2) $L/d_e > 57$
(3) $\text{Re} > 10{,}000$

For values of L/d_e less than 57 these workers recommend Eq. (14-12) with d_w replaced by d_e.

BIBLIOGRAPHY

1. Aladyev, I. T.: *NACA TM* 1356, 1954.
2. Boelter, L. M. K., G. Young, and H. W. Iverson: *NACA TN* 1451, 1948.
3. Boelter, L. M. K., V. D. Sanders, and F. E. Romie: *NACA TN* 2524, 1951.
4. Clark, L., and R. E. Winston: *Chem. Eng. Progr.*, **51**:147 (1955).
5. Colburn, A. P.: *Trans. AIChE*, **29**:174 (1933).
6. Colburn, A. P., and W. J. King: *Trans. AIChE*, **26**:166 (1931).
7. Cope, W. F.: *Proc. Inst. Mech. Engrs.* (*London*), **145**:99 (1941).
8. Deissler, R. G.: *NACA TN* 3016, 1953.
9. Dittus, F. W., and L. M. K. Boelter: *Univ. Calif.* (*Berkeley*) *Publs. Eng.*, **2**:443 (1930).
10. Drexel, R. E., and W. H. McAdams: *NACA WR* W-108 (formerly *ARR* 4F28, February, 1945).
11. Grele, M. D., and L. Gedeon: *NACA RM* E52L09, 1953.
12. Gunter, A. Y., and W. A. Shaw: *Trans. ASME*, **64**:795 (1942).
13. Hobson, M., and J. H. Weber: *Ind. Eng. Chem.*, **46**:2290 (1954).
14. Humble, L. V., W. H. Lowdermilk, and L. G. Desmon: *NACA Rept.* 1020, 1951.
15. Kaufman, S. J., and F. D. Isley: *NACA RM* E50G31, 1950.
16. Kaufman, S. J., and R. W. Henderson: *NACA RM* E51I18, 1951.
17. Knudsen, J. G., and D. L. Katz: *Chem. Eng. Progr.*, **46**:490 (1950).
18. Lorenzo, B. de, and E. D. Anderson: *Trans. ASME*, **67**:697 (1945).
19. Lowdermilk, W. H., W. F. Weiland, and J. N. B. Livingood: *NACA RM* E53J07,
20. Martinelli, R. C., and L. M. K. Boelter: *Proc. Intern. Congr. Appl. Mech., 5th Congr., Cambridge, Mass., 1938*, p. 578.
21. Miller, P., J. J. Byrnes, and D. M. Benforado: *J. AIChE*, **1**:501 (1955).
22. Monrad, C. C., and J. F. Pelton: *Trans. AIChE*, **38**:593 (1942).
23. Nusselt, W.: *Z. Ver. deut. Ing.*, **61**:685 (1917).
24. Royds, R.: "Heat Transmission by Radiation, Conduction, and Convection," Constable & Co., Ltd., London, 1921.
25. Sams, E. W.: *NACA RM* E52D17, 1952.
26. Seigel, L. G.: *Heating, Piping, Air Conditioning*, **18**:111, June (1946).
27. Sieder, E. N., and G. E. Tate: *Ind. Eng. Chem.*, **28**:1429 (1936).
28. Stanton, T. E.: *Trans. Roy. Soc.* (*London*), **190A**:67 (1897).
29. Weiland, W. F., and W. H. Lowdermilk: *NACA RM* E53E04, 1953.
30. West, F. B., and A. T. Taylor: *Chem. Eng. Progr.*, **48**:39 (1952).
31. Wiegand, J. H.: Discussion of Paper by McMillen and Larson, *Trans. AIChE*, **41**:147 (1945).

CHAPTER 15

THE ANALOGY BETWEEN MOMENTUM AND HEAT TRANSFER

15-1. The Relation between Forced-convection Heat Transfer and Friction

When heat is transferred between a solid and a turbulent fluid, it has been observed that the heat-transfer rate may be increased by supplying more energy to the pump handling the fluid. From the results indicated in Chaps. 7 and 14 for turbulent flow in smooth circular tubes (considering only velocity as variable)

$$-\Delta P_f \propto U^{1.8} \qquad \text{(approx.)}$$

and

$$h \propto U^{0.8} \qquad \text{(approx.)}$$

Therefore $-\Delta P_f$ and h increase with increased velocity.

The role of turbulence promoters in increasing heat-transfer coefficients has been discussed in Sec. 14-6. The presence of baffles or spirals in tubes increases the turbulence of the turbulent core and thereby reduces the effective thermal resistance of the laminar layer. The turbulence promoters, however, increase the friction loss for flow through the tube, so the increase in heat transfer has been brought about only by supplying more energy to the pump. The increase in friction loss is manifested by an increase in both the wall shear stress and the velocity gradient at the wall. The increase in rate of heat transfer shows up as an increase in the temperature gradient at the wall. Thus the increase in wall shear stress (skin friction) is accompanied by an increase in rate of heat transfer. The qualitative effect of turbulence promoters is shown in Fig. 15-1.

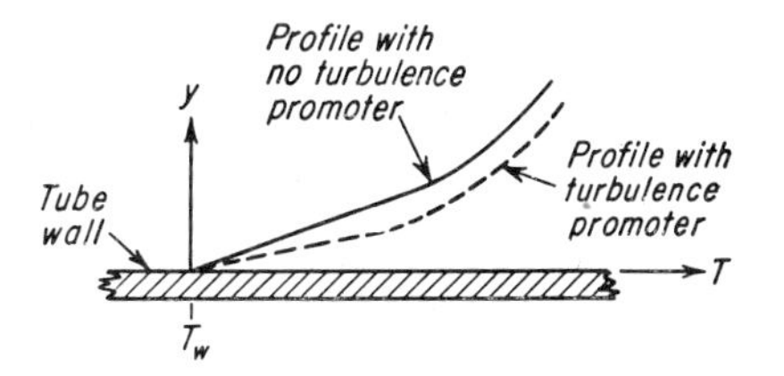

FIG. 15-1. Effect of turbulence promoters on the temperature profile adjacent to the wall of a tube.

Most of the experimental studies of the effect of turbulence promoters on the rate of heat transfer have shown that the actual heat transference per unit of pumping power expended is less for tubes with turbulence promoters than for tubes without them. Similar results have been obtained for heat transfer in rough pipe (see Sec. 14-7). From an economic standpoint, turbulence promoters or rough pipes are not recommended. Heat-transfer equipment is a capital investment to be paid for over a period of

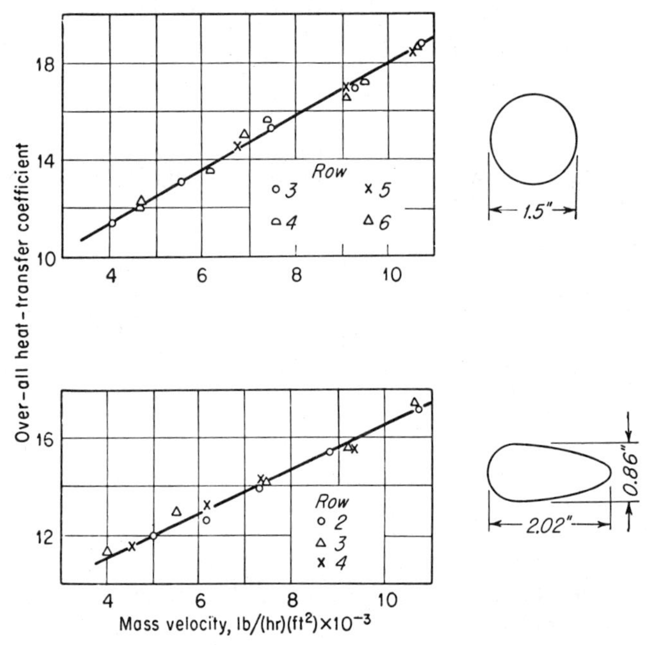

FIG. 15-2. Effect of tube shape on heat transfer in tube banks. [*From C. C. Winding, Ind. Eng. Chem.*, **30**:942 (1938).]

time, and increased pumping costs include both extra equipment and increased operating expenses. Thus, the increase in energy required when the turbulence promoter is used soon costs more than the saving in heat-transfer equipment.

Winding [28] investigated the rates of heat transfer from banks of staggered tubes and determined the influence of tube shape on the heat-transfer coefficient. The curves in Fig. 15-2 depict Winding's data plotted as the over-all heat-transfer coefficient versus the mass velocity of the fluid through the tube bank. The over-all coefficients for the third, fourth, fifth, and sixth rows of circular tubes are shown on the upper plot of Fig. 15-2, while the lower plot shows the coefficients for the streamlined tubes. At low mass velocities the heat-transfer coefficient is nearly the same for streamlined as for circular tubes, while at high mass velocities the coefficient is about 5 per cent higher for circular tubes. The pressure drop across circular tubes would be much greater than across streamlined tubes; so the latter would be the more economical for carrying out heat transfer. The saving in power is achieved not by increasing turbulence and subsequently increasing the heat transfer but by decreasing only that turbulence which was useless for heat transfer. The turbulent wake does not help heat transfer for the row of tubes behind which it is formed.

Knudsen and Katz [11] studied the heat transfer from helical-fin tubes in an annulus. The turbulence between the fins of such fin tubes is illustrated in Fig. 7-33, which is a flow pattern obtained by these workers. Various fin tubes were compared with each other on the basis of heat transferred per unit of power needed to force the fluid through the annulus. The results indicated that there is an optimum fin height and fin spacing from the standpoint of heat transfer. Some fin tubes were more efficient and others were less efficient than smooth tubes. The investigation showed, however, that a certain increase in turbulence due to roughness was beneficial. For some fin tubes turbulent eddies appeared to move the fluid particles away from the heat-transfer surface, while for other tubes this was not true.

These experimental results indicate that an increase in turbulence will bring about an increase in heat transfer, but often an increase in energy loss also occurs, with the result that less heat is transferred per unit of power consumed. If power is not a serious economic factor in a heat-transfer process, then greater heat transfer may be accomplished by the use of turbulence promoters or rough surfaces. However, if power is an economic factor, any increase in turbulence should be more effective in increasing the rate of heat transfer than it is in increasing the energy loss.

15-2. Historical Background of the Analogy between Momentum and Heat Transfer

As pointed out above, a relation exists between heat transfer and skin friction. The determination of this relationship, both theoretically and experimentally, has been the concern of many investigators in the field of fluid flow and heat transfer. A thorough knowledge of the relationship would allow prediction of rates of heat transfer from friction-loss data.

Reynolds,[21] in 1874, was one of the first scientists to recognize the existence of a relationship between heat transfer and skin friction. He did considerable theoretical work on the subject and obtained a simple relationship between the two processes. In 1910 Prandtl [18] did further theoretical work on the relationship and refined Reynolds's formula. Much of the work for the next twenty years was of an experimental nature. Numerous data on heat transfer and friction losses in systems were obtained and used in an attempt to verify the formulas developed by Reynolds and Prandtl. In 1933 Colburn [2] presented j-factor correlations, which represented the empirical relationship between fluid flow and heat transfer. Since that time, most heat-transfer and fluid-flow data have been correlated empirically using the j factor. In 1939 von Kármán [10] analyzed the turbulent heat-transfer and momentum-transfer processes occurring in tubes and proposed a relationship that considerably modified Prandtl's. After 1945 the theoretical studies of turbulent heat and momentum transfer increased rapidly. Some of the notable investigations are those of Martinelli,[15] Seban and Shimazaki,[24] Deissler,[4] and Lyon.[14]

15-3. Comparison of Turbulent Transfer Processes in Circular Tubes

When turbulent flow occurs in circular tubes, momentum is transferred between layers of fluid. This momentum transfer manifests itself as frictional resistance and at the wall of the tube is expressed as τ_w, the shear stress at the wall, which is equivalent to the time rate of momentum transfer per unit area at the wall. Heat and mass transfer may occur during

TABLE 15-1. COMPARISON OF TURBULENT TRANSFER PROCESSES IN CIRCULAR TUBES

	Momentum transfer	Heat transfer	Mass transfer
Quantity transferred	Momentum τ_w	Heat $\dfrac{q_w}{A_w}$	Mass $\dfrac{N_w}{A_w}$
Unit for molecular transfer	Viscosity μ	Thermal conductivity k	Diffusivity D
Coefficients	Fanning friction factor f	Heat-transfer coefficient h	Mass-transfer coefficient k_m
Driving force	Velocity difference U	Temperature difference ΔT	Concentration difference Δc_m
Rate of transfer as a function of gradients at the wall	$\tau_w = \dfrac{\mu}{g_c}\left(\dfrac{\partial u}{\partial y}\right)_{y=0}$	$\dfrac{q_w}{A_w} = -k\left(\dfrac{\partial T}{\partial y}\right)_{y=0}$	$\dfrac{N_w}{A_w} = -D\left(\dfrac{\partial c_m}{\partial y}\right)_{y=0}$
Rate of transfer in terms of coefficients	$\tau_w = f\dfrac{\rho U^2}{2g_c}$	$\dfrac{q_w}{A_w} = h\,\Delta T$	$\dfrac{N_w}{A_w} = k_m\,\Delta c_m$

turbulent flow, taking place by virtue of a temperature or concentration difference, respectively, between the wall and the fluid. Table 15-1 shows a comparison between the processes of momentum, mass, and heat transfer during fluid flow. Included in the table are driving potentials, quantities transferred, and rates of transfer. For momentum transfer, the driving force is the average velocity of the fluid U, which is the average relative velocity of the fluid with respect to the tube wall.

15-4. Definition of the Analogy between Momentum and Heat Transfer

The analogy between momentum and heat transfer is best defined by a statement of Reynolds's theory.[21] According to this theory, the movement of heat between a surface and a fluid follows the same laws as the movement of momentum between the surface and a fluid, whether by conduction or

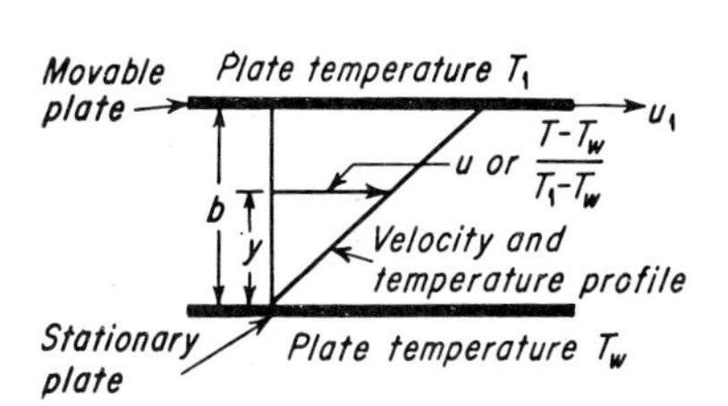

FIG. 15-3. The analogy between momentum transfer and heat transfer for laminar flow.

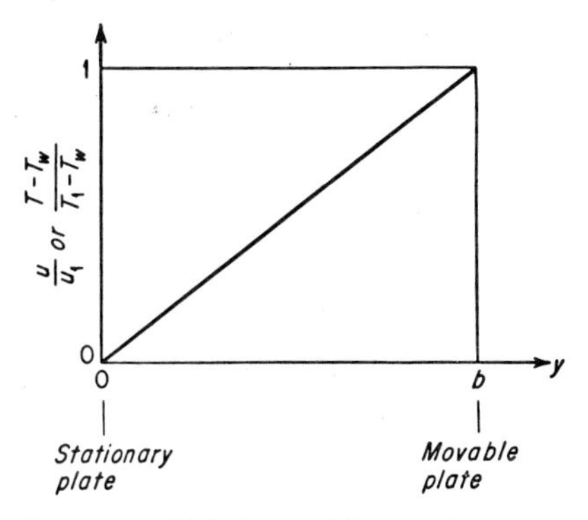

FIG. 15-4. Velocity and temperature profile for laminar motion between a stationary and movable plate (see Fig. 15-3).

convection (radiation is neglected). The similarity of the equations describing the three transfer processes is obvious in Table 15-1, and, at least for molecular transfer, the laws governing the movements are the same. If Reynolds's theory is correct, and if, in fact, heat and momentum transfer follow the same laws, it will be possible to predict rates of heat transfer from rates of momentum transfer and to predict temperature profiles from velocity profiles.

Reynolds's theory states that the analogy between heat and momentum transfer applies both for laminar and turbulent flow. This is most easily demonstrated by considering the laminar motion of a fluid between two parallel plates of which one is stationary and the other moves at a velocity u_1. The stationary plate is maintained at a temperature T_w and the moving plate at a temperature T_1 (see Fig. 15-3). The velocity profile and temperature profile between the two plates are identical when compared on a relative basis. This is illustrated by Fig. 15-4, showing u/u_1 and $(T - T_w)/(T_1 - T_w)$ plotted versus y, where y is the distance from the stationary plate, u is the velocity at y, and T is the temperature at y. The ratios u/u_1 and $(T - T_w)/(T_1 - T_w)$ are zero when $y = 0$; they vary linearly with y and have a value of unity at the moving plate. For this case, the laws governing momentum and heat transfer are the same. From the plot given in Fig. 15-4 both rates of heat transfer and momentum transfer may be calculated if the viscosity and thermal conductivity of the fluid are known.

Rayleigh [20] pointed out that Reynolds's theory was true for laminar flow, as demonstrated above, but indicated that there is question regarding its

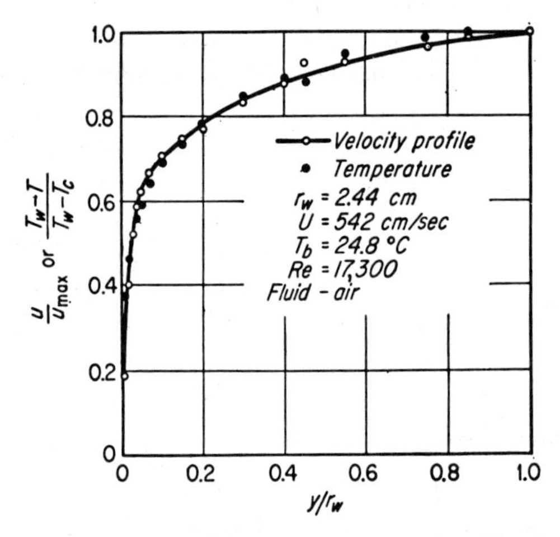

Fig. 15-5. Velocity and temperature profiles for turbulent flow in a circular tube. (*Data from J. R. Pannell, Brit. Aeronaut. Research Comm. R. & M.* 243*ii*, 1916.)

application to turbulent flow. Stanton [27] indicated in an appendix to Rayleigh's article that the theory was applicable to turbulent flow as well as to laminar flow if time-average values of velocity and temperature are used. To substantiate this, he presents some data of Pannell's [17] for turbulent heat transfer in a smooth tube. These data are shown in Fig. 15-5, where $u/u_{\max}$ and $(T_w - T)/(T_w - T_c)$ are plotted versus y/r_w for air flowing in a tube at a Reynolds number of 17,300. The two profiles are nearly coincident, indicating that, for air, the laws governing momentum and heat transfer are the same. It is shown in following sections that, since temperature and velocity profiles are the same only for fluids having a Prandtl number of unity, this group must be considered when predicting heat-transfer rates from friction data.

The analogy between heat and momentum transfer may be shown by considering the mathematical equations describing the two processes. In

Sec. 5-6 it was shown that the total shear stress at a point in a turbulent fluid is composed of a laminar and a turbulent portion, where

$$\tau_l = \frac{\mu}{g_c}\frac{du}{dy} \tag{5-21}$$

$$\tau_t = \frac{\rho\overline{u'v'}}{g_c} \tag{5-19}$$

where the term $\rho\overline{u'v'}/g_c$ is the mean turbulent shear stress. Prandtl [19] expressed this turbulent shear stress in terms of the mixing length in such a way that the total shear could be expressed as

$$\tau g_c = \mu\frac{du}{dy} + \rho l^2 \frac{du}{dy}\frac{du}{dy} \tag{5-23}$$

which becomes

$$\tau g_c = (\mu + \mathrm{E}_M)\frac{du}{dy} \tag{5-24}$$

where E_M is the eddy viscosity. Dividing Eq. (5-24) by the density ρ,

$$\frac{\tau g_c}{\rho} = (\nu + \epsilon_M)\frac{du}{dy} \tag{15-1}$$

where ϵ_M is the eddy diffusivity of momentum and is related to the Prandtl mixing length by

$$\epsilon_M = l^2\frac{du}{dy} \tag{15-2}$$

A similar analysis may be applied to the transfer of heat across a turbulent stream. The sketch in Fig. 15-6 shows two sections of a fluid in turbulent motion, the distance between these two sections being l, the Prandtl mixing length. At section 1 the mean velocity is u, and the mean temperature is T. At section 2 the velocity and temperature are $u + l(du/dy)$ and $T + l(dT/dy)$. The rate of turbulent heat transfer per unit area is the product of the mass flow in the y direction and the temperature difference between the two sections.

$$\left(\frac{q}{A_q}\right)_t = -C_p\rho v' l\frac{dT}{dy} \tag{15-3}$$

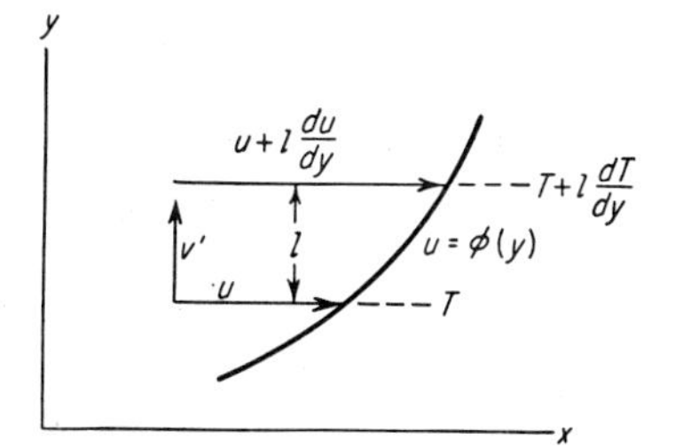

FIG. 15-6. Two sections in a flowing fluid at a different velocity and temperature.

Prandtl assumes v' to be of the order of $l(du/dy)$ (see Sec. 5-6). Thus Eq. (15-3) becomes

$$\left(\frac{q}{A_q}\right)_t = -C_p\rho l^2 \frac{du}{dy}\frac{dT}{dy} \tag{15-4}$$

Combining the turbulent heat transfer with that due to conduction, the total rate of heat transfer becomes

$$\frac{q}{A_q} = -\left(k + C_p\rho l^2 \frac{du}{dy}\right)\frac{dT}{dy} \tag{15-5}$$

and dividing by $C_p\rho$,

$$\frac{q}{C_p\rho A_q} = -\left(\frac{k}{C_p\rho} + l^2 \frac{du}{dy}\right)\frac{dT}{dy} \tag{15-6}$$

which becomes

$$\frac{q}{C_p\rho A_q} = -\left(\alpha + \epsilon_H\right)\frac{dT}{dy} \tag{15-7}$$

where ϵ_H is the eddy diffusivity of heat and is related to the Prandtl mixing length by

$$\epsilon_H = l^2 \frac{du}{dy} \tag{15-8}$$

Equations (15-2) and (15-8) indicate that ϵ_H and ϵ_M are equal. This results from the various assumptions Prandtl made in the analysis. He assumed that the mixing length for heat transfer is the same as the mixing length for momentum transfer; i.e., a particle of fluid travels the same distance before losing its identity and before attaining a different temperature. Under these conditions, a true analogy exists between momentum and heat transfer for turbulent flow.

The use of the analogy between momentum transfer and heat transfer for predicting heat transfer from momentum transfer depends on the relationship between ϵ_H and ϵ_M. Most investigators assume them to be equal, as Prandtl did; however, recent experimental work has shown they are not equal. Therefore, a general relationship between ϵ_H and ϵ_M is needed so that the analogy between heat transfer and momentum transfer will be most useful. Experimental values of ϵ_H and ϵ_M are presented in a subsequent section.

For mass transfer a relationship similar to that for heat transfer may be derived using the molecular diffusivity of mass D, the eddy diffusivity of mass ϵ_m, and the concentration gradient dc_m/dy; i.e.,

$$\frac{N_m}{A_{N_m}} = -\left(D + \epsilon_m\right)\frac{dc_m}{dy} \tag{15-9}$$

15-5. The Momentum and Energy Equations for Turbulent Flow in Circular Tubes

The Reynolds momentum equations for three-dimensional flow of a turbulent fluid were derived in Sec. 5-4, using a rectangular coordinate system. For steady flow in a circular tube in which the mean radial velocity is zero the Reynolds momentum equations (when cylindrical coordinates are used) reduce to

$$\frac{g_c}{\rho}\frac{\partial P}{\partial x} = \frac{\nu}{r}\frac{\partial}{\partial r}\left(r\frac{\partial u}{\partial r}\right) - \frac{1}{r}\frac{\partial}{\partial r}\overline{(ru'v')} \tag{15-10}$$

where u' is the turbulent velocity fluctuation in the x direction and v' is the fluctuation in the radial direction. The last term of Eq. (15-10) represents the turbulent shear in the fluid.

By methods similar to those used in Sec. 5-4 the three-dimensional energy equation may be modified to include both velocity and temperature fluctuations (see Prob. 1 of Chap. 5). For steady turbulent flow in a tube the energy equation becomes

$$u\frac{\partial T}{\partial x} = \frac{\alpha}{r}\frac{\partial}{\partial r}\left(r\frac{\partial T}{\partial r}\right) - \frac{1}{r}\frac{\partial}{\partial r}\overline{(rv'T')} \tag{15-11}$$

Utilizing the concept of the Prandtl mixing length and the eddy diffusivities, Eqs. (15-10) and (15-11) may be written as

$$\frac{g_c}{\rho}\frac{\partial P}{\partial x} = \frac{1}{r}\frac{\partial}{\partial r}\left[r(\nu + \epsilon_M)\frac{\partial u}{\partial r}\right] \tag{15-12}$$

$$u\frac{\partial T}{\partial x} = \frac{1}{r}\frac{\partial}{\partial r}\left[r(\alpha + \epsilon_H)\frac{\partial T}{\partial r}\right] \tag{15-13}$$

The simultaneous solution of Eqs. (15-12) and (15-13), with the appropriate boundary conditions, will give the velocity and temperature distribution in the pipe. The solution involves knowing values of ϵ_M and ϵ_H as a function of the radius. From Eqs. (4-13) and (4-17) and (4-21) for a horizontal tube

$$\frac{g_c}{\rho}\frac{\partial P}{\partial x} = \frac{g_c}{\rho}\frac{\partial P_f}{\partial x} = -\frac{2\tau_w g_c}{\rho r_w} \tag{15-14}$$

Thus Eq. (15-12) becomes

$$-\frac{2\tau_w g_c}{\rho r_w} = \frac{1}{r}\frac{\partial}{\partial r}\left[r(\nu + \epsilon_M)\frac{\partial u}{\partial r}\right] \tag{15-15}$$

and integrating with respect to r from $r = 0$ to $r = r$, Eq. (15-15) becomes

$$-\frac{\tau_w g_c r}{\rho r_w} = (\nu + \epsilon_M)\frac{du}{dr} \qquad (15\text{-}16)$$

But from Eq. (4-14)

$$\tau = \tau_w \frac{r}{r_w} \qquad (4\text{-}14)$$

$$\frac{\tau g_c}{\rho} = -(\nu + \epsilon_M)\frac{du}{dr} \qquad (15\text{-}17)$$

and since $y = r_w - r$ and $dy = -dr$,

$$\frac{\tau g_c}{\rho} = (\nu + \epsilon_M)\frac{du}{dy} \qquad (15\text{-}1)$$

In Eq. (15-13) the term $u(\partial T/\partial x)$ may be expressed in terms of the rate of heat transfer at radius r. Considering an annular element of fluid of

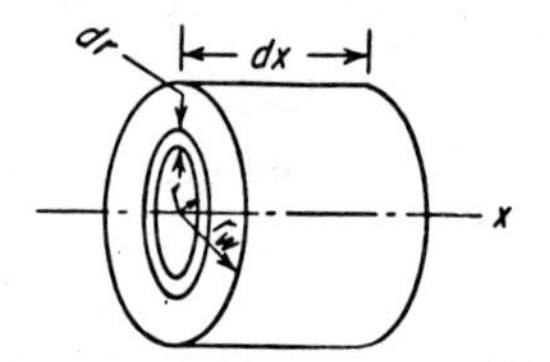

FIG. 15-7. Differential element for energy balance in Eq. (15-18).

length dx, as shown in Fig. 15-7, an energy balance may be made as follows:

$$\text{Input} = \text{output} + \text{accumulation} \qquad (2\text{-}1)$$

$$q_r + \frac{\partial q_r}{\partial r}dr = q_r + \left(2\pi r\, dr\, C_p\rho u \frac{\partial T}{\partial x}\right)dx \qquad (15\text{-}18)$$

from which

$$u\frac{\partial T}{\partial x} = \frac{1}{2\pi r C_p \rho\, dx}\frac{\partial q_r}{\partial r} \qquad (15\text{-}19)$$

where q_r is the rate of heat transfer at radius r. Thus Eq. (15-13) becomes

$$\frac{1}{2\pi r C_p\rho\, dx}\frac{\partial q_r}{\partial r} = \frac{1}{r}\frac{\partial}{\partial r}\left[r(\alpha + \epsilon_H)\frac{\partial T}{\partial r}\right] \qquad (15\text{-}20)$$

Integrating Eq. (15-20) from 0 to r ($q_r = 0$ at $r = 0$),

$$\frac{q_r}{2\pi r C_p\rho\, dx} = (\alpha + \epsilon_H)\frac{dT}{dr} \qquad (15\text{-}21)$$

Since $2\pi r\, dx = A_r$, the area for radial flow of heat

$$\frac{q_r}{C_p \rho A_r} = (\alpha + \epsilon_H) \frac{dT}{dr} \tag{15-22}$$

$$\frac{q_r}{C_p \rho A_r} = -(\alpha + \epsilon_H) \frac{dT}{dy} \tag{15-23}$$

Equation (15-23) is the same as Eq. (15-7). Equations (15-1) and (15-23) are used in the various analogies between heat transfer and momentum transfer to predict rates of heat transfer from friction losses and velocity profiles.

15-6. The Reynolds Analogy

In 1874 Reynolds [21] postulated that the laws governing momentum and heat transfer were the same. In 1883 he [22] discovered the laws of friction resistance for turbulent flow in conduits and was thereby able to express the analogy between heat transfer and momentum transfer in mathematical form.

Consider a fluid flowing in a pipe of length L, the walls of which are maintained at a constant temperature T_w (see Fig. 15-8). The fluid enters at

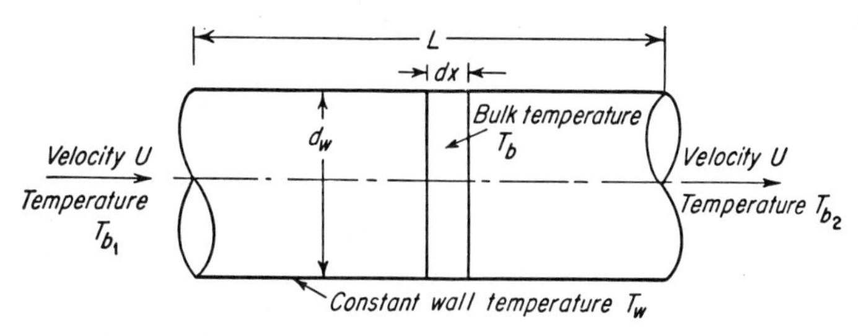

FIG. 15-8. Definition of terms for the derivation of the Reynolds analogy for turbulent flow in a circular tube.

a temperature T_{b_1} and leaves at T_{b_2}. Considering a differential element of length dx, the temperature of the fluid is T_b, and the friction loss over this differential length of pipe is expressed as

$$\frac{g_c}{\rho} \frac{-dP_f}{dx} = \frac{2fU^2}{d_w} \tag{15-24}$$

Letting the mass rate of flow through the pipe be m, Eq. (15-24) is rearranged as follows:

$$\frac{\pi d_w^{\,2}}{4} g_c \frac{-dP_f}{dx} = \frac{\pi d_w^{\,2}}{4m} \frac{2fU\rho}{d_w} mU \tag{15-25}$$

In Eq. (15-25) the left-hand term is the rate of momentum transfer per unit length of pipe. The term mU is the momentum of the fluid with respect to the tube wall. According to Reynolds's theory, $mC_p\,dT_b/dx$ may be substituted for $(\pi d_w^2/4)g_c(-dP_f)/dx$, and $mC_p(T_w - T_b)$ may be substituted for mU, where $mC_p\,dT_b/dx$ is the rate of heat transfer per unit length of pipe and $mC_p(T_w - T_b)$ is the sensible heat in the fluid using the tube-wall temperature as a basis; Eq. (15-25) becomes

$$mC_p\frac{dT_b}{dx} = \frac{\pi d_w^2}{4m}\frac{2fU\rho}{d_w}mC_p(T_w - T_b) \tag{15-26}$$

Since $m = (\pi d_w^2/4)\rho U$, Eq. (15-26) becomes

$$\frac{dT_b}{T_w - T_b} = \frac{2f}{d_w}dx \tag{15-27}$$

Integrating Eq. (15-27) between $T_b = T_{b_1}$ and $T_b = T_{b_2}$ and between $x = 0$ and $x = L$,

$$\ln\frac{T_w - T_{b_1}}{T_w - T_{b_2}} = \frac{2fL}{d_w} \tag{15-28}$$

The result expressed by Eq. (15-28) relates the temperature change in the fluid to the friction factor, the tube length, tube diameter, tube-wall temperature, and fluid inlet temperature. The outlet temperature can be calculated from the friction factor, and the heat-transfer coefficient need not be considered. Stanton [26] was one of the first investigators to attempt to verify the Reynolds analogy with experimental data. He measured rates of heat transfer to water flowing in tubes. A plot of Stanton's data is shown in Fig. 15-9, where $\ln[(T_w - T_{b_1})/(T_w - T_{b_2})]$ is plotted against $2fL/d_w$. The values of the friction factor were not reported by Stanton but are determined from Stanton's data, using Fig. 7-21. The straight line is a plot of Eq. (15-28), and the experimental data (open circles) do not agree well with this line.

An expression for the heat-transfer coefficient and Nusselt number may also be obtained from the Reynolds analogy. A heat balance on the element of length dx in Fig. 15-8 may be expressed as

$$\frac{\pi d_w^2}{4}UC_p\rho\,dT_b = h\pi d_w(T_w - T_b)\,dx \tag{15-29}$$

Combining Eqs. (15-27) and (15-29) and solving for h gives

$$h = \frac{f}{2}UC_p\rho \tag{15-30}$$

or

$$\frac{h}{UC_p\rho} = \mathrm{St} = \frac{f}{2} \tag{15-31}$$

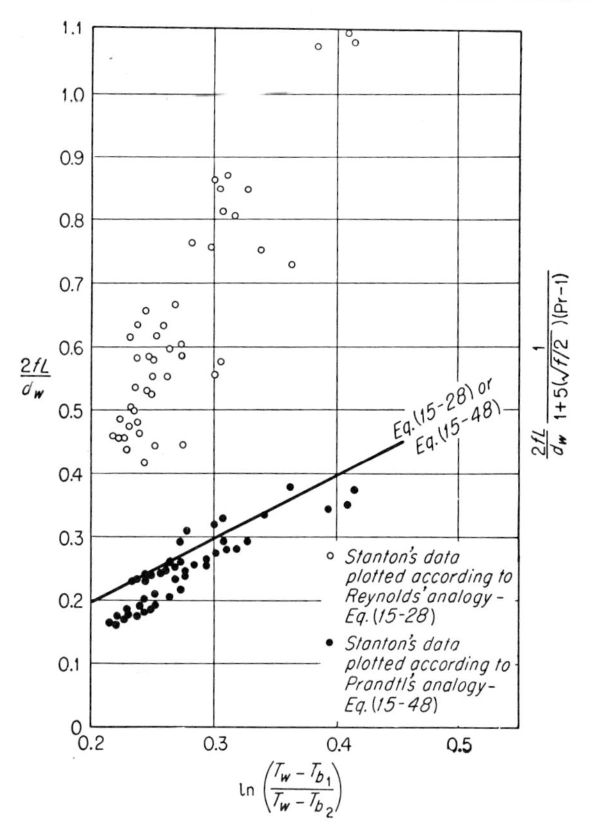

FIG. 15-9. The Reynolds and Prandtl analogies for water flowing in tubes. [*Data from T. E. Stanton, Trans. Roy. Soc. (London),* **190A**:67 (1897).]

Equation (15-30) may be expressed in another way to show the Reynolds analogy. Multiplying each side by U and replacing h by $q_w/[A_w(T_w - T_b)]$ [Eq. (12-2)],

$$\frac{q_w U}{A_w(T_w - T_b)} = \frac{f}{2} U^2 C_p \rho \tag{15-32}$$

and from Eqs. (7-12) and (7-50)

$$u^* = \sqrt{\frac{\tau_w g_c}{\rho}} = U\sqrt{\frac{f}{2}}$$

Thus Eq. (15-32) becomes

$$\frac{C_p \rho (T_w - T_b)}{q_w/A_w} = \frac{U}{\tau_w g_c/\rho} \tag{15-33}$$

which is another way of expressing the Reynolds analogy. The Reynolds analogy predicts that the Stanton number is one-half the Fanning friction factor. Since

$$\mathrm{St} = \frac{\mathrm{Nu}}{\mathrm{RePr}} \tag{15-34}$$

the Nusselt number becomes

$$\mathrm{Nu} = \frac{f}{2}\,\mathrm{RePr} \tag{15-35}$$

or, when the Prandtl number is unity,

$$\mathrm{Nu} = \frac{f}{2}\,\mathrm{Re} \tag{15-36}$$

Equation (15-36) gives the Nusselt number as predicted by the Reynolds analogy. It agrees well with turbulent heat-transfer data on fluids which have a Prandtl number close to 1. For mass transfer the Reynolds analogy gives the relation

$$\frac{k_m d_w}{D} = \frac{f}{2}\,\mathrm{Re} \tag{15-37}$$

which also agrees best with data on systems where the Schmidt number $\mu/\rho D$ is close to 1.

15-7. The Prandtl Analogy

The Reynolds analogy does not take into account the velocity distribution across the tube. In 1910 Prandtl [18] extended Reynolds's work by considering the velocity distribution in the laminar sublayer and obtained a relationship which involved the ratio of the average velocity to the velocity *at the edge of the laminar sublayer*.

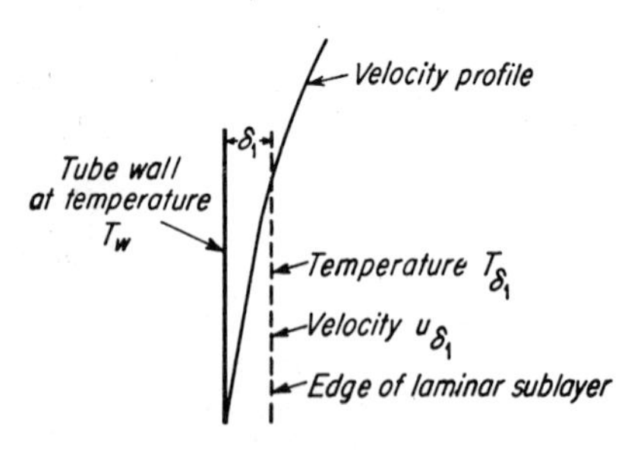

Fig. 15-10. Velocity and temperature at the edge of the laminar sublayer.

The sketch in Fig. 15-10 shows the velocity profile adjacent to the tube wall. The thickness of the laminar sublayer is δ_1, and the velocity at the edge of the laminar layer is $u\delta_1$. The temperature of the wall is T_w, and the temperature of the fluid at the edge of the laminar layer is $T\delta_1$. In the laminar sublayer ϵ_H and ϵ_M are both zero, and they are assumed to be equal in the turbulent core. Equations (15-1) and (15-23) are used.

$$\frac{\tau g_c}{\rho} = (\nu + \epsilon_M)\frac{du}{dy} \tag{15-1}$$

$$\frac{q_r}{A_r C_p \rho} = -(\alpha + \epsilon_H)\frac{dT}{dy} \tag{15-23}$$

Integrating these equations from $y = 0$ to $y = \delta_1$ ($\epsilon_H = \epsilon_M = 0$ and also at $y = 0$, $\tau = \tau_w$, and $q_r/A_r = q_w/A_w$),

$$\int_0^{u_{\delta_1}} du = \frac{\tau_w g_c}{\rho}\int_0^{\delta_1}\frac{dy}{\nu} \tag{15-38}$$

$$\int_{T_w}^{T_{\delta_1}} dT = -\frac{q_w}{A_w C_p \rho}\int_0^{\delta_1}\frac{dy}{\alpha} \tag{15-39}$$

which become

$$u_{\delta_1} = \frac{\tau_w g_c \delta_1}{\rho \nu} \tag{15-40}$$

$$T_w - T_{\delta_1} = \frac{q_w \delta_1}{A_w C_p \rho \alpha} \tag{15-41}$$

Combining Eqs. (15-40) and (15-41),

$$\frac{u_{\delta_1}\rho\nu}{\tau_w g_c \alpha} = \frac{A_w C_p \rho}{q_w}(T_w - T_{\delta_1}) \tag{15-42}$$

Applying the Reynolds analogy for the turbulent core [Eq. (15-33)],

$$\frac{(U - u_{\delta_1})\rho}{\tau_w g_c} = \frac{A_w C_p \rho}{q_w}(T_{\delta_1} - T_b) \tag{15-43}$$

Eliminating T_{δ_1} between Eqs. (15-42) and (15-43),

$$[U + u_{\delta_1}(\mathrm{Pr} - 1)]\frac{\rho}{\tau_w g_c} = \frac{A_w C_p \rho}{q_w}(T_w - T_b) \tag{15-44}$$

where $\mathrm{Pr} = \nu/\alpha$. Introducing

$$U^2\frac{f}{2} = \frac{\tau_w g_c}{\rho} \qquad \text{and} \qquad h = \frac{q_w}{A_w(T_w - T_b)}$$

gives

$$\frac{U + u_{\delta_1}(\mathrm{Pr} - 1)}{U^2(f/2)} = \frac{C_p \rho}{h} \tag{15-45}$$

from which

$$\frac{h}{C_p \rho U} = \mathrm{St} = \frac{f/2}{1 + (u_{\delta_1}/U)(\mathrm{Pr} - 1)} \tag{15-46}$$

Equation (15-46) predicts the Stanton number from the friction factor, the

Prandtl number, and the ratio u_{δ_1}/U. It reduces to Eq. (15-31) when the Prandtl number is unity. From Eq. (8-5) the velocity at the edge of the laminar sublayer is

$$\frac{u_{\delta_1}}{U} = 5\sqrt{\frac{f}{2}} \tag{8-5}$$

Thus Eq. (15-46) becomes, after replacing St by Nu/RePr,

$$\mathrm{Nu} = \frac{(f/2)\mathrm{RePr}}{1 + 5\sqrt{f/2}(\mathrm{Pr} - 1)} \tag{15-47}$$

Equation (15-47) is the relation obtained by Prandtl's analogy taking the thickness of the laminar sublayer into consideration. From Eqs. (15-29) and (15-47) it may be shown that

$$\ln\frac{T_w - T_{b_1}}{T_w - T_{b_2}} = \frac{2fL}{d_w}\frac{1}{1 + 5\sqrt{f/2}(\mathrm{Pr} - 1)} \tag{15-48}$$

Stanton's[26] data are calculated to obtain values of the right-hand side of Eq. (15-48). The data (solid circles) are plotted in Fig. 15-9 and show good agreement with the straight line, which represents Eq. (15-48).

For mass transfer Prandtl's analogy becomes

$$\frac{k_m d_w}{D} = \frac{(f/2)\mathrm{ReSc}}{1 + 5\sqrt{f/2}(\mathrm{Sc} - 1)} \tag{15-49}$$

where Sc is the Schmidt number $\mu/\rho D$.

Example 15-1

Estimate the temperature rise for water flowing in a heat-exchanger tube 10 ft long, 1.000 in. OD with a 16 BWG wall. The water flows at the rate of 15 gal/min and enters the tube at a temperature of 60°F. Neglect entrance effects and use the properties of water evaluated at the arithmetic average bulk temperature. The tube-wall temperature is 212°F.

Solution

Kinematic viscosity of water at 60°F

$$\nu = 1.21 \times 10^{-5}\ \mathrm{ft^2/sec}$$

$$\mathrm{Pr} = 7.5$$

$$\text{Tube ID} = 0.870\ \text{in.}$$

$$\text{Water velocity} = \frac{15}{60}\left|\frac{}{7.48}\right|\frac{144}{(0.870)^2}\left|\frac{4}{\pi}\right. = 8.10\ \mathrm{ft/sec}$$

$$\mathrm{Re} = \frac{0.870}{12}\left|\frac{8.10}{1.21}\right|\frac{}{10^{-5}} = 4.85 \times 10^4$$

From Fig. 7-21

$$f = 0.0051 \qquad \sqrt{\frac{f}{2}} = 0.0505$$

From Eq. (15-48)

$$\ln \frac{212 - 60}{212 - T_{b_2}} = \frac{(2)(0.0051)(10)(12)}{0.870} \frac{1}{1 + (5)(0.0505)(7.5 - 1)}$$

$$= 0.532$$

from which

$$T_{b_2} = 122°\text{F}$$

$$\text{Average water temperature} = \frac{60 + 122}{2} = 91°\text{F}$$

$$\nu = 0.82 \times 10^{-5} \text{ ft}^2/\text{sec}$$

$$\text{Pr} = 5.5$$

$$\text{Re} = \frac{0.870}{12} \left| \frac{8.10}{0.82} \right| \frac{}{10^{-5}} = 7.1 \times 10^4$$

$$f = 0.0048 \qquad \sqrt{\frac{f}{2}} = 0.049$$

$$\ln \frac{212 - 60}{212 - T_{b_2}} = \frac{(2)(0.0048)(10)(12)}{0.870} \frac{1}{1 + (5)(0.049)(5.5 - 1)}$$

$$= 0.630$$

from which

$$T_{b_2} = 131°\text{F}$$

Further calculation gives a value for T_{b_2} of 132°F.

From Eq. (14-9) the value of the Nusselt number may be calculated at the average bulk temperature.

$$\text{Average temperature} = \frac{60 + 132}{2} = 96°\text{F}$$

From Eq. (14-9)

$$h = 1780 \text{ Btu/(hr)(ft}^2\text{)(°F)}$$

With this value of h the value of T_{b_2} may also be calculated.

$$mC_p(T_{b_2} - T_{b_1}) = h\pi d_w L \frac{T_w - T_{b_1} - (T_w - T_{b_2})}{\ln [(T_w - T_{b_1})/(T_w - T_{b_2})]}$$

from which

$$\ln \frac{T_w - T_{b_1}}{T_w - T_{b_2}} = \frac{h\pi d_w L}{mC_p}$$

Thus

$$\ln \frac{212 - 60}{212 - T_{b_2}} = \frac{(1780)(\pi)(10)(0.870)}{(15)(8.33)(1)(60)(12)} = 0.541$$

giving

$$T_{b_2} = 123°\text{F}$$

Prandtl's analogy predicts a higher temperature rise in the tube than Eq. (14-9) does by about 14 per cent.

15-8. The von Kármán Analogy

Von Kármán[10] obtained a further improvement of Prandtl's analogy by integrating Eqs. (15-1) and (15-23) and making use of the universal velocity distribution shown in Fig. 7-10. The resulting equation for the Nusselt number is

$$\mathrm{Nu} = \frac{(f/2)\mathrm{RePr}}{1 + 5\sqrt{f/2}\{\mathrm{Pr} - 1 + \ln[1 + \frac{5}{6}(\mathrm{Pr} - 1)]\}} \tag{15-50}$$

15-9. Colburn's Analogy; The j Factor

A useful empirical means of correlating convection-heat-transfer data and showing the relationship to the friction factor was developed by Colburn,[2] who proposed the j factor. The correlation of heat-transfer data obtained by Colburn has been mentioned [Eq. (14-10)].

$$\left(\frac{h_m}{GC_p}\right)_b \left(\frac{C_p\mu}{k}\right)_{T0.5}^{2/3} = 0.023 \left(\frac{d_w G}{\mu}\right)_{T0.5}^{-0.2} \tag{14-10}$$

When the right side of Eq. (14-10) is compared with Eq. (7-64), it is seen to be equivalent to one-half the friction factor; i.e.,

$$\left(\frac{h_m}{GC_p}\right)_b \left(\frac{C_p\mu}{k}\right)_{T0.5}^{2/3} = \frac{f}{2} = j_H \tag{15-51}$$

where j_H is the j factor for heat transfer and is the product of the Stanton number and the Prandtl number to the two-thirds power.

Since St = Nu/RePr, Eq. (15-51) may be written

$$\mathrm{Nu} = \frac{f}{2}\mathrm{Re}(\mathrm{Pr})^{1/3} \tag{15-52}$$

For mass transfer one obtains

$$\frac{k_m d_w}{D} = \frac{f}{2}\mathrm{Re}(\mathrm{Sc})^{1/3} \tag{15-53}$$

Colburn investigated a large number of convection-heat-transfer and pressure-drop data and found that a correlation in the form of Eq. (15-51) was obtainable, thus making it possible to predict the heat-transfer coefficient from the friction factor. A graphical comparison of Eqs. (15-36), (15-47), (15-50), (15-52), and (14-11) is shown in Fig. 15-11, where the Nusselt number is plotted versus the Reynolds number at constant values of the Prandtl number. The top curves of Fig. 15-11 are for a Prandtl number of 10.0 and the lower curves for a Prandtl number of 0.01. When the

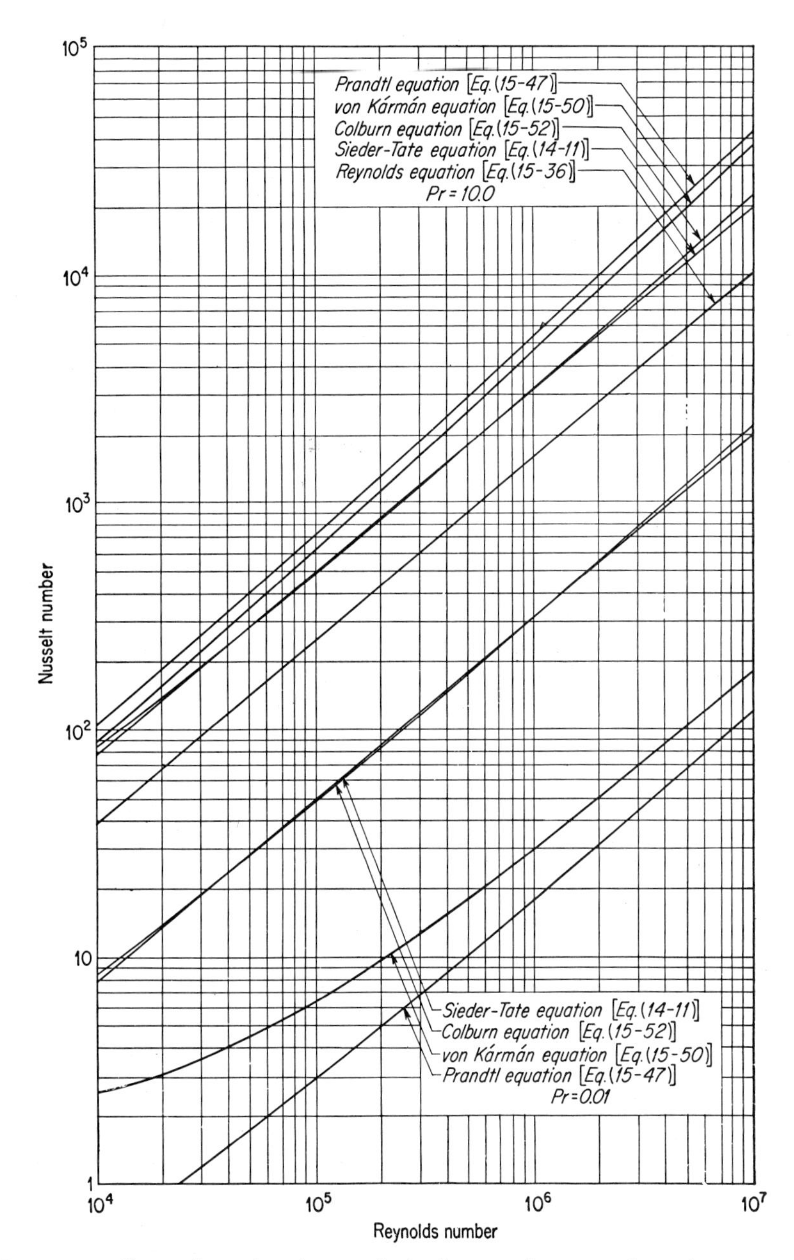

FIG. 15-11. Comparison of various analogies between heat transfer and momentum transfer.

Prandtl number is unity, all equations become identical. All equations agree well at high Prandtl numbers. At low Prandtl numbers there is wide divergence of the various equations, those of Prandtl and von Kármán giving much lower values of the Nusselt number. These results would indicate that for fluids with a high Prandtl number the empirical equations satisfactorily predict the heat-transfer coefficient. For low-Prandtl-number fluids the large difference between the theoretical and empirical equations suggests that the present empirical relationships are not suitable. New equations have been developed for predicting heat-transfer coefficients for such fluids.

15-10. The Martinelli Analogy; Uniform Heat Flux

In 1947 Martinelli [15] presented a thorough analysis of turbulent-flow heat transfer in circular tubes. He emphasized the calculation of Nusselt numbers for liquid metals but presented relationships which apply to all fluids at all Reynolds numbers. The conditions of flow and heat transfer for which Martinelli developed relationships are as follows:

(1) Velocity and temperature profiles are fully developed. There are no entrance effects.
(2) The fluid properties are independent of temperature.
(3) There is uniform heat flux along the tube wall.

Starting with Eqs. (15-1) and (15-23),

$$\frac{\tau g_c}{\rho} = (\nu + \epsilon_M)\frac{du}{dy} \tag{15-1}$$

$$\frac{q_r}{C_p \rho A_r} = -(\alpha + \epsilon_H)\frac{dT}{dy} \tag{15-23}$$

The shear at any point y is related to the shear at the wall by

$$\tau = \frac{r}{r_w}\tau_w \tag{4-14}$$

and, since $r = r_w - y$,

$$\tau = \left(1 - \frac{y}{r_w}\right)\tau_w \tag{15-54}$$

Martinelli assumes the same linear distribution for the rate of heat transfer across the tube as exists for the shear; thus

$$\frac{q_r}{C_p \rho A_r} = \frac{q_w}{C_p \rho A_w}\left(1 - \frac{y}{r_w}\right) \tag{15-55}$$

Equations (15-1) and (15-23) become

$$\frac{\tau_w g_c}{\rho}\left(1 - \frac{y}{r_w}\right) = (\nu + \epsilon_M)\frac{du}{dy} \tag{15-56}$$

$$\frac{q_w}{C_p \rho A_w}\left(1 - \frac{y}{r_w}\right) = -(\alpha + \epsilon_H)\frac{dT}{dy} \tag{15-57}$$

The Nusselt number may be calculated from the temperature distribution by Eq. (12-12).

$$\mathrm{Nu} = \frac{hd_w}{k} = -\frac{d_w}{T_w - T_b}\left(\frac{\partial T}{\partial y}\right)_{y=0} \tag{12-12}$$

The temperature distribution as a function of y is obtained by integrating Eq. (15-57). To carry out this integration ϵ_H must be known as a function of y. Martinelli assumed that ϵ_H was proportional to ϵ_M and determined ϵ_M from Eq. (15-56), using the universal velocity distribution given in Fig. 7-10 and described by Eqs. (7-35), (7-37), and (7-38).

The important relationships derived by Martinelli are as follows:

1. Temperature distribution:

a. Laminar sublayer ($0 < y^+ < 5$):

$$\frac{T_w - T}{T_w - T_c} = \frac{\dfrac{\epsilon_H}{\epsilon_M}\mathrm{Pr}\dfrac{y}{y_1}}{\dfrac{\epsilon_H}{\epsilon_M}\mathrm{Pr} + \ln\left(1 + 5\dfrac{\epsilon_H}{\epsilon_M}\mathrm{Pr}\right) + 0.5F_1 \ln\dfrac{\mathrm{Re}}{60}\sqrt{\dfrac{f}{2}}} \tag{15-58}$$

where y_1 is the value of y at $y^+ = 5$.

b. Buffer layer ($5 < y^+ < 30$):

$$\frac{T_w - T}{T_w - T_c} = \frac{\dfrac{\epsilon_H}{\epsilon_M}\mathrm{Pr} + \ln\left[1 + \dfrac{\epsilon_H}{\epsilon_M}\mathrm{Pr}\left(\dfrac{y}{y_1} - 1\right)\right]}{\dfrac{\epsilon_H}{\epsilon_M}\mathrm{Pr} + \ln\left(1 + 5\dfrac{\epsilon_H}{\epsilon_M}\mathrm{Pr}\right) + 0.5F_1 \ln\dfrac{\mathrm{Re}}{60}\sqrt{\dfrac{f}{2}}} \tag{15-59}$$

c. Turbulent core $y^+ > 30$:

$$\frac{T_w - T}{T_w - T_c} = \frac{\dfrac{\epsilon_H}{\epsilon_M}\mathrm{Pr} + \ln\left(1 + 5\dfrac{\epsilon_H}{\epsilon_M}\mathrm{Pr}\right) + 0.5F_1 \ln\dfrac{\mathrm{Re}}{60}\sqrt{\dfrac{f}{2}}\dfrac{y}{r_w}}{\dfrac{\epsilon_H}{\epsilon_M}\mathrm{Pr} + \ln\left(1 + 5\dfrac{\epsilon_H}{\epsilon_M}\mathrm{Pr}\right) + 0.5F_1 \ln\dfrac{\mathrm{Re}}{60}\sqrt{\dfrac{f}{2}}} \tag{15-60}$$

2. Heat-transfer coefficient:

$$\mathrm{Nu} = \frac{\dfrac{\epsilon_H}{\epsilon_M}\sqrt{\dfrac{f}{2}}\dfrac{T_w - T_c}{T_w - T_b}\mathrm{RePr}}{5\left[\dfrac{\epsilon_H}{\epsilon_M}\mathrm{Pr} + \ln\left(1 + 5\dfrac{\epsilon_H}{\epsilon_M}\mathrm{Pr}\right) + 0.5F_1 \ln\dfrac{\mathrm{Re}}{60}\sqrt{\dfrac{f}{2}}\right]} \tag{15-61}$$

where values of F_1 and $(T_w - T_b)/(T_w - T_c)$ are given in Tables 15-2 and 15-3. The calculations to determine F_1 and $(T_w - T_b)/(T_w - T_c)$ are given in detail by Martinelli. The values of F_1 given in Table 15-2 apply for ϵ_H/ϵ_M equal to unity.

TABLE 15-2. VALUES OF F_1 TO BE USED IN EQ. (15-61) †

Pe \ Re	10^4	10^5	10^6
10^2	0.18	0.098	0.052
10^3	0.65	0.45	0.29
10^4	0.92	0.83	0.65
10^5	0.99	0.985	0.980
10^6	1.00	1.00	1.00

† From W. H. McAdams, "Heat Transmission," McGraw-Hill Book Company, Inc., New York, 1954.

TABLE 15-3. VALUES OF $\dfrac{T_w - T_b}{T_w - T_c}$ TO BE USED IN EQ. (15-61) †

Pr \ Re	10^4	10^5	10^6	10^7
0	0.564	0.558	0.553	0.550
10^{-4}	0.568	0.560	0.565	0.617
10^{-3}	0.570	0.572	0.627	0.728
10^{-2}	0.589	0.639	0.738	0.813
10^{-1}	0.692	0.761	0.823	0.864
1.0	0.865	0.877	0.897	0.912
10	0.958	0.962	0.963	0.966
10^2	0.992	0.993	0.993	0.994
10^3	1.00	1.00	1.00	1.00

† From W. H. McAdams, "Heat Transmission," McGraw-Hill Book Company, Inc., New York, 1954.

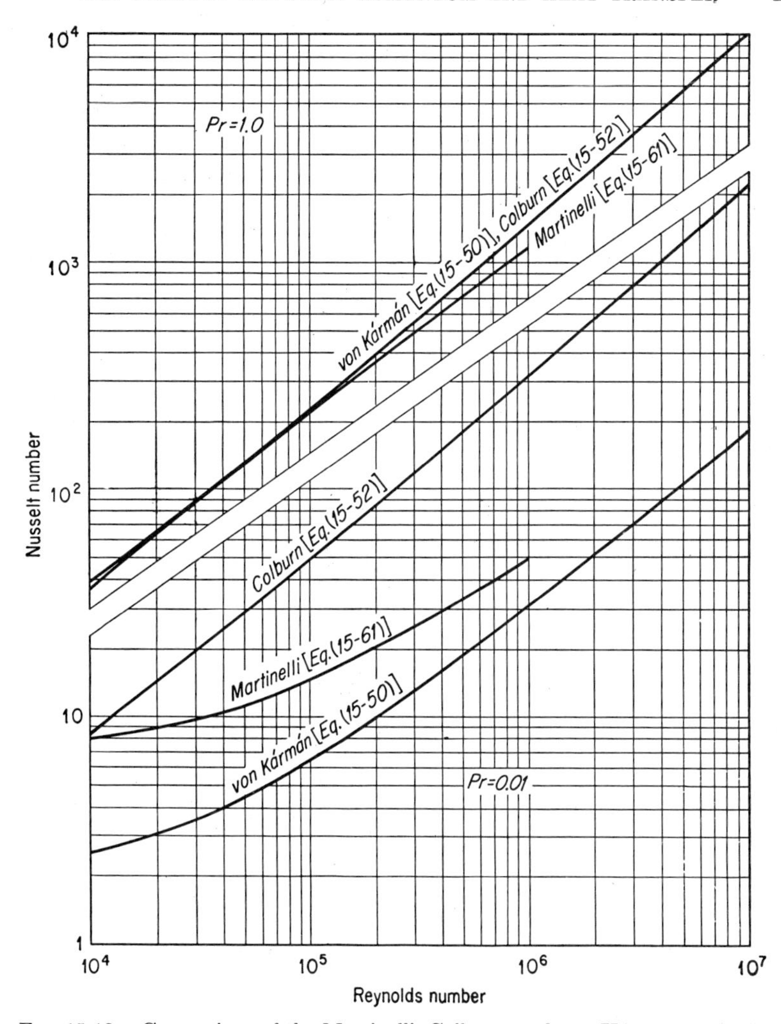

FIG. 15-12. Comparison of the Martinelli, Colburn, and von Kármán analogies.

Equations (15-58) to (15-60) may be used to calculate the temperature distribution for any value of Prandtl number and Reynolds number. Equation (15-61) may be used in conjunction with Tables 15-1 and 15-2 to predict the Nusselt number. All the above equations depend on the value of ϵ_H/ϵ_M, which, lacking actual experimental data, is usually taken as unity. In deriving these relationships, Martinelli assumed ϵ_H/ϵ_M to be independent of the distance from the wall. This assumption is now known to be incorrect. Figure 15-12 shows a comparison of Martinelli's analogy [Eq. (15-61) for $\epsilon_H/\epsilon_M = 1$], von Kármán's analogy [Eq. (15-50)], and Colburn's analogy [Eq. (15-52)] at Prandtl numbers of 1 and 0.01.

15-11. Seban and Shimazaki's Analogy; Constant Wall Temperature

Seban and Shimazaki [24] made an analysis similar to that of Martinelli for turbulent flow in circular tubes. They did not, however, assume a linear radial distribution of heat flux, as Martinelli did [see Eqs. (15-55) and (15-57)], and they considered the case of constant wall temperature. They determined the radial temperature distribution by solving Eq. (15-13) after obtaining an expression for the term $\partial T/\partial x$. Differentiating the expression $(T_w - T)/(T_w - T_b)$ with respect to x (this derivative is assumed to be zero since the temperature profile is considered to be fully developed; i.e., entrance effects are absent),

$$\frac{\partial}{\partial x}\left(\frac{T_w - T}{T_w - T_b}\right) = \frac{1}{T_w - T_b}\left(-\frac{\partial T}{\partial x}\right) - \frac{T_w - T}{(T_w - T_b)^2}\left(\frac{\partial T_w}{\partial x} - \frac{\partial T_b}{\partial x}\right) = 0 \tag{15-62}$$

Since the wall temperature is constant, $\partial T_w/\partial x = 0$; therefore

$$\frac{\partial T}{\partial x} = \frac{T_w - T}{T_w - T_b}\frac{\partial T_b}{\partial x} \tag{15-63}$$

Substituting Eq. (15-63) into Eq. (15-13),

$$u\,\frac{T_w - T}{T_w - T_b}\frac{\partial T_b}{\partial x} = \frac{1}{r}\frac{\partial}{\partial r}\left[r(\alpha + \epsilon_H)\frac{\partial T}{\partial r}\right] \tag{15-64}$$

Seban and Shimazaki integrated Eq. (15-64) twice to get the radial temperature distribution and the Nusselt number. Their results are summarized below. Two functions of r/r_w are obtained.

$$\phi_1\left(\frac{r}{r_w}\right) = \int_0^{r/r_w} \frac{u}{U}\frac{T_w - T}{T_w - T_c}\frac{r}{r_w}\,d\frac{r}{r_w} \tag{15-65}$$

$$\phi_2\left(\frac{r}{r_w}\right) = \int_0^{r/r_w} \frac{\phi_1(r/r_w)}{(r/r_w)(\epsilon_H + \alpha)/\nu}\,d\frac{r}{r_w} \tag{15-66}$$

The temperature distribution becomes

$$\frac{T_w - T}{T_w - T_c} = 1 - \frac{\phi_2(r/r_w)}{\phi_2(1)} \tag{15-67}$$

and the Nusselt number

$$\mathrm{Nu} = \frac{\mathrm{Pr}}{\phi_2(1)} \tag{15-68}$$

To solve the above equation the temperature distribution is first calculated using Martinelli's equations (15-58) to (15-60). This distribution is used in Eq. (15-65) to evaluate $\phi_1(r/r_w)$, which, in turn, is used in Eq. (15-66) to get $\phi_2(r/r_w)$. The temperature distribution is then calculated from Eq. (15-67), this distribution is put back in Eq. (15-65), and the procedure is repeated to obtain another temperature distribution. This iterative process is continued until the calculated temperature distribution does not change. The Nusselt number is then calculated from Eq. (15-68).

15-12. Lyon's Analogy; Uniform Heat Flux

Lyon [14] derived relationships for the Nusselt number using the analogy between heat transfer and momentum transfer for the case of uniform heat flux. In deriving his relationship, Lyon made the following assumptions:

(1) Velocity and temperature profiles are fully developed. There are no end effects.
(2) Steady state exists. There are no changes in temperature or velocity over a reasonable length of time in a given position in a given fluid. The rapid fluctuations in these conditions due to eddying or molecular movement still exist, and allowance for such fluctuations is made in the usual definitions of eddy diffusivity, viscosity, and thermal conductivity.
(3) At any point in the tube all properties and quantities are in radial symmetry.
(4) Fluid properties are independent of temperature.
(5) The heat flux at the wall of the tube q_w/A_w is uniform along the length of the tube. On this basis $\partial T/\partial x$ and dT_b/dx are constant and assumed equal to each other.

Strictly speaking, the above conditions are met only in the limiting case of no heat transfer. However, in sections where temperature changes of the heated or cooled fluid are small and where the temperature difference between the tube wall and the fluid remains approximately constant, the conditions are fulfilled to the degree required to permit useful application of the results.

The following derivation is based on the work of Lyon.[14] The heat-transfer coefficient is defined as

$$h = \frac{q_w}{A_w(T_w - T_b)} \tag{15-69}$$

The sketch in Fig. 15-13 shows a circular tube of unit length with flow taking place through the tube. The heat flow through the wall per unit time is q_w. In Eq. (15-69) all quantities are easily determined except the mean flow temperature, or bulk temperature, T_b. Determination of this temperature requires the use of the velocity distribution and the analogy between heat transfer and momentum transfer.

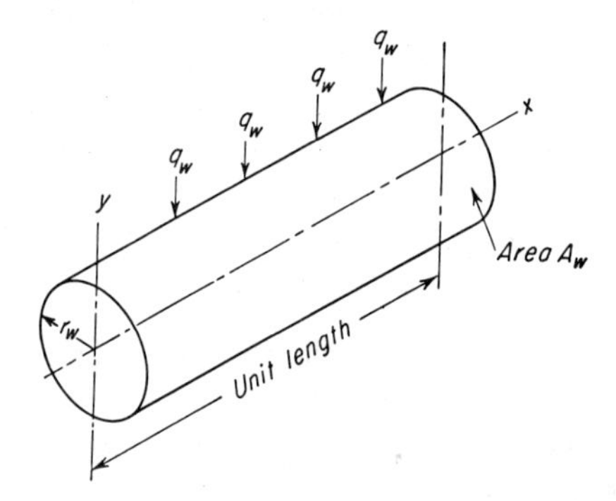

FIG. 15-13. Unit length of tube in which fluid is flowing and heat transfer is taking place.

By a heat balance, q_w is equated to the heat accumulated in the fluid in flowing through the unit length of the tube.

$$q_w = \pi r_w^2 U C_p \rho \frac{dT_b}{dx} \tag{15-70}$$

The area A_w equals $2\pi r_w$. Substituting these quantities in Eq. (15-69),

$$h = \frac{\pi r_w^2 U C_p \rho (dT_b/dx)}{2\pi r_w (T_w - T_b)} = \frac{r_w U C_p \rho (dT_b/dx)}{2(T_w - T_b)} \tag{15-71}$$

The temperature T_b is defined as the bulk temperature of the fluid, i.e., the temperature that would be measured if all liquid flowing through a

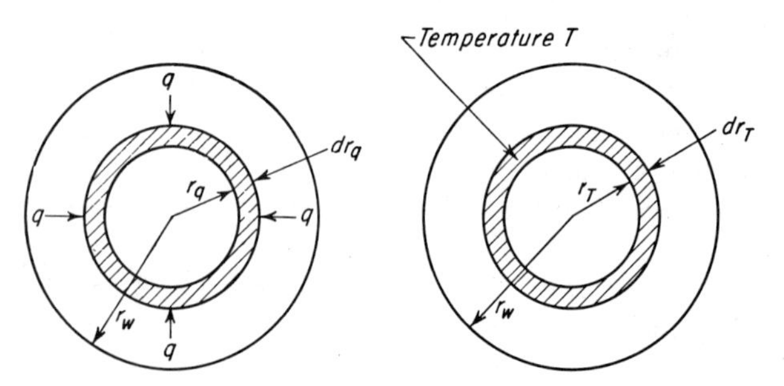

FIG. 15-14. Definition of terms for the derivation of Lyon's analogy.

section were thoroughly mixed. Referring to the right-hand sketch of Fig. 15-14, the amount of fluid flowing through an element of width dr_T is $2\pi r_T u_T \, dr_T$, where r_T and u_T are the radius and velocity, respectively, and the temperature of the fluid is T. Therefore, the bulk temperature of the fluid is

$$T_b = \frac{\int_0^{r_w} (2\pi r_T u_T \, dr_T) T}{\int_0^{r_w} 2\pi r_T u_T \, dr_T} \tag{15-72}$$

Subtracting each side of Eq. (15-72) from T_w results in

$$T_w - T_b = \frac{\int_0^{r_w} 2\pi r_T u_T (T_w - T) \, dr_T}{\int_0^{r_w} 2\pi r_T u_T \, dr_T} \tag{15-73}$$

On integration the denominator becomes $\pi r_w^2 U$, thus resulting in

$$T_w - T_b = \frac{2\int_0^{r_w} \pi r_T u_T (T_w - T) \, dr_T}{\pi r_w^2 U}$$

$$= 2\int_0^1 \frac{r_T}{r_w} \frac{u_T}{U} (T_w - T) \, d\frac{r_T}{r_w} \tag{15-74}$$

A new set of conditions may now be developed independently. At a radius r_q (left-hand sketch of Fig. 15-14) the radial heat flux is q and the total conductivity, including molecular and eddy conductivities, is K.

Then $$\frac{dT}{dr_q} = \frac{q}{2\pi r_q K} \tag{15-75}$$

Equation (15-75) is a conductivity equation, which, upon integration between the limits $r = r_w$ and $r = r_T$, gives

$$T_w - T = \int_{r_T}^{r_w} \frac{q}{2\pi r_q K} \, dr_q \tag{15-76}$$

Just as Eq. (15-70) gives the heat flow at the tube wall by an expression for the temperature rise in the entire tube, so a similar relationship may be set up for the heat flow to a cylinder of radius r_q within the tube. The difference lies in the necessity for finding an expression for the mean velocity in the cylinder of radius r_q. This mean velocity may be expressed as

$$\frac{\int_0^{r_q} 2\pi r u \, dr}{\pi r_q^2} \tag{15-77}$$

Thus the heat transferred into a cylinder of radius r_q becomes

$$q = \pi r_q^{\,2} C_p \rho \frac{\partial T}{\partial x} \frac{\int_0^{r_q} 2\pi r u \, dr}{\pi r_q^{\,2}} \tag{15-78}$$

Substituting Eq. (15-78) into (15-76),

$$T_w - T = \int_{r_T}^{r_w} \frac{C_p \rho (\partial T / \partial x) \int_0^{r_q} 2\pi r u \, dr}{2\pi r_q K} \, dr_q \tag{15-79}$$

Equation (15-79) expresses the temperature distribution across the tube. Substituting this equation into Eq. (15-74) and putting the resulting relationship into Eq. (15-71) gives an expression for the heat-transfer coefficient in terms of the radius, the point velocity, and the total conductivity, as shown in Eq. (15-80). In obtaining Eq. (15-80), $\partial T/\partial x$ is assumed equal to dT_b/dx.

$$h = \frac{r_w U}{4 \int_0^1 \frac{r_T}{r_w} \frac{u_T}{U} \left(\int_{r_T}^{r_w} \frac{\int_0^{r_q} r u \, dr}{r_q K} \, dr_q \right) d \frac{r_T}{r_w}} \tag{15-80}$$

Multiplying each side of Eq. (15-80) by $2r_w/k$ results in an expression for the Nusselt number in terms of the radius, point velocity, and conductivity.

$$\mathrm{Nu} = \frac{1}{2 \int_0^1 \frac{r_T}{r_w} \frac{u_T}{U} \left[\int_{r_T/r_w}^1 \frac{\int_0^{r_q/r_w} (r/r_w)(u/U) \, d(r/r_w)}{(r_q/r_w)(K/k)} \, d \frac{r_q}{r_w} \right] d \frac{r_T}{r_w}} \tag{15-81}$$

The following dimensionless terms are defined:

$$\frac{u}{U} = \overline{V} \tag{15-82}$$

$$\frac{u_T}{U} = \overline{V}_T \tag{15-83}$$

$$\frac{r}{r_w} = R \tag{15-84}$$

$$\frac{r_q}{r_w} = R_q \tag{15-85}$$

$$\frac{r_T}{r_w} = R_T \tag{15-86}$$

The quantities R_T, R_q, and R are different variables. Thus, substituting Eqs. (15-82) to (15-86) in Eq. (15-81) and rearranging the integrals gives

$$\mathrm{Nu} = \frac{1}{2\int_0^1 \int_{R_T}^1 \int_0^{R_q} \frac{R_T R \overline{V}_T \overline{V}}{R_q(K/k)}\, dR\, dR_q\, dR_T} \tag{15-87}$$

which is equivalent to

$$\mathrm{Nu} = \frac{1}{2\int_0^1 \frac{\left(\int_0^{R_q} R\overline{V}\, dR\right)^2}{R_q(K/k)}\, dR_q} \tag{15-88}$$

The equality of the integrals in Eqs. (15-87) and (15-88) may be shown. If I_3 and I_3' are triple integrals expressed as

$$I_3 = \int_0^1 \int_{R_T}^1 \int_0^{R_q} \frac{R_T R \overline{V}_T \overline{V}}{R_q(K/k)}\, dR\, dR_q\, dR_T \tag{15-89}$$

and

$$I_3' = \int_0^1 \int_0^{R_q} \int_0^{R_q} \frac{R_T R \overline{V}_T \overline{V}}{R_q(K/k)}\, dR\, dR_T\, dR_q \tag{15-90}$$

then the first integration of each gives, respectively,

$$I_3 = \int_0^1 \int_{R_T}^1 \phi(R_T,R_q)\, dR_q\, dR_T \tag{15-91}$$

and

$$I_3' = \int_0^1 \int_0^{R_q} \phi(R_T,R_q)\, dR_T\, dR_q \tag{15-92}$$

where

$$\phi(R_T,R_q) = \frac{R_T \overline{V}_T}{R_q(K/k)} \int_0^{R_q} R\overline{V}\, dR \tag{15-93}$$

$\phi(R_T,R_q)$ represents a surface in space, and I_3 and I_3' represent a solid volume under this surface. The limits of the second integration of I_3 are from $R_q = R_T$ to $R_q = 1$. These limits are illustrated in Fig. 15-15 (*top*), which also shows the limits of the third integration of I_3 from $R_T = 0$ to $R_T = 1$. The area thus covered by these two integrations of I_3 is the triangle BOA, and the surface $\phi(R_T,R_q)$ is integrated over this area.

The second integration of I'_3 is from $R_T = 0$ to $R_T = R_q$, and the third integration is from $R_q = 0$ to $R_q = 1$, as shown in Fig. 15-15 (*bottom*).

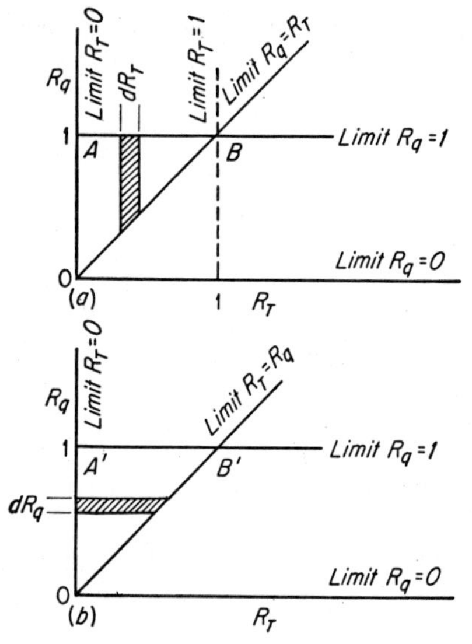

FIG. 15-15. Sketch to show limits of integration of Eqs. (15-87) and (15-88).

The area covered by the two integrations of I_3' is the triangle $B'OA'$, which is the same as the triangle BOA in Fig. 15-15 (*top*). Thus,

$$I_3 = I_3' \tag{15-94}$$

The integral I_3' is then equivalent to the integral in Eq. (15-88).

Equation (15-88) must be integrated in order to determine the Nusselt number. The velocity distribution is generally assumed to be of the form indicated by Fig. 7-10, and integrations are carried out graphically. Values of the total conductivity are determined by methods described in Sec. 15-14.

15-13. Comparison of Martinelli's, Seban and Shimazaki's, and Lyon's Equations for the Nusselt Number

A comparison of the Nusselt number as calculated by Martinelli's equation (15-61), Seban and Shimazaki's equation (15-68), and Lyon's equation (15-88) is shown in Table 15-4.

TABLE 15-4. NUSSELT NUMBER CALCULATED FROM EQS. (15-61), (15-68), AND (15-88) ($\epsilon_H = \epsilon_M$)

Pr	Reynolds number	Eq. (15-61) (Martinelli) uniform heat flux $u = U$	Eq. (15-68) (Seban and Shimazaki) constant wall temp. u variable	Eq. (15-88) (Lyon) uniform heat flux u variable
1.0	10,000	36.7	35.4	
	100,000	217.0	209.0	
	1,000,000	1,416.0	1,360.0	
0.1	10,000	14.0	11.9	15.0
	100,000	53.0	46.0	53.0
	1,000,000	271.0	256.0	265.0
0.01	10,000	7.97	5.97	8.0
	100,000	14.50	11.6	13.7
	1,000,000	48.7	43.1	45.0

15-14. The Calculation of Eddy Diffusivities and Total Conductivities

The eddy diffusivity of momentum may be calculated from Eq. (15-56) if the velocity profile is known.

$$\frac{\tau_w g_c}{\rho}\left(1 - \frac{y}{r_w}\right) = (\nu + \epsilon_M)\frac{du}{dy} \tag{15-56}$$

Solving for ϵ_M and introducing the relation $u^{*2} = \tau_w g_c/\rho$,

$$\epsilon_M = \frac{u^{*2}(1 - y/r_w)}{du/dy} - \nu \tag{15-95}$$

From the relations

$$y^+ = \frac{yu^*}{\nu} = \frac{y}{r_w}\frac{\text{Re}}{2}\sqrt{\frac{f}{2}} \tag{7-52}$$

$$u^+ = \frac{u}{u^*} \tag{7-51}$$

one obtains

$$dy^+ = \frac{u^*\,dy}{\nu} \tag{15-96}$$

$$du^+ = \frac{du}{u^*} \tag{15-97}$$

from which

$$\frac{du^+}{dy^+} = \frac{\nu}{u^{*2}}\frac{du}{dy} \tag{15-98}$$

Combining Eqs. (15-98) and (15-95),

$$\frac{\epsilon_M}{\nu} = \frac{1 - y/r_w}{du^+/dy^+} - 1 \tag{15-99}$$

and from Eq. (7-52)

$$\frac{\epsilon_M}{\nu} = \frac{1 - y^+/[(\text{Re}/2)\sqrt{f/2}\,]}{du^+/dy^+} - 1 \tag{15-100}$$

The eddy diffusivity of momentum may be calculated from Eq. (15-100). It is the usual practice to calculate eddy diffusivities of momentum from the universal velocity distribution equations; i.e.,

$$u^+ = y^+ \qquad y^+ < 5 \tag{7-37}$$

$$u^+ = -3.05 + 5.0 \ln y^+ \qquad 5 < y^+ < 30 \tag{7-38}$$

$$u^+ = 5.5 + 2.5 \ln y^+ \qquad y^+ > 30 \tag{7-35}$$

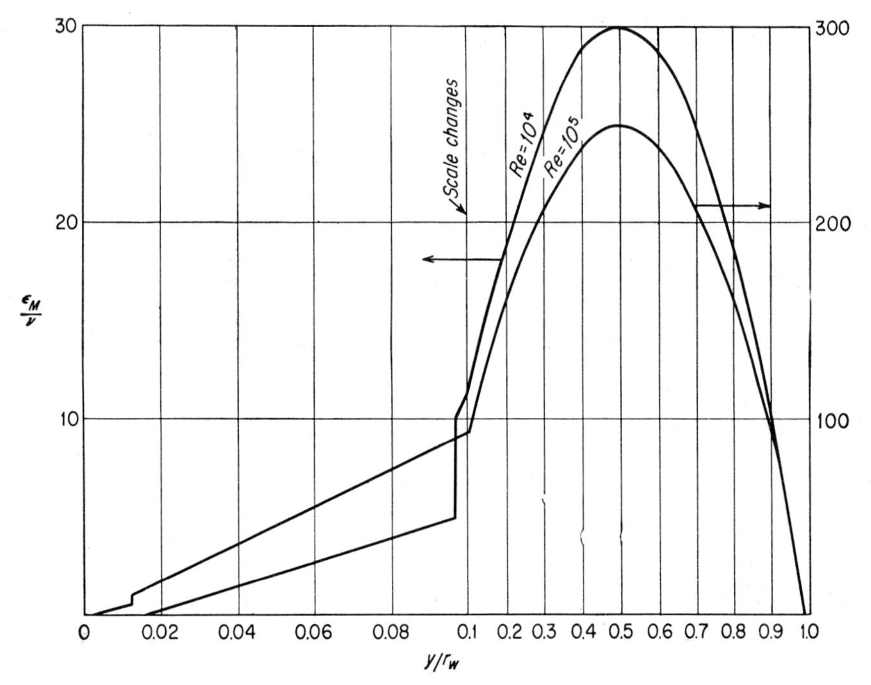

FIG. 15-16. Values of ϵ_M/ν for turbulent flow in tubes calculated from Eq. (15-100).

Equation (15-100) is plotted in Fig. 15-16 for Reynolds numbers of 10^4 and 10^5.

The total conductivity K, which appears in Eq. (15-88), is made up of the molecular conductivity and the eddy conductivity of heat.

$$K = k + \mathrm{E}_H \tag{15-101}$$

Since $\mathrm{E}_H = C_p\rho\epsilon_H$,

$$K = k + C_p\rho\epsilon_H \tag{15-102}$$

which may be rearranged to

$$\frac{K}{k} = 1 + \frac{\epsilon_H}{\nu}\,\mathrm{Pr} \tag{15-103}$$

Equation (15-103) gives the ratio of the total conductivity to the molecular conductivity. If ϵ_H is known as a function of ϵ_M, Eq. (15-100) may be used to obtain values of ϵ_H. In calculating Nusselt numbers ϵ_H and ϵ_M are usually assumed equal.

15-15. Experimental Values of Eddy Diffusivities

Obviously, the equations making use of the analogy between momentum and heat transfer require the relation between ϵ_H and ϵ_M to be known if results are to be accurate. To simplify calculations ϵ_H/ϵ_M is taken to be unity in most work; however, experimental investigation has shown ϵ_H/ϵ_M to vary between 1.0 and 1.6.

Isakoff and Drew [9] determined values of ϵ_M and ϵ_H for the turbulent flow of mercury in a circular tube. Temperature and velocity profiles were measured, and values of the eddy diffusivities were calculated from these data. The results of the work are plotted in Fig. 15-17. The ratio ϵ_H/ϵ_M is a function not only of the Reynolds number but of the position in the cross section of the pipe.

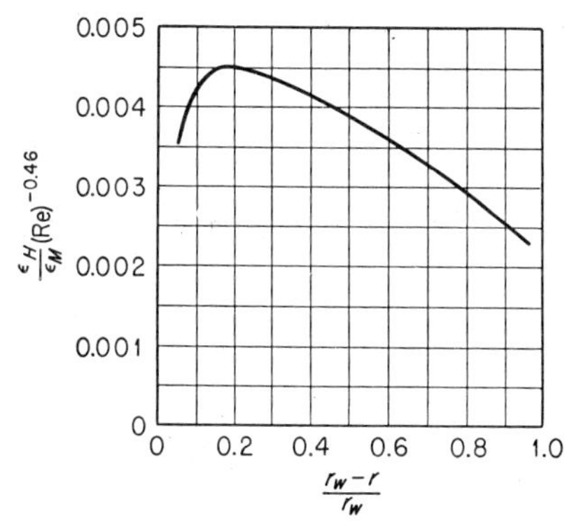

FIG. 15-17. Experimental values of eddy diffusivities. *(From S. E. Isakoff and T. B. Drew, "Proceedings of the General Discussion on Heat Transfer," Institution of Mechanical Engineers, London, and American Society of Mechanical Engineers, New York, 1951, p. 405.)*

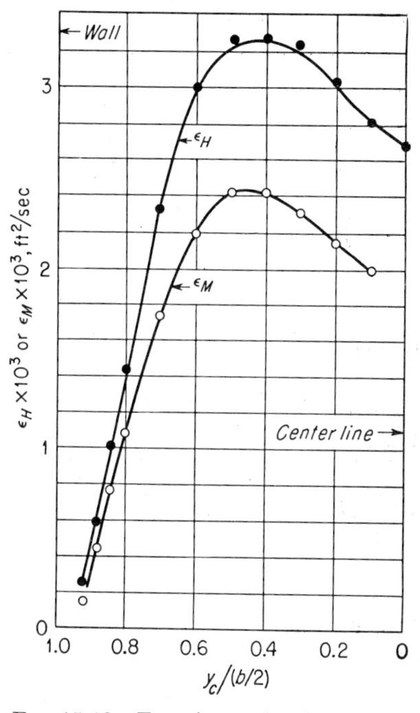

FIG. 15-18. Experimental values of eddy diffusivities of heat and momentum for air flowing between parallel planes. Reynolds number = 9,390. [*From F. Page, W. G. Schlinger, D. K. Breaux, and B. H. Sage, Ind. Eng. Chem.*, **44**:424 (1952).]

Similar results were obtained by Page, Schlinger, Breaux, and Sage,[16] who measured velocity and temperature profiles for air flowing between two parallel plates. Figure 15-18 is a plot representing one test they carried out. It shows a plot of both ϵ_H and ϵ_M as a function of the position in the stream. Both diffusivities reach a maximum value. Throughout the whole cross section ϵ_H is greater than ϵ_M. The ratio ϵ_M/ϵ_H as determined

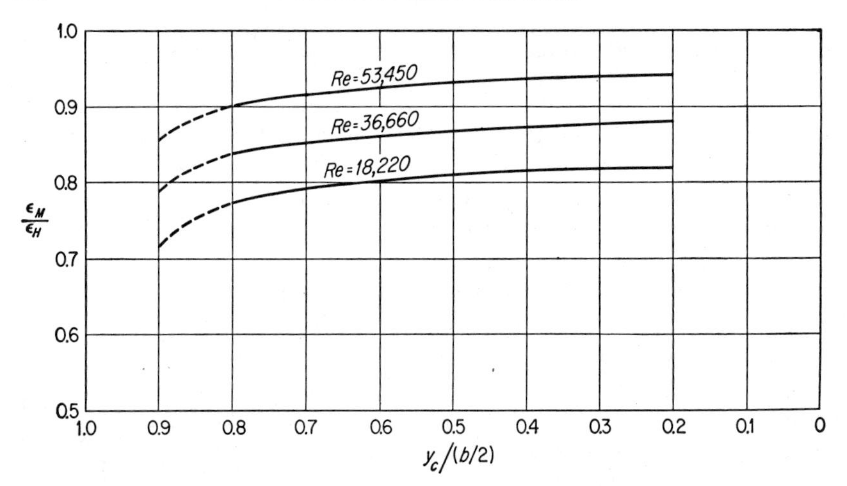

FIG. 15-19. Value of ϵ_M/ϵ_H for air flow between parallel planes. [*Data from F. Page et al., Ind. Eng. Chem.*, **44**:424 (1952).]

by Page et al. for air is shown in Fig. 15-19 for three different Reynolds numbers.

Deissler[5] proposed the following semiempirical equation for the ratio ϵ_H/ϵ_M:

$$\frac{\epsilon_H}{\epsilon_M} = n\mathrm{Pe}[1 - e^{-(1/n\mathrm{Pe})}] \tag{15-104}$$

where n is determined experimentally to be 0.000153. Using Eq. (15-104), Deissler predicted Nusselt numbers for liquid metals which agree well with experimental values (see Sec. 16-8).

15-16. Effect of Prandtl Number on Temperature Distribution in a Tube

The integration of Eq. (15-13) gives the temperature of the fluid as a function of the radius. When the radial temperature distribution is obtained, the heat-transfer coefficient may be calculated from the temperature gradient at the wall, just as the friction factor may be calculated from the velocity gradient at the wall.

The temperature distribution for turbulent flow in tubes is a function of the Reynolds number and the Prandtl number, as indicated by Martinelli's equations (15-58) to (15-60). The temperature profile across the section of a tube as calculated from Martinelli's equations is plotted in Fig. 15-20 for a Reynolds number of 10,000 and for Prandtl numbers of 10.0, 1.0, and 0.01. The line for Pr = 1.0 in Fig. 15-20 is also the velocity profile, u/u_{max}, calculated from Eqs. (7-35), (7-37), and (7-38) for a Reynolds number of 10,000. When the Prandtl number is unity, the velocity

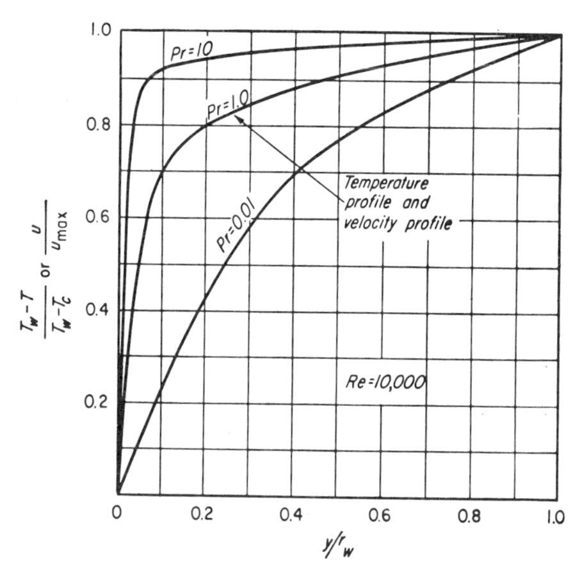

FIG. 15-20. Effect of Prandtl number on temperature profile for turbulent flow in circular tubes. Calculated from Eqs. (15-58) to (15-60).

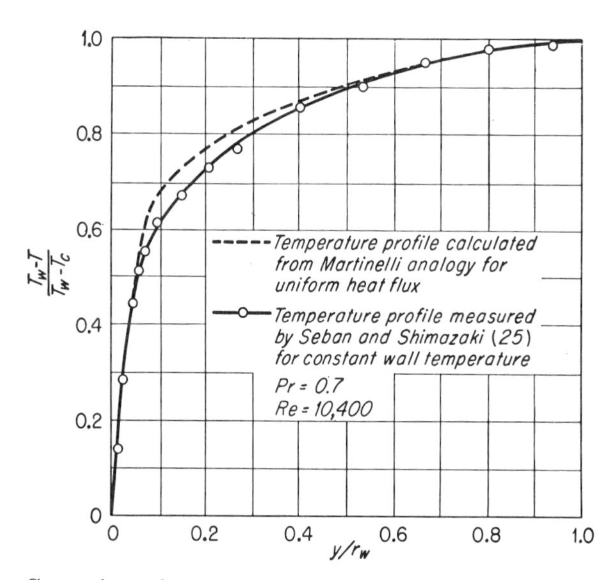

FIG. 15-21. Comparison of measured and calculated temperature profiles in circular tubes.

and temperature profiles are the same when plotted on a dimensionless basis, as in Fig. 15-20. The effect of the Prandtl number on the temperature distribution is clearly indicated in the figure.

Seban and Shimazaki [25] measured temperature profiles for air flowing in a circular tube in which the wall temperature was constant. The experimental data they reported are plotted in Fig. 15-21 for a Reynolds number of 10,400. The temperature profile as calculated from Eqs. (15-58) to (15-60) for a Prandtl number of 0.7 is shown. The difference between Martinelli's temperature profile and Seban and Shimazaki's result is due to two factors:

1. Martinelli's profile is based on $\epsilon_H = \epsilon_M$.
2. Martinelli's profile is based on uniform heat flux, and Seban and Shimazaki's data were obtained for the case of constant wall temperature.

15-17. Limitation of Equations Which Determine the Nusselt Number from Velocity Data

The various equations which have been given for calculating the Nusselt number from the velocity distribution, in addition to being restricted to the conditions for which they were derived, cannot be solved without

1. A knowledge of ϵ_H/ϵ_M, which is generally assumed to be equal to 1, although experimental data have been obtained which indicate that it may have values considerably different from 1, depending on the flow conditions.
2. An accurate relationship between u^+ and y^+.

In the laminar layer adjacent to the wall the total conductivity is equal to the molecular conductivity of the fluid. The integral in Eq. (15-88) is composed of three separate integrals, each representing a certain region in the cross section of the stream, these regions being the laminar, buffer, and turbulent layers. Thus, letting I_3' represent the integral in Eq. (15-88), the three separate integrals may be expressed

$$I_3' = I_{LL} + I_{BL} + I_{TL} \tag{15-105}$$

where I_{LL} is the integral over the laminar layer integrated over the limits $R_q = 1$ and the value of R_q at the edge of the laminar layer, I_{BL} is the integral evaluated over the buffer layer, having limits from the value of R_q at the edge of the laminar layer to the value of R_q at the edge of the buffer layer, and I_{TL} is the integral evaluated for the turbulent core over the limits of the value of R_q at the edge of the buffer layer to the value of R_q at the center of the pipe, i.e., at $R_q = 0$.

When the Prandtl number is large, the total conductivity is much greater than the molecular conductivity in the turbulent portion of the stream. Hence I_{LL}, the contribution of the laminar layer to the heat-transfer coefficient, is large. It is necessary to know the limits of integration of I_{LL} very accurately in order to obtain accurate values of the heat-transfer coefficient. However, owing to the difficulty of measuring the thickness of the laminar layer, it is not possible to determine the value of I_{LL} with any degree of accuracy. Thus, equations which relate the heat-transfer coefficient to the velocity distribution are not suitable for fluids having a Prandtl number greater than 1.

On the other hand, if the Prandtl number is small, indicating that the thermal conductivity is large, the contribution of the integrals I_{BL} and I_{TL} to the complete integral is large compared to that of the integral I_{LL}, so that the thickness of the laminar layer is relatively unimportant. The various equations for calculating heat-transfer coefficients from velocity-distribution data have found extensive application for fluids having a low Prandtl number, and they have been used as a basis for developing simple relationships between the heat-transfer coefficient and the flow conditions.

15-18. Deissler's Analysis; Fluid Properties Variable

Work on the calculation of temperature profiles and heat-transfer coefficients for turbulent flow in tubes has been done by Deissler.[6,8] The theoretical analyses have covered both high- and low-Prandtl-number fluids and have also taken into consideration the effect of variable fluid properties. The conditions of Deissler's analysis are as follows:

(1) Velocity and temperature profiles fully developed; no entrance effects
(2) $\epsilon_H = \epsilon_M$
(3) Prandtl number and heat capacity independent of temperature

The equations which are integrated to give the temperature and velocity distribution are

$$\frac{\tau g_c}{\rho} = (\nu + \epsilon_M)\frac{du}{dy} \qquad (15\text{-}1)$$

$$\frac{q_r}{C_p \rho A_r} = -(\alpha + \epsilon_H)\frac{dT}{dy} \qquad (15\text{-}23)$$

The distinguishing feature of Deissler's work is his assumption of the value of the eddy diffusivity of momentum. In the region close to the wall he assumes

$$\epsilon_M = n^2 u y \qquad y^+ < 26 \qquad (15\text{-}106)$$

and for the turbulent core he assumes the von Kármán relation

$$\epsilon_M = \kappa^2 \frac{(du/dy)^3}{(d^2u/dy^2)^2} \qquad y^+ > 26 \tag{15-107}$$

Experimental values of n are 0.109 and of κ are 0.36.

The velocity distributions Deissler obtained by integrating Eq. (15-1) are discussed in Sec. 7-8, Eqs. (7-44), (7-45), (7-48), and (7-49). In deriving them, the viscosity is assumed to be a function of temperature according to Eq. (7-47) for gases or the relation following Eq. (7-49) for liquids.

The analytical heat-transfer results obtained by Deissler are summarized below.

1. Gases:[8]

$$\mathrm{Nu} = \frac{2r_w^+ \mathrm{Pr}}{T_b^+} \tag{15-108}$$

where

$$r_w^+ = \frac{u^* r_w}{\nu}$$

$$T_b^+ = \frac{1}{\omega}\left(1 - \frac{T_b}{T_w}\right)$$

(All properties evaluated at tube-wall temperature)

and

$$\frac{T_b}{T_w} = \frac{\int_0^{r_w^+} u^+(r_w^+ - y^+)\, dy^+}{\int_0^{r_w^+} [u^+(r_w^+ - y^+)/(1 - \omega T^+)]\, dy^+} \tag{15-109}$$

where ω is defined by Eq. (7-46), and

$$T^+ = \frac{1}{\omega}\left(1 - \frac{T}{T_w}\right)$$

For gases with a Prandtl number of 1 the predicted temperature distribution plotted as T^+ versus y^+ is the same as the generalized velocity distribution given in Fig. 7-13. For air (Pr = 0.73) the generalized temperature distribution predicted by Deissler is shown in Fig. 15-22.

Deissler also shows that when the fluid properties are evaluated at the film temperature $T_{0.4}$ [see Eq. (14-7)], the Nusselt number is independent of the rate of heat addition. For air with a Prandtl number of 0.73, the predicted relation between Nu_m and Re is shown in Fig. 15-23. Deissler's experimental data on the heating of air, which are included, show good

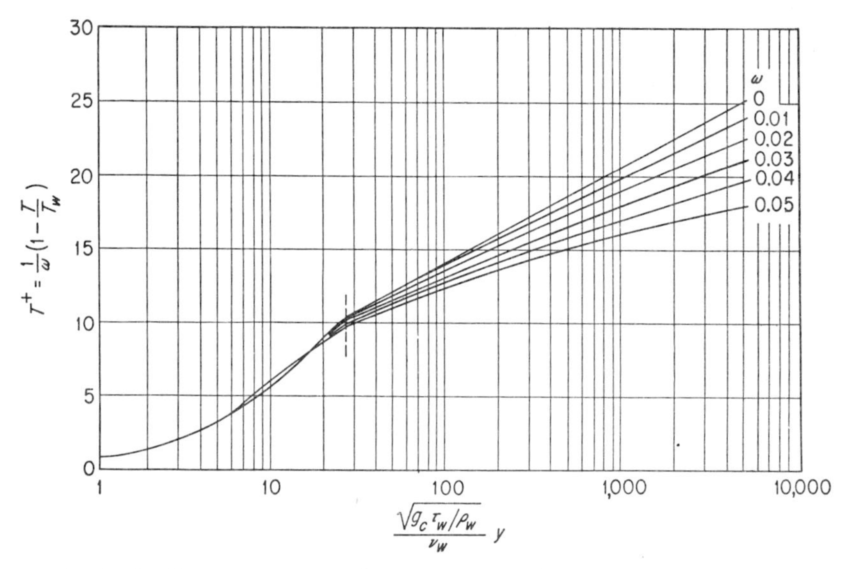

FIG. 15-22. Predicted temperature distribution for the heating of gases by forced convection in a circular tube. (*From R. G. Deissler, NACA TN* 2629, 1952.)

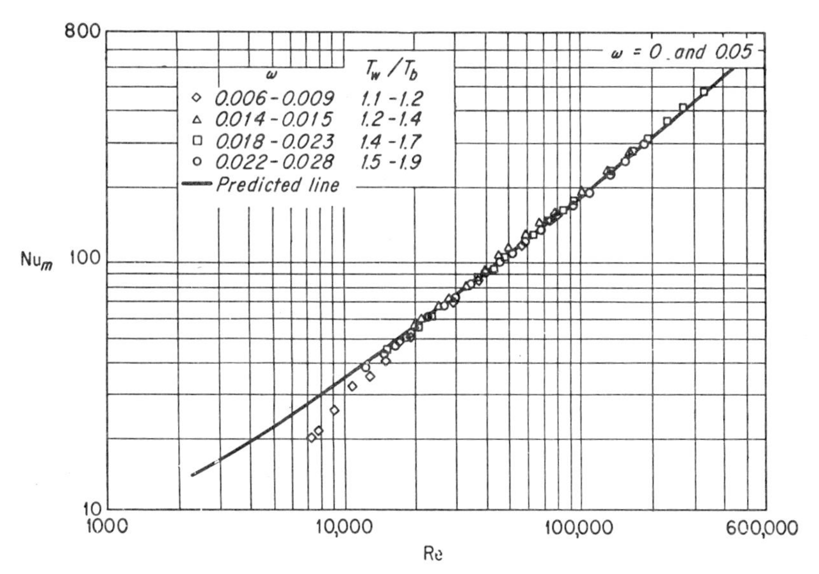

FIG. 15-23. Mean Nusselt numbers for heating gases in circular tubes (properties evaluated at $T_{0.4}$). (*From R. G. Deissler, NACA TN* 2629, 1952.)

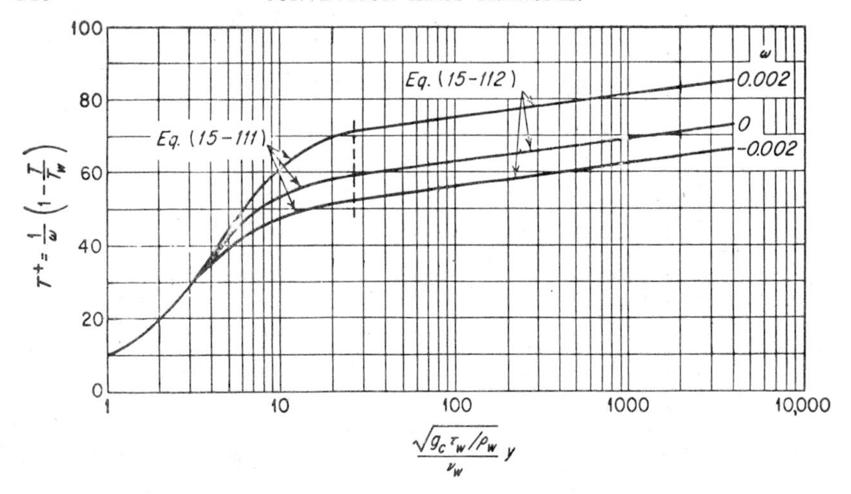

FIG. 15-24. Predicted temperature distribution for the heating and cooling of liquids in a circular tube. Velocity profile fully developed; Pr = 10.0; $\mu/\mu_w = (T/T_w)^{-4}$. (*From R. G. Deissler, NACA TN* 3145, 1954.)

agreement with theoretical results. The approximate equation of the predicted line for Re > 20,000 is

$$\mathrm{Nu}_m = 0.0185(\mathrm{Re})^{0.8} \tag{15-110}$$

(Properties evaluated at $T_{0.4}$)

This compares favorably with Eq. (14-9) with the Prandtl number taken as 0.73.

2. Liquids:

Deissler [6] predicts the temperature distribution for liquids to be

a. For $y < 26$:

$$T^+ = \int_0^{y^+} \frac{dy^+}{1/\mathrm{Pr} + n^2u^+y^+(1 - \exp[-n^2u^+y^+/(\mu/\mu_w)]} \tag{15-111}$$

(Properties evaluated at tube-wall temperature)

b. For $y^+ > 26$:

$$T^+ = c_1 + \frac{1}{\kappa} \ln y^+ \tag{15-112}$$

where $\frac{\mu}{\mu_w} = \left(\frac{T}{T_w}\right)^{d'} = (1 - \omega T^+)^{d'}$

$n = 0.109$

$c_1 =$ constant evaluated at $y^+ = 26$

$\kappa = 0.36$

Equation (15-111), which gives the temperature distribution up to $y^+ = 26$, must be solved in conjunction with Eq. (7-48).

The Nusselt number for heat transfer to liquids is calculated from Eq. (15-108) (for constant properties).

$$\mathrm{Nu} = \frac{2r_w^+ \mathrm{Pr}}{T_b^+} \tag{15-108}$$

where

$$T_b^+ = \frac{\int_0^{r_w^+} T^+ u^+ (r_w^+ - y^+)\, dy^+}{\int_0^{r_w^+} u^+ (r_w^+ - y^+)\, dy^+}$$

Figure 15-24 shows Deissler's predicted temperature distribution for a liquid with Pr = 10.0 and a viscosity variation $\mu/\mu_w = (T/T_w)^{-4}$.

Example 15-2

Calculate the Nusselt number for water flowing in a 2-in.-ID smooth pipe at a Reynolds number of 30,000. The water temperature is 60°F.

(*a*) Use Martinelli's equation (15-61).

(*b*) Use Lyon's equation (15-88) and use the velocity distribution given by Eqs. (7-35), (7-37), and (7-38).

(*c*) Use Eq. (14-11).

Solution

At Re = 30,000

$$f = 0.00585 \qquad \text{(from Fig. 7-21)}$$

$$\sqrt{\frac{f}{2}} = \sqrt{\frac{0.00585}{2}} = 0.0542$$

$$\mathrm{Re}\sqrt{\frac{f}{2}} = 1{,}620$$

At 60°F

$$\nu = 1.218 \times 10^{-5}\ \mathrm{ft^2/sec}$$

$$\mathrm{Pr} = 7.74$$

$$U = \frac{\mathrm{Re}\nu}{d_w} = \frac{(30{,}000)(1.218 \times 10^{-5})(12)}{2} = 2.185\ \mathrm{ft/sec}$$

$$\mathrm{Pe} = \mathrm{RePr} = (7.74)(30{,}000) = 2.32 \times 10^5$$

(*a*) *Martinelli's solution.* From Tables 15-1 and 15-2

$$F_1 = 0.99$$

$$\frac{T_w - T_c}{T_w - T_b} = 1.065$$

Substituting in Eq. (15-61) (with $\epsilon_H = \epsilon_M$),

$$\mathrm{Nu} = \frac{(0.0542)(1.065)(2.32 \times 10^5)}{(5)\{7.74 + \ln [1 + (5)(7.74)] + (0.5)(0.99) \ln [(3 \times 10^4)(0.0542)/60]\}}$$

$$= \frac{13{,}400}{(5)(7.74 + 3.68 + 1.63)} = 205$$

(*b*) By Lyon's method Eq. (15-88) must be integrated.

$$\mathrm{Nu} = \frac{1}{2 \int_0^1 \frac{\left(\int_0^{R_q} R\overline{V}\, dR \right)^2}{R_q(K/k)}\, dR_q}$$

ϵ_M/ν is calculated from Eq. (15-100). For $0 < y^+ < 5$

$$u^+ = y^+$$

$$\frac{du^+}{dy^+} = 1$$

Therefore $$\frac{\epsilon_M}{\nu} = \frac{1 - y^+/(1{,}620/2)}{1} - 1 \approx 0$$

From Eq. (15-103)

$$\frac{K}{k} = 1$$

For $5 < y^+ < 30$

$$u^+ = -3.05 + 5.00 \ln y^+$$

$$\frac{du^+}{dy^+} = \frac{5.00}{y^+}$$

$$\frac{\epsilon_M}{\nu} = \frac{y^+(1 - y^+/810)}{5} - 1$$

and $$\frac{K}{k} = 1 + 7.74 \left[\frac{y^+(1 - y^+/810)}{5} - 1 \right]$$

For $30 > y^+ > 810$ (tube axis)

$$u^+ = 5.5 + 2.5 \ln y^+$$

$$\frac{du^+}{dy^+} = \frac{2.5}{y^+}$$

$$\frac{\epsilon_M}{\nu} = \frac{y^+(1 - y^+/810)}{2.5} - 1$$

$$\frac{K}{k} = 1 + 7.74 \left[\frac{y^+(1 - y^+/810)}{2.5} - 1 \right]$$

Table 15-5. Quantities for the Integration of Eq. (15-88)

y^+	u^+	R or R_q	$\overline{V}$	$R\overline{V}$	$\int_0^{Rq} R\overline{V}\,dR$	$\left(\int_0^{Rq} R\overline{V}\,dR\right)^2$	$\frac{\epsilon_M}{\nu}$	$\frac{K}{k}$	$R_q \frac{K}{k}$	$\frac{\left(\int_0^{Rq} RV\,dR\right)^2}{R_q(K/k)}$
0	0	1	0	0	0.49198	0.24303	0	1	1	0.24303
1	1	0.9988	0.0541	0.0540	0.49195	0.24201	0	1	0.9988	0.24230
2	2	0.9975	0.1082	0.1079	0.49184	0.24191	0	1	0.9975	0.24252
3	3	0.9963	0.1623	0.1617	0.49167	0.24174	0	1	0.9963	0.24264
4	4	0.9951	0.2164	0.2153	0.49144	0.24152	0	1	0.9951	0.24271
5	5	0.9938	0.2705	0.2688	0.49115	0.24123	0	1	0.9938	0.24273
6	5.91	0.9926	0.3197	0.3173	0.49078	0.24086	0.1911	2.479	2.4646	0.09772
8	7.34	0.9901	0.3971	0.3932	0.48993	0.24003	0.5842	5.522	5.4670	0.04391
10	8.45	0.9877	0.4571	0.4515	0.48887	0.24000	0.9753	8.549	8.4435	0.02834
15	10.50	0.9815	0.5681	0.5576	0.48571	0.23591	1.945	16.05	15.775	0.01495
20	11.93	0.9753	0.6454	0.6295	0.48192	0.23225	2.901	23.46	22.877	0.01015
30	13.95	0.9630	0.7547	0.7267	0.47382	0.22451	3.167	55.47	53.415	0.004203
40	14.73	0.9506	0.7969	0.7575	0.46420	0.21548	14.21	111.0	105.51	0.00204
100	17.00	0.8766	0.9197	0.8046	0.40710	0.16573	36.06	264.6	231.97	0.00071
200	18.75	0.7531	1.0144	0.7639	0.30670	0.09407	60.25	467.3	351.92	0.00027
500	21.04	0.3827	1.1383	0.4357	0.08400	0.00714	75.54	585.7	224.17	0.000032
700	21.89	0.1358	1.1842	0.1608	0.01100	0.00012	37.02	287.6	39.07	0.000003
810	22.25	0	1.2037	0	0	0	−1	−6.74	0	0

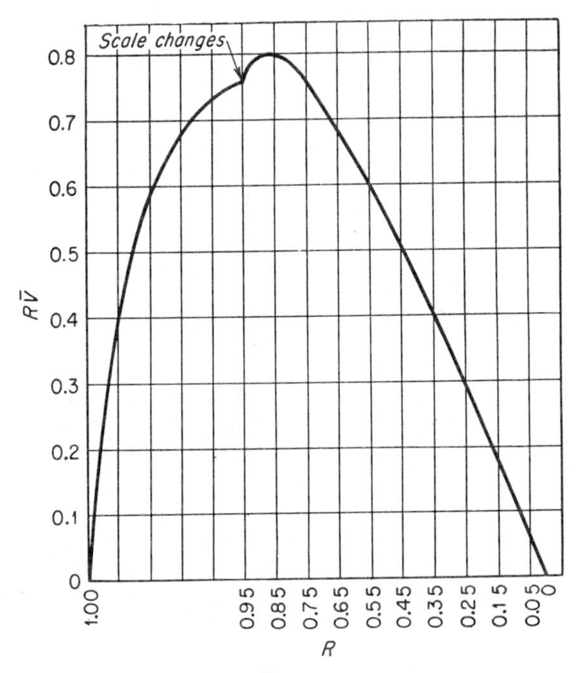

FIG. 15-25. Plot of $R\overline{V}$ versus R for Example 15-2.

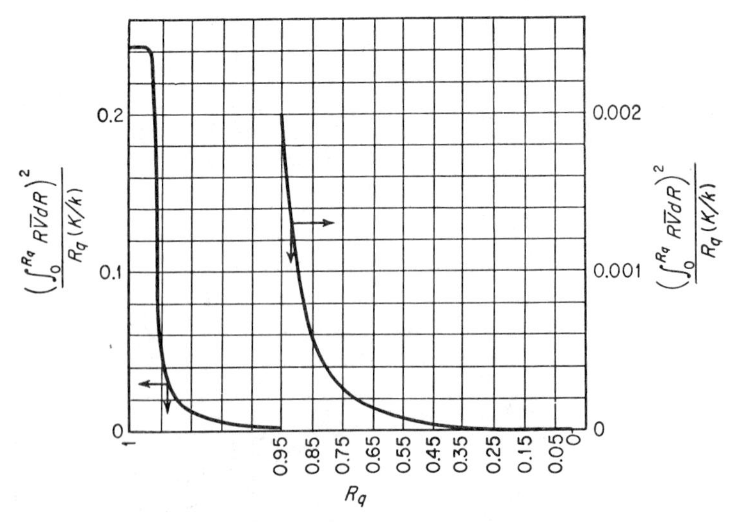

FIG. 15-26. Plot of $\left(\int_o^{R_q} R\overline{V}\, dR\right)^2/[R_q(K/k)]$ versus R_q for Example 15-2.

To express R in terms of y^+

$$R = \frac{r}{r_w} = 1 - \frac{y}{r_w}$$

$$= 1 - \frac{y^+}{(\mathrm{Re}/2)\sqrt{f/2}}$$

$$= 1 - \frac{y^+}{810}$$

To express $\overline{V}$ in terms of u^+

$$V = \frac{u}{U} = \frac{u^+u^*}{U} = \frac{u^+U\sqrt{f/2}}{U} = u^+\sqrt{\frac{f}{2}} = 0.0542u^+$$

The various quantities for the integration of Eq. (15-88) are shown in Table 15-5. In Fig. 15-25 $R\overline{V}$ is plotted versus R, and integration of the resulting curve gives values of $\int_0^{R_q} RV\,dR$. These values are then used to calculate values of $\left(\int_0^{R_q} R\overline{V}\,dR\right)^2/[R_q(K/k)]$, which is plotted versus R_q in Fig. 15-26. Integration of the curve in Fig. 15-26 from $R_q = 0$ to $R_q = 1$ gives

$$\int_0^1 \frac{\left(\int_0^{R_q} R\overline{V}\,dR\right)^2}{R_q(K/k)}\,dR_q = 2.70 \times 10^{-3}$$

Substituting in Eq. (15-88),

$$\mathrm{Nu} = \frac{1}{2(2.7 \times 10^{-3})} = 185$$

The value calculated by Lyon's equation is about 10 per cent below that calculated by Martinelli's method.

(c) By Eq. (14-11), assuming $\mu_b/\mu_w = 1$,

$$\mathrm{Nu} = (0.023)(30{,}000)^{0.8}(7.74)^{1/3} = 174$$

Values of the Nusselt number obtained in parts (a), (b), and (c) compare well with each other.

15-19. Lyon's Analogy for Heat Transfer in Smooth Annuli

Theoretical equations for the calculation of heat-transfer coefficients in annuli are not so numerous as for circular tubes. This is due to the fact that the velocity distribution for circular tubes is well known, except in the region immediately adjacent to the wall. Only a few velocity-distribution data are available for annuli, some work having been done by Rothfus [23] and Knudsen and Katz.[12]

Following the procedure Lyon used for circular tubes, Knudsen and Katz developed an analogous equation for calculating the heat-transfer coefficient in annuli.

$$h = \frac{(r_2^2 - r_1^2)^2 U^2}{4r_1 \int_{r_1}^{r_2} r_T u_T \int_{r_1}^{r_T} \frac{\int_{r_q}^{r_2} ru\, dr}{r_q K}\, dr_q\, dr_T} \tag{15-113}$$

Bailey [1] also derived the same equation for the heat-transfer coefficient in annuli but proceeded further, to include dimensionless quantities so that the coefficient could be expressed in terms of the fluid properties, the velocity distribution, and the diameter ratio of the annulus. Bailey defined the quantity R, where $R = (r - r_1)/(r_2 - r_1)$, so that, when $r = r_1$, $R = 0$, and when $r = r_2$, $R = 1$. Also, $dr = dR/(r_2 - r_1)$. Substituting these values into Eq. (15-113) results in the following expression for the Nusselt number:

$$\text{Nu} = \frac{(r_2^2 - r_1^2)^2}{2(r_2 - r_1)^2 r_1 \int_0^1 r_T \overline{V}_T \int_0^{R_T} \frac{\int_{R_q}^1 r\overline{V}\, dR}{r_q(K/k)}\, dR_q\, dR_T} \tag{15-114}$$

The same difficulty is encountered in the solution of Eq. (15-114) as in that of (15-88); i.e., for fluids having large Prandtl numbers it is necessary to know the thickness of the laminar layer very accurately, and existing data for this region in annuli are even less definite than the corresponding data for circular tubes. Another difficulty is encountered in the solution of Eq. (15-114) because the integration must be carried out over the whole profile in the annular space. This requires an accurate knowledge of the position of maximum velocity.

For circular tubes the conductivity ratio K/k may be obtained from Eq. (15-103).

$$\frac{K}{k} = 1 + \frac{\epsilon_H}{\nu}\,\text{Pr} \tag{15-103}$$

The shear at any point on the outer and inner portions of the annular velocity profile may be obtained from Eqs. (4-68), (4-69), and (4-72).

$$\left(\frac{\tau g_c}{\rho}\right)_2 = U^2 \frac{f}{2} \frac{r^2 - r_{\max}^2}{r(r_2 - r_1)} \tag{15-115}$$

and

$$\left(\frac{\tau g_c}{\rho}\right)_1 = U^2 \frac{f}{2} \frac{r_{\max}^2 - r^2}{r(r_2 - r_1)} \tag{15-116}$$

where the subscript 1 refers to the inner portion of the velocity profile and the subscript 2 refers to the outer portion. Equations (15-115) and (15-116) give the shear at any point on the annular velocity profile. From these two relationships and Eq. (15-1) it may be shown that

$$\left(\frac{\epsilon_M}{\nu}\right)_2 = \left[\frac{fU^2}{2\nu r}\,\frac{(r^2 - r_{\max}^2)/(r_2 - r_1)}{(du/dy)_2}\right] - 1 \qquad (15\text{-}117)$$

and

$$\left(\frac{\epsilon_M}{\nu}\right)_1 = \left[\frac{fU^2}{2\nu r}\,\frac{(r_{\max}^2 - r^2)/(r_2 - r_1)}{(du/dy)_1}\right] - 1 \qquad (15\text{-}118)$$

Equations (15-117) and (15-118) may be used to determine the value of K/k in Eq. (15-114).

Equation (15-114) has been evaluated for water flowing in an annulus by means of the velocity distributions obtained by Knudsen and Katz.[12, 13] Since water has a relatively large Prandtl number, it is necessary to know the thickness of the laminar layer adjacent to the wall. This thickness was calculated from the turbulent-velocity-distribution equation given by Eqs. (7-90) and (7-91). It may also be shown from Eqs. (4-64), (4-65), (15-115), and (15-116) that the velocity in the laminar layer may be expressed as

$$u_2 = \frac{fU^2}{2\nu}\,\frac{r_2{}^2 - r_{\max}^2}{r_2(r_2 - r_1)}\,y_2 \qquad (15\text{-}119)$$

and

$$u_1 = \frac{fU^2}{2\nu}\,\frac{r_{\max}^2 - r_1{}^2}{r_1(r_2 - r_1)}\,y_1 \qquad (15\text{-}120)$$

Assuming that no buffer zone exists, the velocity at the edge of the laminar layer given by the turbulent-core equations is the same as the velocity given by the laminar-layer equations. Hence, equating Eqs. (7-91) and (15-120) and solving for y_1, the calculated thickness of the laminar layer is obtained. Equation (15-114) was solved using this calculated value of the laminar-layer thickness, and it was also solved using one-half of the calculated thickness.

It is found that the value of K/k calculated from Eq. (15-103) has maximum values on either side of the point of maximum velocity and becomes zero at the point of maximum velocity. This is probably not true physically, so that values of K/k in the region of the point of maximum velocity are determined by linear interpolation between the two maximum points.† The sketch given in Fig. 15-27 illustrates this procedure, where K/k is plotted against the radius. The solid line represents calculated values of

† Little error is involved in this assumption.

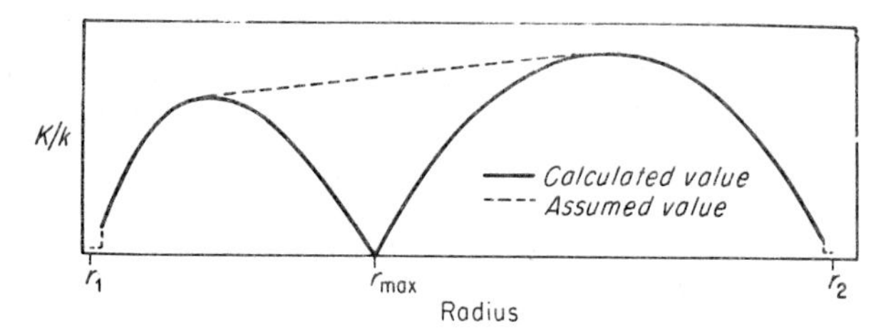

FIG. 15-27. Value of K/k for smooth concentric annuli.

K/k. The broken line represents assumed values of K/k which were used in the solution of Eq. (15-114).

Table 15-6 shows the results obtained from solving Eq. (15-114) for water flowing in an annulus.

TABLE 15-6. CALCULATED NUSSELT NUMBERS FOR WATER FLOWING IN AN ANNULUS ($r_1 = 0.311$ in.; $r_2 = 1.120$ in.)

Reynolds number	Velocity, ft/sec	Friction factor	Prandtl number	Calculated laminar-layer thickness, in.	Calculated Nusselt number		
					Using calculated laminar-layer thickness	Using one-half of calculated laminar-layer thickness	Using Eq. (14-17)
9,190	1.032	0.00882	10.0	0.0268	47	84	121
35,700	2.93	0.00652	7.28	0.0092	146	234	323
69,900	4.91	0.0058	6.10	0.0052	227	394	507

All the Nusselt numbers calculated by means of Eq. (15-114) are lower than those predicted by the empirical equation (14-17), indicating that the laminar-layer thicknesses used in solving Eq. (15-114) were too great. It is reasonable to assume that the thickness calculated by Eqs. (7-91) and (15-120) would be too great, since the process involves the extrapolation of both the laminar profile and the turbulent profile into the buffer zone. It appears from the results given in Table 15-6 that the actual thickness of the laminar layer is about one-third the calculated value given in the table. Considering the small thicknesses involved, however, the results given in the table are reasonable.

BIBLIOGRAPHY

1. Bailey, R. V.: *Oak Ridge Natl. Lab. Tech. Div. Eng. Research Sec. ORNL* 521, 1950.
2. Colburn, A. P.: *Trans. AIChE*, **29**:174 (1933).
3. Deissler, R. G.: *NACA TN* 2138, 1950.
4. Deissler, R. G.: *Trans. ASME*, **73**:101 (1951).
5. Deissler, R. G.: *NACA RM* E52F05, 1952.
6. Deissler, R. G.: *NACA TN* 3145, 1954.
7. Deissler, R. G.: *NACA Rept.* 1210, 1955.
8. Deissler, R. G., and C. S. Eian: *NACA TN* 2629, 1952.
9. Isakoff, S. E., and T. B. Drew: "Proceedings of the General Discussion on Heat Transfer," Institution of Mechanical Engineers, London, and American Society of Mechanical Engineers, New York, 1951, p. 405.
10. Kármán, T. von: *Trans. ASME*, **61**:705 (1939).
11. Knudsen, J. G., and D. L. Katz: *Chem. Eng. Progr.*, **46**:490 (1950).
12. Knudsen, J. G., and D. L. Katz: *Proc. Midwest. Conf. Fluid Dynamics, 1st Conf., Univ. Illinois*, 1950, p. 175.
13. Knudsen, J. G., and D. L. Katz: unpublished work.
14. Lyon, R. N.: *Chem. Eng. Progr.*, **47**:75 (1951).
15. Martinelli, R. C.: *Trans. ASME*, **69**:947 (1947).
16. Page, F., W. G. Schlinger, D. K. Breaux, and B. H. Sage: *Ind. Eng..Chem.*, **44**:424 (1952).
17. Pannell, J. R.: *Brit. Aeronaut. Research Comm. R. & M.* 243ii, 1916.
18. Prandtl, L.: *Physik. Z.*, **11**:1072 (1910).
19. Prandtl, L.: *Z. angew. Math. u. Mech.*, **5**:136 (1925); see also *NACA TM* 1231, 1949.
20. Rayleigh, Lord: *Brit. Aeronaut. Research Comm. R. & M.* 497, 1917.
21. Reynolds, O.: *Proc. Manchester Lit. Phil. Soc.*, **14**:7 (1874).
22. Reynolds, O.: *Trans. Roy. Soc. (London)*, **174A**:935 (1883).
23. Rothfus, R. R., C. C. Monrad, and V. E. Senecal: *Ind. Eng. Chem.*, **42**:2511 (1950).
24. Seban, R. A., and T. T. Shimazaki: *Trans. ASME*, **73**:803 (1951).
25. Seban, R. A., and T. T. Shimazaki: "Proceedings of the General Discussion on Heat Transfer," Institution of Mechanical Engineers, London, and American Society of Mechanical Engineers, New York, 1951, p. 122.
26. Stanton, T. E.: *Trans. Roy. Soc. (London)*, **190A**:67 (1897).
27. Stanton, T. E.: *Brit. Aeronaut. Research Comm. R. & M.* 497, appendix, 1917.
28. Winding, C. C.: *Ind. Eng. Chem.*, **30**:942 (1938).

CHAPTER 16

HEAT TRANSFER WITH LIQUID METALS

16-1. Liquid Metals as Heat-transfer Media

In recent years there has been a growing interest in the use of low-melting metals as heat-transfer media. Liquid metals have had limited application in the field of heat transfer for a number of years, but developments in handling and metering liquid metals and their suitability for high-temperature, high-heat-flux applications have led to a considerable amount of theoretical and empirical investigation in this field. Liquid metals have been used extensively to cool valves in aircraft engines. Power generation using mercury instead of water as the working fluid in boilers has been carried out commercially since 1922. Small quantities of liquid metals have found numerous applications in research work as heat-transfer media. The present interest in liquid metals in the field of heat transfer stems from their use in atomic reactors. In this application large quantities of heat must be removed rapidly at very high temperatures and at relatively low temperature differences. Under these conditions liquid metals have certain advantages over water at high pressure.

The common low-melting metals which have proved suitable for heat-transfer purposes are lithium, sodium, potassium, mercury, bismuth, and lead. In addition, sodium-potassium and lead-bismuth alloys have received considerable attention. In the present chapter some of the theoretical and experimental investigations on liquid metals are discussed. The purpose is to show the differences between liquid-metal heat transfer (low Prandtl number) and heat transfer with the common gases and liquids (high Prandtl number). In Chaps. 12 to 15 consideration has been restricted mainly to high-Prandtl-number fluids.

It is beyond the scope of this discussion to cover all the work, both theoretical and experimental, that has been done recently on liquid-metal heat transfer. (For a complete discussion of liquid metals see refs. 7 and 13.) A thorough summary of experimental investigations of liquid-metal heat transfer is given by Lubarsky and Kaufman.[11]

TABLE 16-1. PHYSICAL PROPERTIES OF SOME COMMON LOW-MELTING-POINT METALS

Metal	Melting point, °F	Normal boiling point, °F	Temperature, °F	Density, lb_m/ft^3	Viscosity, $lb_m/(ft)(sec)$	Heat capacity, $Btu/(lb_m)(°F)$	Thermal conductivity, $Btu/(hr)(ft^2)(°F)/ft$	Prandtl number
Bismuth	520	2691	600	625	1.09	0.0345	9.5	0.014
			1400	591	0.53	0.0393	9.0	0.0084
Lead	621	3159	700	658	1.61	0.038	9.3	0.024
			1300	633	0.92	0.037	8.6	0.016
Lithium	354	2403	400	31.6	0.40	1.0	22.0	0.065
			1800	27.6	0.28	1.0		
Mercury	−38.0	675	50	847	1.07	0.033	4.7	0.027
			600	802	0.58	0.032	8.1	0.0084
Potassium	147	1400	300	50.4	0.25	0.19	26.0	0.0066
			1300	42.1	0.09	0.18	19.1	0.0031
Sodium	208	1621	400	56.3	0.29	0.32	46.4	0.0072
			1300	48.6	0.12	0.30	34.5	0.0038
Sodium-Potassium, 22% Na	66.2	1518	200	53.0	0.330	0.226	14.1	0.019
			1400	43.1	0.0981	0.211		
56% Na	12	1443	200	55.4	0.390	0.270	14.8	0.026
			1400	46.2	0.108	0.249	16.7	0.058
Lead-Bismuth, 44.5% Pb	257	3038	550	646	1.18	0.035	6.20	0.024
			1200	614	0.772			

16-2. Properties of Liquid Metals

With liquid metals, as with other fluids, the density, viscosity, thermal conductivity, and specific gravity are the properties to consider in heat transfer. Liquid metals are unique among fluids because of their high thermal conductivity and consequently low Prandtl number. Their viscosity is comparable to that of water. A summary of the properties of the common liquid metals used for heat-transfer purposes is given in Table 16-1. Complete information on physical properties is available elsewhere.[13]

16-3. Advantages and Disadvantages of Liquid Metals as Heat-transfer Media

One major advantage of liquid metals as heat-transfer media is the fact that they are liquid over a wide temperature range. An inspection of Table 16-1 shows that for the metals listed the difference between the normal boiling point and the melting point is greater than 1000°F (except for mercury) and may be over 2000°F. This extremely wide temperature range is possible at pressures up to 1 atm. If water is used as a heat-transfer medium, the total pressure must be as high as 3,000 psi to ensure that the water will remain liquid over a temperature range of about 650°F. The liquid metals can also be used at very high temperatures. An additional advantage of liquid metals is their high heat-transfer coefficient; thus very high heat fluxes, such as are necessary in atomic reactors, may be handled.

The main disadvantage of using liquid metals for heat-transfer purposes is the difficulty of handling them, but techniques have been developed which virtually eliminate this disadvantage. Sodium and sodium-potassium alloys, which have the most desirable properties because of their high heat-transfer capacity and low pumping power, are dangerous to handle, especially if pumping and storing equipment should fail. They react violently with water and chlorinated hydrocarbons, and they also ignite in air at elevated temperatures. Even from the handling standpoint, liquid metals possess some advantages over other liquids, one being the fact that electromagnetic pumps and metering equipment may be used conveniently. Such equipment has no moving parts and is superior to conventional equipment from this standpoint.

I. LIQUID-METAL HEAT TRANSFER IN CIRCULAR TUBES

16-4. Laminar-flow Heat Transfer with Liquid Metals

Very few empirical data exist on heat transfer during laminar flow of liquid metals in circular tubes. It is generally stated [13] that the relationships derived in Chap. 13 for laminar flow in tubes are suitable for liquid metals as well as for other fluids. Results in Sec. 13-4 indicated that limiting Nusselt numbers Nu_∞ were 3.66 for constant wall temperature and 4.36 for uniform heat flux and uniform wall-temperature gradient. These limiting Nusselt numbers are presumed to apply for liquid metals also.

Johnson, Hartnett, and Clabaugh [9] measured Nusselt numbers for mercury and the lead-bismuth eutectic during laminar and transition flow in circular tubes and obtained Nusselt numbers considerably below the limiting value of 4.36 predicted in Chap. 13. The solid curve shown in Fig. 16-1 represents the data of Johnson, Hartnett, and Clabaugh within ±20

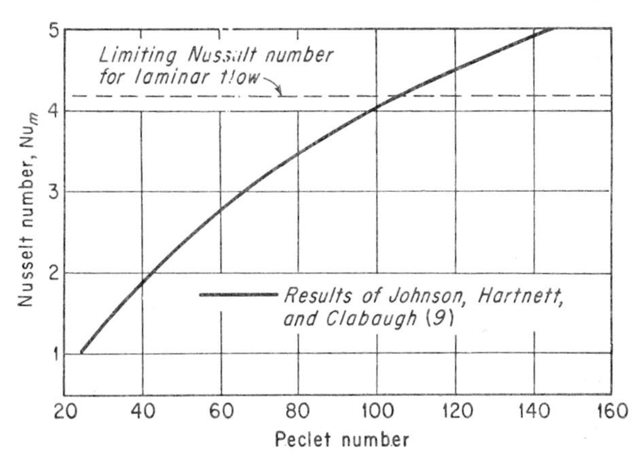

FIG. 16-1. Heat transfer for laminar and transition flow of liquid metals in circular tubes.

per cent. The data cover a range of Peclet numbers between 25 and 150, with Reynolds numbers ranging from 1,000 to 10,000. The laminar range is at Peclet numbers below 40, and the measured Nusselt numbers here are much below the theoretical value of 4.36 indicated by the broken line. These investigators could find no reason (such as oxide or gaseous films on the pipe wall or nonwetting of the wall by the molten metal) to account for the low values of Nusselt number obtained.

16-5. The Analogy between Momentum and Heat Transfer Applied to Liquid Metals in Turbulent Flow

When engineers began designing heat-exchange equipment for use with liquid metals, the usual empirical equations for predicting heat-transfer coefficients were found to be unsuitable. Such relationships as Eqs. (14-9) to (14-11) were developed for fluids with Prandtl numbers greater than 0.7. These equations do not apply for fluids with Prandtl numbers of the order of 0.001 to 0.1, which is in the range for most liquid metals.

The effect of low Prandtl number on the Nusselt number is shown in Fig. 15-11, where the von Kármán, Colburn, and Prandtl analogies are compared with Eq. (14-11). The von Kármán analogy predicts Nusselt numbers (for Pr = 0.01) much lower than Colburn's analogy or Eq. (14-11). It also predicts Nusselt numbers much closer to those measured experimentally for liquid metals. Consequently, there has been extensive application of the analogies between heat transfer and momentum transfer, particularly those of Martinelli, Lyon, and Seban and Shimazaki, to turbulent liquid-metal heat transfer in circular tubes.

16-6. Lyon's Analogy Applied to Liquid-metal Heat Transfer; Uniform Heat Flux

The various relationships given in Chap. 15 for predicting Nusselt numbers from velocity-distribution data are well suited for application to liquid metals. Since liquid metals have a high thermal conductivity, the total conductivity in turbulent flow (molecular conductivity + eddy conductivity) is nearly equal to the molecular conductivity. In other words, the eddy conductivity does not become so large (relatively) as it does for fluids like water (see Example 15-2). The following example will illustrate these statements.

Example 16-1

Calculate the Nusselt number for liquid sodium flowing in a 2-in.-ID smooth pipe at a Reynolds number of 30,000. The metal temperature is 1300°F.

(*a*) Use Martinelli's analogy (Table 15-4).

(*b*) Use Lyon's equation (15-88).

Solution

From Example 15-2

$$\sqrt{\frac{f}{2}} = 0.0542 \qquad \text{Re}\sqrt{\frac{f}{2}} = 1{,}620$$

At 1300°F, from Table 16-1,

$$\text{Pr} = 0.0038$$
$$\text{Pe} = \text{PrRe} = (30{,}000)(0.0038) = 114$$

(*a*) *Martinelli's solution.* From Table 15-4 by graphical interpolation

$$\text{Nu} = 8.1$$

(*b*) By Lyon's method Eq. (15-88) must be integrated.

$$\text{Nu} = \frac{1}{2\displaystyle\int_0^1 \frac{\left(\int_0^{R_q} RV\,dR\right)^2}{R_q(K/k)}\,dR_q}$$

The quantities for the integration of this equation are computed as in Example 15-2 and are tabulated in Table 16-2.

Table 16-2. Data for Integration of Eq. (15-88) (Example 16-1)

y^+	u^+	R or R_q	V	$R\overline{V}$	$\int_0^{R_q} R\overline{V}\,dR$	$\left(\int_0^{R_q} R\overline{V}\,dR\right)$	$\frac{\epsilon_M}{\nu}$	$\frac{K}{k}$	$R_q \frac{K}{k}$	$\frac{\left(\int_0^{R_q} R\overline{V}\,dR\right)^2}{R_q(K/k)}$
0	0	1	0	0	0.49198	0.24303	0	1	1	0.24303
1	1	0.99877	0.0541	0.0540	0.49195	0.24201	0	1	0.99877	0.24231
2	2	0.99753	0.1082	0.1079	0.49184	0.24191	0	1	0.99753	0.24251
3	3	0.99630	0.1623	0.1617	0.49167	0.24174	0	1	0.99630	0.24264
4	4	0.99506	0.2164	0.2153	0.49144	0.24151	0	1	0.99506	0.24271
5	5	0.99380	0.2705	0.2688	0.49115	0.24123	0	1	0.99380	0.24273
6	5.91	0.99259	0.3197	0.3173	0.49078	0.24086	0.1911	1.0007	0.99328	0.24249
8	7.34	0.99012	0.3971	0.3932	0.48993	0.24003	0.5842	1.0022	0.99230	0.24189
10	8.45	0.98766	0.4571	0.4515	0.48887	0.23899	0.9753	1.0037	0.99131	0.24109
15	10.50	0.98148	0.5681	0.5576	0.48571	0.23591	1.945	1.0074	0.98874	0.23860
20	11.93	0.97531	0.6454	0.6295	0.48192	0.23225	2.901	1.0110	0.98604	0.23554
30	13.95	0.96296	0.7547	0.7267	0.47382	0.22451	3.167	1.0272	0.98915	0.22697
40	14.73	0.95062	0.7969	0.7575	0.46420	0.21548	14.21	1.0540	1.00195	0.21506
100	17.00	0.87655	0.9197	0.8046	0.40710	0.16573	36.06	1.1370	0.99664	0.16629
200	18.75	0.75309	1.0144	0.7639	0.30670	0.09406	60.25	1.2290	0.92555	0.10163
500	21.04	0.38272	1.1383	0.4357	0.08400	0.00714	75.54	1.2870	0.49256	0.003480
700	21.89	0.13580	1.1842	0.1608	0.01100	0.000121	37.02	1.1407	0.15491	0.000732
810	22.25	0	1.2037	0	0	0	−1	0.9962	0	0

In Fig. 16-2 $\left(\int_0^{R_q} RV\, dR\right)^2/[R_q(K/k)]$ is plotted versus R_q, and the area under the curve is determined between $R_q = 0$ and $R_q = 1$.

From this graphical integration

$$\int_0^1 \frac{\left(\int_0^{R_q} R\overline{V}\, dR\right)^2}{R_q(K/k)}\, dR_q = 0.0568$$

Thus, from Eq. (15-88),

$$\text{Nu} = \frac{1}{(2)(0.0568)} = 8.80$$

This value agrees well with that predicted by Martinelli's equation.

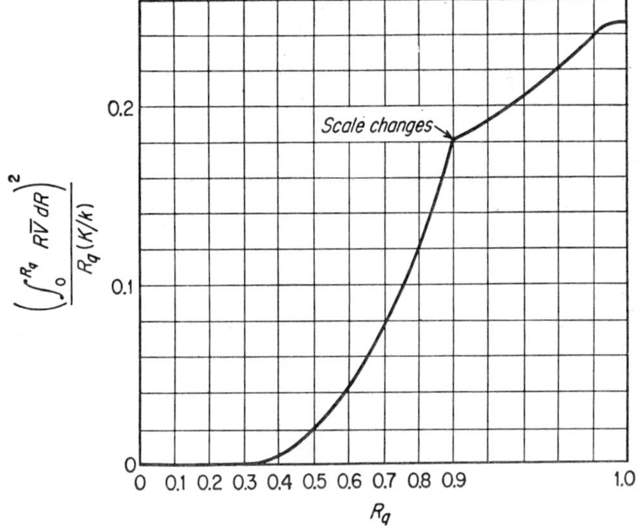

FIG. 16-2. Plot of $\left(\int_0^{R_q} R\overline{V}\, dR\right)^2/[R_q(K/k)]$ versus R_q for Example 16-1.

It is interesting to note the values of K/k in Table 16-2 and compare them with the values of K/k calculated for water in Table 15-5. For the liquid sodium considered in Table 16-2 the total conductivity K is never more than 1.3 times as great as the thermal conductivity k, which means that molecular heat transfer is predominant throughout the cross section of the tube. For this reason, the laminar boundary thickness is not important, since the resistance of the laminar layer is only a small fraction of the total resistance to heat transfer. In Table 15-5 for water the value of K is as much as 600 times the value of k. In this case the thickness of the laminar layer is important, since it offers the major portion of the total resistance to heat transfer.

Equation (15-88) may be used to calculate the Nusselt numbers for slug flow of liquid metals in tubes. In slug flow the velocity profile is uniform, and hence $\overline{V}$ becomes unity. Assuming $K = k$ (a valid assumption, as indicated by Table 16-2), Eq. (15-88) becomes

$$(\mathrm{Nu}_s)_\infty = \frac{1}{2\int_0^1 \frac{\left(\int_0^{R_q} R\,dR\right)^2}{R_q}\,dR_q} \tag{16-1}$$

which upon integration gives a value of $(\mathrm{Nu}_s)_\infty$ as follows:

$$(\mathrm{Nu}_s)_\infty = 8.0 \tag{16-2}$$

Equation (16-2) gives the value of the Nusselt number for slug flow of fluids in which total conductivity can be assumed equal to the molecular conductivity.

Lyon [12] used Eq. (15-88) to develop a relatively simple expression for predicting Nusselt numbers for heat transfer with liquid metals. He integrated Eq. (15-88) using Nikuradse's [15] velocity-distribution data for smooth tubes. Assuming $K = k$, Lyon obtained Nusselt numbers between 6.84 and 7.15 in the range of Reynolds numbers investigated by Nikuradse. On the basis of these results, Lyon proposed the following equation for predicting Nusselt numbers for liquid metals:

$$\mathrm{Nu}_m = 7 + 0.025(\mathrm{Pe})^{0.8} \tag{16-3}$$

Conditions:

(1) Properties evaluated at bulk temperature
(2) $\mathrm{Pe} > 100$
(3) $L/d_w > 60$
(4) Uniform heat flux

Equation (16-3) is recommended for calculating Nusselt numbers for sodium-potassium alloys (see Sec. 16-9).

16-7. Seban and Shimazaki's Equation; Constant Wall Temperature

Seban and Shimazaki [18] proposed an equation for predicting Nusselt numbers for liquid metals. This relation was obtained from their analytical equations (15-65) to (15-68). For low-Prandtl-number fluids, with heat transfer taking place at constant wall temperature, these authors propose the relation

$$\mathrm{Nu}_m = 5.0 + 0.025(\mathrm{Pe})^{0.8} \tag{16-4}$$

Conditions:

(1) Properties evaluated at bulk temperature
(2) $\mathrm{Pe} > 100$
(3) $L/d_w > 60$
(4) Constant wall temperature

The difference between the conditions of constant wall temperature and uniform heat flux is evident in Eqs. (16-3) and (16-4) (also compared in Table 15-4). For low Peclet numbers, lower Nusselt numbers are obtainable with constant wall temperature than with uniform heat flux. At high Peclet numbers there is little difference between the two conditions.

16-8. Deissler's Equation for Liquid Metals

Deissler [2] proposed a relation for predicting Nusselt numbers for liquid metals based on the assumptions indicated in Eqs. (15-106) and (15-107) and taking the relation between ϵ_H and ϵ_M to be

$$\frac{\epsilon_H}{\epsilon_M} = n\mathrm{Pe}[1 - e^{-(1/n\mathrm{Pe})}] \qquad (15\text{-}104)$$

where $n = 0.000153$. Deissler's analytical results can be represented very closely by the equation

$$\mathrm{Nu}_m = 6.3 + 0.000222(\mathrm{Pe})^{1.3} \qquad (16\text{-}5)$$

This equation agrees favorably with a large number of heat-transfer data on liquid metals.

16-9. Experimental Heat-transfer Data on Liquid Metals and Comparison with Analytical Results

1. *Uniform Heat Flux.* Considerable experimental work has been done on the determination of heat-transfer coefficients for liquid metals, most of it being on sodium-potassium alloys, mercury, and lead-bismuth alloys. In general, the empirical data for sodium-potassium alloys agree well with Lyon's equation (16-3) for uniform heat flux. Most of the experimental coefficients on mercury and lead-bismuth alloys are considerably below those predicted by Lyon's equation, and an empirical relationship is recommended for these liquid metals. A thorough review of the available experimental data on liquid-metal heat transfer is given by Lubarsky and Kaufman.[11]

Lyon measured over-all heat-transfer coefficients in a double-tube heat exchanger with sodium-potassium alloy (50 per cent sodium) as both the hot and cold fluids. He also predicted over-all coefficients from Eq. (16-3)

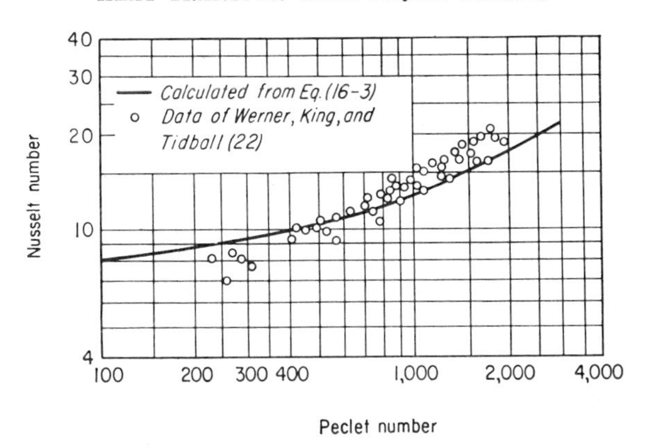

FIG. 16-3. Nusselt number as a function of Peclet number for the turbulent flow of liquid metals in circular tubes.

for circular tubes and from Eq. (16-10) for annuli. Reasonable agreement is obtained between predicted and measured over-all coefficients. Werner, King, and Tidball [22] report heat-transfer coefficients for sodium-potassium alloys flowing in circular tubes; their data are shown in Fig. 16-3. The experimental data compare favorably with the solid line, which represents Eq. (16-3).

In Fig. 16-4 the experimental data of Johnson, Hartnett, and Clabaugh [8] on the lead-bismuth eutectic and of Trefethen [20] on mercury are plotted.

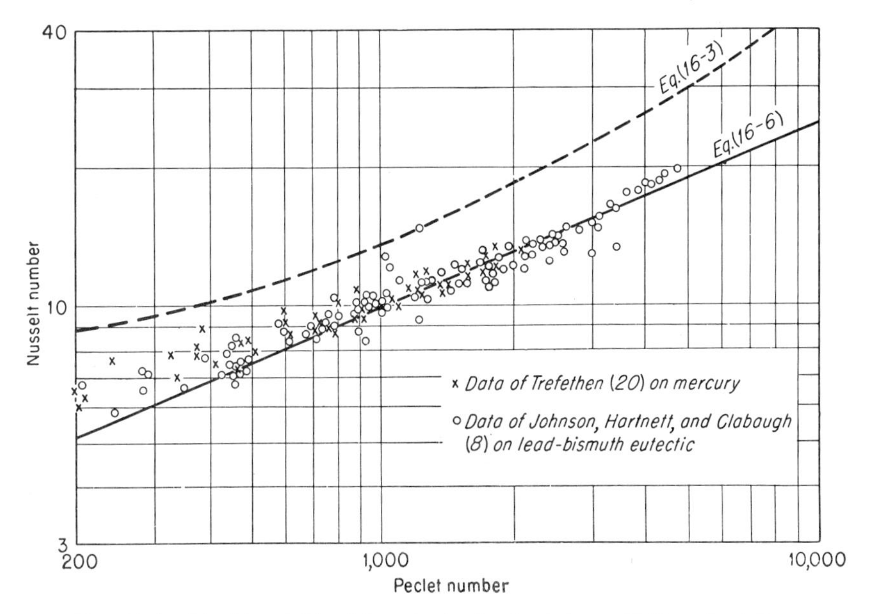

FIG. 16-4. Liquid-metal heat transfer in circular tubes. Uniform heat flux.

Most available experimental data on mercury and lead-bismuth alloys agree with those of the above workers. The broken line represents Lyon's equation (16-3), and is seen to lie somewhat above the experimental points. The solid line represents the empirical relation recommended by Lubarsky and Kaufman.[11]

$$\mathrm{Nu}_m = 0.625(\mathrm{Pe})^{0.4} \tag{16-6}$$

Conditions:

(1) Properties evaluated at bulk temperature
(2) $10^2 < \mathrm{Pe} < 10^4$
(3) Uniform heat input
(4) $L/d_w > 60$

2. *Uniform Wall Temperature.* Several investigations of heat transfer with liquid metals have been carried out under conditions of constant wall

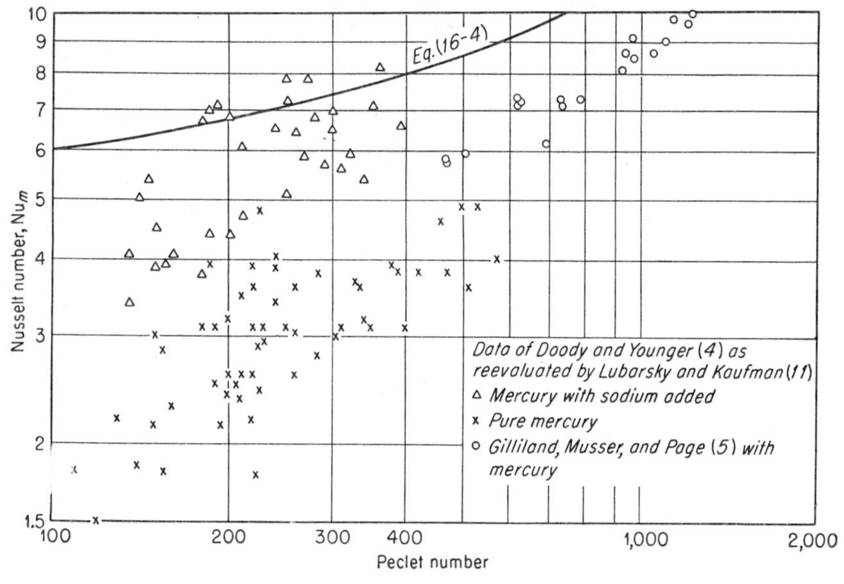

FIG. 16-5. Liquid-metal heat transfer in circular tubes. Constant wall temperature.

temperature. Gilliland, Musser, and Page[5] measured heat-transfer coefficients for mercury flowing in a circular tube surrounded by condensing steam. Doody and Younger[4] studied heat transfer to mercury and mercury-sodium using hot water as a heating medium, which represents a condition between uniform heat flux and constant wall temperature. The experimental data of these workers are plotted in Fig. 16-5. The curve represents Eq. (16-4), which is Seban and Shimazaki's theoretical equation for heat transfer to liquid metals at constant wall temperature. The data of Gilliland et al. fall below the curve, while those of Doody and

Younger for mercury-sodium alloys come near the curve. The data of the latter workers for pure mercury fall much below that for mercury-sodium-alloy solutions. They attribute the low heat-transfer coefficients of pure mercury to the fact that the fluid does not wet the tube wall.

16-10. The Effect of Wetting in Liquid-metal Heat Transfer

Experimental liquid-metal heat-transfer coefficients vary widely. Most data on sodium and sodium-potassium alloys agree well with Lyon's theoretical equation (16-3). On the other hand, data on mercury and lead-bismuth alloys fall considerably below Lyon's equation. Many investigators have attributed the low coefficients of mercury and the lead-bismuth alloys to the fact that, since these metals do not wet the surface, there is a contact resistance which is measured along with the heat-transfer coefficient. The data on these metals are therefore greatly influenced by the condition of the tube wall.

It is not yet established definitely whether this wetting or nonwetting effect is the complete cause of the wide variation of data. Many investigators claim to have observed a physical effect of wetting. The data of Doody and Younger shown in Fig. 16-5 indicate much higher coefficients for mercury-sodium alloys, which wet the wall, than for mercury, which does not wet the wall. On the other hand, an almost equal number of investigators report no difference between wetting and nonwetting metals. Figures 16-6 and 16-7 are plots of data obtained by Lubarsky [10] and Stromquist,[19] respectively. The former studied heat transfer with the lead-bismuth eutectic with and without magnesium, which was used as a wetting agent. Some small difference can be seen between the wetting and nonwetting data. Similar results were obtained by Stromquist, who investigated heat transfer with mercury with and without sodium as a wetting agent.

Other factors besides the wetting effect have been suggested as possible reasons for the wide variation of data, e.g., gas entrainment, insulating gas layer, and oxide films. Johnson, Hartnett, and Clabaugh [9] report that these factors do not account for the low observed heat-transfer coefficients in laminar flow.

MacDonald and Quittenton [14] attribute the wide scattering of liquid-metal heat-transfer data to gas entrainment in the fluid. They studied heat transfer with liquid sodium and observed a large variation in the heat-transfer coefficient at the same conditions of temperature and flow rate. Their experimental equipment was designed in such a way that there was a definite possibility of entraining varying amounts of argon gas in the liquid sodium. They report that gas concentrations as low as 0.1 per cent (by volume) produced marked reductions in heat-transfer rates.

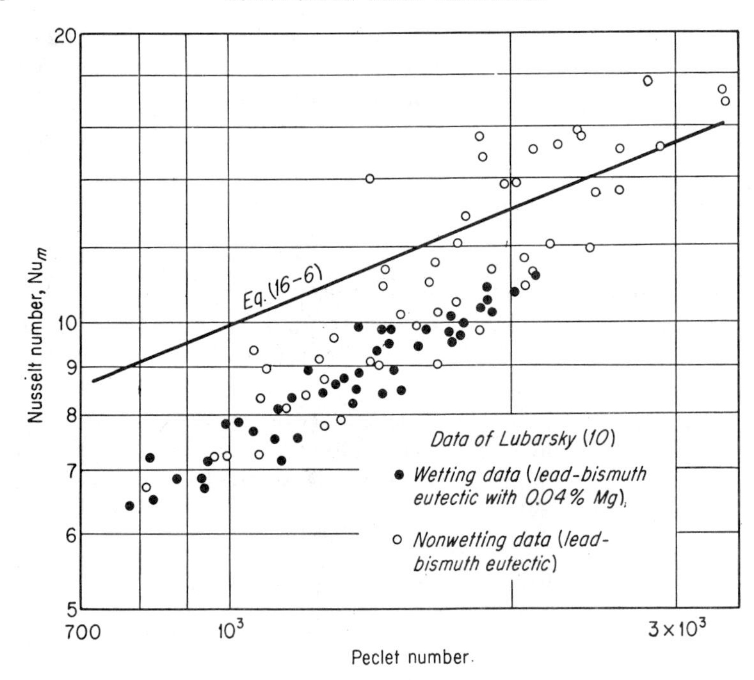

FIG. 16-6. Liquid-metal heat transfer in circular tubes under wetting and nonwetting conditions. (*Results from B. Lubarsky, NACA RM* E51G02, 1951.)

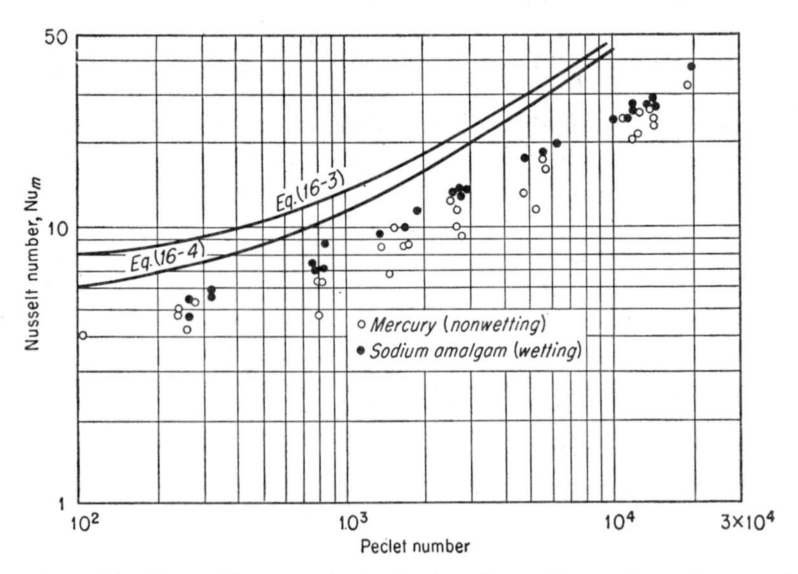

FIG. 16-7. Liquid-metal heat transfer in circular tubes under wetting and nonwetting conditions. (*Results from W. K. Stromquist, USAEC Tech. Inform. Service* ORO-93, *March*, 1953.)

16-11. Liquid-metal Heat Transfer in Entrance Regions; Velocity and Temperature Profile Developing

The theoretical and empirical equations presented above for the prediction of liquid-metal heat-transfer coefficients apply to long tubes with values of L/d_w greater than 60. Many of the experimental data have been obtained on long tubes, and no extensive data are available on the local coefficients in the entrance region or on average coefficients in the entrance region. Poppendiek and Harrison [16] mathematically analyzed liquid-metal heat transfer in the entrance section of a circular tube. They propose the following equations for the local and mean Nusselt numbers respectively:

$$\mathrm{Nu} = \frac{1}{\Gamma[(n+3)/(n+2)]}\left[\frac{n+1}{2^{1-n}(n+2)}\right]^{1/(n+2)}\left(\mathrm{Pe}\frac{d_w}{x}\right)^{1/(n+2)} \tag{16-7}$$

$$\mathrm{Nu}_m = \frac{n+2}{(n+1)\Gamma[(n+3)/(n+2)]}\left[\frac{n+1}{2^{1-n}(n+2)}\right]^{1/(n+2)}\left(\mathrm{Pe}\frac{d_w}{L}\right)^{1/(n+2)} \tag{16-8}$$

where Nu_m is the mean Nusselt number over a length of tube L. Equations (16-7) and (16-8) are based on a velocity distribution which follows a power law, and n is the exponent in the power-law expression. The value of n is $1/7$ for turbulent flow in circular tubes. The solid line in Fig. 16-8 is a

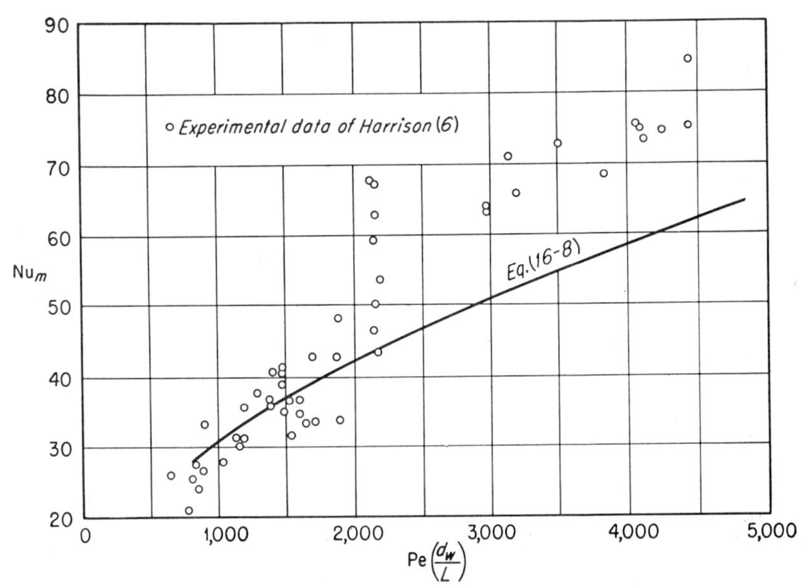

FIG. 16-8. Liquid-metal heat transfer in the entrance region of a circular tube. (*From W. B. Harrison, Oak Ridge Natl. Lab. Reactor Exptl. Eng. Div. ORNL* 915, 1954.)

plot of Eq. (16-8). The data points represent the results obtained by Harrison [6] on heat transfer with mercury. The experimental results agree reasonably well with the theoretical line. Harrison also obtained data on sodium, but, because of nonwetting conditions, they were very erratic.

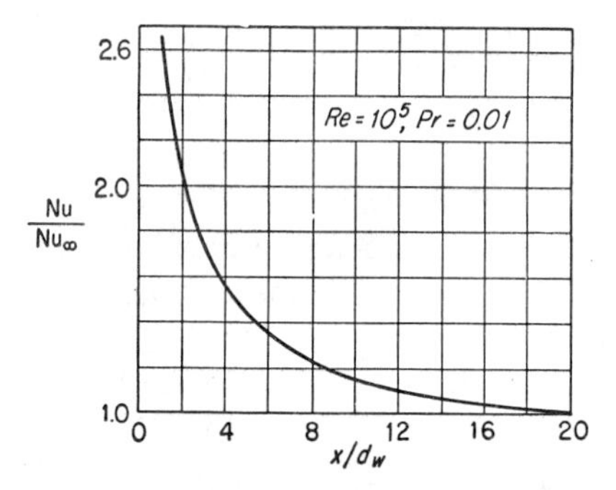

FIG. 16-9. Liquid-metal heat transfer in the entrance region of a circular tube. (*From R. G. Deissler, NACA TN* 3016, 1953.)

An analysis of liquid-metal heat transfer in entrance regions was carried out by Deissler,[3] whose results for a Prandtl number of 0.01 and a Reynolds number of 10^5 are shown by the curve in Fig. 16-9, in which $\mathrm{Nu}/\mathrm{Nu}_\infty$ is plotted versus x/d_w.

II. LIQUID-METAL HEAT TRANSFER IN ANNULI AND BETWEEN PARALLEL PLANES

16-12. Analytical and Experimental Results of Liquid-metal Heat Transfer in Annuli and between Parallel Planes

As with liquid-metal heat transfer in circular tubes, Eqs. (15-113) and (15-114) may be simplified by assuming slug flow ($\overline{V} = 1$) and taking $K = k$. Under these conditions the Nusselt number may be expressed by Eq. (13-40), which was derived by Trefethen [21] and Bailey.[1]

$$(\mathrm{Nu}_s)_\infty = \frac{h(d_2 - d_1)}{k}$$

$$= \frac{8(d_2/d_1 - 1)[(d_2/d_1)^2 - 1]^2}{4(d_2/d_1)^4 \ln (d_2/d_1) - 3(d_2/d_1)^4 - 4(d_2/d_1)^2 - 1} \tag{13-40}$$

Table 13-3 gives values of $(\mathrm{Nu}_s)_\infty$ as a function of d_2/d_1.

Seban [17] developed an expression for liquid-metal heat transfer between parallel plates with heat transferred through one plate only. This relationship is recommended for use with annuli when d_2/d_1 is less than 1.4.

$$\mathrm{Nu}_m = 5.8 + 0.020(\mathrm{Pe})^{0.8} \tag{16-9}$$

Conditions:

(1) Pe > 50
(2) Nu_m and Pe based on equivalent diameter of system
(3) Applies for parallel plates with heat transfer through one side or annuli with $d_2/d_1 < 1.4$
(4) Uniform heat flux
(5) Sodium, potassium, and their alloys

Werner, King, and Tidball [22] developed an equation for liquid-metal heat transfer in annuli which applies a correction of $0.70(d_2/d_1)^{0.53}$ to the circular-tube equation (16-3).

$$Nu_m = 0.70\left(\frac{d_2}{d_1}\right)^{0.53}[7.0 + 0.025(Pe)^{0.8}] \tag{16-10}$$

The conditions are the same as for Eq. (16-9), but (16-10) applies to annuli only. This correction to the circular-tube equation to predict heat-

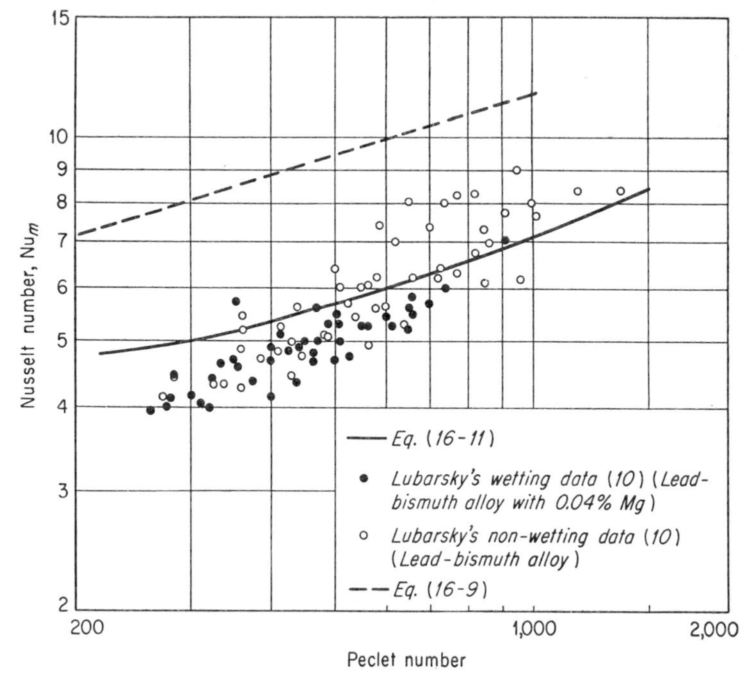

FIG. 16-10. Liquid-metal heat transfer in annuli.

transfer coefficients for annuli is suggested by the fact that Eq. (14-18) for annuli is obtained by multiplying Eq. (14-9), for circular tubes, by the factor $0.87(d_2/d_1)^{0.53}$.

Lubarsky [10] obtained heat-transfer data on a double-pipe heat exchanger using the lead-bismuth eutectic as both the hot and cold fluid. He also studied the effect of wetting by adding 0.04 per cent magnesium to the eutectic. The results are shown in Fig. 16-10 for both the wetting and nonwetting tests. Little difference is detectable between the two sets of data, both being satisfactorily represented by the relation

$$\mathrm{Nu}_m = 3.80 + 0.0133(\mathrm{Pe})^{0.8} \tag{16-11}$$

which is shown as a solid curve. Equation (16-9) is also plotted in Fig. 16-10 and lies somewhat above Lubarsky's data. Equation (16-9) is recommended for sodium-potassium alloys and Eq. (16-11) for mercury and the lead-bismuth eutectic. Equation (16-11) also represents Lubarsky's circular-tube data satisfactorily (see Fig. 16-6).

BIBLIOGRAPHY

1. Bailey, R. V.: *Oak Ridge Natl. Lab. Tech. Div. Eng. Research Sec. ORNL* 521, 1950.
2. Deissler, R. G.: *NACA RM* E52F05, 1952.
3. Deissler, R. G.: *NACA TN* 3016, 1953.
4. Doody, T. C., and A. H. Younger: *Chem. Eng. Progr. Symposium Ser.*, [5] **49**:33 (1953).
5. Gilliland, E. R., R. J. Musser, and W. R. Page: "Proceedings of the General Discussion on Heat Transfer," Institution of Mechanical Engineers, London, and American Society of Mechanical Engineers, New York, 1951, p. 402.
6. Harrison, W. B.: *Oak Ridge Natl. Lab. Reactor Exptl. Eng. Div. ORNL* 915, 1954.
7. Jackson, C. B. (ed.): "Liquid Metals Handbook, Sodium and NaK Supplement," 3d ed., USAEC and U.S. Dept. of the Navy, Washington, D.C., 1955.
8. Johnson, H. A., J. P. Hartnett, and W. J. Clabaugh: *Trans. ASME*, **75**:1191 (1953).
9. Johnson, H. A., J. P. Hartnett, and W. J. Clabaugh: paper 53-A-188, presented at annual meeting of the ASME, New York, 1953.
10. Lubarsky, B.: *NACA RM* E51G02, 1951.
11. Lubarsky, B., and S. J. Kaufman: *NACA TN* 3336, 1955.
12. Lyon, R. N.: *Chem. Eng. Progr.*, **47**:75 (1951).
13. Lyon, R. N. (ed.): "Liquid Metals Handbook," 2d ed., USAEC and U.S. Dept. of the Navy, Washington, D.C., 1952.
14. MacDonald, W. C., and R. C. Quittenton: *Chem. Eng. Progr. Symposium Ser.*, [9] **50**:59 (1954).
15. Nikuradse, J.: *VDI-Forschungsheft* 356, 1932.
16. Poppendiek, H. F., and W. B. Harrison: *Chem. Eng. Progr. Symposium Ser.*, [17] **51**:49 (1955).
17. Seban, R. A.: *Trans. ASME*, **72**:789 (1950).
18. Seban, R. A., and T. T. Shimazaki: *Trans. ASME*, **73**:803 (1951).
19. Stromquist, W. K.: *USAEC Tech. Inform. Service* ORO-93, March, 1953.
20. Trefethen, L. M.: *USAEC Tech. Inform. Service* NP1788, July 1, 1950.
21. Trefethen, L. M.: "Proceedings of the General Discussion on Heat Transfer," Institution of Mechanical Engineers, London, and American Society of Mechanical Engineers, New York, 1951, p. 436.
22. Werner, R. C., E. C. King, and R. A. Tidball: paper presented at annual meeting of the AIChE, Pittsburgh, Pa., Dec. 5, 1949.

CHAPTER 17

HEAT TRANSFER DURING INCOMPRESSIBLE FLOW PAST IMMERSED BODIES

17-1. The Thermal Boundary Layer

When fluids flow past immersed bodies, such as plates, cylinders, and spheres, heat is often transferred between the boundary and the fluid. This transfer of heat must take place through the hydrodynamical boundary layer which forms on the immersed object (see Sec. 10-1). For heat

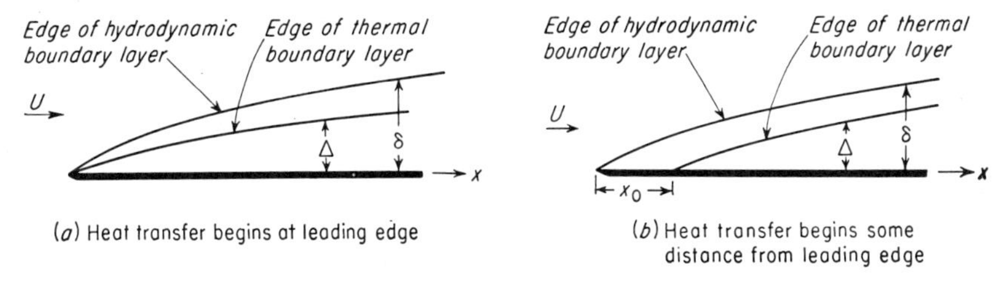

FIG. 17-1. Thermal and hydrodynamical boundary layers on a flat plate.

transfer to take place the temperature of the immersed body must be different from that of the flowing fluid. As the fluid flows past the immersed solid, a hydrodynamical boundary layer, as well as a thermal boundary layer, forms. In the thermal boundary layer the fluid temperature at the solid surface equals that of the surface. On moving away from the surface, the fluid temperature changes and becomes equal to the temperature of the undisturbed stream at the edge of the thermal boundary layer. The thickness of the thermal boundary layer may be different from that of the hydrodynamical boundary layer, and it starts at the point where the heat transfer starts. The thermal and hydrodynamical boundary layers on a flat plate are shown schematically in Fig. 17-1. In Fig. 17-1*a* the heat transfer starts at the leading edge of the plate, while in Fig. 17-1*b* the heat transfer begins a distance x_0 from the leading edge.

17-2. The Equations of Two-dimensional Flow and Heat Transfer in the Boundary Layer; Fluid Properties Constant

The momentum and continuity equations for two-dimensional flow past immersed bodies are

$$u \frac{\partial u}{\partial x} + v \frac{\partial u}{\partial y} = -\frac{g_c}{\rho} \frac{\partial P}{\partial x} + \frac{\mu}{\rho} \frac{\partial^2 u}{\partial y^2} \tag{10-3}$$

$$\frac{\partial u}{\partial x} + \frac{\partial v}{\partial y} = 0 \tag{10-4}$$

The two-dimensional energy equation (neglecting viscous friction) is

$$C_p \rho \left(u \frac{\partial T}{\partial x} + v \frac{\partial T}{\partial y} \right) = k \left(\frac{\partial^2 T}{\partial x^2} + \frac{\partial^2 T}{\partial y^2} \right) \tag{2-58}$$

In the boundary layer $\partial^2 T/\partial x^2$ is much less than $\partial^2 T/\partial y^2$, and so it may be neglected. Equation (2-58) becomes

$$u \frac{\partial T}{\partial x} + v \frac{\partial T}{\partial y} = \alpha \frac{\partial^2 T}{\partial y^2} \tag{17-1}$$

where $\alpha = k/C_p\rho$, the thermal diffusivity.

Equations (10-3), (10-4), and (17-1) are based on the assumption that the fluid properties are independent of temperature. Solution of Eqs. (10-3) and (10-4) gives the velocity distribution in the hydrodynamical boundary layer. Equation (17-1) may then be solved to give the temperature distribution in the thermal boundary layer; i.e., it gives the temperature T as a function of the space coordinates x and y.

17-3. The Integral Energy Equation

In Sec. 10-5 the two-dimensional momentum and continuity equations (10-3) and (10-4) were integrated to give von Kármán's integral momentum equation. In a similar way, the energy equation (17-1) may be integrated to give the integral energy equation.

Equation (17-1) is considered to be valid throughout the thickness Δ of the thermal boundary layer, and it is integrated from $y = 0$ to $y = \Delta$, giving

$$\int_0^\Delta u \frac{\partial T}{\partial x} dy + \int_0^\Delta v \frac{\partial T}{\partial y} dy = \alpha \int_0^\Delta \frac{\partial^2 T}{\partial y^2} dy \tag{17-2}$$

Equation (17-2) becomes

$$\frac{\partial}{\partial x}\int_0^\Delta uT\,dy - \int_0^\Delta \frac{\partial u}{\partial x} T\,dy + [vT]_0^\Delta - \int_0^\Delta T\frac{\partial v}{\partial y}\,dy = -\alpha\left(\frac{\partial T}{\partial y}\right)_{y=0} \qquad (17\text{-}3)$$

From the continuity equation (10-4) and the conditions

At $y = 0$
$v = 0$

At $y = \Delta$

$$v = -\int_0^\Delta \frac{\partial u}{\partial x}\,dy$$

$T = T_\infty$

Eq. (17-3) becomes

$$\frac{\partial}{\partial x}\int_0^\Delta uT\,dy - \int_0^\Delta T\frac{\partial u}{\partial x}\,dy - T_\infty\int_0^\Delta \frac{\partial u}{\partial x}\,dy + \int_0^\Delta T\frac{\partial u}{\partial x}\,dy = -\alpha\left(\frac{\partial T}{\partial y}\right)_{y=0} \qquad (17\text{-}4)$$

which reduces to

$$\frac{\partial}{\partial x}\int_0^\Delta uT\,dy - T_\infty\frac{\partial}{\partial x}\int_0^\Delta u\,dy = -\alpha\left(\frac{\partial T}{\partial y}\right)_{y=0} \qquad (17\text{-}5)$$

The temperature T_∞ of the undisturbed stream is independent of x, so Eq. (17-5) becomes

$$\frac{\partial}{\partial x}\int_0^\Delta u(T - T_\infty)\,dy = -\alpha\left(\frac{\partial T}{\partial y}\right)_{y=0} \qquad (17\text{-}6)$$

Equation (17-6) is the integral energy equation. It may be integrated only if $T - T_\infty$ is known as a function of y. The resulting integration will give the boundary-layer thickness as a function of x and will yield the local heat-transfer coefficient as a function of x.

The local heat-transfer coefficient for flow past immersed bodies is defined by Eq. (12-10)

$$h = -\frac{k}{T_w - T_\infty}\left(\frac{\partial T}{\partial y}\right)_{y=0} \qquad (12\text{-}10)$$

from which

$$-\left(\frac{\partial T}{\partial y}\right)_{y=0} = \frac{h}{k}(T_w - T_\infty) \qquad (17\text{-}7)$$

Substituting Eq. (17-7) into Eq. (17-6),

$$\frac{\partial}{\partial x}\int_0^{\Delta} u(T - T_\infty)\,dy = \frac{h}{C_p\rho}(T_w - T_\infty) \tag{17-8}$$

When T_w is independent of x, and dividing each side by U, the mainstream velocity (also independent of x), the Stanton number becomes

$$\frac{h}{C_p\rho U} = \mathrm{St} = \frac{\partial}{\partial x}\int_0^{\Delta} \frac{u}{U}\frac{T - T_\infty}{T_w - T_\infty}\,dy \tag{17-9}$$

Equation (17-9) gives the local Stanton number, which may be evaluated when u/U and $(T - T_\infty)/(T_w - T_\infty)$ are known as a function of y.

I. LAMINAR FLOW PARALLEL TO FLAT PLATES

17-4. Pohlhausen's Solution; Constant Wall Temperature

The case of heat transfer from an isothermal flat plate has been studied analytically by Pohlhausen,[39] who solved the energy equation (17-1) to obtain the temperature distribution in the boundary layer as well as the local heat-transfer coefficient on the plate.

The momentum and continuity equations (10-3) and (10-4) have been solved by Blasius [2] (Sec. 10-4), and Pohlhausen used Blasius' solution to solve the energy equation.

Consider a fluid with velocity U and temperature T_∞ flowing past a flat plate maintained at a temperature T_w (see Fig. 17-2). The temperature

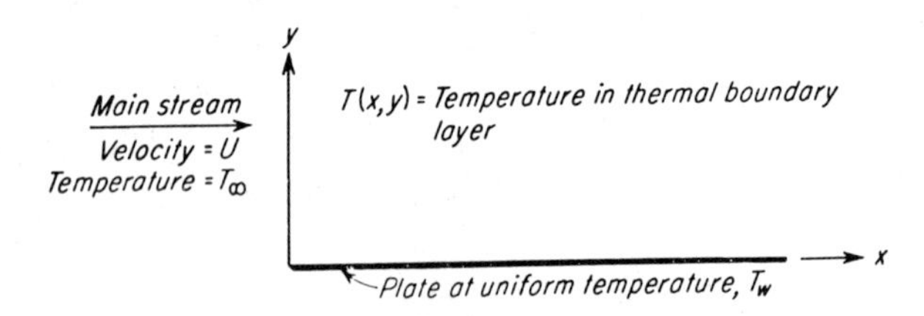

FIG. 17-2. Boundary conditions for Pohlhausen's solution.

T of the fluid at any point beyond the leading edge of the plate is a function of x and y and is related to these independent variables through the energy equation

$$u\frac{\partial T}{\partial x} + v\frac{\partial T}{\partial y} = \alpha\frac{\partial^2 T}{\partial y^2} \tag{17-1}$$

with boundary conditions

At $y = 0$ all x
$T = T_w$

At $y = \infty$ all x
$T = T_\infty$

At $x = 0$ all y
$T = T_\infty$

In Blasius' solution for the velocity distribution in the laminar boundary layer on a flat plate the following variables were defined (see Sec. 10-4):

$$\eta = \frac{y}{2}\left(\frac{U\rho}{\mu x}\right)^{1/2} \tag{10-15}$$

$$\psi = \left(\frac{\mu U x}{\rho}\right)^{1/2} \phi \tag{10-16}$$

from which were obtained

$$u = \frac{U}{2}\phi' \tag{10-17}$$

$$v = \frac{1}{2}\left(\frac{\mu U}{\rho x}\right)^{1/2} (\eta\phi' - \phi) \tag{10-21}$$

Defining the dimensionless temperature ratio

$$\bar{T} = \frac{T_w - T}{T_w - T_\infty} \tag{17-10}$$

and making use of Eqs. (10-15) to (10-17) and (10-21), Eq. (17-1) becomes

$$\frac{d^2\bar{T}}{d\eta^2} + \mathrm{Pr}\phi\frac{d\bar{T}}{d\eta} = 0 \tag{17-11}$$

with boundary conditions

At $\eta = 0$
$\bar{T} = 0$

At $\eta = \infty$
$\bar{T} = 1$

Equation (17-11) is an ordinary, linear differential equation which may be easily solved. Letting $\bar{T}' = d\bar{T}/d\eta$, Eq. (17-11) becomes

$$\frac{d\bar{T}'}{d\eta} + \mathrm{Pr}\phi\bar{T}' = 0 \tag{17-12}$$

which rearranges to

$$\frac{d\bar{T}'}{\bar{T}'} = -\mathrm{Pr}\phi\, d\eta \tag{17-13}$$

Integrating,

$$\ln \bar{T}' = -\mathrm{Pr}\int_0^\eta \phi\, d\eta + C_1 \tag{17-14}$$

or

$$\bar{T}' = \frac{d\bar{T}}{d\eta} = C_2 \exp\left(-\mathrm{Pr}\int_0^\eta \phi\, d\eta\right) \tag{17-15}$$

Integration of Eq. (17-15) gives

$$[\bar{T}]_0^\eta = C_2\int_0^\eta \exp\left(-\mathrm{Pr}\int_0^\eta \phi\, d\eta\right) d\eta \tag{17-16}$$

Introducing the boundary conditions gives the value of C_2:

$$C_2 = \frac{1}{\displaystyle\int_0^\infty \exp\left(-\mathrm{Pr}\int_0^\eta \phi\, d\eta\right) d\eta} \tag{17-17}$$

So the complete solution of Eq. (17-11) with boundary conditions inserted is

$$\bar{T} = \frac{\displaystyle\int_0^\eta \exp\left(-\mathrm{Pr}\int_0^\eta \phi\, d\eta\right) d\eta}{\displaystyle\int_0^\infty \exp\left(-\mathrm{Pr}\int_0^\eta \phi\, d\eta\right) d\eta} \tag{17-18}$$

Equation (17-18) gives the temperature distribution in the boundary layer as a function of the Prandtl number and of η, which is a function of x and y. The values of ϕ in Eq. (17-18) may be obtained from Table 10-1. Pohlhausen has tabulated values of the integral $\left[\int_0^\infty \exp\left(-\mathrm{Pr}\int_0^\eta \phi\, d\eta\right) d\eta\right]^{-1}$ as a function of the Prandtl number. Values of the integral are tabulated in Table 17-1.

It will be seen from Eqs. (17-15) and (17-17) that

$$\left(\frac{d\bar{T}}{d\eta}\right)_{\eta=0} = C_2 = \left[\int_0^\infty \exp\left(-\mathrm{Pr}\int_0^\eta \phi\, d\eta\right) d\eta\right]^{-1} \tag{17-19}$$

From the definition of $\bar{T}$ and η [Eqs. (10-15) and (17-10)],

$$\left(\frac{d\bar{T}}{d\eta}\right)_{\eta=0} = \frac{(\partial\bar{T}/\partial y)_{y=0}}{(\partial\eta/\partial y)_{y=0}} = -\frac{1}{T_w - T_\infty}\frac{(\partial T/\partial y)_{y=0}}{\frac{1}{2}(U\rho/\mu x)^{\frac{1}{2}}} \tag{17-20}$$

TABLE 17-1. VALUES OF $\left[\int_0^\infty \exp\left(-\Pr\int_0^\eta \phi\, d\eta\right) d\eta\right]^{-1}$ AND THE LOCAL NUSSELT NUMBER FOR POHLHAUSEN'S FLAT-PLATE SOLUTION

Pr	$\left[\int_0^\infty \exp\left(-\Pr\int_0^\eta \phi\, d\eta\right) d\eta\right]^{-1}$	$\frac{\mathrm{Nu}_x}{(\mathrm{Re}_x)^{1/2}}$	$\frac{\mathrm{Nu}_L}{(\mathrm{Re}_L)^{1/2}}$
0.001	0.0346	0.0173	0.0346
0.01	0.103	0.0516	0.103
0.1	0.280	0.140	0.280
0.6	0.552	0.276	0.552
0.7	0.585	0.292	0.585
0.8	0.614	0.307	0.614
0.9	0.640	0.320	0.640
1.0	0.664	0.332	0.664
1.1	0.687	0.343	0.687
7.0	1.29	0.695	1.29
10.0	1.46	0.730	1.46
15.0	1.67	0.835	1.67

From Eq. (12-10),

$$h = -\frac{k}{T_w - T_\infty}\left(\frac{\partial T}{\partial y}\right)_{y=0} \tag{12-10}$$

and the local Nusselt number Nu_x is

$$\mathrm{Nu}_x = \frac{hx}{k} = -\frac{x}{T_w - T_\infty}\left(\frac{\partial T}{\partial y}\right)_{y=0} \tag{17-21}$$

Combining Eqs. (17-19) to (17-21),

$$\mathrm{Nu}_x = \frac{1}{2}\left(\frac{Ux\rho}{\mu}\right)^{1/2}\left[\int_0^\infty \exp\left(-\Pr\int_0^\eta \phi\, d\eta\right) d\eta\right]^{-1} \tag{17-22}$$

Conditions:

(1) $\mathrm{Re}_x < 300{,}000$
(2) Fluid properties evaluated [14] at $0.58(T_w - T_\infty) + T_\infty$

Equation (17-22) gives the predicted local Nusselt number for laminar flow over a flat plate. The value of the integral in Eq. (17-22) is given in Table 17-1. Also tabulated are values of $\mathrm{Nu}_x/(\mathrm{Re}_x)^{1/2}$ as a function of the Prandtl number. Heat-transfer coefficients may be calculated from these quantities.

The average heat-transfer coefficient for a plate of length L is

$$h_m = \frac{1}{L}\int_0^L h\, dx \tag{17-23}$$

Thus from Eqs. (17-22) and (17-23)

$$\frac{h_m L}{k} = \text{Nu}_L = \left(\frac{UL\rho}{\mu}\right)^{1/2} \left[\int_0^\infty \exp\left(-\text{Pr}\int_0^\eta \phi\, d\eta\right) d\eta\right]^{-1} \tag{17-24}$$

The values of $\text{Nu}_L/(\text{Re}_L)^{1/2}$ given in Table 17-1 may be used to calculate average Nusselt numbers for laminar flow over flat plates.

17-5. Solution of the Integral Energy Equation; Constant Wall Temperature

A solution for the thermal-boundary-layer thickness, temperature distribution, and local Nusselt number may be obtained using the integral energy equation (17-6). This solution involves the assumption of the form of the temperature and velocity profiles. In solving von Kármán's integral momentum equation for the hydrodynamical boundary layer the velocity distribution was assumed to be of the form

$$\frac{u}{U} = 1.5\frac{y}{\delta} - \frac{1}{2}\left(\frac{y}{\delta}\right)^3 \tag{10-44}$$

The temperature distribution is assumed to be of the same form.

$$\frac{T_w - T}{T_w - T_\infty} = 1.5\frac{y}{\Delta} - \frac{1}{2}\left(\frac{y}{\Delta}\right)^3 \tag{17-25}$$

Since

$$\frac{T - T_\infty}{T_w - T_\infty} = 1 - \frac{T_w - T}{T_w - T_\infty} \tag{17-26}$$

$$\frac{T - T_\infty}{T_w - T_\infty} = 1 - 1.5\frac{y}{\Delta} + \frac{1}{2}\left(\frac{y}{\Delta}\right)^3 \tag{17-27}$$

Substituting Eqs. (10-44) and (17-27) into (17-6),

$$-\alpha\left(\frac{\partial T}{\partial y}\right)_{y=0} = U(T_w - T_\infty)\frac{\partial}{\partial x}\int_0^\Delta \left[1.5\frac{y}{\delta} - \frac{1}{2}\left(\frac{y}{\delta}\right)^3\right] \times \left[1 - 1.5\frac{y}{\Delta} + \frac{1}{2}\left(\frac{y}{\Delta}\right)^3\right] dy \tag{17-28}$$

From Eq. (17-27)

$$\left(\frac{\partial T}{\partial y}\right)_{y=0} = -\frac{1.5(T_w - T_\infty)}{\Delta} \tag{17-29}$$

Integrating Eq. (17-28) and making use of Eq. (17-29),

$$\frac{1.5\alpha(T_w - T_\infty)}{\Delta} = U(T_w - T_\infty)\frac{\partial}{\partial x}\left[\delta\left(\tfrac{3}{20}\zeta^2 - \tfrac{3}{280}\zeta^4\right)\right] \tag{17-30}$$

where ζ is the ratio of the two boundary-layer thicknesses Δ/δ. It is assumed that Δ/δ is less than unity, so the term $\frac{3}{280}\zeta^4$ may be neglected. It will be seen later that Δ/δ is always less than unity if the Prandtl number is greater than 1. Equation (17-30) becomes

$$\alpha = \frac{U\Delta}{10}\frac{d}{dx}(\zeta^2\delta) \tag{17-31}$$

Carrying out the differentiation in Eq. (17-31) and noting that $\Delta = \zeta\delta$,

$$\alpha = \frac{U}{10}\left(\zeta^3\delta\frac{d\delta}{dx} + 2\zeta^2\delta^2\frac{d\zeta}{dx}\right) \tag{17-32}$$

which becomes, after substituting Eq. (10-49) for $\delta(d\delta/dx)$ and Eq. (10-51) for δ^2,

$$\tfrac{14}{13}\left(\zeta^3 + 4x\zeta^2\frac{d\zeta}{dx}\right) = \frac{1}{\mathrm{Pr}} \tag{17-33}$$

The ratio $\frac{14}{13}$ may be considered to be unity, so Eq. (17-33) becomes

$$\zeta^3 + \tfrac{4}{3}x\frac{d(\zeta^3)}{dx} = \frac{1}{\mathrm{Pr}} \tag{17-34}$$

Equation (17-34) is a first-order linear differential equation the solution of which gives ζ^3 as a function of x. The boundary conditions on Eq. (17-34) are:

At $x = x_0$ starting point of heat transfer
$\zeta = 0$

The solution of Eq. (17-34) is given by Eckert [14] thus

$$\zeta = \frac{\Delta}{\delta} = \frac{1}{(\mathrm{Pr})^{1/3}}\left[1 - \left(\frac{x_0}{x}\right)^{3/4}\right]^{1/3} \tag{17-35}$$

If $x_0 = 0$,

$$\frac{\Delta}{\delta} = \frac{1}{(\mathrm{Pr})^{1/3}} \tag{17-36}$$

It will be recalled that the analysis was based on the presumption that Δ/δ has a value less than 1. From Eq. (17-36) this is seen to be true if the Prandtl number is greater than unity. Equation (17-36) gives the relation between the thickness of the hydrodynamical boundary layer and the thermal boundary layer. For high-Prandtl-number fluids Δ is less than δ. For gases with Prandtl number as low as 0.6, Δ is only 1.19 times as great as δ, and Eq. (17-36) is valid for this case; it is not valid for the liquid metals, which have low Prandtl numbers. From the definition of the local Nusselt number given by Eq. (17-21) and from Eq. (17-29)

$$\mathrm{Nu}_x = \frac{1.5x}{\Delta} \tag{17-37}$$

But from Eq. (10-52)

$$\delta = 4.64\left(\frac{\nu x}{U}\right)^{1/2} \tag{10-52}$$

Thus from Eq. (17-36) (with $x_0 = 0$)

$$\Delta = \frac{4.64}{(\mathrm{Pr})^{1/3}}\left(\frac{\nu x}{U}\right)^{1/2} \tag{17-38}$$

Equation (17-38) may be used to calculate the thickness of the thermal boundary layer. Substituting Eq. (17-38) into (17-37),

$$\mathrm{Nu}_x = 0.324(\mathrm{Re}_x)^{1/2}(\mathrm{Pr})^{1/3} \tag{17-39}$$

Conditions:

(1) $\mathrm{Re}_x < 300{,}000$
(2) $\mathrm{Pr} > 0.6$
(3) Fluid properties evaluated at $0.58(T_w - T_\infty) + T_\infty$
(4) Heating starts at leading edge

Equation (17-39) may be used to calculate local Nusselt numbers for laminar flow past flat plates. It is derived by means of the integral energy equation, the form of the temperature and velocity profiles being assumed.

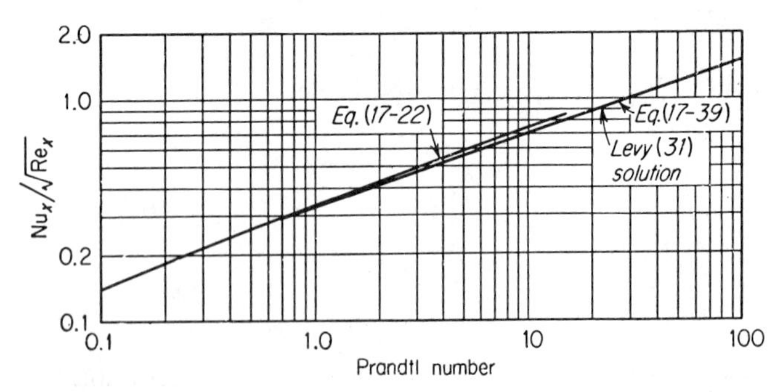

FIG. 17-3. Nusselt numbers for laminar flow over flat plates.

If heat transfer starts at a distance x_0 from the leading edge, Eq. (17-39) becomes

$$\mathrm{Nu}_x = 0.324(\mathrm{Re}_x)^{1/2}(\mathrm{Pr})^{1/3}\frac{1}{[1 - (x_0/x)^{3/4}]^{1/3}} \tag{17-40}$$

From Eqs. (17-23) and (17-39) the mean Nusselt number for a plate of length L is

$$\mathrm{Nu}_L = 0.648(\mathrm{Re}_L)^{1/2}(\mathrm{Pr})^{1/3} \tag{17-41}$$

Two relationships are now available for predicting local Nusselt numbers for laminar flow on flat plates, namely, Eqs. (17-22) and (17-39). They are compared graphically in Fig. 17-3, where $\mathrm{Nu}_x/(\mathrm{Re}_x)^{1/2}$ is plotted versus the Prandtl number. At Prandtl numbers greater than 0.6 both relationships agree well with each other, and either one may be used. Below Prandtl numbers of 0.6 Eq. (17-39) is not valid, and therefore Eq. (17-22) must be used.

Example 17-1

A thin copper cooling fin is 2 in. square, and its temperature is essentially uniform at 180°F. Air at 60°F flows parallel to the fin at a velocity of 40 ft/sec. Pressure is atmospheric.

(*a*) Determine the temperature gradient at the surface and the local rate of heat transfer at the center of the fin.

(*b*) Determine the total rate of heat transfer from the fin.

(*c*) Determine the thermal-boundary-layer thickness for a point at the center of the fin.

Solution

Use fluid properties evaluated at $0.58(T_w - T_\infty) + T_\infty = 130°\mathrm{F}$.

$$\mathrm{Pr} = 0.70$$

$$\nu = 1.98 \times 10^{-4}\ \mathrm{ft^2/sec}$$

$$k = 0.0165\ \mathrm{Btu/(hr)(ft^2)(°F)/ft}$$

(*a*) At the center of the fin

$$\mathrm{Re}_x = \frac{1}{12}\left|\frac{40}{1.98}\right|\frac{}{10^{-4}} = 1.69 \times 10^4$$

From Eq. (17-39)

$$\mathrm{Nu}_x = (0.324)(1.69 \times 10^4)^{1/2}(0.70)^{1/3} = 37.4$$

so

$$h = \frac{(37.4)(0.0165)(12)}{1} = 7.42\ \mathrm{Btu/(hr)(ft^2)(°F)}$$

The local rate of heat transfer is

$$h(T_w - T_\infty) = (7.42)(180 - 60) = 890\ \mathrm{Btu/(hr)(ft^2)}$$

From Eq. (12-8)

$$-0.0165\left(\frac{dT}{dy}\right)_{y=0} = 890$$

giving

$$\left(\frac{dT}{dy}\right)_{y=0} = -\frac{890}{0.0165} = -54{,}000°\mathrm{F/ft}$$

which is the temperature gradient at the surface.

(b)

$$\mathrm{Re}_L = \frac{2}{12}\left|\frac{40}{1.98}\right|\frac{}{10^{-4}} = 3.36 \times 10^4$$

From Eq. (17-41)

$$\mathrm{Nu}_L = (0.648)(3.36 \times 10^4)^{1/2}(0.70)^{1/3} = 105.2$$

giving

$$h_m = \frac{(105.2)(0.0165)(12)}{2} = 10.4 \text{ Btu/(hr)(ft}^2\text{)(°F)}$$

The rate of heat loss from the whole plate is

$$\frac{(10.4)(180 - 60)(2)(2)(2)}{144} = 69.4 \text{ Btu/hr}$$

(c) From Eq. (17-38)

$$\frac{\Delta}{x} = \frac{4.64}{(\mathrm{Pr})^{1/3}(\mathrm{Re}_x)^{1/2}}$$

$$= \frac{4.64}{(0.70)^{1/3}(1.69 \times 10^4)^{1/2}} = 0.0401$$

Thus, when $x = 1$ in.,

$$\Delta = 0.0401 \text{ in.}$$

which is the thickness of the thermal boundary layer at the center of the plate.

Example 17-2

Compare the rate of heat transfer from six individual fins similar to that described in Example 17-1 and from one fin 2 by 12 in. oriented with the long side parallel to the flow.

Solution

From Example 17-1 the rate of heat transfer from a single small fin is 69.4 Btu/hr. Thus for six identical fins the rate of heat transfer is 416 Btu/hr. For a single fin 2 by 12 in.

$$\mathrm{Re}_L = \frac{12}{12}\left|\frac{40}{1.98}\right|\frac{}{10^{-4}} = 2.02 \times 10^5$$

From Eq. (17-41)

$$\mathrm{Nu}_L = (0.648)(2.02 \times 10^5)^{1/2}(0.70)^{1/3} = 258$$

$$h_m = \frac{(258)(0.0165)(12)}{12} = 4.26 \text{ Btu/(hr)(ft}^2\text{)(°F)}$$

$$\text{Rate of heat transfer} = \frac{(4.26)(180 - 60)(2)(12)(2)}{144} = 171 \text{ Btu/hr}$$

A much higher rate of heat transfer is obtained using a number of small fins than using one long fin of equal area.

17-6. Laminar Flow over Flat Plates; Wall Temperature Variable

The case of laminar-flow heat transfer from a flat plate with wall temperature variable is considered in Sec. 17-16, where the general case of two-dimensional flow and heat transfer is discussed.

II. TURBULENT FLOW PARALLEL TO FLAT PLATES

17-7. Solution of the Integral Energy Equation; Constant Wall Temperature

The integral energy equation may be used to determine the local Nusselt number for turbulent flow parallel to flat plates. To solve this equation the temperature and velocity profiles are assumed to have the same form. This assumption is, in effect, an application of the Reynolds analogy [44] (see Sec. 15-6). The velocity distribution is assumed to be of the form given by Eq. (10-66).

$$\frac{u}{U} = \left(\frac{y}{\delta}\right)^{1/7} \tag{10-66}$$

Thus the temperature distribution is assumed to be

$$\frac{T_w - T}{T_w - T_\infty} = \left(\frac{y}{\Delta}\right)^{1/7} \tag{17-42}$$

For fluids with Prandtl numbers close to unity Δ/δ may also be assumed to be unity. Thus, substituting Eqs. (17-42) and (10-66) into (17-9),

$$\frac{h}{C_p\rho U} = \frac{\partial}{\partial x}\int_0^\delta \left(\frac{y}{\delta}\right)^{1/7}\left[1 - \left(\frac{y}{\delta}\right)^{1/7}\right] dy \tag{17-43}$$

and carrying out the integration

$$\frac{h}{C_p\rho U} = \frac{7}{72}\frac{\partial\delta}{\partial x} \tag{17-44}$$

but from Eq. (10-71)

$$\frac{\delta}{x} = 0.376(\mathrm{Re}_x)^{-1/5} \tag{10-71}$$

so

$$\frac{h}{C_p\rho U} = 7/72(0.376)\frac{d}{dx}\left[x\left(\frac{\mu}{U\rho x}\right)^{1/5}\right] \tag{17-45}$$

which becomes

$$\frac{h}{C_p\rho U} = 0.0292\left(\frac{\mu}{U\rho x}\right)^{1/5} \tag{17-46}$$

and since the Prandtl number is unity,

$$\mathrm{Nu}_x = 0.0292(\mathrm{Re}_x)^{4/5} \tag{17-47}$$

Conditions:

(1) Boundary layer turbulent over whole plate
(2) Pr = 1
(3) Heat transfer takes place over whole plate
(4) Fluid properties evaluated [14] at

$$T_\infty - \frac{0.1\text{Pr} + 40}{\text{Pr} + 72}(T_\infty - T_w)$$

Equation (17-47) gives the local Nusselt number for turbulent flow parallel to flat plates as a function of the local Reynolds number. The average Nusselt number on a plate of length L is

$$\text{Nu}_L = 0.0366(\text{Re}_L)^{4/5} \tag{17-48}$$

The conditions are the same as for Eq. (17-47).

17-8. The Prandtl Analogy Applied to Flat Plates; Constant Wall Temperature

For fluids which have Prandtl numbers much greater than 1, Eqs. (17-47) and (17-48) are not applicable. The resistance of the laminar sublayer must be considered, and Prandtl's analogy [41] may be applied to determine an expression for the local Nusselt number. In Fig. 17-4 the approximate temperature distribution on a flat plate is shown. The temperature distribution is approximately linear throughout the laminar sublayer, which has a thickness δ_1. At the edge of the laminar sublayer the velocity is u_{δ_1}, and the temperature is T_{δ_1}. The Prandtl analogy, which was derived for circular tubes, is equally applicable for flat plates if T_b is replaced by T_∞. Thus Eq. (15-44) becomes

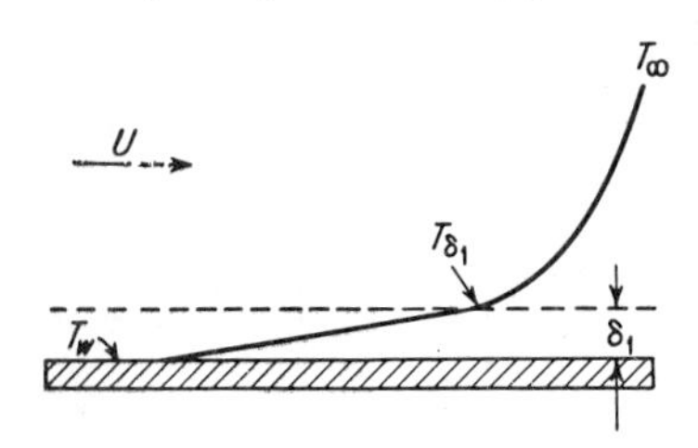

FIG. 17-4. Approximate turbulent temperature profile over a flat plate for the derivation of the Prandtl analogy.

$$\frac{\rho}{\tau_w g_c}[U + u_{\delta_1}(\text{Pr} - 1)] = \frac{A_w C_p \rho}{q_w}(T_w - T_\infty) \tag{17-49}$$

Introducing from Eq. (10-69) $\tau_w g_c/\rho = 0.0228U^2(\nu/U\delta)^{1/4}$ and from Eq. (12-6) $h = q_w/[A_w(T_w - T_\infty)]$,

$$\frac{U + u_{\delta_1}(\text{Pr} - 1)}{0.0228U^2(\nu/U\delta)^{1/4}} = \frac{C_p\rho}{h} \tag{17-50}$$

Replacing δ by Eq. (10-71),

$$\frac{h}{C_p\rho U} = \frac{0.0292(\text{Re}_x)^{-1/5}}{1 + (u_{\delta_1}/U)(\text{Pr} - 1)} \tag{17-51}$$

Eckert [14] reports that

$$\frac{u_{\delta_1}}{U} = 1.3(\text{Pr})^{-1/6}(\text{Re}_x)^{-1/10} \tag{17-52}$$

Since

$$\frac{h}{C_p\rho U} = \frac{\text{Nu}_x}{\text{Re}_x\text{Pr}}$$

Equation (17-51) becomes

$$\text{Nu}_x = \frac{0.0292(\text{Re}_x)^{4/5}\text{Pr}}{1 + (u_{\delta_1}/U)(\text{Pr} - 1)} \tag{17-53}$$

Conditions:

(1) Turbulent boundary layer starts at leading edge
(2) Heat transfer starts at leading edge
(3) Properties evaluated at [14]

$$T_\infty - \frac{0.1\text{Pr} + 40}{\text{Pr} + 72}(T_\infty - T_w)$$

(4) $\text{Pr} > 1.0$

Equation (17-53) may be used to calculate local Nusselt numbers for turbulent flow parallel to flat plates.

17-9. The Colburn Analogy Applied to Flat Plates; Constant Wall Temperature

The Colburn analogy may be easily applied for turbulent flow parallel to flat plates. Defining the heat-transfer j factor as

$$j_H = \frac{h}{C_p\rho U}\left(\frac{C_p\mu}{k}\right)^{2/3} \tag{17-54}$$

and equating j_H to one-half the local drag coefficient [which is given as a function of Re_x in Eq. (10-75)],

$$\frac{h}{C_p\rho U}\left(\frac{C_p\mu}{k}\right)^{2/3} = 0.0292(\text{Re}_x)^{-1/5} \tag{17-55}$$

from which

$$\text{Nu}_x = 0.0292(\text{Re}_x)^{4/5}(\text{Pr})^{1/3} \tag{17-56}$$

Conditions:

(1) $\mathrm{Pr} > 0.6$
(2) Heat transfer and turbulent boundary layer start at leading edge
(3) Fluid properties evaluated at [14]

$$T_\infty - \frac{0.1\,\mathrm{Pr} + 40}{\mathrm{Pr} + 72}(T_\infty - T_w)$$

From Eq. (17-56) the average Nusselt number for a plate of length L is

$$\mathrm{Nu}_L = 0.0366(\mathrm{Re}_L)^{4/5}(\mathrm{Pr})^{1/3} \tag{17-57}$$

Example 17-3

The leading edge of the cooling fin in Example 17-1 is rough, so that the boundary layer is completely turbulent. Determine the rate of heat transfer from the fin.

Solution

From Eq. (17-57)

$$\mathrm{Nu}_L = (0.0366)(3.36 \times 10^4)^{4/5}(\mathrm{Pr})^{1/3} = 136$$

$$h_m = \frac{(136)(0.0165)(12)}{2} = 13.5\ \mathrm{Btu/(hr)(ft^2)(°F)}$$

$$\text{Rate of heat loss} = \frac{(13.5)(180 - 60)(2)(2)(2)}{144} = 90\ \mathrm{Btu/hr}$$

The rate of heat loss for laminar flow was 69.4 Btu/hr. At larger values of Re_L the difference between the heat transferred for laminar flow and for turbulent flow is much greater than shown in this example.

17-10. Combined Laminar and Turbulent Flow Parallel to a Flat Plate

Equations (17-47), (17-53), and (17-56) are valid if the turbulent boundary layer begins at the leading edge of the flat plate. If both a laminar and a turbulent boundary layer exist on the plate, a correction must be applied to the above-mentioned relationships. Eckert [14] reported the following relationships for the average Nusselt number in cases where the laminar boundary layer must be considered:

1. For transition Reynolds number of 10^5:

$$\mathrm{Nu}_L = 0.0366(\mathrm{Pr})^{1/3}[(\mathrm{Re}_L)^{4/5} - 4{,}200] \tag{17-58}$$

2. For transition Reynolds number of 5×10^5:

$$\mathrm{Nu}_L = 0.0366(\mathrm{Pr})^{1/3}[(\mathrm{Re}_L)^{4/5} - 23{,}100] \tag{17-59}$$

17-11. Effect of Unheated Starting Length

The relationships for the local Nusselt number given by Eqs. (17-47), (17-53), and (17-56) are applicable only if the heat transfer takes place over the entire plate. A correction must be made if the forward portion of the plate is unheated for a distance of x_0 from the leading edge. This situation is illustrated schematically in Fig. 17-5. At the point $x = x_0$ there is essentially a stepwise change in temperature. For $x < x_0$ the plate temperature is T_∞, and for $x > x_0$ the plate temperature is T_w.

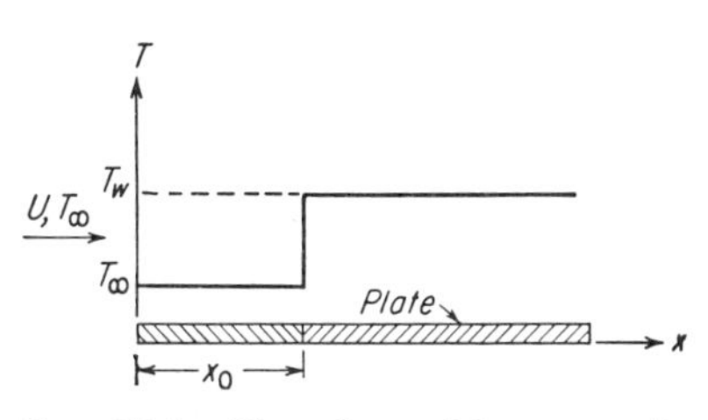

FIG. 17-5. Flat plate with a stepwise temperature change at a point x_0 from the leading edge.

Rubesin [46] developed an analytical expression for the local Nusselt number on a plate having such a stepwise surface-temperature discontinuity.

$$\mathrm{Nu}_x = 0.0292(\mathrm{Re}_x)^{4/5}(\mathrm{Pr})^{1/3}\left[1 - \left(\frac{x_0}{x}\right)^{39/40}\right]^{-7/39} \qquad (17\text{-}60)$$

where x_0 is the point where the stepwise discontinuity in temperature occurs. Equation (17-60) has been substantiated experimentally by Scesa and Sauer,[47] who studied heat transfer from plates at values of x_0/L of 0.104, 0.208, 0.316, and 0.528. Step increases in the temperature of the plate surface were 10 and 22°F. The last term of Eq. (17-60) represents the correction for unheated starting lengths. Scesa and Levy [48] presented an analytical relationship to account for the unheated portion of the plate. This relationship gives local Nusselt numbers somewhat higher than those predicted by Eq. (17-60).

17-12. Experimental Heat-transfer Data for Laminar and Turbulent Flow past Flat Plates

There has been considerable investigation of heat transfer during laminar and turbulent flow parallel to flat plates. Many of the data have been obtained for the case where the plate temperature is a constant. The more important investigations are those of Juerges,[24] Slegel and Hawkins,[53] and Elias.[16] The data of these workers cover both the laminar and turbulent flow regimes, although most of the observations were made on the latter. The data essentially confirm Eq. (17-41) for completely laminar flow and Eq. (17-57) for completely turbulent flow. These equations are shown in

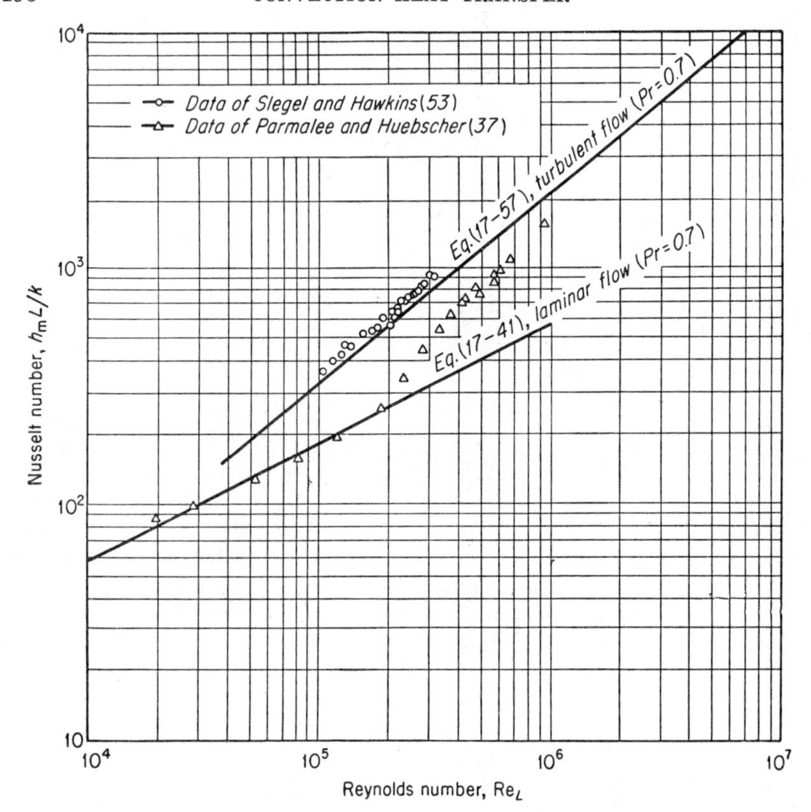

FIG. 17-6. Total Nusselt numbers for the flow of air parallel to flat plates (constant plate temperature).

Fig. 17-6, where Nu_L is plotted versus Re_L for a Prandtl number of 0.7. Experimental data are included.

Drake [12] investigated heat transfer to air from a plate having variable surface temperature. His measured Nusselt numbers agree with those predicted by Eq. (17-57), and his results indicate that for air and higher-Prandtl-number fluids the local Nusselt number is not affected by the temperature variation of the plate surface. A large effect might be noted, however, for low-Prandtl-number fluids, such as the liquid metals.

17-13. Nusselt Numbers for Turbulent Flow past Rough Plates

There has been little investigation of heat transfer from rough plates. Stanton and Booth [55] could find no effect of surface roughness on the rate of heat transfer from a plate. If the surface roughness is less than the admissible roughness [see Eq. (10-83)], the plate may be considered smooth,

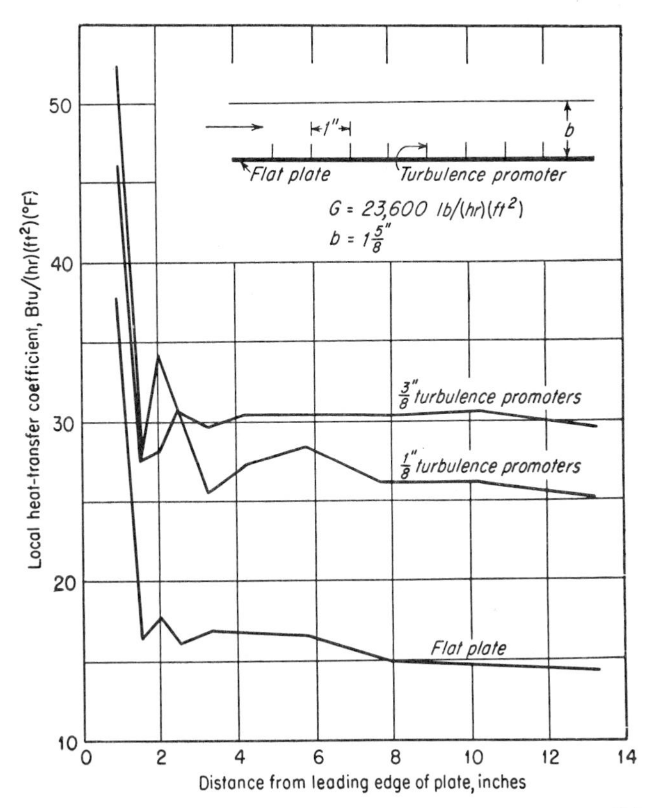

FIG. 17-7. Effect of turbulence promoters on the local heat-transfer coefficient for air flowing parallel to a flat plate. (*From L. M. K. Boelter et al., NACA TN* 2517, 1951.)

and Eqs. (17-47), (17-53), and (17-56) may be used. If the roughness is greater than the admissible roughness, the total drag coefficient may be determined from Eq. (10-86), and Colburn's analogy can be employed to determine an approximate heat-transfer coefficient.

17-14. Effect of Turbulence Promoters on Flat Plates

Boelter and coworkers [3] determined local heat-transfer coefficients on a flat plate containing turbulence promoters, which were vertical strips affixed to the plate at 1-in. intervals. Strips ⅛ and ⅜ in. high were studied. These investigators found that the local heat-transfer coefficient on the plate increased by as much as 200 per cent, the increase being also a function of the width of the duct in which the plate was located. They found that the local heat-transfer coefficient was approximately constant beyond a point 2 in. from the leading edge. The ⅜-in. turbulence promoters brought about an increase of about 30 per cent in the local coefficient over that for the ⅛-in. promoters. Figure 17-7 is a plot of some data obtained by Boelter and coworkers on the effect of turbulence promoters on flat plates. From the standpoint of power consumed per unit of heat transferred the plate with turbulence promoters offers no advantage over a smooth flat plate.

III. HEAT TRANSFER DURING FLOW NORMAL TO TWO-DIMENSIONAL BODIES

17-15. The Two-dimensional Momentum and Energy Equations

When heat transfer takes place during flow past two-dimensional bodies, such as cylinders or airfoils, the velocity distribution and temperature distribution in the boundary layer are obtained by the simultaneous solution of the continuity, momentum, and energy equations, which are as follows:

1. Momentum:

$$u\frac{\partial u}{\partial x_1} + v\frac{\partial u}{\partial y} = u_{\max}\frac{\partial u_{\max}}{\partial x_1} + \frac{\mu}{\rho}\frac{\partial^2 u}{\partial y^2} \tag{10-89}$$

2. Continuity:

$$\frac{\partial u}{\partial x_1} + \frac{\partial v}{\partial y} = 0 \tag{10-90}$$

3. Energy:

$$u\frac{\partial T}{\partial x_1} + v\frac{\partial T}{\partial y} = \frac{k}{C_p\rho}\frac{\partial^2 T}{\partial y^2} \tag{17-61}$$

where x_1 is defined in Fig. 10-26.

By means of the several methods described in Sec. 10-20, Eqs. (10-89) and (10-90) may be solved to give the velocity distribution in the laminar boundary layer. The velocities obtained from the solution of Eqs. (10-89) and (10-90) may be used in Eq. (17-61) to obtain the temperature distribution in the boundary layer. In Sec. 17-4 the analytical solution of Eqs. (10-89), (10-90), and (17-61) is described for the case of a flat plate at constant wall temperature. Of much greater interest is the solution of these equations for flow past any two-dimensional immersed body with any arbitrary surface-temperature distribution. Since the introduction of the electronic digital computer and other high-speed computational methods it has been possible to obtain such a solution of these three differential equations. Important analytical investigations have been those of Chapman and Rubesin,[6] Levy,[31] Klein and Tribus,[28] Seban,[50] and Levy and Seban.[32]

Chapman and Rubesin [6] solved the case for which the surface-temperature distribution is arbitrary (in the form of a polynomial) and the pressure gradient $u_{\max}(\partial u_{\max}/\partial x_1)$ in the direction of flow is zero. The latter condition gives the solution of Eqs. (10-89) and (10-90) in the form of Blasius' solution (see Sec. 10-4). These workers' results are readily applicable to practical cases. Seban [50] solved Eqs. (10-89), (10-90), and (17-61) for the general case of arbitrary free-stream velocity and surface-temperature distribution. Klein and Tribus [28] presented an excellent review of the analytical work done on forced-convection heat transfer from nonisothermal surfaces. In addition, they provided solutions for the case where the surface temperature contains discontinuities, e.g., those which would be present if heating elements in the form of wires or strips were imbedded in the surface. Levy [31] dealt with the problem of heat transfer from two-dimensional bodies where the velocity u_{max} and surface temperature T_w are power functions. His solution,[31] which is described in more detail in the following section, is applicable to flow over the forward part of cylinders and wedges and reduces to Pohlhausen's solution for flow over flat plates.

17-16. Levy's Solution for Heat Transfer from Two-dimensional Bodies

Levy's solution for forced-convection heat transfer from two-dimensional bodies is based on the following variation of the velocity at the edge of the boundary layer and surface temperature:

$$u_{max} = cx_1^{n'} \tag{10-91}$$

$$T_w - T_\infty = Ax_1^{m'} \tag{17-62}$$

where c and A are constants.

By defining the stream function ψ according to Eqs. (3-16) and (3-17) and defining the functions η_1 and ϕ according to Eqs. (10-93) and (10-94) the momentum equation was transformed to the ordinary differential equation

$$\phi''' + \phi''\phi + \lambda(1 - \phi'^2) = 0 \tag{10-95}$$

where

$$\lambda = \frac{2n'}{1 + n'} \tag{10-96}$$

and ϕ is a function of η_1 only. The energy equation (17-61) may be transformed into

$$\frac{d^2\overline{T}}{d\eta_1^2} + \phi\text{Pr}\frac{d\overline{T}}{d\eta_1} - \text{Pr}(2 - \lambda)m'\phi'\overline{T} = 0 \tag{17-63}$$

where

$$\overline{T} = \frac{T_w - T}{T_w - T_\infty} \tag{17-10}$$

The solution of Eqs. (10-95) and (17-63) gives the boundary-layer velocity and temperature profile respectively. These equations may be solved analytically for only two cases: that for which $m' = 0$ (constant surface temperature) and that for which $(2 - \lambda)m' = -1$. For all other cases they must be solved by numerical methods, which give only approximate results. Levy solved Eqs. (10-95) and (17-63) numerically for various values of λ and m and considers his results to be in error by not more than 4 per cent. He shows a number of temperature-distribution curves and reports values of the temperature gradient at the surface. The latter quantity is required in determining local heat-transfer coefficients on the two-dimensional body by means of the relationship

$$\frac{-(d\overline{T}/d\eta_1)_{\eta_1=0}}{\sqrt{2-\lambda}} = \frac{\mathrm{Nu}_{x_1}}{\sqrt{\mathrm{Re}_{x_1}}} \tag{17-64}$$

Values of $(d\overline{T}/d\eta_1)_{\eta_1=0}$ reported by Levy are shown in Table 17-2 for $\lambda = 0$, Table 17-3 for $\lambda = 1$, and Table 17-4 for $\lambda = 1.6$. For flow parallel

TABLE 17-2. VALUES OF $\left(\frac{d\overline{T}}{d\eta_1}\right)_{\eta_1=0}$ FOR $\lambda = 0$ (FLAT PLATE) †

m' \ Pr	0.7	2.0	10.0	20.0
−0.50	0	0	0	0
−0.25	......	−0.3789	−0.6257	−0.7668
0	−0.4065	−0.5822	−0.9863	−1.230
0.25	−0.4989	−0.7130	−1.210	−1.513
0.50	−0.5690	−0.8112	−1.377	−1.721
1.00	−0.6746	−0.9593	−1.625	−2.024
2.00	−0.8218	−1.165	−1.965	−2.445
3.00	−0.9296	−1.316	−2.211	−2.741
4.00	−1.017	−1.437	−2.406	−2.974

† From S. Levy, *J Aeronaut. Sci.*, **19**:341 (1952).

to a flat plate $\lambda = 0$, and for flow over the forward portion of a circular or elliptical cylinder $\lambda = 1$. Various values of the Prandtl number and of m' are shown in the tables. The data in these tables may be used to calculate local Nusselt numbers and local heat-transfer coefficients by means of Eq. (17-64). In Fig. 17-8 values of $\mathrm{Nu}_{x_1}/\sqrt{\mathrm{Re}_{x_1}}$ are plotted versus m' for a Prandtl number of 0.70 (air at moderate temperatures). When $m' = 0$, the surface temperature is constant. Levy's solution for $m' = 0$ and $\lambda = 0$ agrees with Eqs. (17-22) and (17-39) for laminar-flow heat transfer on a flat plate as shown in Fig. 17-3.

TABLE 17-3. VALUES OF $\left(\frac{d\bar{T}}{d\eta_1}\right)_{\eta_1=0}$ FOR $\lambda = 1$ †

m' \ Pr	0.7	1.0	5.0	10.0
−1.0	0	0	0	0
−0.75	−0.1755	−0.2001	.....	−0.4062
−0.50	......	......	.....	−0.8094
−0.25	−0.4093	−0.4708	.....	−1.081
0	−0.4879	−0.5603	−1.011	−1.286
0.25	−0.5535	−0.6345	−1.141	−1.451
0.50	−0.6094	−0.6979	−1.251	−1.590
1.00	−0.7033	−0.8116	−1.432	−1.818
2.00	−0.8461	−0.9647	.....	−2.159
2.40	......	......	−1.795	
3.00	−0.9567	−1.089	−1.914	−2.418
4.00	−1.048	−1.192	−2.086	−2.630

† From S. Levy, *J. Aeronaut. Sci.*, **19**:341 (1952).

TABLE 17-4. VALUES OF $\left(\frac{d\bar{T}}{d\eta_1}\right)_{\eta_1=0}$ FOR $\lambda = 1.6$ †

m' \ Pr	0.7	1.0	5.0	10.0
−2.50	0	0	0	0
−1.50	−0.2687	−0.3101	−0.5587	−0.7005
−0.50	−0.4413	−0.5085	−0.9303	−1.186
0.00	−0.5062	−0.5828	−1.064	−1.357
0.25	−0.5353	−0.6161	−1.122	−1.432
0.50	−0.5626	−0.6468	−1.176	−1.501
1.00	−0.6120	−0.7031	−1.275	−1.626
2.00	−0.6975	−0.7995	−1.442	−1.836
3.00	−0.7692	−0.8808	−1.581	−2.010
4.00	−0.8315	−0.9512	−1.701	−2.159

† From S. Levy, *J. Aeronaut. Sci.*, **19**:341 (1952).

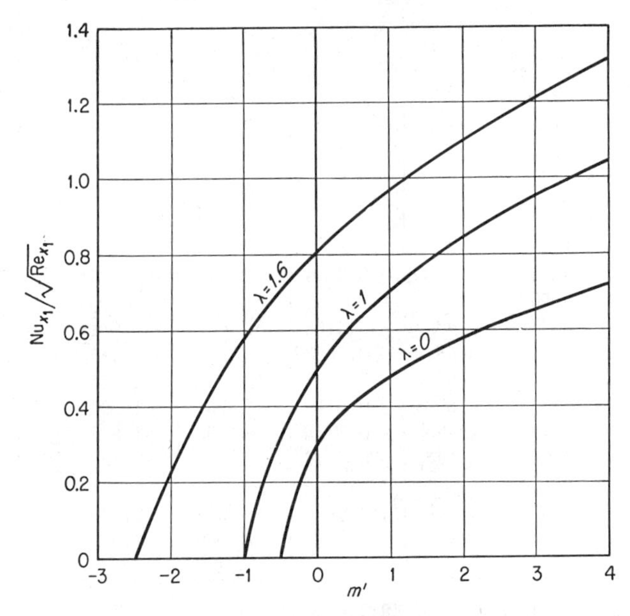

FIG. 17-8. Levy's solution for laminar-flow heat transfer from immersed bodies (Pr = 0.7). [*From S. Levy, J. Aeronaut. Sci.*, **19**:341 (1952).]

17-17. Local Heat-transfer Coefficients for Flow past Circular Cylinders; Experimental Values

Considerable experimental work has been done on heat transfer between a circular cylinder and a fluid flowing perpendicular to its axis. Although a large amount of the investigation has been on the determination of average heat-transfer coefficients, there has been rather extensive study of local heat-transfer coefficients on cylinders, which have been correlated, at least semiquantitatively, with the mechanism of flow which exists.

As might be expected, the local heat-transfer coefficient at the stagnation point of cylinders is fairly high. It decreases with increasing distance from the stagnation point and would continue to do so if it were not for the occurrence of transition from a laminar to a turbulent boundary layer or the separation of the boundary layer. In the following discussion the effect of these factors on the local heat-transfer coefficient is noted.

1. *Low-Reynolds-number Range* ($20 < Re_0 < 500$). There has been little investigation of local coefficients on cylinders in the low-Reynolds-number range. The major experimental contribution is the work of Eckert and Soehngen,[15] who studied the flow of air past heated cylinders at Reynolds numbers d_0U/ν ranging from 20 to 500. They determined local heat-transfer coefficients and also reported a correlation between the average

Nusselt number $h_m d_0/k$ on the cylinder and the Reynolds number. Figure 17-9 is a plot of the local Nusselt number hd_0/k versus θ, the angle measured from the stagnation point for three of the Reynolds numbers studied by Eckert and Soehngen. These curves demonstrate the variation of the local Nusselt number over the cylindrical surface. At the low Reynolds numbers for which the curves are drawn the minimum occurs between 120 and

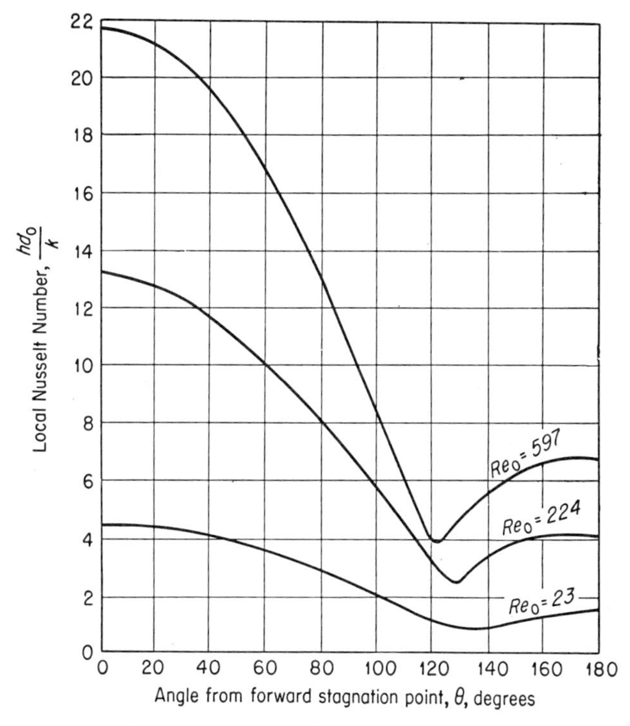

FIG. 17-9. Local Nusselt numbers for the flow of air past cylinders at low Reynolds numbers. [*From E. R. G. Eckert and E. Soehngen, Trans. ASME,* **74**:343 (1952).]

130° from the forward point of stagnation. Beyond the minimum, the local Nusselt number rises slightly to the trailing edge ($\theta = 180°$). The highest rate of heat transfer is obtained on the forward stagnation point.

2. *High-Reynolds-numbers Range* ($\mathrm{Re}_0 > 1{,}000$). The most extensive investigations of local heat-transfer coefficients on cylinders at relatively high Reynolds numbers are those of Lorisch,[33] Krujilin,[30] Schmidt and Wenner,[49] Comings, Clapp, and Taylor,[9] Giedt,[18, 19] and Zapp.[61] All the above studies, except that of Giedt in 1949, were restricted to cylinders with isothermal surfaces. The work of Comings, Clapp, and Taylor, Zapp, and Giedt considered the effect of turbulence on local coefficients.

The manner in which the local heat-transfer coefficient varies over the cylindrical surface changes drastically as the Reynolds number is increased up to and beyond the critical value. When the turbulence of the main stream is low (about 1 per cent), the plot of local Nusselt number versus the angle measured from the forward stagnation point has much the same shape as the curves shown in Fig. 17-9 for low Reynolds numbers. A series of

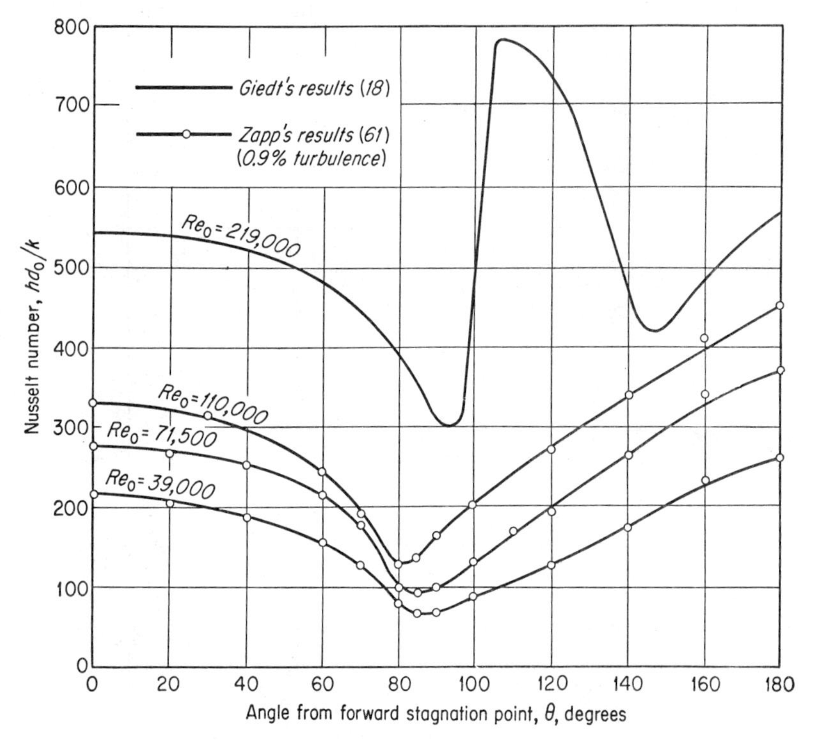

FIG. 17-10. Local Nusselt numbers for air flowing past circular cylinders. [*Results from W. H. Giedt, Trans. ASME,* **71**:375 (1949); *G. M. Zapp, M.S. Thesis, Oregon State College,* 1950.]

curves obtained by Zapp at Reynolds numbers of 39,000, 71,500, and 110,000 is shown in Fig. 17-10. All curves have minimum values at θ slightly greater than 80°, which is about at the point of separation of the boundary layer. Beyond the minimum point, the curves rise, reaching values of the Nusselt number at the trailing edge which are higher than those at the forward stagnation point. The curves obtained by Zapp in Fig. 17-10 were for a main-stream turbulence intensity of 0.9 per cent. The upper curve in Fig. 17-10 was obtained by Giedt (probably near the same percentage turbulence) for a Reynolds number of 219,000. The shape of this curve is considerably different from the lower three. It contains two minimum points and between them is a maximum point that is much higher than the local

Nusselt number at the forward point of stagnation. The first minimum in the local-Nusselt-number curve (at about 95°) is thought to be due to the transition from a laminar boundary layer to a turbulent boundary layer. At this transition the wall shear stress also begins to increase. The second minimum (at about 140°) is thought to lie where the boundary layer separates, since this minimum coincides with the point of zero shear stress at the surface (see Fig. 10-35). A Reynolds number of 219,000 is in the vicinity of the abrupt drop in the total drag coefficient, this drop being caused by the transition from a laminar to a turbulent boundary layer (see Sec. 10-27).

Martinelli, Guibert, Morrin, and Boelter[35] studied the data of Schmidt and Wenner and proposed the following empirical equation for predicting the local Nusselt number on a cylinder up to $\theta = 80°$:

$$\frac{hd_0}{k} = 1.14(\mathrm{Pr})^{0.4}\left(\frac{d_0 U \rho}{\mu}\right)^{0.5}\left[1 - \left(\frac{\theta}{90}\right)^3\right] \tag{17-65}$$

For air with a Prandtl number of 0.74 Eq. (17-65) becomes

$$\mathrm{Nu} = 1.01(\mathrm{Re}_0)^{0.5}\left[1 - \left(\frac{\theta}{90}\right)^3\right] \tag{17-66}$$

Conditions:

(1) θ measured in degrees
(2) $0 < \theta < 80°$
(3) Main-stream turbulence less than 1 per cent

Equations (17-65) and (17-66) may be used to calculate local Nusselt numbers on cylinders.

17-18. The Effect of Turbulence Intensity on Local Heat-transfer Coefficients on Circular Cylinders

Comings, Clapp, and Taylor[9] studied the effect of the intensity of turbulence of the main stream on heat transfer from a cylinder. In the relatively low range of Reynolds numbers that was studied (400 to 20,000) these workers found no large effect of turbulence intensity on heat-transfer rates. Average heat-transfer coefficients on the cylinder at turbulence levels above 7 per cent were somewhat higher than those obtained at levels below 3 per cent, the difference being greater at higher Reynolds numbers. Figure 17-11 is a plot of the mean Nusselt number as a function of percentage turbulence for a Reynolds number of 5,800. Most of the 30 per cent increase in Nusselt number occurs at a turbulence intensity between 2 and 4 per cent.

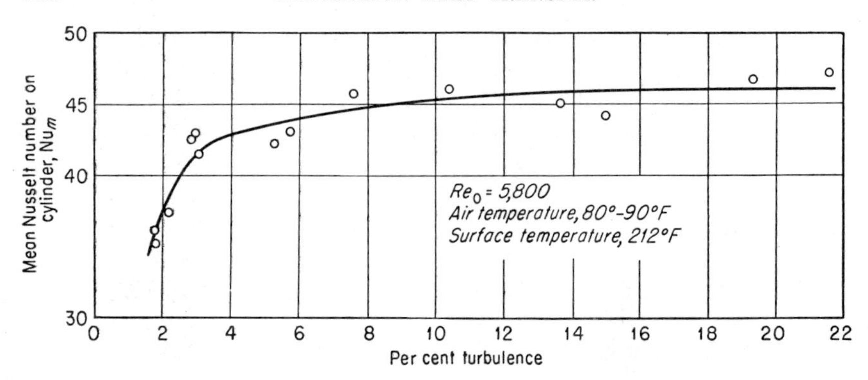

FIG. 17-11. Effect of turbulence intensity on the rate of heat transfer for air flowing past a circular cylinder. [*From E. W. Comings et al., Ind. Eng. Chem.*, **40**:1076 (1948).]

The results obtained by Zapp [61] and Giedt [19] on the effect of turbulence intensity are essentially the same; both investigated relatively high Reynolds numbers. The results obtained by Zapp are shown in Fig. 17-12, which is a plot of the local Nusselt number versus θ at a Reynolds number of 39,000 and at turbulence intensities of 0.9, 3, and 11.5 per cent, and in

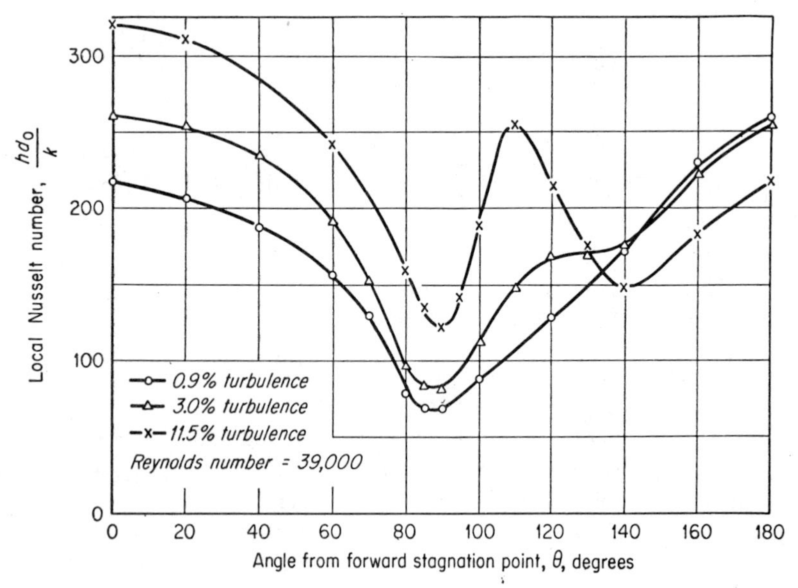

FIG. 17-12. Effect of percentage turbulence on local Nusselt number for air flowing past a circular cylinder, Re_0 = 39,000. (*From G. M. Zapp, M.S. Thesis, Oregon State College,* 1950.)

Fig. 17-13, which is a similar plot for a Reynolds number of 110,000. The effect of increased turbulence level is to increase the local heat-transfer coefficient over most of the cylinder and to bring about transition from a laminar to a turbulent boundary layer at much lower Reynolds numbers. In Fig. 17-12 the turbulence level of 11.5 per cent was sufficient to cause the formation of a turbulent boundary layer even at the relatively low Reynolds number of 39,000. The presence of the turbulent boundary layer is indicated by the fact that the curve for the 11.5 per cent turbulence contains two minimum points.

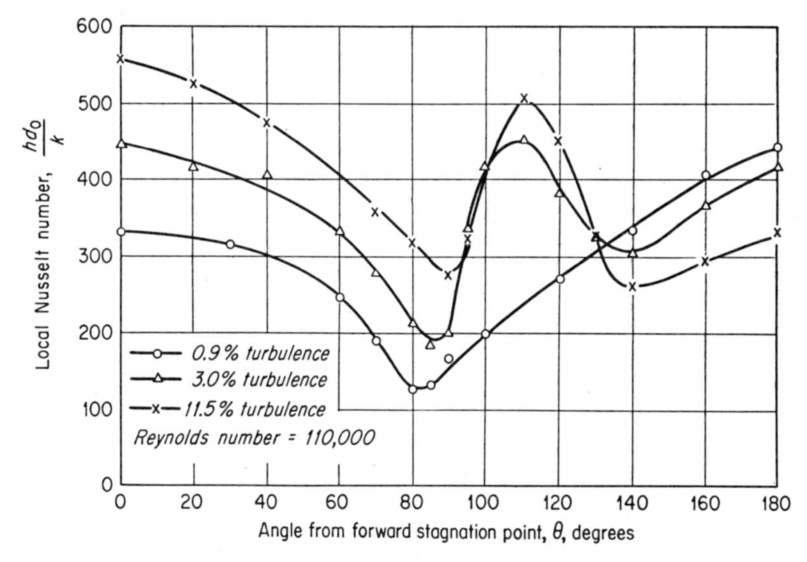

FIG. 17-13. Effect of percentage turbulence on local Nusselt number for air flowing past a circular cylinder, Re_0 = 110,000. (*From G. M. Zapp, M.S. Thesis, Oregon State College*, 1950.)

17-19. The Effect of a Single Roughness Element on the Cylindrical Surface

Schmidt and Wenner [49] determined the effect of a single 1.5-mm-diameter wire on the surface of a 100-mm-diameter cylinder. The wire was placed 77.5° from the forward point of stagnation. The effect of the wire is shown schematically in Fig. 17-14. At the point where the wire is located on the surface a maximum in the curve occurs, on either side of which are two minimum points. The wire evidently causes transition from laminar to turbulent flow. The over-all effect of the wire was a slight increase in the average heat-transfer coefficient on the cylinder.

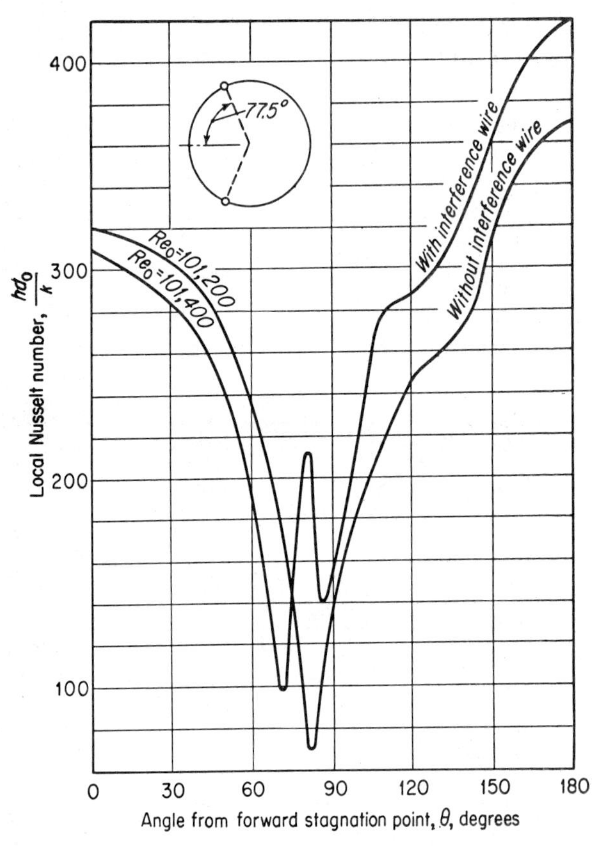

FIG. 17-14. Effect of an interference wire on the rate of heat transfer from a cylinder to air. (*From E. Schmidt and K. Wenner, NACA TM* 1050, 1943.)

17-20. Local Nusselt Numbers at the Forward Point of Stagnation of Circular Cylinders

The local Nusselt number at the forward point of stagnation ($\theta = 0$) of a circular cylinder immersed in an air stream was given by Goldstein [20] as

$$\mathrm{Nu}_{\theta=0} = 1.14(\mathrm{Pr})^{0.4}(\mathrm{Re}_0)^{0.5} \tag{17-67}$$

and for air with Pr = 0.74

$$\mathrm{Nu}_{\theta=0} = 1.01\sqrt{\mathrm{Re}_0} \tag{17-68}$$

Equation (17-68) is plotted as a solid line in Fig. 17-15, and it shows good agreement with the experimental data of Eckert and Soehngen,[15] Zapp,[61] Giedt,[18] and Schmidt and Wenner.[49] Robinson and Han [45] studied heat transfer to circular cylinders in ducts to find the effect of duct size. They recommend the use of an additional factor $\sqrt{1 + d_0/b}$, by which the right side of Eq. (17-68) should be multiplied. The term b is the duct width, and as it increases indefinitely the term $\sqrt{1 + d_0/b}$ becomes zero.

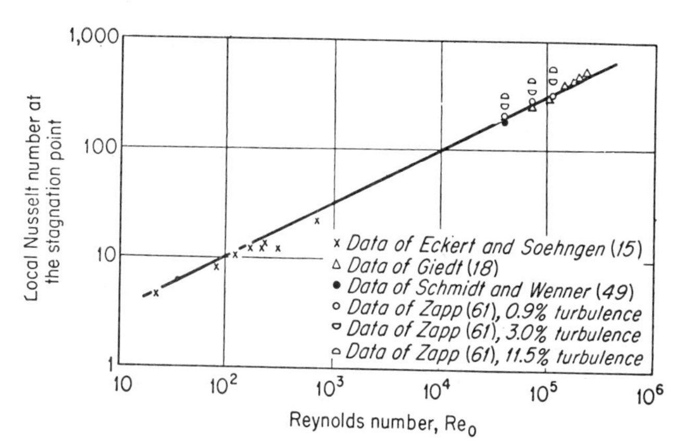

FIG. 17-15. Local Nusselt number at the forward stagnation point for the flow of air past circular cylinders.

Example 17-4

From Fig. 17-10 the local Nusselt number at a point 30° from the forward stagnation point is 316 for an air Reynolds number d_0U/ν of 110,000. Compare this value with that predicted by Levy's solution. Assume constant surface temperature [$m' = 0$ in Eq. (17-62)].

Solution

On the forward portion of a cylinder

$$n' \approx 1 \qquad \text{[in Eq. (10-91)]}$$

Therefore from Eq. (10-96)

$$\lambda = \frac{2n'}{1 + n'} = 2/2 = 1$$

From Table 17-3, for $\lambda = 1$, $m' = 0$, $\text{Pr} = 0.7$,

$$\left(\frac{dT}{d\eta_1}\right)_{\eta_1 = 0} = -0.488$$

From Eq. (17-64)

$$\frac{hx_1/k}{\sqrt{Ux_1/\nu}} = 0.488$$

x_1 is the distance measured from the forward stagnation point.

$$x_1 = r_0\theta = \frac{d_0\theta}{2}$$

Thus
$$\frac{hd_0/k}{\sqrt{Ud_0/\nu}} = \frac{0.488\sqrt{2}}{\sqrt{\theta}}$$

When
$$\theta = 30° = \pi/6 \text{ radians}$$

and
$$Ud_0/\nu = 110{,}000$$

$$\frac{hd_0}{k} = \frac{(0.488)(\sqrt{2})(\sqrt{6})(\sqrt{110{,}000})}{\sqrt{\pi}}$$

$$= 316$$

The value of the Nusselt number predicted by Levy's solution agrees with the measured value.

17-21. Average Nusselt Numbers on Circular Cylinders

Extensive data are available on average Nusselt numbers for flow past circular cylinders. Much of the work has been done with air, the Reynolds numbers investigated for this fluid ranging from 1.0 to 4×10^5. Cylinders ranging from 0.0004 in. diameter (fine wires) to 3.25 in. diameter have been studied. McAdams [36] gives a good summary of the various work that has been done on heat transfer from circular cylinders. He recommends the correlation reported by Hilpert [22] as being most reliable and representative of most of the available data for circular cylinders. Some of the data, however, do not agree with Hilpert's correlation. Douglas and Churchill [11] reanalyzed the data studied by McAdams and suggest that some of the data failed to correlate with Hilpert's curve because of the high temperature differences involved. They recalculated the data, using the physical properties determined at the film temperature $0.5(T_w + T_\infty)$. In McAdams's plot the density in the Reynolds number is evaluated at the free-stream temperature instead of the film temperature. The relation for predicting the average heat-transfer coefficient for gases flowing past cylinders is

$$\left(\frac{h_m d_0}{k}\right)_{T_{0.5}} = c\left(\frac{d_0 U \rho}{\mu}\right)^n_{T_{0.5}} \tag{17-69}$$

Conditions:

(1) All fluid properties evaluated at $T_{0.5} = 0.5(T_w + T_\infty)$
(2) Applies for gases and liquids
(3) Turbulence intensity 1 to 2 per cent
(4) Values of c and n are as given in Table 17-5

To consider the effect of duct width, the factor c should be multiplied by $\sqrt{1 + d_0/b}$. Equation (17-69) is plotted in Fig. 17-16, which also includes the data of Zapp,[61] Schmidt and Wenner,[49] Giedt,[18] and Comings, Clapp, and Taylor.[9] The data of Zapp and of Comings, Clapp, and Taylor clearly

TABLE 17-5

$\left(\frac{d_0 U \rho}{\mu}\right)_{T_{0.5}}$	n	c	
		For gases	For liquids
1–4	0.330	0.891	$0.989(\mathrm{Pr})^{1/3}$
4–40	0.385	0.821	$0.911(\mathrm{Pr})^{1/3}$
40–4,000	0.466	0.615	$0.683(\mathrm{Pr})^{1/3}$
4,000–40,000	0.618	0.174	$0.193(\mathrm{Pr})^{1/3}$
40,000–250,000	0.805	0.0239	$0.0266(\mathrm{Pr})^{1/3}$

show the effect of turbulence intensity on the average Nusselt number. At low turbulence intensities the data of all the above-mentioned investigators agree with the curve representing Eq. (17-69).

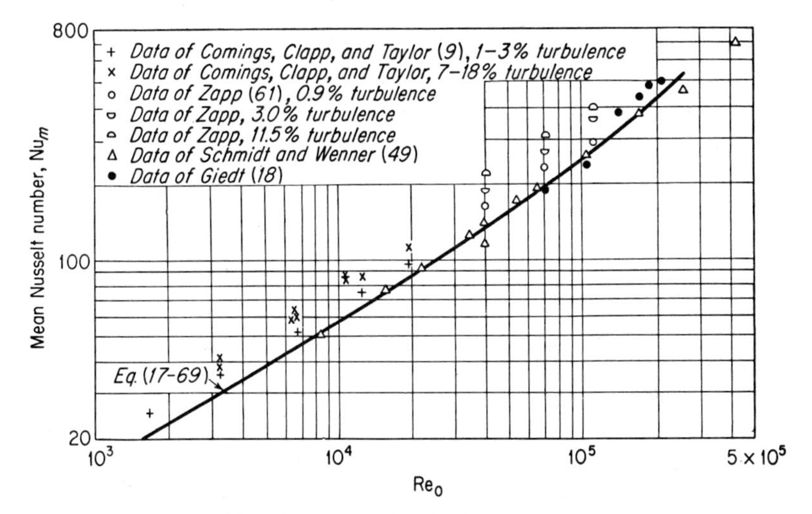

FIG. 17-16. Mean Nusselt numbers for the flow of air past circular cylinders.

Churchill and Brier[7] studied heat transfer from cylinders at very high temperature differences and propose the following relationship for predicting the average Nusselt number:

$$\mathrm{Nu}_m = 0.60(\mathrm{Pr})^{1/3}(\mathrm{Re}_0)^{1/2}(T_\infty/T_w)^{0.12} \qquad (17\text{-}70)$$

Conditions:

(1) All fluid properties evaluated at bulk temperature
(2) $300 < \mathrm{Re}_0 < 2{,}250$

Their data correlate well with Eq. (17-69) when the fluid properties are evaluated at the film temperature.

17-22. Local Heat-transfer Coefficients for Flow past Elliptical Cylinders

Local heat-transfer coefficients on elliptical cylinders with flow parallel to the major axis and for the case of uniform heat flux have been studied by Seban and Drake [51] and Drake, Seban, Doughty, and Levy.[13] The latter investigators studied an elliptical cylinder with axis ratio of 1:3. The

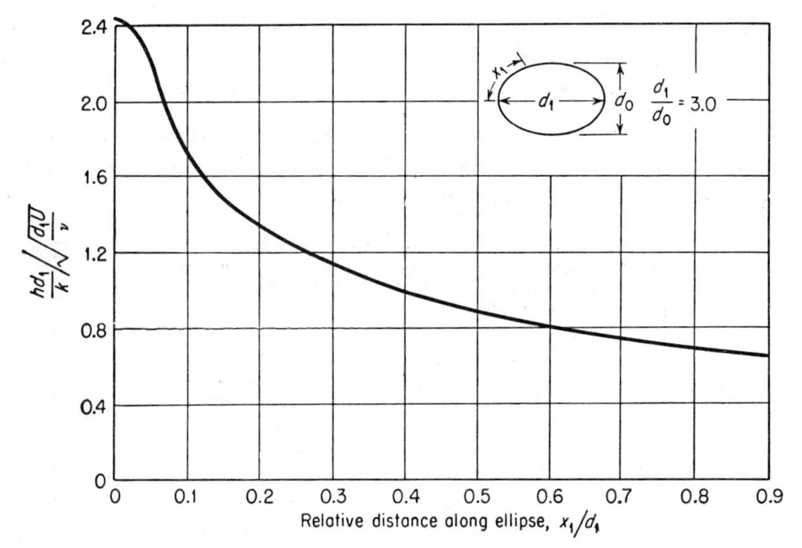

FIG. 17-17. Local Nusselt numbers for air flowing past an elliptical cylinder. [*From R. M. Drake et al., Trans. ASME,* **75**:1291 (1953).]

experimental results obtained are shown in Fig. 17-17, where $(hd_1/k)/\sqrt{d_1U/\nu}$ is plotted versus x_1/d_1, where d_1 is the axis parallel to the flow and x_1 is measured from the forward stagnation point. The experimental curve shown in Fig. 17-17 agrees well with the curve predicted by Seban [50] up to $x_1/d_1 = 0.4$. At the point of separation ($x_1/d_1 = 0.9$) the experimental curve is about 20 per cent above the predicted value of the coefficient. As the ratio of the axes of the cylinder becomes greater than 3:1, it behaves more like a flat plate, and the flat-plate equations may be used to predict approximate local coefficients on elliptical cylinders except in the vicinity of the leading edge. At the forward stagnation point the following approximate relationship holds for gases:

$$\frac{hd_1}{k} = 2.4\left(\frac{d_1U}{\nu}\right)^{0.5} \tag{17-71}$$

17-23. Local Heat-transfer Coefficients on Streamlined Cylinders

Martinelli et al.[35] recommend the following procedure for calculating local heat-transfer coefficients on streamlined cylinders. The forward portion is assumed to be a circular cylinder, and the trailing portion is assumed

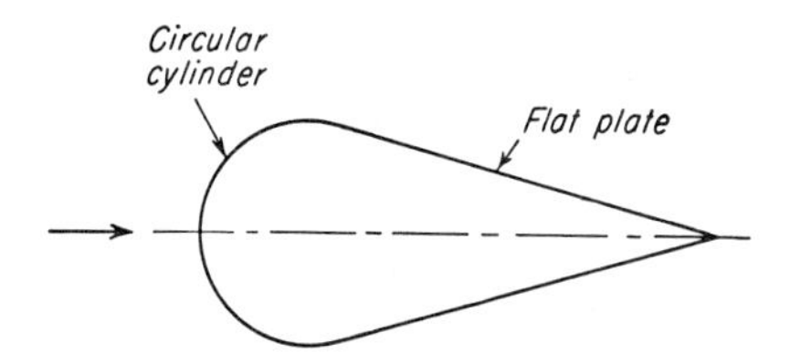

FIG. 17-18. Approximation of an airfoil so that Eqs. (17-39) and (17-65) may be used to predict local rates of heat transfer.

to be a flat plate, as shown in Fig. 17-18. Equation (17-65) is recommended for the forward portion.

$$\frac{hd_0}{k} = 1.14(\mathrm{Pr})^{0.4}\left(\frac{d_0U\rho}{\mu}\right)^{0.5}\left[1-\left(\frac{\theta}{90}\right)^3\right] \qquad 0 < \theta < 80° \qquad (17\text{-}65)$$

For the trailing portion of the streamlined cylinder Eq. (17-39) is recommended for laminar flow

$$\mathrm{Nu}_x = 0.324(\mathrm{Re}_x)^{1/2}(\mathrm{Pr})^{1/3} \qquad (17\text{-}39)$$

and Eq. (17-56) for the turbulent boundary layer

$$\mathrm{Nu}_x = 0.0292(\mathrm{Re}_x)^{4/5}(\mathrm{Pr})^{1/3} \qquad (17\text{-}56)$$

The distance x is measured from the forward stagnation point along the surface of the streamlined cylinder.

17-24. Average Nusselt Numbers on Cylinders of Noncircular Cross Section

Eckert [14] reports the empirical results of Hilpert [22] for average Nusselt numbers on cylinders of noncircular cross section. The form of the relationship for air is

$$\left(\frac{h_m d_0}{k}\right)_{T_{0.5}} = c\left(\frac{d_0U\rho}{\mu}\right)^n_{T_{0.5}} \qquad (17\text{-}69)$$

Conditions:

(1) Applies for gases only
(2) Fluid properties evaluated at $0.5(T_w + T_\infty)$
(3) Values of c and n are as given in Table 17-6

TABLE 17-6

Cross section	$\frac{d_0 U \rho}{\mu}$	c	n
→ d_0	5×10^3–10^5	0.0921	0.675
→ d_0	5×10^3–10^5	0.222	0.588
→ d_0	5×10^3–10^5	0.138	0.638
→ d_0	5×10^3–1.95×10^4 1.95×10^4–10^5	0.144 0.0347	0.638 0.782

IV. HEAT TRANSFER DURING FLOW PAST SPHERES

17-25. Transfer Processes on Spheres

There has been considerable interest in heat-, momentum-, and mass-transfer processes on spheres. In many operations, such as spray drying, humidification, and water cooling, heat and mass transfer occur simultaneously at the surface of a sphere or particle. Such simultaneous heat and mass transfer also occurs at the surface of a solid catalyst pellet on which a chemical reaction is occurring in either the liquid or gas phase. Liquid-liquid extraction involves mass transfer at the surface of spherical droplets of the discontinuous phase. The heat- and mass-transfer processes occurring in beds of fluidized catalyst take place at the surface of spheres or particles.

Much of the experimental work on spheres has involved the study of mass-transfer rates, particularly evaporation, from the surface. As a result, most of the data, including those obtained on heat transfer, have been correlated using Colburn's j-factor analogy.

17-26. Local Heat-transfer Coefficients on Spheres

Very little has been done either theoretically or experimentally on the determination of local heat-transfer coefficients on spheres. Carey [5] determined local heat-transfer coefficients for air flowing past spheres at Reynolds numbers from 44,000 to 150,000. He compared his experimental results with theoretical relationships in which many simplifying assumptions had been made. A plot of the local coefficients measured by Carey

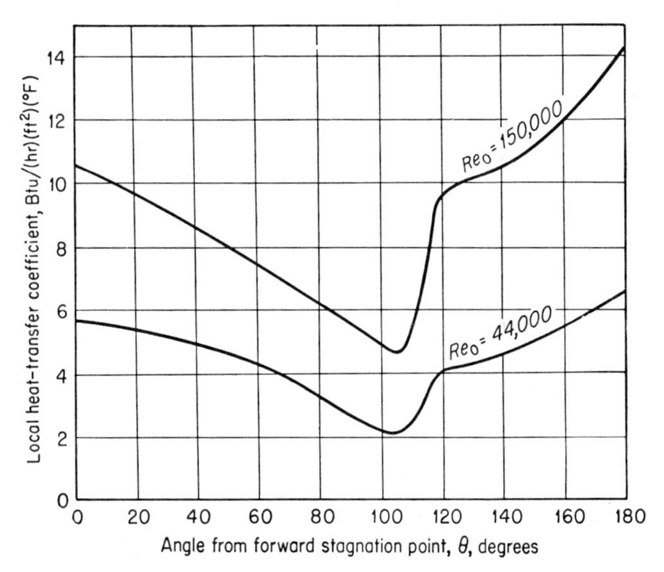

FIG. 17-19. Local heat-transfer coefficients for air flowing past a sphere. [*From J. R. Carey, Trans. ASME*, **75**:483 (1953).]

is shown in Fig. 17-19 for two Reynolds numbers, 44,000 and 150,000. At all Reynolds numbers Carey observed a minimum at $\theta = 105°$. Between $\theta = 105$ and $120°$ the coefficient rises abruptly and then rises more slowly between 120 and 180°. The region between 90 and 120° is the region of transition and separation.

Carey correlated his data on local coefficients by plotting $\mathrm{Nu}/\sqrt{\mathrm{Re}_0}$ versus θ, and curve I of Fig. 17-20 represents the correlation for values of θ from 0 to 105°. Curve II of Fig. 17-20 is a curve predicted from the theoretical equation derived by Pohlhausen [40]

$$\frac{h\delta}{k} = 0.765 \tag{17-72}$$

where δ is the thickness of the laminar boundary layer on the forward portion of the sphere. Values of δ may be determined from the "experimental" curve of Fig. 10-41. The theoretical curve in Fig. 17-20 is somewhat below the line representing Carey's experimental values.

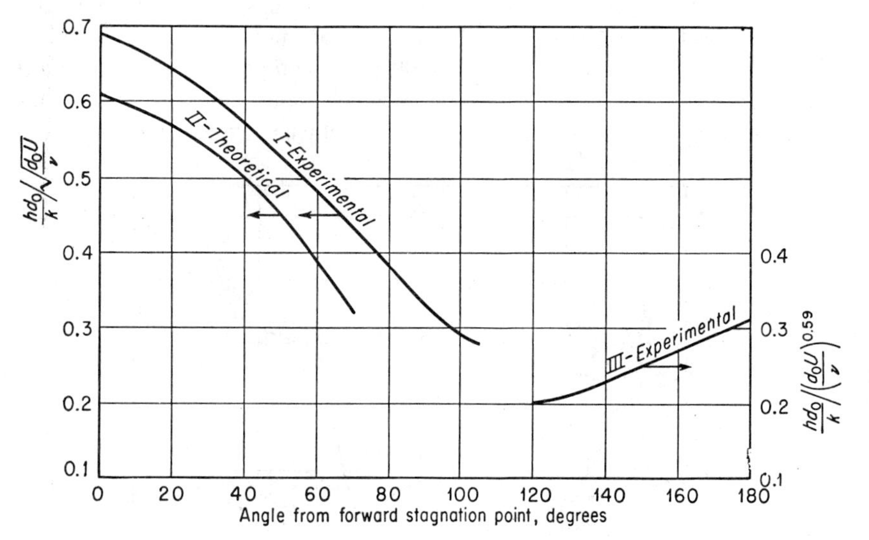

FIG. 17-20. Heat transfer for the flow of air past spheres. [*From J. R. Carey, Trans. ASME*, **75**:483 (1953).]

Curve III of Fig. 17-20 shows the empirical relation between $\text{Nu}/(\text{Re}_0)^{0.59}$ and θ for values of θ between 120 and 180°. Curves I and III are recommended for predicting local heat-transfer coefficients for the flow of air past spheres.

Use of the Reynolds analogy for relating the local heat- or mass-transfer coefficient to the local coefficient of friction does not appear possible for the case of spheres, i.e., the following do not apply.

$$\frac{h}{C_p \rho U} = \frac{f'}{2} \tag{17-73}$$

$$\frac{k_m}{U} = \frac{f'}{2} \tag{17-74}$$

At the forward point of stagnation f' is zero but neither h nor k_m are zero. The variation of h or k_m is much different from f' up to a value of $\theta = 50°$. There has been some success, however, in relating mean heat-transfer coefficients on a sphere to a mean value of the coefficient of friction through the Reynolds analogy. (See Sec. 17-29.)

17-27. Local Heat-transfer Coefficients at the Forward Stagnation Point of Spheres

Carey measured local heat-transfer coefficients at the stagnation point of spheres and obtained the following empirical relationship:

$$\mathrm{Nu}_{\theta=0} = 0.37(\mathrm{Re}_0)^{0.53} \qquad 44{,}000 < \mathrm{Re}_0 < 150{,}000 \qquad (17\text{-}75)$$

17-28. Average Heat-transfer Coefficients on Spheres

Extensive experimental work has been done on both mass and heat transfer from spheres in order to determine average coefficients.

A general correlation obtained by Froessling [17] on the rates of evaporation of water drops appears to be the best available for predicting heat- and mass-transfer coefficients for spheres. This correlation is:

1. For heat transfer:

$$\mathrm{Nu}_m = 2.0 + 0.60(\mathrm{Pr})^{1/3}(\mathrm{Re}_0)^{1/2} \qquad (17\text{-}76)$$

2. For mass transfer:

$$\left(\frac{k_m d_w}{D}\right)_m = 2.0 + 0.60(\mathrm{Sc})^{1/3}(\mathrm{Re}_0)^{1/2} \qquad (17\text{-}77)$$

Conditions:

(1) $1 < \mathrm{Re}_0 < 70{,}000$
(2) $0.6 < \mathrm{Pr} < 400$
(3) $0.6 < \mathrm{Sc} < 400$
(4) Dilute solutions

Ranz [42] compared the data of Froessling,[17] Maisel and Sherwood,[34] and Ranz and Marshall [43] on mass transfer and those of Kramers [29] on heat transfer and found Eqs. (17-76) and (17-77) to be representative of the experimental data. Equation (17-76) is plotted as the solid curve in Fig. 17-21 (curve I), where Nu_m is shown as a function of $(\mathrm{Pr})^{2/3}\mathrm{Re}_0$. The data points shown on the graph represent the heat-transfer results of Kramers [29] for oil, water, and air and those of Tang, Duncan, and Schweyer [56] for air. Kramers's studies covered a Prandtl-number range of 0.7 to 380. His data may be represented by the relation

$$\mathrm{Nu}_m = 2.0 + 1.3(\mathrm{Pr})^{0.15} + 0.66(\mathrm{Pr})^{0.31}(\mathrm{Re}_0)^{0.5} \qquad (17\text{-}78)$$

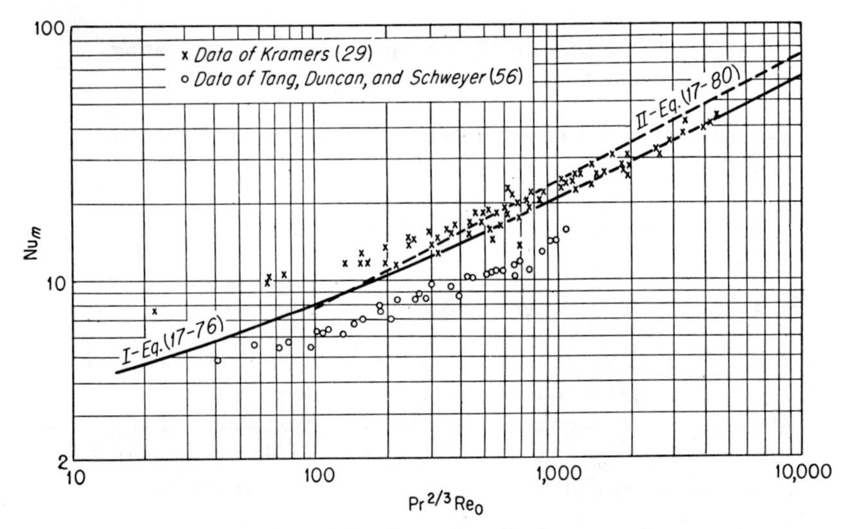

FIG. 17-21. Mean Nusselt numbers for flow past spheres.

Equation (17-78) is recommended by Kramers for the range $10 < \mathrm{Re}_0 < 10^5$. Tang, Duncan, and Schweyer[56] obtained the following relation from their data:

$$\frac{h_m}{C_p \rho U} = \frac{3.10}{\mathrm{Re}_0} + \frac{0.55}{\sqrt{\mathrm{Re}_0}} \tag{17-79}$$

Conditions:

(1) $50 < \mathrm{Re}_0 < 1{,}000$
(2) Applicable to gases ($\mathrm{Pr} = 0.7$)

Curve II represents a correlation obtained by Carey[5] for heat transfer from spheres to air. It is expressed by the following relation:

$$\mathrm{Nu}_m = 0.69\sqrt{\mathrm{Re}_0} \tag{17-80}$$

Conditions:

(1) $40{,}000 < \mathrm{Re}_0 < 150{,}000$
(2) $\mathrm{Pr} = 0.7$ (gases)

Equation (17-80) is essentially the same as the correlation obtained by Yuge[60] for a Reynolds-number range from 6.5×10^2 to 1.3×10^5. Equation (17-76) is the best one to use (for the complete Reynolds-number range) for the prediction of average Nusselt numbers for flow of fluids past spheres.

17-29. Use of Reynolds' Analogy for Predicting Average Heat-transfer Coefficients on Spheres

Tang, Duncan, and Schweyer [56] made use of Reynolds' analogy to derive a relationship for predicting average heat-transfer coefficients on spheres. The Reynolds analogy is expressed by

$$\frac{h}{C_p \rho U} = \frac{f'}{2} \qquad \text{Pr} = 1 \tag{17-73}$$

or

$$\frac{h_m}{C_p \rho U} = \frac{f'_m}{2} \qquad \text{Pr} = 1 \tag{17-81}$$

where f'_m is the average value of the local coefficient of friction; i.e.,

$$f'_m = \frac{\int f' \, dA}{A} \tag{17-82}$$

where A = the area of the sphere. These workers solved Millikan's equation (10-119) using the pressure distribution over the cylinder as given in Fig. 10-40 for Re_0 = 157,200 (below critical) and from the solution obtained

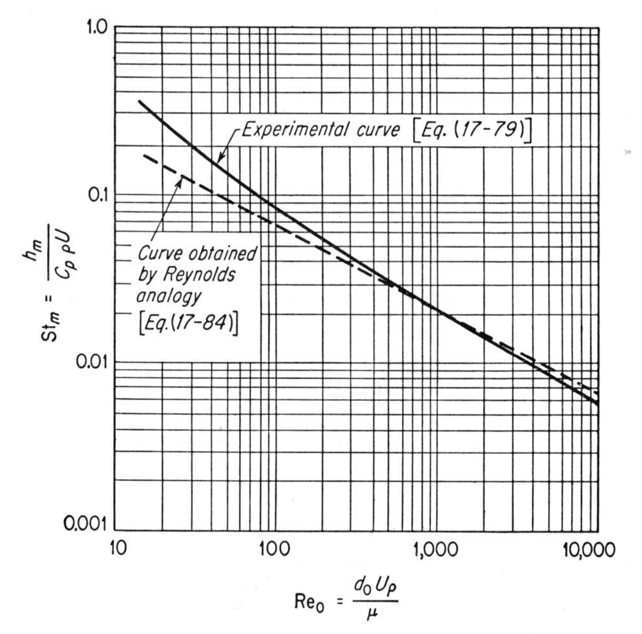

FIG. 17-22. Heat transfer from spheres. Comparison of experimental results with those obtainable from the Reynolds analogy. (*Based on the work of Y. S. Tang, J. M. Duncan, and H. E. Schweyer, NACA TN* 2867, 1953.)

values of f' up to the point of separation ($\theta = 86°$). Beyond the point of separation the local coefficient of friction is small, and the value of f'_m from $\theta = 86°$ to $\theta = 180°$ was considered to be 5 per cent of the value of f'_m from $\theta = 0°$ to $\theta = 86°$. The value of f'_m obtained by these calculations is

$$f'_m = 1.32/\sqrt{\text{Re}_0} \tag{17-83}$$

Substituting Eq. (17-83) into (17-81),

$$\text{St}_m = \frac{h_m}{C_p \rho U} = 0.66/\sqrt{\text{Re}_0} \qquad \text{Pr} = 1 \tag{17-84}$$

Equation (17-84) gives the relation between the average Stanton number and the Reynolds number using the Reynolds analogy. Figure 17-22 shows plots of Eqs. (17-79), the empirical equation obtained by Tang, Duncan, and Schweyer, and Eq. (17-84), which was derived using the Reynolds analogy based on the average value of the local coefficient of friction. The equations agree well with each other in the Reynolds-number range from 200 to 10,000.

V. HEAT TRANSFER ON THE SHELL SIDE OF TUBULAR HEAT EXCHANGERS

17-30. Nusselt Numbers for Flow across Banks of Circular Tubes

As pointed out in Sec. 11-1, most room heaters and radiators are so arranged so that the air to be heated flows perpendicular to a bank of tubes. A large portion of the flow in baffled heat exchangers is perpendicular to the tubes. For this reason it is necessary to be able to predict heat-transfer coefficients for flow across tube banks.

One of the earliest and simplest correlations of heat-transfer data for flow across banks of staggered tubes was proposed by Colburn.[8]

$$\left(\frac{h_{av} d_0}{k}\right)_{T0.5} = 0.33 \left(\frac{d_0 G_{max}}{\mu}\right)_{T0.5}^{0.6} \left(\frac{C_p \mu}{k}\right)_{T0.5}^{1/3} \tag{17-85}$$

where h_{av} is the average heat-transfer coefficient for the tube bank.

Conditions:

(1) $10 < \text{Re}_0 < 40{,}000$
(2) Mass velocity G_{max} is based on the minimum free-flow area
(3) Fluid properties evaluated at $0.5(T_w + T_b)$
(4) Staggered tube arrangement
(5) Ten or more rows of tubes

Equation (17-85) represents the simplest correlation of heat-transfer data for flow across banks of tubes. It is based on the pertinent data available up to 1933. In 1937 Pierson [38] and Huge [23] investigated the heat-transfer characteristics of a large number of tube banks. Grimison [21] correlated the data obtained by these workers and gave an equation of the form

$$\left(\frac{h_{av}d_0}{k}\right)_{T0.5} = c\left(\frac{d_0 G_{max}}{\mu}\right)^n_{T0.5} \qquad 2{,}000 < Re_0 < 40{,}000 \qquad (17\text{-}86)$$

Equation (17-86) is applicable for banks having 10 or more transverse rows of tubes. The values of c and n for the heating of air are given in Table 17-7. For fluids other than air the factor c should be multiplied by

TABLE 17-7. VALUES OF c AND n TO BE USED IN EQ. (17-86) FOR BANKS OF TUBES CONTAINING MORE THAN TEN TRANSVERSE ROWS †

Arrangement	S_T/d_0 → / S_L/d_0 ↓	1.25		1.50		2.0		3.0	
		c	n	c	n	c	n	c	n
Staggered	0.6							0.213	0.636
	0.9					0.446	0.571	0.401	0.581
	1.0			0.497	0.558				
	1.125					0.478	0.565	0.518	0.560
	1.250	0.518	0.556	0.505	0.554	0.519	0.556	0.522	0.562
	1.50	0.451	0.568	0.460	0.562	0.452	0.568	0.488	0.568
	2.0	0.404	0.572	0.416	0.568	0.482	0.556	0.449	0.570
	3.0	0.310	0.592	0.356	0.580	0.440	0.562	0.421	0.574
In-line	1.25	0.348	0.592	0.275	0.608	0.100	0.704	0.0633	0.752
	1.50	0.367	0.586	0.250	0.620	0.101	0.702	0.0678	0.744
	2.0	0.418	0.570	0.299	0.602	0.229	0.632	0.198	0.648
	3.0	0.290	0.601	0.357	0.584	0.374	0.581	0.286	0.608

† From E. D. Grimison, *Trans. ASME*, **59**:583 (1937).

$1.13(Pr)^{1/3}$. Nusselt numbers predicted by Eq. (17-86) are somewhat higher than those predicted by Eq. (17-85). Equation (17-86) is recommended for predicting heat-transfer coefficients for turbulent flow past tube banks.

Snyder [54] measured average heat-transfer coefficients on individual tubes in a bank of staggered tubes for which $S_L/d_0 = S_T/d_0 = 1.8$. He determined average heat-transfer coefficients for each tube row and obtained a correlation in the form of Eq. (17-86) for air, the values of the c and n

TABLE 17-8. VALUES OF c AND n TO BE USED IN EQ. (17-86) ACCORDING TO SNYDER'S CORRELATION †

$$\left(8{,}000 < \mathrm{Re}_0 < 20{,}000;\ \frac{S_L}{d_0} = \frac{S_T}{d_0} = 1.8\right)$$

Tube row	n	c
1	0.588	0.200
2	0.608	0.190
3	0.638	0.162
4	0.640	0.169
5	0.654	0.150
6	0.640	0.178
7	0.625	0.207
8	0.624	0.213
9	0.622	0.216
10	0.620	0.223

† From N. W. Snyder, *Chem. Eng. Progr. Symposium Ser.*, [5] **49**:11 (1953).

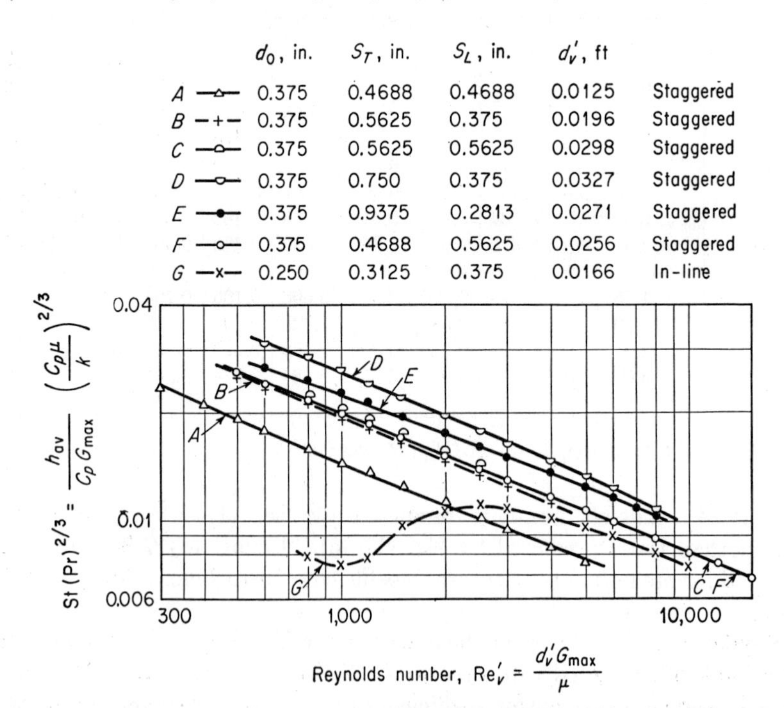

FIG. 17-23. Heat-transfer coefficients for flow across banks of circular tubes. [*Data from W. M. Kays et al., Trans. ASME,* **76**:387 (1954).]

being given in Table 17-8. These values may be used to predict average Nusselt numbers for each row of tubes. For fluids other than air multiply c by $1.13(\text{Pr})^{1/3}$.

For the laminar and transition region of flow across banks of tubes experimental results have been reported by Bergelin, Brown, and Doberstein [1] and Kays, London, and Lo.[27] The results of the former group are shown in the lower part of Fig. 11-11, where $(h_{av}/C_pG_{max})(\text{Pr})^{2/3}(\mu_w/\mu_b)^{0.14}$ is plotted versus Re_0 in the range from 10 to 4,000. Curves are shown for three staggered and two in-line arrangements, and these are recommended for predicting heat-transfer coefficients in tube banks of similar dimensions. The experimental results of Kays, London, and Lo are shown in Fig. 17-23 for six different staggered arrangements and one in-line arrangement (see Sec. 11-8). In this figure $(h_{av}/C_pG_{max})(\text{Pr})^{2/3}$ is plotted versus Re'_v. These curves may also be used to predict heat-transfer coefficients for laminar and transition flow across tube banks.

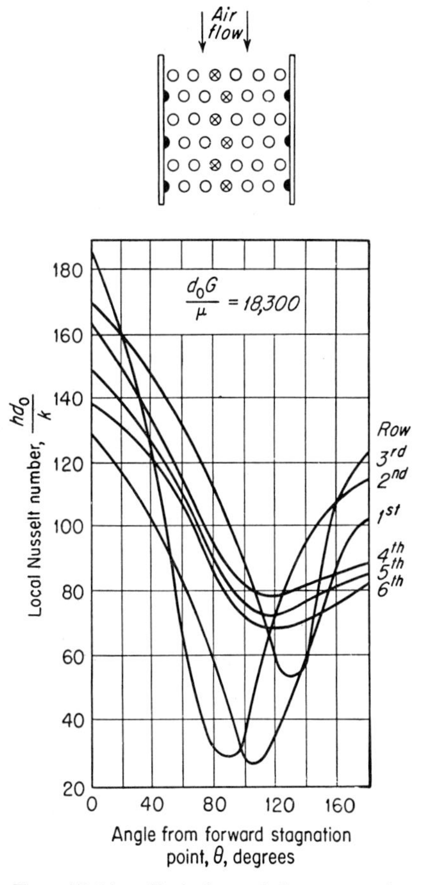

FIG. 17-24. Variation of heat transfer around tubes in a tube bank. (*Data from A. S. T. Thompson et al., "Proceedings of the General Discussion on Heat Transfer," Institution of Mechanical Engineers, London, and American Society of Mechanical Engineers, New York,* 1951, *p.* 177.)

17-31. Local Heat-transfer Coefficients on Tubes in Tube Banks

Robinson and Han [45] and Thompson, Scott, Laird, and Holden [57] have reported measurements of local heat-transfer coefficients on tubes in a tube bank. The latter group obtained coefficients on a single tube in each transverse row of a bank containing six transverse rows. Their results are shown in Fig. 17-24, where the local Nusselt number on each tube is plotted versus the angle from the forward stagnation point. All curves shown were obtained at a Reynolds number ($d_0U\rho/\mu$, where U is the superficial air velocity) of 18,300. A sketch shows the tube bank and the single tubes which were studied. All curves are similar in shape, the curve for the first row being more nearly like the curves obtained for flow past single circular cylinders (see Fig. 17-10). The local Nusselt numbers for tubes in the second row are generally higher than those in the first row. This is undoubtedly due to the turbulence caused by the first row. The local coefficients for tubes in the third and succeeding rows progressively decrease, becoming nearly constant (except at the forward stagnation point) for rows five and six. It appears that the second and third rows are most affected by the turbulence created by the first row.

17-32. Average Nusselt Numbers on Single Tubes in a Tube Bank

Snyder [54] measured average heat-transfer coefficients on a single tube for air flowing across a bank of staggered tubes. His results are in general agreement with those of Thompson et al. which are described above.

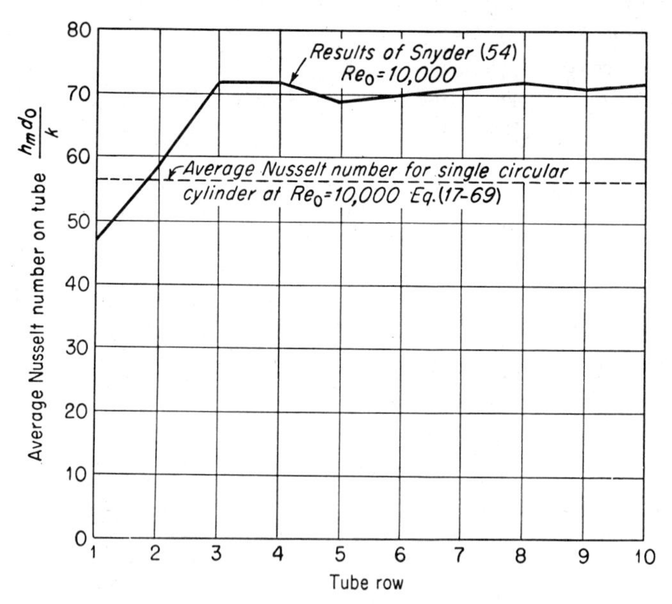

FIG. 17-25. Variation of average Nusselt number on tubes for crossflow of air through a tube bank. [*Results from N. W. Snyder, Chem. Eng. Progr. Symposium Ser.*, [5] **49**:11 (1953).]

Snyder's results for a Reynolds number (Re_0 based on maximum mass velocity) of 10,000 are shown in Fig. 17-25, where $h_m d_0/k$ is plotted versus the number of the tube row (h_m is the average heat-transfer coefficient on the tube). The average Nusselt number increases up to the third row, decreases slightly, and then remains essentially constant beyond the fifth row. The value of the average Nusselt number beyond the fifth row is about 80 per cent over that existing at the first row. Figure 17-25 shows clearly the effect of the front rows of tubes on the succeeding rows.

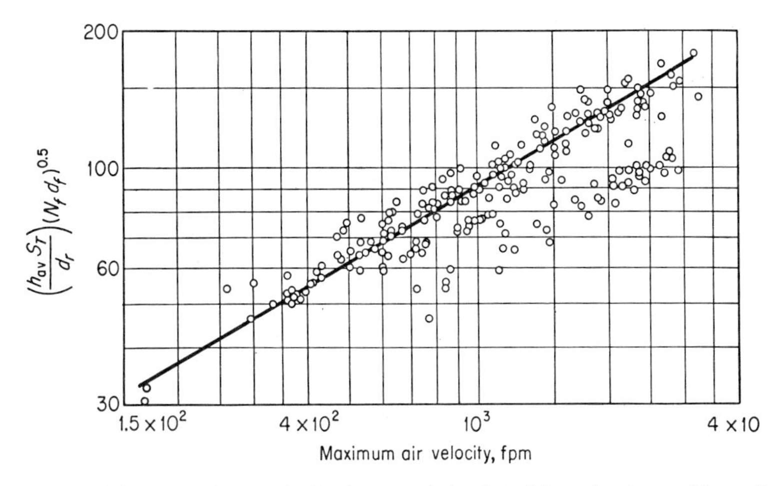

FIG. 17-26. Heat transfer to air flowing across banks of finned tubes. (*From D. L. Katz et al., Univ. Mich. Eng. Research Inst. Rept. on Project* M592, *August*, 1952.)

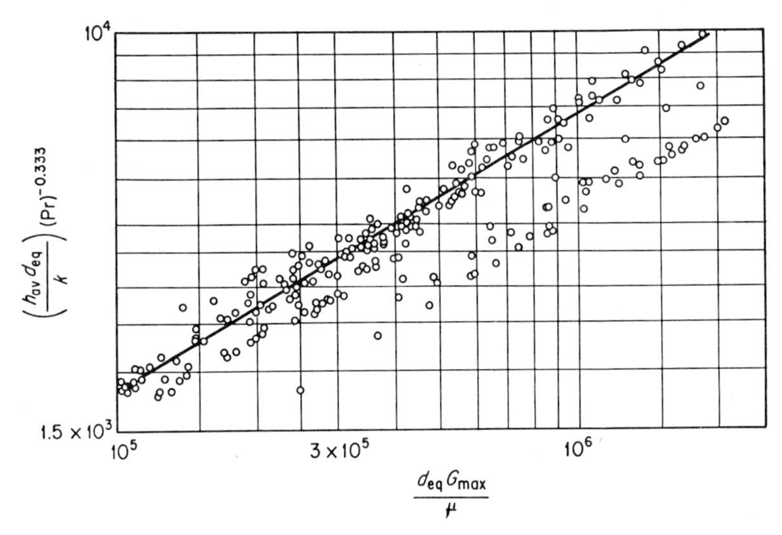

FIG. 17-27. Generalized correlation for heat transfer from banks of finned tubes (*From D. L. Katz et al., Univ. Mich. Eng. Research Inst. Rept. on Project* M592, *August*, 1952.)

17-33. Heat Transfer for Flow across Banks of Finned Tubes

Katz et al.[25] correlated the heat-transfer data for air flowing across banks of transverse-finned tubes. Their results are plotted in Fig. 17-26 for air, and a generalized correlation for all fluids is given in Fig. 17-27. Figure 17-26 shows the group $(h_{av}S_T/d_r)(N_f d_f)^{0.5}$ plotted versus the maximum air velocity U_{max} through the finned tube bank. The heat-transfer coeffi-

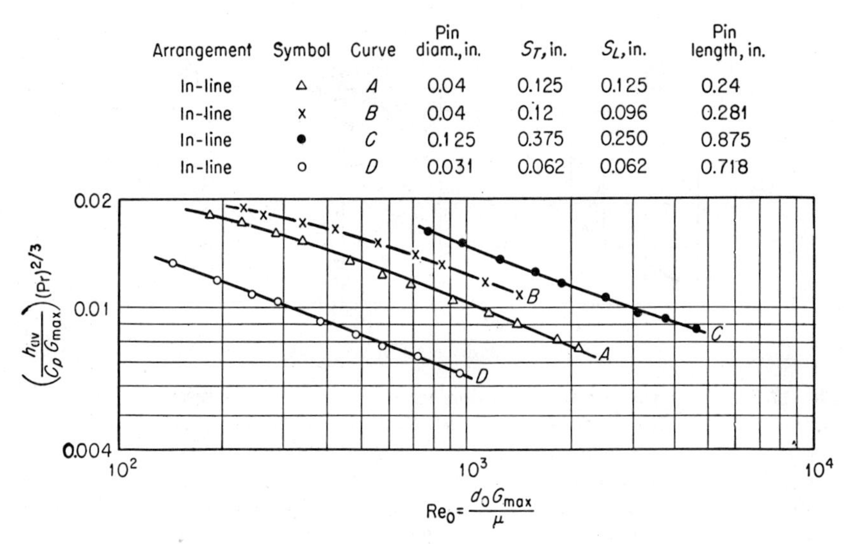

Arrangement	Symbol	Curve	Pin diam., in.	S_T, in.	S_L, in.	Pin length, in.
In-line	△	A	0.04	0.125	0.125	0.24
In-line	x	B	0.04	0.12	0.096	0.281
In-line	●	C	0.125	0.375	0.250	0.875
In-line	o	D	0.031	0.062	0.062	0.718

FIG. 17-28. Heat transfer in pin-fin heat exchangers. (*Data from W. M. Kays, Stanford Univ. Dept. Mech. Eng. Tech. Rept.* 19, *Navy Contract* N6-onr-251, *Task Order* 6, *August,* 1953.)

cient is a function of the tube pitch, root diameter, fin diameter, and number of fins per inch. The generalized correlation shown in Fig. 17-27 is a plot of $(h_{av}d_{eq}/k)(\text{Pr})^{-1/3}$ versus $d_{eq}G_{max}/\mu$, where d_{eq} is defined by Eq. (11-29).

An investigation of heat transfer in pin-fin heat exchangers (see Sec. 11-11) was reported by Kays.[26] A plot of the group $(h_{av}/C_pG_{max})(\text{Pr})^{2/3}$ versus Re_0 (based on d_0 and G_{max}) is shown in Fig. 17-28 for four different arrangements of the pins in the exchangers. The curves in Fig. 17-28 may be used to predict heat-transfer rates in such heat exchangers.

17-34. Heat-transfer Coefficients on the Shell Side of Heat Exchangers

The most extensive analysis of momentum and heat transfer on the shell side of baffled heat exchangers has been carried out by Tinker,[58] whose analysis includes a consideration of the effects of dimensional characteristics and mechanical clearances on shell-side performance. The reader is referred to the original articles for details concerning Tinker's work.

A correlation of shell-side heat-transfer coefficients was proposed by Donohue,[10] based on data he obtained as well as that of Short[52] and Bowman.[4] The following equations are recommended:

1. For unbaffled heat exchangers:

$$\frac{h_{av}d_0}{k} = 0.128{d_e}^{0.6}\left(\frac{d_0G_u}{\mu}\right)^{0.6}(\mathrm{Pr})^{1/3}\left(\frac{\mu_b}{\mu_w}\right)^{0.14} \qquad 100 < d_e\frac{d_0G_u}{\mu} < 20{,}000 \tag{17-87}$$

where d_e is defined in Table 11-1 and must be expressed in *inches* and G_u is the mass velocity in the unbaffled exchanger based on the free-flow area.

2. For disk-and-doughnut baffles:

$$\frac{h_{av}d_0}{k} = 0.23{d_e}^{0.6}\left(\frac{d_0G_e}{\mu}\right)^{0.6}(\mathrm{Pr})^{1/3}\left(\frac{\mu_b}{\mu_w}\right)^{0.14} \tag{17-88}$$

where d_e is in *inches*

$G_e = \sqrt{G_wG_c}$

G_w = mass velocity through baffle window

G_c = crossflow mass velocity based on flow area at diameter of shell

3. For segmental baffles:

$$\frac{h_{av}d_0}{k} = 0.25\left(\frac{d_0G_e}{\mu}\right)^{0.6}(\mathrm{Pr})^{1/3}\left(\frac{\mu_b}{\mu_w}\right)^{0.14} \tag{17-89}$$

For Eqs. (17-87) to (17-89) the fluid properties (except μ_w) are evaluated at the bulk temperature of the fluid. These relationships are recommended for predicting shell-side heat-transfer coefficients in commercial heat exchangers.

Williams and Katz [59] investigated shell-side heat-transfer coefficients in baffled heat exchangers containing plain and finned tubes. They obtained a heat-transfer correlation in the form of Eq. (17-89).

$$\frac{h_{av}d_0}{k} = c\left(\frac{d_0G_e}{\mu}\right)^{0.6}(\mathrm{Pr})^{1/3}\left(\frac{\mu_b}{\mu_w}\right)^{0.14} \qquad \mathrm{Re}_0 > 200 \tag{17-90}$$

Values of c are given in Table 11-5 for six different tube banks. Equation (17-90) is recommended for baffled heat exchangers containing transverse-finned tubes.

BIBLIOGRAPHY

1. Bergelin, O. P., G. A. Brown, and S. C. Doberstein: *Trans. ASME*, **74**:953 (1952).
2. Blasius, H.: *Z. Math. u. Phys.*, **56**:1 (1908).
3. Boelter, L. M. K., G. Young, M. L. Greenfield, V. D. Sanders, and M. Morgan: *NACA TN* 2517, 1951.

4. Bowman, R. A.: *ASME Misc. Paper* 28, p. 75, 1936.
5. Carey, J. R.: *Trans. ASME,* **75**:483 (1953).
6. Chapman, D. R., and M. W. Rubesin: *J. Aeronaut. Sci.,* **16**:547 (1949).
7. Churchill, S. W., and J. C. Brier: *Chem. Eng. Progr. Symposium Ser.,* [17] **51**:57 (1955).
8. Colburn, A. P.: *Trans. AIChE,* **29**:174 (1933).
9. Comings, E. W., J. T. Clapp, and J. F. Taylor: *Ind. Eng. Chem.,* **40**:1076 (1948).
10. Donohue, D. A.: *Ind. Eng. Chem.,* **41**:2499 (1949).
11. Douglas, W. J. M., and S. W. Churchill: *Heat Transfer Symposium, AIChE National Meeting, Louisville, Ky., 1955, Preprint* 16.
12. Drake, R. M.: *J. Appl. Mechanics,* **16**:1 (1949).
13. Drake, R. M., R. A. Seban, D. L. Doughty, and S. Levy: *Trans. ASME,* **75**:1291 (1953).
14. Eckert, E. R. G.: "Introduction to the Transfer of Heat and Mass," McGraw-Hill Book Company, Inc., New York, 1950.
15. Eckert, E. R. G., and E. Soehngen: *Trans. ASME,* **74**:343 (1952).
16. Elias, F.: *Abhandl. Aerodyn. Inst. Tech. Hochschule Aachen* 9, 1930; see also *NACA TM* 614, 1931.
17. Froessling, N.: *Gerlands Beitr. Geophys.,* **52**:170 (1938).
18. Giedt, W. H.: *Trans. ASME,* **71**:375 (1949).
19. Giedt, W. H.: *J. Aeronaut. Sci.,* **18**:725 (1951).
20. Goldstein, S. (ed.): "Modern Developments in Fluid Dynamics," vol. 1, Oxford University Press, London, 1938.
21. Grimison, E. D.: *Trans. ASME,* **59**:583 (1937).
22. Hilpert, R.: *Forsch. Gebiete Ingenieurw.,* **4**:215 (1933).
23. Huge, E. C.: *Trans. ASME,* **59**:573 (1937).
24. Juerges, W.: *Gesundh.-Ing.,* Reihe I, Beiheft 19, p. 1, 1924.
25. Katz, D. L., E. H. Young, R. B. Williams, G. Balekjian, and R. P. Williamson: Correlation of Heat Transfer and Pressure Drop for Air Flowing across Banks of Finned Tubes, *Univ. Mich. Eng. Research Inst. Rept. on Project* M592, August, 1952.
26. Kays, W. M.: *Stanford Univ. Dept. Mech. Eng. Tech. Rept.* 19, Navy Contract N6-onr-251, Task Order 6, August, 1953.
27. Kays, W. M., A. L. London, and R. K. Lo: *Trans. ASME,* **76**:387 (1954).
28. Klein, J., and M. Tribus: Forced Convection from Nonisothermal Surfaces, *Univ. Mich. Eng. Research Inst. Rept. on Project* M992-B, August, 1952.
29. Kramers, H.: *Physica,* **12**:61 (1946).
30. Krujilin, G. J.: *Tech. Phys. U.S.S.R.,* **5**:289 (1938).
31. Levy, S.: *J. Aeronaut. Sci.,* **19**:341 (1952).
32. Levy, S., and R. A. Seban: *J. Appl. Mechanics,* **20**:415 (1953).
33. Lorisch, W.: *VDI-Forschungsheft* 322, 1929.
34. Maisel, D. S., and T. K. Sherwood: *Chem. Eng. Progr.,* **46**:131 (1950).
35. Martinelli, R. C., A. G. Guibert, E. H. Morrin, and L. M. K. Boelter: *NACA WR* W-14 (formerly *ARR,* March, 1943).
36. McAdams, W. H.: "Heat Transmission," 3d ed., McGraw-Hill Book Company, Inc., New York, 1954.
37. Parmalee, G. V., and R. G. Huebscher: *Heating, Piping, Air Conditioning,* **19**:115 (Aug., 1947).
38. Pierson, O. L.: *Trans. ASME,* **59**:563 (1937).
39. Pohlhausen, E.: *Z. angew. Math. u. Mech.,* **1**:115 (1921).
40. Pohlhausen, K.: *Z. angew. Math. u. Mech.,* **1**:252 (1921).
41. Prandtl, L.: *Physik. Z.,* **11**:1072 (1910).
42. Ranz, W. E.: *Chem. Eng. Progr.,* **48**:247 (1952).

43. Ranz, W. E., and W. R. Marshall: *Chem. Eng. Progr.*, **48**:141 (1952).
44. Reynolds, O.: *Proc. Manchester Lit. Phil. Soc.*, **14**:7 (1874).
45. Robinson, W., and L. S. Han: *Proc. Midwest. Conf. Fluid Mechanics, 2nd Conf., Ohio State University, 1952*, p. 349.
46. Rubesin, M. W.: An Analytical Investigation of Convective Heat Transfer from a Flat Plate Having a Stepwise Discontinuous Surface Temperature, M.S. Thesis, University of California, 1947; see also *NACA TN* 2345, 1951.
47. Scesa, S., and F. M. Sauer: *Trans. ASME*, **74**:1251 (1952).
48. Scesa, S., and S. Levy: *Trans. ASME*, **76**:279 (1954).
49. Schmidt, E., and K. Wenner: *Forsch. Gebiete Ingenieurw.*, **12**(2):65 (1941); see also *NACA TM* 1050, 1943.
50. Seban, R. A.: *University Calif. (Berkeley) Inst. Eng. Research Rept.* 12, ser. 2, 1950.
51. Seban, R. A., and R. M. Drake: *Trans. ASME*, **75**:235 (1953).
52. Short, B. E.: *Univ. Texas Publ.* 4324, pp. 1–55, 1943.
53. Slegel, L. G., and G. A. Hawkins: *Purdue Univ. Eng. Bull. Research Ser.* 97, 1946.
54. Snyder, N. W.: *Chem. Eng. Progr. Symposium Ser.*, [5] **49**:11 (1953).
55. Stanton, T. E., and H. C. Booth: *British Aeronaut. Research Comm. R. & M.* 271, 1917.
56. Tang, Y. S., J. M. Duncan, and H. E. Schweyer: *NACA TN* 2867, 1953.
57. Thompson, A. S. T., A. W. Scott, A. M. Laird, and H. S. Holden: "Proceedings of the General Discussion on Heat Transfer," Institution of Mechanical Engineers, London, and American Society of Mechanical Engineers, New York, 1951, p. 177.
58. Tinker, T.: "Proceedings of the General Discussion on Heat Transfer," Institution of Mechanical Engineers, London, and American Society of Mechanical Engineers, New York, 1951, pp. 89, 97, 110.
59. Williams, R. B., and D. L. Katz: *Trans. ASME*, **74**:1307 (1952).
60. Yuge, T.: *Repts. Inst. High Speed Mechanics*, **5**(49):175 (1955).
61. Zapp, G. M.: The Effect of Turbulence on Local Heat Transfer Coefficients around a Cylinder Normal to an Air Stream, M.S. Thesis, Oregon State College, 1950.

APPENDIX I

MATHEMATICAL TERMS AND VECTOR NOTATION

I-1. Vectors and Scalars

The physical quantities appearing in problems are either vectors or scalars. Scalar quantities, such as density, viscosity, thermal conductivity, energy, etc., are represented by a number which signifies their magnitude. Scalar quantities can be either dimensional or dimensionless. Vector quantities are those quantities which are not completely described by magnitude alone. A velocity having a magnitude of 10 ft/sec is not completely described until its direction is also specified. Other vector quantities occurring in physical problems are acceleration and force.

I-2. The Velocity Vector

In dealing with the arbitrary motion of fluid in space one is concerned with the velocity of the fluid at various points. This velocity is represented by the velocity

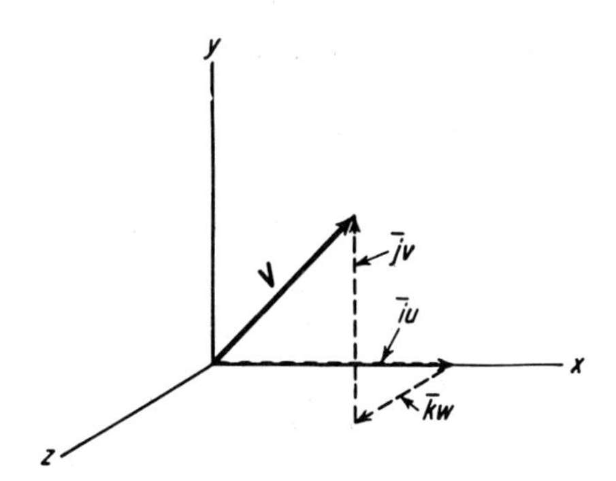

Fig. I-1. Components of an arbitrary velocity vector.

vector **V** in Fig. I-1. As shown, the velocity **V** is equivalent to three components of magnitude, u, v, and w, normal to each other and in the direction of the x, y, and z coordinates respectively; i.e.,

$$\mathbf{V} = \bar{\imath}u + \bar{\jmath}v + \bar{k}w \tag{I-1}$$

where $\bar{\imath}$, $\bar{\jmath}$, and $\bar{k}$ are unit vectors.

I-3. The Derivative of a Vector

The derivative of a vector is equivalent to the derivative of the components of the vector and is also a vector quantity. For example, taking the partial derivative of each side of Eq. (I-1) with respect to x gives

$$\frac{\partial \mathbf{V}}{\partial x} = \frac{\partial}{\partial x}(\bar{\imath}u + \bar{\jmath}v + \bar{k}w) = \bar{\imath}\frac{\partial u}{\partial x} + \bar{\jmath}\frac{\partial v}{\partial x} + \bar{k}\frac{\partial w}{\partial x} \qquad \text{(a vector)} \tag{I-2}$$

I-4. The Divergence of a Vector

The divergence div $\mathbf{V}$ of the vector $\mathbf{V}$ is defined as the scalar product of the two vectors $\mathbf{V}$ and $\bar{\imath}(\partial/\partial x) + \bar{\jmath}(\partial/\partial y) + \bar{k}(\partial/\partial z)$. The scalar product of two vectors results in a scalar quantity.

$$\begin{aligned} \text{div } \mathbf{V} &= \mathbf{V} \cdot \left(\bar{\imath}\frac{\partial}{\partial x} + \bar{\jmath}\frac{\partial}{\partial y} + \bar{k}\frac{\partial}{\partial z}\right) = (\bar{\imath}u + \bar{\jmath}v + \bar{k}w)\left(\bar{\imath}\frac{\partial}{\partial x} + \bar{\jmath}\frac{\partial}{\partial y} + \bar{k}\frac{\partial}{\partial z}\right) \\ &= \frac{\partial u}{\partial x} + \frac{\partial v}{\partial y} + \frac{\partial w}{\partial z} \qquad \text{(a scalar)} \end{aligned} \tag{I-3}$$

$$\text{div } (\rho\mathbf{V}) = \frac{\partial(\rho u)}{\partial x} + \frac{\partial(\rho v)}{\partial y} + \frac{\partial(\rho w)}{\partial z} \tag{I-4}$$

where ρ is a scalar. Carrying out the differentiation in Eq. (I-4),

$$\begin{aligned} \text{div } (\rho\mathbf{V}) &= u\frac{\partial \rho}{\partial x} + \rho\frac{\partial u}{\partial x} + v\frac{\partial \rho}{\partial y} + \rho\frac{\partial v}{\partial y} + w\frac{\partial \rho}{\partial z} + \rho\frac{\partial w}{\partial z} \\ &= \rho \text{ div } \mathbf{V} + \frac{D\rho}{Dt} - \frac{\partial \rho}{\partial t} \end{aligned} \tag{I-5}$$

where $D\rho/Dt$ is the substantial derivative of ρ with respect to t (see Sec. I-8).

I-5. The Derivative of the Divergence of a Vector

The derivative of the divergence of a vector is equal to the divergence of the derivative of the vector. The result is a scalar quantity.

$$\begin{aligned} \frac{\partial}{\partial x}(\text{div } \mathbf{V}) &= \frac{\partial}{\partial x}\left(\frac{\partial u}{\partial x} + \frac{\partial v}{\partial y} + \frac{\partial w}{\partial z}\right) \qquad \text{(a scalar)} \\ &= \frac{\partial^2 u}{\partial x^2} + \frac{\partial^2 v}{\partial x\, \partial y} + \frac{\partial^2 w}{\partial x\, \partial z} \\ &= \text{div}\left(\bar{\imath}\frac{\partial u}{\partial x} + \bar{\jmath}\frac{\partial v}{\partial x} + \bar{k}\frac{\partial w}{\partial x}\right) \\ &= \text{div}\frac{\partial \mathbf{V}}{\partial x} \end{aligned} \tag{I-6}$$

I-6. The Gradient of a Scalar

The gradient is a term applied to the operation on scalar quantities which consists of multiplying the scalar by the vector $\bar{\imath}(\partial/\partial x) + \bar{\jmath}(\partial/\partial y) + \bar{k}(\partial/\partial z)$. The result is a vector quantity; i.e.,

$$\text{grad } \rho = \left(\bar{\imath}\frac{\partial}{\partial x} + \bar{\jmath}\frac{\partial}{\partial y} + \bar{k}\frac{\partial}{\partial z}\right)\rho = \bar{\imath}\frac{\partial \rho}{\partial x} + \bar{\jmath}\frac{\partial \rho}{\partial y} + \bar{k}\frac{\partial \rho}{\partial z} \quad \text{(a vector)} \tag{I-7}$$

I-7. The Laplacian of a Scalar

The Laplacian of a scalar is the operation on a scalar with the quantity $\partial^2/\partial x^2 + \partial^2/\partial y^2 + \partial^2/\partial z^2$, which is generally called the Laplacian operator ∇^2. The Laplacian operation on a scalar also gives a scalar quantity. The Laplacian operation is equivalent to the divergence of the gradient of a scalar.

$$\begin{aligned}\nabla^2 u &= \frac{\partial^2 u}{\partial x^2} + \frac{\partial^2 u}{\partial y^2} + \frac{\partial^2 u}{\partial z^2} \\ &= \text{div grad } u \\ &= \left(\bar{\imath}\frac{\partial}{\partial x} + \bar{\jmath}\frac{\partial}{\partial y} + \bar{k}\frac{\partial}{\partial z}\right)\left(\bar{\imath}\frac{\partial u}{\partial x} + \bar{\jmath}\frac{\partial u}{\partial y} + \bar{k}\frac{\partial u}{\partial z}\right) \\ &= \frac{\partial^2 u}{\partial x^2} + \frac{\partial^2 u}{\partial y^2} + \frac{\partial^2 u}{\partial z^2}\end{aligned} \tag{I-8}$$

I-8. The Total Differential and Total Derivative

In Fig. I-1 the component of **V** in the x direction is u. In the arbitrary motion of a fluid

$$u = F_1(x,y,z,t) \tag{I-9}$$

For a differential change of dx, dy, dz, and dt the total change in u is du, which is the total differential defined as

$$du = \frac{\partial u}{\partial x}dx + \frac{\partial u}{\partial y}dy + \frac{\partial u}{\partial z}dz + \frac{\partial u}{\partial t}dt \tag{I-10}$$

However, in moving the distances dx, dy, dz there has been a certain elapsed time dt; or one might consider that x, y, and z are also functions of t. Dividing Eq. (I-10) by dt results in

$$\frac{du}{dt} = \frac{\partial u}{\partial x}\frac{dx}{dt} + \frac{\partial u}{\partial y}\frac{dy}{dt} + \frac{\partial u}{\partial z}\frac{dz}{dt} + \frac{\partial u}{\partial t} \tag{I-11}$$

dx/dt, dy/dt, and dz/dt are respectively u, v, and w, the components of the velocity **V**. Thus

$$\frac{du}{dt} = u\frac{\partial u}{\partial x} + v\frac{\partial u}{\partial y} + w\frac{\partial u}{\partial z} + \frac{\partial u}{\partial t} \tag{I-12}$$

du/dt is the total derivative of u with respect to t and represents the total acceleration of the fluid due to changes of velocity in both space and time. The total derivative is often given the symbol Du/Dt. Following are two more examples of total derivatives:

$$\frac{D\rho}{Dt} = u\frac{\partial \rho}{\partial x} + v\frac{\partial \rho}{\partial y} + w\frac{\partial \rho}{\partial z} + \frac{\partial \rho}{\partial t} \tag{I-13}$$

$$\frac{DT}{Dt} = u\frac{\partial T}{\partial x} + v\frac{\partial T}{\partial y} + w\frac{\partial T}{\partial z} + \frac{\partial T}{\partial t} \tag{I-14}$$

APPENDIX II

COMPLEX NUMBERS AND CONFORMAL MAPPING

II-1. Complex Numbers

The quantity $x + iy$ is a complex number, where x and y are real numbers and i is $\sqrt{-1}$. A complex number consists of two parts; a real part given by x and an imaginary part given by iy. Complex numbers may be represented graphically on two-dimensional rectangular graph paper. The x axis represents the *real* axis and the y axis represents the *imaginary* axis. The two-dimensional plane of the

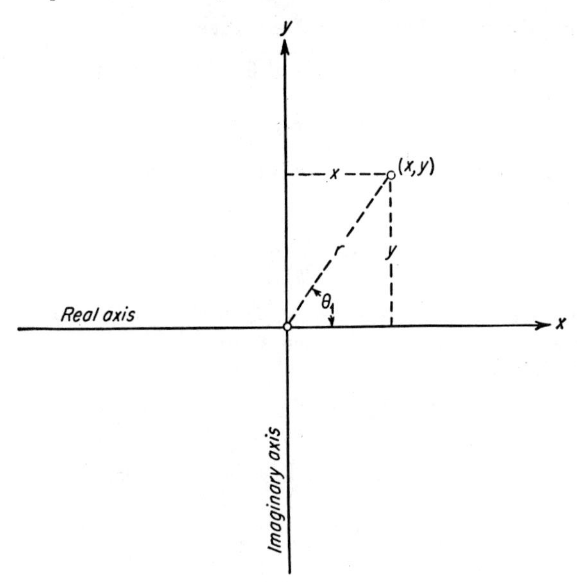

FIG. II-1. Graphical representation of the complex number $x + iy$.

graph is the *complex* plane. Figure II-1 shows the graphical representation of the complex number $x + iy$, the point (x,y) representing the complex number.

An alternative expression for complex numbers is

$$r(\cos \theta_1 + i \sin \theta_1)$$

where r is the radial distance from the origin of the x and y coordinates and θ_1 is the angle made with the x axis.

$$r = \sqrt{x^2 + y^2} \tag{II-1}$$

$$x = r \cos \theta_1 \tag{II-2}$$

$$y = r \sin \theta_1 \tag{II-3}$$

In the above expression $r \cos \theta_1$ is the real part, and $ri \sin \theta_1$ is the imaginary part. It may be shown by expansion into series that

$$r(\cos \theta_1 + i \sin \theta_1) = re^{i\theta_1} \tag{II-4}$$

The right side of Eq. (II-4) is the exponential expression for a complex number.

The operations of addition, subtraction, multiplication, and division of complex numbers follow the usual rules of real numbers. Table II-1 shows the results of carrying out these operations.

TABLE II-1. OPERATIONS WITH COMPLEX NUMBERS

Operation	Algebraic representation	Result
Addition	$x_1 + iy_1 + x_2 + iy_2$	$x_1 + x_2 + i(y_1 + y_2)$
Subtraction	$(x_1 + iy_1) - (x_2 + iy_2)$	$(x_1 - x_2) + i(y_1 - y_2)$
Multiplication	$(x_1 + iy_1)(x_2 + iy_2)$	$x_1x_2 - y_1y_2 + i(x_1y_2 + x_2y_1)$
	$r_1e^{i\theta_1}r_2e^{i\theta_2}$	$r_1r_2e^{i(\theta_1+\theta_2)}$
Division	$\dfrac{x_1 + iy_1}{x_2 + iy_2}$	$\dfrac{x_1x_2 + y_1y_2}{x_2^2 + y_2^2} + \dfrac{i(y_1x_2 - x_1y_2)}{x_2^2 + y_2^2}$
	$\dfrac{r_1e^{i\theta_1}}{r_2e^{i\theta_2}}$	$\dfrac{r_1}{r_2} e^{i(\theta_1-\theta_2)}$
Powers of numbers	$(x + iy)^n$	$x^n + inx^{n-1}y - \dfrac{n(n-1)}{2!} x^{n-2}y^2 + \cdots$
	$(re^{i\theta_1})^n$	$r^ne^{in\theta_1}$

II-2. Functions of Complex Variables

Consider two complex numbers $\phi + i\psi$ and $x + iy$ such that

$$\phi + i\psi = F_1(x + iy) \tag{II-5}$$

In these complex numbers ϕ, ψ, x, and y are all real variables. The functional relationship given in Eq. (II-5) also requires that ϕ, ψ, x, and y be related; i.e., $\phi = \phi(x,y)$ and $\psi = \psi(x,y)$.

Let

$$w = \phi + i\psi \tag{II-6}$$

and

$$z = x + iy \tag{II-7}$$

Differentiating w with respect to x.

$$\frac{\partial w}{\partial x} = \frac{dw}{dz}\frac{\partial z}{\partial x} \quad \text{(II-8)}$$

From Eqs. (II-6) and (II-7)

$$\frac{\partial w}{\partial x} = \frac{\partial \phi}{\partial x} + i\frac{\partial \psi}{\partial x} \quad \text{(II-9)}$$

$$\frac{\partial z}{\partial x} = 1 \quad \text{(II-10)}$$

Substituting Eqs. (II-9) and (II-10) into (II-8),

$$\frac{dw}{dz} = \frac{\partial \phi}{\partial x} + i\frac{\partial \psi}{\partial x} \quad \text{(II-11)}$$

Differentiating w with respect to y,

$$\frac{\partial w}{\partial y} = \frac{dw}{dz}\frac{\partial z}{\partial y} \quad \text{(II-12)}$$

From Eqs. (II-6) and (II-7)

$$\frac{\partial w}{\partial y} = \frac{\partial \phi}{\partial y} + i\frac{\partial \psi}{\partial y} \quad \text{(II-13)}$$

$$\frac{\partial z}{\partial y} = i \quad \text{(II-14)}$$

Substituting Eqs. (II-14) and (II-13) into (II-12),

$$\frac{dw}{dz} = \frac{1}{i}\frac{\partial \phi}{\partial y} + \frac{\partial \psi}{\partial y} = \frac{\partial \psi}{\partial y} - i\frac{\partial \phi}{\partial y} \quad \text{(II-15)}$$

Thus, combining Eqs. (II-11) and (II-15),

$$\frac{\partial \phi}{\partial x} + i\frac{\partial \psi}{\partial x} = \frac{\partial \psi}{\partial y} - i\frac{\partial \phi}{\partial y} \quad \text{(II-16)}$$

Equating the real and imaginary parts of Eq. (II-16),

$$\frac{\partial \phi}{\partial x} = \frac{\partial \psi}{\partial y}$$

$$\frac{\partial \phi}{\partial y} = -\frac{\partial \psi}{\partial x} \quad \text{(II-17)}$$

Equations (II-17) are known as the Cauchy-Riemann equations. Any functional relationship defined throughout a region, such as the function described in Eq. (II-5), satisfies Eqs. (II-17) provided all partial derivatives are continuous.

II-3. Conformal Mapping

The complex number $\phi + i\psi$ may be represented graphically on the $\phi\psi$ complex plane just as $x + iy$ is represented on the xy complex plane (see Fig. II-1). If a functional relationship exists between $\phi + i\psi$ and $x + iy$, i.e.,

$$\phi + i\psi = F_1(x + iy) \quad \text{(II-5)}$$

then for every point on the $\phi\psi$ plane there will be a corresponding point on the xy plane, the position of the point depending upon the functional relationship in Eq. (II-5). This is most easily illustrated if one considers a line at constant ϕ on the $\phi\psi$ plane. In Fig. II-2a the vertical line labeled ϕ_1 represents a constant value of ϕ. Corresponding to this line will be a line on the xy plane, the shape of which is determined by the functional relationship between the two complex numbers. This line is labeled ϕ_1 on the xy plane shown in Fig. II-2b. Similarly the line ϕ_2 on the $\phi\psi$ plane corresponds to the curve ϕ_2 on the xy plane, as is also the case for the lines labeled ψ_1 and ψ_2. In Fig. II-2a the lines ϕ_1 and ϕ_2 intersect the lines ψ_1 and ψ_2

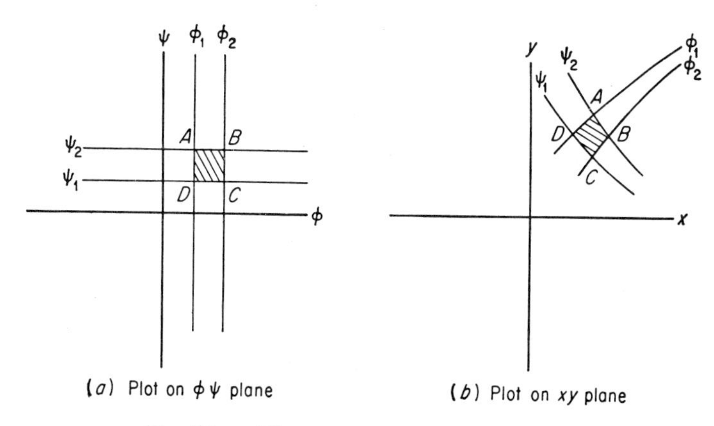

FIG. II-2. Illustration of conformal mapping.

forming the square $ABCD$. In Fig. II-2b the lines intersect to form the mesh $ABCD$; i.e., the square $ABCD$ on the $\phi\psi$ plane has been mapped onto the xy plane as the mesh $ABCD$. This process is *conformal mapping*. The lines of constant ϕ and ψ on the $\phi\psi$ plane all intersect at right angles and form a grid of square meshes. The corresponding lines mapped onto the xy plane also intersect at right angles and form a grid of meshes which are square if they are infinitesimal in size.

As an example of conformal mapping consider that the functional relationship of Eq. (II-5) is as follows:

$$\phi + i\psi = (x + iy)^2 \tag{II-18}$$

or

$$\phi + i\psi = x^2 - y^2 + 2ixy \tag{II-19}$$

Equating the real and imaginary parts of Eq. (II-19),

$$\phi = x^2 - y^2 \tag{II-20}$$

$$\psi = 2xy \tag{II-21}$$

Equations (II-20) and (II-21) may now be used for conformal mapping. For example, the line corresponding to $\phi = 1$ on the $\phi\psi$ plane will map into the curve $y = \pm\sqrt{x^2 - 1}$ on the xy plane. Figure II-3 shows the results of the conformal mapping.

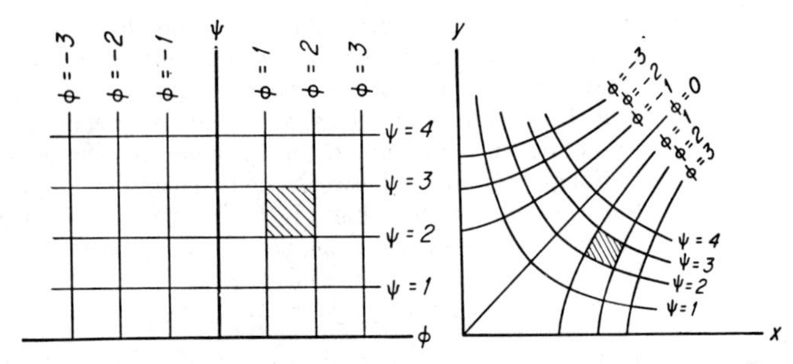

Fig. II-3. Conformal map of the function $\phi + i\psi = (x + iy)^2$.

II-4. The Application of Conformal Mapping to Nonviscous Fluid Flow

Differentiating the first of Eqs. (II-17) with respect to x and the second with respect to y and adding the results gives

$$\frac{\partial^2 \phi}{\partial x^2} + \frac{\partial^2 \phi}{\partial y^2} = 0 \tag{II-22}$$

Likewise differentiating the first of Eqs. (II-17) with respect to y and the second with respect to x and subtracting, one obtains

$$\frac{\partial^2 \psi}{\partial x^2} + \frac{\partial^2 \psi}{\partial y^2} = 0 \tag{II-23}$$

Equations (II-22) and (II-23) are the differential equations describing two-dimensional, incompressible, nonviscous flow of fluids, where ϕ is the velocity potential and ψ is the stream function. ϕ and ψ are defined as follows:

$$\begin{aligned} \frac{\partial \phi}{\partial x} &= u \\ \frac{\partial \phi}{\partial y} &= v \\ \frac{\partial \psi}{\partial y} &= u \\ \frac{\partial \psi}{\partial x} &= -v \end{aligned} \tag{II-24}$$

The above relationships also satisfy the Cauchy-Riemann conditions.

Therefore, the constant ϕ lines shown in Fig. II-3 (either on the $\phi\psi$ plane or the xy plane) may be considered to be the equipotential lines, or lines of constant velocity, in the flowing fluid. Also, the lines of constant ψ may be considered to be streamlines in the flowing fluid. The system of intersecting lines on the xy plane in Fig. II-3 satisfies all the conditions of a flow net in that all the lines intersect at right angles and the meshes are approximately square. The meshes would be square if the lines were located an infinitesimal distance apart.

In systems where flow takes place past solid boundaries, the limiting streamline coincides with the shape of the solid boundary. This fact is helpful in interpreting flow nets which have been obtained by conformal mapping.

This may be illustrated by the use of Fig. II-3. The conformal mapping of the net on the $\phi\psi$ plane to that on the xy plane was a purely mathematical procedure. No thought was given to any physical significance of the nets thus obtained. On the $\phi\psi$ plane the line corresponding to $\psi = 0$ is the ϕ axis. On the xy plane (referring to Fig. II-3) the line corresponding to $\psi = 0$ is the positive x and y axis. This line may be considered to be the solid boundaries of a flow system. Hence the flow net shown on the xy plane of Fig. II-3 describes the ideal flow along two planes which intersect each other at right angles. Similarly, the streamline corresponding to $\psi = 1$ could have been considered to be the solid boundary of the system, and the flow net then would represent the flow past a two-dimensional solid having a shape corresponding to the curve for $\psi = 1$ on the xy plane.

The use of conformal mapping in ideal-fluid flow problems then consists of determining the functional relationship between $\phi + i\psi$ and $x + iy$ such that a streamline or streamlines mapped on the xy plane coincide with the solid boundaries of the given system. Mathematically, what has been obtained is a solution of the differential equations of ideal-fluid flow [Eqs. (II-22) and (II-23)]. The boundary conditions are obtained from the shape of the solid surfaces past which flow is taking place.

APPENDIX III

TABLE OF NOMENCLATURE

The fundamental dimensions are represented by the following letters:

F = force L = length

m = mass t = time

T = temperature

Symbol	Meaning	Dimensions
A	Constant	Variable
A	Cross-sectional area; A_P, projected area of immersed body perpendicular to direction of flow; A_q, area through which rate of heat transfer is q; A_r, area through which rate of heat transfer is q_r; A_w, area of surface in contact with fluid; area of wall	L^2
A_{N_m}	Area through which rate of mass transfer is N_m	L^2
a	Constant	Variable
a	Length of short side of rectangular duct; dimensions on noncircular ducts shown in Fig. 4-21	L
b	Distance between parallel planes; length of long side of rectangular duct; dimensions on noncircular ducts shown in Fig. 4-21	L
$C, C_0, C_1, \ldots, C_n$	Constant or coefficient	Variable
C_p	Heat capacity at constant pressure	FL/mT
C_v	Heat capacity at constant volume	FL/mT
$c, c_0, c_1, \ldots, c_n$	Constant or coefficient	Variable
c_m	Concentration	m/L^3
D	Quantity defined by Eq. (10-128)	None
D	Molecular diffusivity of mass	L^2/t
D/Dt	Substantial or total derivative with respect to t	$1/t$

Symbol	Meaning	Dimensions
d	Diameter; d_0, sphere or cylinder diameter; projected diameter of immersed bodies perpendicular to direction of flowing fluid; d_1, outside diameter of inner tube of annulus; d_2, inside diameter of outer tube of annulus; d_c, clearance between tubes in transverse rows; d_e, equivalent diameter of closed conduit [Eq. (11-2)]; d_{eq}, equivalent diameter of bank of finned tubes defined by Eq. (11-29); d_f, fin diameter; d_{max}, diameter of annulus where maximum point velocity occurs; d_r, root diameter of finned tube; d_v, volumetric equivalent diameter defined by Eq. (11-4); d_v', modified volumetric equivalent diameter defined by Eq. (11-5); d_w, inside diameter of circular tube	L
d'	Exponent giving relation of gas or liquid viscosity to temperature	None
div	Divergence of a vector	
E	Internal energy	FL/m or FL
e	2.718, base of natural logarithms	
e	Fin height; height of roughness projections; height of threads; e_{admiss}, admissible roughness [Eq. (10-83)]	L
e_I, e_{II}, e_n	Voltage across hot-wire anemometer	
exp	Denotes 2.718 raised to a certain exponent	
F	Force; F_{edge}, force of resistance due to edges of flat plate	F
F_1, F_2, F_3, ...	Function or coefficient	
FF	Field force	F
$F(n)$	Fraction of total turbulent energy due to frequencies between n and $n + dn$	None
f	Friction factor for flow in closed conduits; drag coefficient [defined by Eq. (10-5)] for flow over immersed bodies; f_1, f_2, friction factors for inner and outer walls of annulus, respectively, as defined by Eqs. (7-94) and (7-95); f', local drag coefficient or skin-friction coefficient defined by Eq. (10-7); f_m', mean value of local drag coefficient over immersed body; f_{av}, average friction factor for finite length of tube; f_B, Blasius friction factor defined by Eq. (4-22); f_{CG}, friction factor for tube banks defined by Eq. (11-10); f_D, drag coefficient for immersed bodies defined by Eq. (10-6); f_f, friction factor defined after Eq. (14-14); f_{GS}, f_{GS}', friction factor for tube banks defined by Eqs. (11-13) and (11-14); f_{KL}, friction factor for tube banks defined by Eq. (11-17); f_{TB}, friction factor for tube banks defined by Eq. (11-7); f_{TB}', friction factor for tube banks defined by Eq. (11-31)	None
G	Mass velocity; G_c, crossflow mass velocity in heat exchangers based on minimum flow area at shell diameter; G_e, geometric mean mass velocity $\sqrt{G_cG_w}$; G_{max}, maximum mass velocity through tube banks; G_u, mass velocity in unbaffled heat exchanger based on free-flow area; G_w, mass velocity through baffle window based on free-flow area	m/L^2t

Symbol	Meaning	Dimensions
Gr	Gas gravity	None
g	Acceleration of gravity	L/t^2
g_c	Gravitational constant	mL/Ft^2
grad	Gradient of a scalar	
H	Enthalpy; ΔH_v, latent heat of vaporization	FL/m
h	Local heat-transfer coefficient defined by Eq. (12-2); $h_{\text{a.m.}}$, arithmetic mean heat-transfer coefficient [Eq. (12-5)]; h_{av}, average heat-transfer coefficient for shell side of a tube bank; h_m, mean heat-transfer coefficient [Eqs. (12-3) and (12-4)]; mean heat-transfer coefficient on immersed bodies; h_∞, local heat-transfer coefficient a long distance from an entrance	F/LtT
I_0	Bessel function of the first kind of order zero; I_2, Bessel function of the first kind of order two	
I_3, I_3'	Triple integral	
I_{BL}	Triple integral over buffer layer; I_{LL}, triple integral over laminar sublayer; I_{TL}, triple integral over turbulent core	
IF	Inertial force	F
i	Imaginary number, $\sqrt{-1}$	
i	Number of independent dimensionless groups	
$\bar{\imath}$	Unit vector in x direction	
J_0	Bessel function of the first kind of order zero; $J_{1/3}$, Bessel function of the first kind of order one-third	
j	Number of fundamental dimensions	
$\bar{\jmath}$	Unit vector in the y direction	
j_H	j factor for heat transfer defined by Eq. (15-51)	None
K	Total conductivity of heat	F/tT
K'	$N_f d_f / d_r^{0.2}$, parameter defined in Fig. 11-14	
k	Thermal conductivity	F/tT
$\bar{k}$	Unit vector in z direction	
k_m	Mass-transfer coefficient	L/t
L	Tube length; plate length; length of immersed body; heat-transfer length; flow length of a tube bank; L_1, L_2, etc., length dimensions; L_e, entrance length required for development of velocity profile; L_x, L_y, scale of turbulence defined by Eqs. (5-33) and (5-34)	L
l	Time-average Prandtl mixing length; l_i, instantaneous Prandtl mixing length; l_1, scale of turbulence defined by Eq. (5-32)	L
ln	Logarithm to base e	
log	Logarithm to base 10	
lw	Energy lost because of friction	FL
$\overline{lw}$	Energy lost because of friction per unit mass of flowing fluid	FL/m
M	Molecular weight	m/m
M	Mesh size of turbulence-producing screen	L
m	Mass	m

Symbol	Meaning	Dimensions
m	Mass rate of flow	m/t
m'	Exponent showing variation of surface temperature of immersed body [Eq. (17-62)]	
N_1	Number of tube rows traversed between baffle windows (Fig. 11-16)	
N_f	Number of fins per inch on a transverse-finned tube	L^{-1}
N_m	Rate of mass transfer	m/t
N_T	Number of transverse rows of tubes	
N_w	Rate of mass transfer at the wall	m/t
N'	Number of major restrictions in flow through a tube bank	
n	Constant or exponent used in velocity-distribution equations	
n	Exponent	
n	Frequency	
n	Moles per unit mass	m/m
n	Number of dimensional variables	
n'	Exponent showing variation of u_{max} on immersed body [Eq. (10-91)]	
P	Static pressure; ΔP, increase in static pressure over a finite length of conduit; P_0, reference pressure; $-\Delta P_f$, pressure decrease equivalent to energy loss due to friction; $(-\Delta P_f)_w$, friction loss through baffle window; P_I, impact pressure; $(P_I)_{av}$, average or measured impact pressure; $(P_I)_c$, impact pressure at center of pitot-tube opening; P_{st}, stagnation pressure; P_θ, static pressure at the angle θ; P_c, critical pressure	F/L^2
P_r	Reduced pressure	None
p	Perimeter	L
p'	Fluctuating component of static pressure	F/L^2
p_i	Instantaneous static pressure	F/L^2
p_x	Normal stress in the x direction	F/L^2
p_y	Normal stress in the y direction	F/L^2
p_z	Normal stress in the z direction	F/L^2
Q	Volumetric rate of flow	L^3/t
$Q_1, Q_2, Q_3, \ldots$	Dimensional variable	Variable
q	Heat absorbed by flowing fluid; rate of heat transfer through area A_q; q_r, rate of heat transfer at radius r; q_w, rate of heat transfer at a solid boundary	FL/t
q'	Rate of energy generation in a fluid or solid	F/L^2t
R	Gas constant	FL/mT
R	Ratio of radii as given by Eq. (15-84); R_q, R_T, ratios of radii defined by Eqs. (15-85) and (15-86)	None
R_t	Correlation coefficient defined by Eq. (5-25); $R_{xu'}$, $R_{yu'}$, $R_{zu'}$, $R_{u'v'}$, $R_{v'w'}$, $R_{u'w'}$, correlation coefficients defined by Eqs. (5-26) to (5-31)	

Symbol	Meaning	Dimensions
r	Radial distance from axis or center; radius of transverse section at point (x_1,y) above surface of a body of revolution (Fig. 10-39); r_0, outside tube radius; radius of circular cylinder; radius of bodies of revolution measured normal to direction of fluid flow; one-half the minor axis of ellipse; r_1, outside radius of inner tube of annulus; r_2, inside radius of outer tube of annulus; r_c, radius of curvature (Fig. 10-39); r_f, outside radius of fin; r_H, hydraulic radius of annulus defined by Eq. (7-75); $(r_h)_{max}$, hydraulic radius at maximum cross section; $(r_h)_{min}$, hydraulic radius at minimum cross section; r_{max}, radius where maximum point velocity occurs in annulus; r_q, radius where rate of heat transfer is q; r_T, radius where temperature of flowing fluid is T; r_t, radius of body of revolution at any transverse cross section (Fig. 10-39); r_v, volumetric hydraulic radius; r_w, inside radius of circular tube	L
$r'_2, r'_3, r'_4, \ldots, r'_n$	Dimensionless length ratios	None
r_w^+	Dimensionless radius $r_w u^*/\nu$	None
S	Entropy	FL/T
S_L	Distance between tubes in longitudinal rows; S'_L, distance between tubes in adjacent transverse rows (staggered arrangement); S_m, minimum center-to-center spacing of adjacent tubes in a tube bank; S_T, distance between tubes in transverse rows	L
ΣF	External forces	F
s	Fin spacing; thread spacing	L
sp. gr.	Specific gravity	None
T	Absolute temperature; T_0, reference temperature; $T_{0.4}$, $T_{0.5}$, $T_{0.6}$, film temperatures defined by Eqs. (14-6) to (14-8); T_1, temperature of inner wall of annulus; T_∞, temperature of undisturbed flowing stream; temperature of fluid at entrance to conduit; T_a, air temperature; T_B, normal boiling point; T_b, bulk temperature of fluid; T_C, critical temperature; T_c, temperature at tube axis; T_f, film temperature defined by Eq. (7-72); $(T_w - T_b)_{l.m.}$, logarithmic mean temperature difference between conduit wall and flowing fluid; T_w, temperature of wall or at boundary between a solid and a fluid; T_{δ_1}, temperature at edge of laminar sublayer	T
$\bar{T}$	Dimensionless temperature defined by Eq. (17-10); $\bar{T}'$, $d\bar{T}/d\eta$	None
T^+	Dimensionless temperature $(1/\omega)(1 - T/T_w)$	None
T_b^+	Dimensionless temperature $(1/\omega)(1 - T_b/T_w)$	None
t	Time	t
U	Time-average velocity of undisturbed flowing stream; average velocity in a closed conduit; U_{crit}, velocity at transition from laminar to turbulent flow in a conduit; U_{max}, maximum velocity through tube bank	L/t

Symbol	Meaning	Dimensions
u	Time-average velocity in x direction; u_1, point velocity in inner portion of annular velocity profile; u_2, point velocity in outer portion of annular velocity profile; u', fluctuating component of velocity in x direction; $\overline{u'}$, mean value of the velocity fluctuation; $\sqrt{\overline{u'^2}}$, rms of velocity fluctuation; $\overline{u'v'}$, $\overline{u'w'}$, and $\overline{v'w'}$, turbulent shear components; u_i, instantaneous velocity in the x direction; u_{max}, time-average maximum point velocity; point velocity at edge of boundary layer; u_T, time-average point velocity where temperature is T; u_δ, point velocity at a distance δ from a solid boundary; u_{δ_1}, velocity at edge of laminar sublayer	L/t
u^*	Friction velocity defined by Eq. (7-12); $\overset{*}{u_1}$, $\overset{*}{u_2}$, friction velocity for inner and outer portions of annular velocity profile, respectively, defined by Eqs. (7-78) and (7-79); $\overset{*}{u_f}$, friction velocity for flow in rough pipes defined after Eq. (14-14)	L/t
u^+	Dimensionless velocity parameter u/u^*; u_1^+, u_2^+, dimensionless velocity parameter for inner and outer portions respectively of annular velocity profile [Eqs. (7-80) and (7-81)]	None
V	Volume of mass m	L^3
$\mathbf{V}$	Velocity vector; $\mathbf{V}_\theta$, velocity vector at angle θ	L/t
V_C	Critical molal volume; V_m, molal volume	L^3/m
$\overline{V}$	Dimensionless quantity defined by Eq. (15-82); $\overline{V}_T$, dimensionless quantity defined by Eq. (15-83)	None
v	Time-average velocity in y direction; v', fluctuating velocity in y direction; $\overline{v'}$, mean value of velocity fluctuation; $\sqrt{\overline{v'^2}}$, rms of velocity fluctuation; v_i, instantaneous velocity in y direction; v_δ, value of v at a distance δ from a solid boundary	L/t
W	Weight	F
W	Thread thickness	L
w	Complex number	
w	Time-average velocity in z direction; w', fluctuating velocity in the z direction; $\overline{w'}$, mean value of velocity fluctuation; $\sqrt{\overline{w'^2}}$, rms of velocity fluctuation; w_i, instantaneous velocity in z direction	L/t
w''	Work done by fluid on surroundings	FL
$\bar{w}$	Work done per unit mass of fluid	FL/m
X	Quantity defined by Eq. (13-4)	None
x	Cartesian coordinate; distance from leading edge of a flat plate; distance from entrance of closed conduit; x_0, distance from leading edge of flat plate where heat transfer starts; x_1, distance measured along surface from forward stagnation point of an immersed body	L

Symbol	Meaning	Dimensions
y	Cartesian coordinate; distance measured normal to solid boundary; y_0, distance of center of impact tube opening from wall; y_1, distance from inner wall of annulus; y_2, distance from outer wall of annulus; y_c, distance measured normal to center line between two parallel planes; $(y_{max})_1$, distance from inner wall of annulus to point of maximum velocity; $(y_{max})_2$, distance from outer wall of annulus to point of maximum velocity	L
y^+	Dimensionless distance yu^*/ν; y_1^+, y_2^+, dimensionless distances for annuli defined by Eqs. (7-82) and (7-83)	None
Z	Vertical distance above a datum plane	L
z	Complex number	
z	Cartesian coordinate; width of immersed flat plate	L

Greek Symbols

Symbol	Meaning	Dimensions
α	Angle between two intersecting planes; half the angle of a wedge	None
α	Thermal diffusivity $k/C_p\rho$	L^2/t
α	Factor in kinetic-energy term to correct for velocity distribution	None
β	Coefficient of thermal expansion	$1/T$
β_n	Coefficients in Graetz solution [Eq. (13-22)]	
Γ	Gamma function	
Γ	Circulation	L^2/t
γ	Quantity defined by Eq. (10-104); γ_1, quantity defined by Eq. (10-110)	L
Δ	Final value minus initial value; increase	
Δ	Thickness of thermal boundary layer	L
δ	Thickness of hydrodynamical boundary layer; δ_1, thickness of laminar sublayer; δ^*, displacement thickness defined by Eq. (10-54)	L
E_H	Eddy conductivity of heat	F/tT
E_M	Eddy viscosity	m/Lt
ϵ_H	Eddy diffusivity of heat; ϵ_M, eddy diffusivity of momentum; ϵ_m, eddy diffusivity of mass	L^2/t
ζ	Ratio of thermal-boundary-layer thickness to hydrodynamical-boundary-layer thickness Δ/δ	
η	Function defined by Eq. (10-15); η_1, η_2, η_3, functions defined by Eqs. (10-93), (10-105), and (10-109) respectively	None
Θ	Polar coordinate for definition of complex numbers	None
θ	Angle measured from forward stagnation point; cylindrical coordinate; θ_1, angle measured from trailing edge of ellipse or cylinder; elliptical coordinate; polar coordinate for definition of complex numbers; θ_2, polar coordinate for definition of complex numbers	None

Symbol	Meaning	Dimensions
κ	Universal constant appearing in velocity-distribution equations	None
λ	Bulk modulus of viscosity	m/Lt
λ	Quantity defined by Eq. (10-96); λ_1, λ_2, quantities defined by Eqs. (10-103) and (10-111) respectively	None
μ	Viscosity; μ_1, viscosity at temperature T_1; μ_b, viscosity at temperature T_b; μ_c, viscosity at critical point; μ_w, viscosity at temperature T_w; μ_∞, viscosity at temperature T_∞	m/Lt
ν	Kinematic viscosity	L^2/t
ξ	Elliptical coordinate [Eqs. (10-107) and (10-108)]	
Π	Dimensionless group	None
π	3.1416	None
ρ	Density	m/L^3
σ	Surface tension	F/L
τ	Shear stress in a flowing fluid; τ_0, yield stress for a plastic; τ_1, τ_2, shear stresses at inner and outer walls respectively, of an annulus; τ_l, laminar shear stress or shear stress due to molecular viscosity; τ_t, time-average turbulent shear stress; $(\tau_t)_i$, instantaneous turbulent shear stress; τ_w, shear stress at boundary between a fluid and a solid; τ_{xy}, τ_{yx}, τ_{yz}, τ_{zy}, τ_{zx}, τ_{xz}, shear stresses (first subscript refers to plane parallel to stress, and last subscript gives direction of stress)	F/L^2
Φ	Viscous-dissipation function	F/L^2t
ϕ	Velocity potential	L^2/t
ϕ, ϕ_0, ϕ_1, . . ., ϕ_n	Function; ϕ', ϕ'', etc., first, second, and higher derivatives of the function ϕ; $\phi(0)$, $\phi(1)$, value of ϕ when independent variable has values of 0 and 1 respectively	
ψ	Stream function	L^2/t
Ω	Force potential	FL/m
ω	Dimensionless heat-transfer factor defined by Eq. (7-46)	

Miscellaneous symbols

$\lvert\ \rvert$	Absolute value
∇^2	Laplacian operator
Z	Compressibility factor

Dimensionless groups

Eu	Euler number $\Delta P g_c/\rho U^2$
Fr	Froude number U^2/Lg
Gr	Grashof number $(d_w{}^3\rho^2 g\beta\ \Delta T)/\mu^2$

Dimensionless groups (*Continued*)

Gz	Graetz number $\pi \mathrm{Pe}/(4x/d_w)$; Gz′, modified Graetz number $\mathrm{Pe}/(x/d_w)$ or $\mathrm{Pe}/(x/d_e)$; $\mathrm{Gz}_{\mathrm{a.m.}}$, Graetz number based on fluid properties evaluated at arithmetic mean bulk temperature of fluid
Nu	Nusselt number based on local heat-transfer coefficient hd_w/k, hd_e/k, hd_0/k, etc.; Nu_∞, asymptotic value of Nusselt number; $\mathrm{Nu}_{\mathrm{a.m.}}$, Nusselt number based on arithmetic mean heat-transfer coefficient $h_{\mathrm{a.m.}}$; Nu_L, total Nusselt number for a flat plate h_mL/k; Nu_m, mean Nusselt number based on mean heat-transfer coefficient h_m; Nu_s, Nusselt number for slug flow; $(\mathrm{Nu}_s)_\infty$, asymptotic value of Nu_s; Nu_x, local Nusselt number hx/k; Nu_{x_1}, local Nusselt number on immersed body hx_1/k; $\mathrm{Nu}_{\theta=0}$, value of the local Nusselt number hd_0/k at forward stagnation point
Pe	Peclet number RePr
Pr	Prandtl number $C_p\mu/k$
Re	Reynolds number $d_wU\rho/\mu$; Reynolds number based on equivalent diameter defined in Eqs. (11-2) and (11-3); $\mathrm{Re}_{\max}$, Reynolds number in circular tubes based on maximum velocity; Re_c, Re_0, Re_v, Re'_v, Reynolds number based on d_c, d_0, d_v, d'_v respectively (for tube banks all Reynolds numbers use $G_{\max}$); Re_2, Reynolds number for annuli defined by Eq. (7-74); Re_L, total Reynolds number on a flat plate LU/ν; Re_x, local Reynolds number on a flat plate xU/ν; Re_{x_1}, local Reynolds number for immersed body x_1U/ν
Sc	Schmidt number $\mu/\rho D$
St	Stanton number based on local heat-transfer coefficient h/C_pG; St_m, Stanton number based on mean heat-transfer coefficient h_m/C_pG
We	Weber number $\rho U^2L/\sigma g_c$

PROBLEMS

CHAPTER 1

1-1. Castor oil at 59°F fills the space between two parallel horizontal plates which are ⅜ in. apart. If the upper plate moves with a velocity of 5 ft/sec and the lower one is stationary, what is the shear stress in the oil?

1-2. A long vertical cylinder 3.00 in. in diameter slides concentrically inside a fixed tube having a diameter of 3.02 in. The uniform annular space between tube and cylinder is filled with water at 59°F. Assuming a linear velocity distribution, what is the resistance to motion at a relative velocity of 3 ft/sec for a 6-in. length of cylinder? What would be the resistance if the fluid were SAE 30 oil at 59°F?

1-3. Two concentric cylinders are 10.00 in. long and have diameters 4.52 and 4.50 in. What is the viscosity of a liquid which produces a torque of 0.45 ft-lb_f upon the inner cylinder when the outer one rotates at the rate of 128 rpm? What would the radial spacing of the cylinders have to be to give the same torque speed ratio with carbon tetrachloride at 50°F? Mercury at 60°F? Pure glycerin at 80°F?

1-4. A viscometer consists of two concentric cylinders. The inner stationary cylinder has an outside diameter of 12.00 in. and a length of 6.00 in. The outer cylinder has an inside diameter of 12.10 in. In determining the viscosity of a fluid at constant temperature, the following data were obtained:

Outer cylinder, rpm	*Torque on inner cylinder, ft-lb$_f$*
0	0
10	0.25
20	0.43
30	0.58
40	0.70
50	0.82

(*a*) Show from the above data that the fluid is non-Newtonian.
(*b*) Plot the viscosity [in $(lb_m)/(ft)(hr)$] as a function of the rate of shear.
(*c*) Determine the viscosity at zero rate of shear by two methods.
(*d*) What type of non-Newtonian fluid is being investigated?

1-5. A flat circular disk is separated from a flat stationary surface by a layer of oil 0.01 in. thick. The disk is 6 in. in diameter. Determine the power input to the shaft rotating the disk as a function of the rotational speed. The viscosity of the oil is 100 centipoises.

1-6. A crankshaft on an automobile is 2.000 in. in diameter. The connecting rod bearing on the shaft is 2.005 in. in diameter and 1.50 in. long. The bearing is lubricated with

SAE 10 oil, and the temperature of the bearing is at 235°F. The oil is distributed evenly between the shaft and the bearing. How much heat must be removed per hour so that the bearing remains at a constant temperature? The shaft is rotating at 2,500 rpm. Viscosity of SAE 10 oil at 235°F is 4.5 centipoises.

1-7. A 3- by 3-in. circular piston has a stroke of 4 in. The clearance between the piston and cylinder wall is 0.001 in. and can be assumed to be uniform. If oil (viscosity = 5 centipoises) is to be used as a lubricant, what is the maximum crankshaft rpm if the heat load at the cylinder wall cannot exceed 15,000 Btu/hr?

CHAPTER 2

2-1. For steady laminar flow in a circular tube with axis parallel to the x axis, $v = 0$ and $w = 0$. Derive Eq. (8) of Table 2-2 from Eq. (3) (x direction) of the same table. Make use of the relationship between cartesian (x,y,z) and cylindrical (x,r,θ) coordinates

$$x = x$$

$$y = r \cos \theta$$

$$z = r \sin \theta$$

2-2. Derive Eq. (2-60) from Eq. (2-59).

2-3. A two-dimensional velocity field is described by

$$u = 2xy^2$$

$$v = 2x^2y$$

Under what flow condition does this satisfy the equation of continuity?

2-4. A three-dimensional velocity field is described by

$$u = xyz$$

$$v = -y$$

$$w = -\tfrac{1}{2} yz^2 + z$$

(a) Under what flow condition is the equation of continuity satisfied?

(b) Determine an expression for the absolute value of the velocity in space.

(c) What are the magnitude and direction of the velocity at points (0,0,0), (1,2,3), and (−4,5,1)?

2-5. The temperature in a material is given by the relationship

$$T = (x^2y + xy^2)10$$

where x and y are in units of feet and T is the temperature (°F).

(a) Obtain the equation for the isotherm passing through the point (1,2).

(b) Plot the equation obtained in (a) for $0.5 < x < 5$.

(c) What is the rate of heat conduction in the x direction and y direction at the point (1,2) if the material is water?

(d) Could this temperature field exist in a solid material?

2-6. The velocity u in a two-dimensional field for an incompressible fluid is expressed as:

$$u = 3x^2y + y^3$$

(a) Use the continuity equation to determine v.

(*b*) In the determination of v, a constant of integration appears. Can this constant of integration be a function of x?

(*c*) Assuming the constant of integration to be zero, obtain the equation for the constant velocity line passing through the point (3,4).

(*d*) Draw the velocity vectors at points (0,0), (1,2), (−3,4), (3,4).

CHAPTER 3

3-1. The velocity in a two-dimensional velocity field for an incompressible fluid is expressed as

$$u = x^2y - 2x - \frac{y^3}{3}$$

$$v = \frac{x^3}{3} + 2y - xy^2$$

(*a*) Is the equation of continuity satisfied?

(*b*) Is the flow irrotational?

(*c*) Determine the magnitude of the velocity at points (0,0), (1,1), (2,2).

3-2. A stream function is given by the equation

$$\psi = x^3 - 3y^2$$

(*a*) What are the magnitude and direction of the velocity at points (0,0), (2,3), (−3,6)?

(*b*) Draw the streamlines corresponding to $\psi = 0$, 1, 2, 3, and 4.

3-3. The velocity potential and stream function for a flow system are given by

$$x = \phi + e^{\phi} \cos \psi$$

$$y = \psi + e^{\phi} \sin \psi$$

Draw the flow net for this set of equations and suggest the solid boundaries that would give this type of flow net.

3-4. A point source in a two-dimensional infinite body of water supplies water at the rate of 20 ft^3/(sec) (ft). Calculate the speed of the water at distances 1, 5, and 10 ft from the source. The gauge pressure 1 ft from the source is 200 lb_f/ft^2. What is the pressure 10 ft from the source?

3-5. A point source discharges water into a tank at the rate of 50 ft^3/(min)(ft). The tank is 40 ft long and 20 ft wide. The point source is located at the middle of one end of the tank and a point sink is located at the middle of the other end. Draw the flow net for this flow.

3-6. A very large rectangular tank is divided into two parts by a vertical partition. The partition contains a vertical slot 2 in. wide and 5 ft long. Water at 60°F flows through the slot at the rate of 25 ft^3/sec. Assume two-dimensional, incompressible, nonviscous flow.

(*a*) Draw the flow net for the flow through the orifice. Make a full-scale drawing and let the distance between the streamlines in the orifice be 0.2 in.

(*b*) What is the fluid velocity at a point 4 in. from the orifice? One foot from the orifice?

(*c*) The vapor pressure of water at 60°F is 0.26 psia. What minimum pressure at a point 1 ft from the orifice is required so that the water in the orifice will not vaporize?

3-7. Water is flowing through an open rectangular channel at 20 ft/sec. The channel is circular with a radius of 50 ft. The channel is 2 ft wide. The constant flow rate is 2,000,000 gal/hr. Determine the minimum height of the sides of the channel.

3-8. Compare the shape of the surface of a liquid in a cylindrical tank 3 ft in diameter when (*a*) the tank is rotated at 100 rpm and (*b*) the tank contains an agitator which rotates at 100 rpm. (In the latter case assume that the velocity of the fluid varies inversely as the distance from the center of the tank. Assume also that the fluid in contact with the 2-in.-diameter shaft of the agitator has the same velocity as the shaft.)

3-9. An airplane traveling 400 miles/hr goes into a turn of 14,700-ft radius banking 30°. What is the slope of the gasoline in its tank with respect to the bottom of the tank?

3-10. A parabolic reflector, whose cross section can be represented by $y = 2x^2$, is to be coated by centrifugation. What would be the optimum speed of rotation for a uniform coat?

3-11. The velocity potential and stream function of a flow system are given by

$$x = -\phi + \ln\sqrt{\phi^2 + \psi^2}$$

$$y = -\psi + \tan^{-1}\frac{\psi}{\phi}$$

Draw the flow net for this system and suggest the solid boundaries that would give this type of flow net.

3-12. Construct the flow net for flow along two planes intersecting at an angle of 135°.

3-13. Given the following data on a gas well:

Wellhead pressure	2600 psia
Depth of well	7500 ft
Gas gravity	0.744 (Air = 1)
Pseudocritical pressure	663.8 psia
Pseudocritical temperature	385.6°R

Depth, ft	*Temperature, °F*
0	77
2000	144
4000	174
6000	206
7500	228

Compute the pressure at the bottom of the well by two methods:

(*a*) Using average values for the temperature and compressibility factor.

(*b*) Using an average well-fluid density.

3-14. Derive the barometric formula

$$P = P_0 \exp\left(-\frac{Zg}{RTg_c}\right)$$

where P = barometric pressure at a height Z above sea level
P_0 = barometric pressure at sea level
R = gas constant per unit mass of air
T = absolute temperature

What assumptions are involved in deriving this formula?

3-15. When the barometer at sea level is normal determine the pressure at an altitude of 10,000 ft when the temperature of the atmosphere varies as follows:

$$T°\text{F} = 59 - 0.0036Z$$

where Z is the height above sea level.

3-16. Metal cylinders 6 in. high are cast by a method of centrifugal casting. The molten metal is put into a mold and the mold is rotated so that the molten metal assumes the shape of a cylinder. At what rate must the mold be rotated in order that the inside diameter of a cylinder will vary not more than 0.010 in. from top to bottom? The average inside diameter of the cylinders is 3 in. The molten metal has a density of 8.7 g/cc and a viscosity of 2.5 centipoises.

3-17. A vertical cylindrical tank 2 ft in diameter and 4 ft deep contains water to a depth of 2 ft and oil (specific gravity = 0.86) to a depth of 1 ft. At what speed can the tank be rotated so the oil just reaches the top rim of the tank? At this speed what will be the pressure at the center of the bottom of the tank?

3-18. Water is flowing at the rate of 1 $ft^3/(sec)(ft)$ in a two-dimensional duct having parallel boundaries 5.00 in. apart. An orifice having a width of 2.50 in. is in the duct. Draw the flow net for the flow through the orifice. Plot the pressure profile along the center streamline for the flow across the orifice and compare this with the pressure profile actually obtained for flow across such a sharp-edged circular orifice in a pipeline. Water temperature is 60°F. [See H. Judd, *Trans. ASME,* **38**:331 (1916).]

CHAPTER 4

4-1. Show that the average kinetic energy per unit mass of a fluid in laminar flow in a circular tube is U^2/g_c where U is the average velocity. Perform this calculation for laminar flow in an annulus.

4-2. For laminar flow in a circular tube, determine the point in the cross section of the tube where the rate of change of the kinetic energy is a maximum.

4-3. Pure glycerin at 150°F is flowing from a tank through a 2-in. schedule 40 pipe. The pipe is vertical and connected to the bottom of the tank. The pressure in the tank is atmospheric and the depth of liquid in the tank is 3 ft. Assume Re = 1,000.

(*a*) Determine the static pressure in the pipe a distance of 30 ft from the pipe entrance.

(*b*) What is the difference in static pressure between two points in the pipe located a distance 30 ft apart?

(*c*) What is the friction loss over 30 ft of pipe:

(1) in terms of energy loss per lb_m of fluid?

(2) in terms of pressure loss?

(*d*) Is the pressure loss due to friction between two points equal to the difference in static pressure between the same two points? Explain.

4-4. The joints in a pipeline are of the bell and spigot type as shown in the accompanying illustration. After the joint was sealed and in operation, a crack developed in

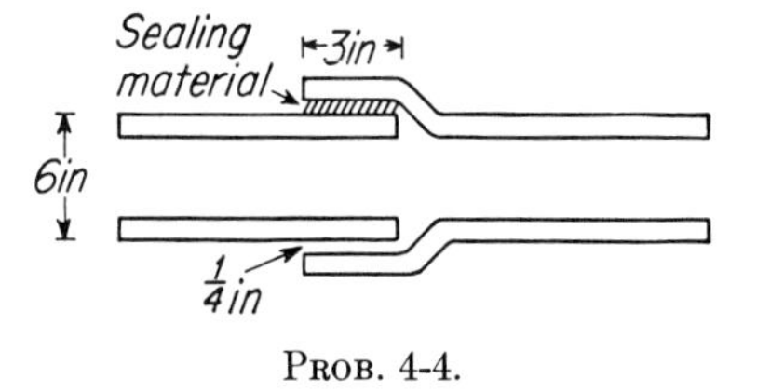

PROB. 4-4.

the sealing material. This crack was 1/16 in. wide and extended the length of the joint. The line is carrying crude oil (viscosity = 150 centipoises, specific gravity = 1.0) at a pressure of 30 psig. Estimate the loss of oil through the crack. Neglect entrance effects.

4-5. A tank contains water at 50°F. A capillary tube 24 in. long with a bore of $\frac{1}{16}$ in. is mounted horizontally on the tank a distance of 5 ft below the water surface. How long will it take for the water surface to fall 2 ft? The tank has a cross-sectional area of 2 ft^2.

4-6. Oil is flowing in a 4-in. schedule 40 steel pipeline. The oil flow rate is 18,000 bbl/day. Given the following data:

Average properties of oil:

Viscosity	300 centipoises
Density	1.09 g/cc
Heat capacity	0.5 Btu/(lb$_m$)(°F)

Properties of steel:

Heat capacity	6 cal/(mole)(°C)

(*a*) Determine the temperature rise in the oil per mile of pipeline if the line is completely insulated. The temperature of the pipe wall is always the same as that of the oil.

(*b*) Determine the heat loss in Btu/hr per 100 ft of pipe if flow is isothermal.

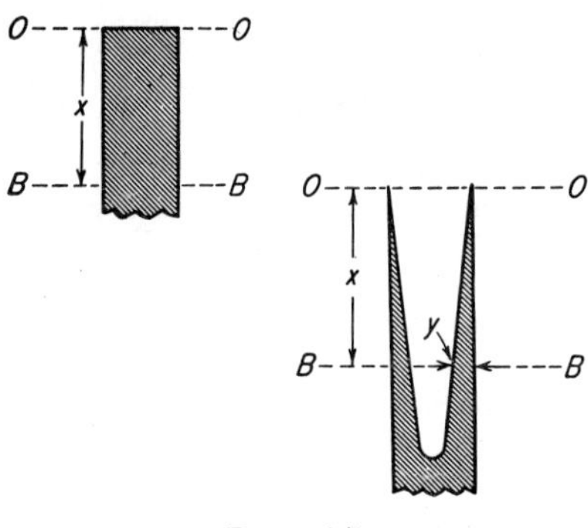

PROB. 4-7.

4-7. Ferrell, Richardson, and Beatty [*Ind. Eng. Chem.*, **47**:29 (1955)] used a dye-displacement technique to measure the velocity distribution for laminar flow in a tube. A tube is filled with liquid containing dye up to a point *O–O*. Clear liquid fills the tube above this point. Laminar flow is started and the thickness y of the dye solution at point *B–B* is determined with a spectrophotometer.

Typical data which might be obtained from such an experiment are tabulated below for the flow of water in a 0.500-in.-ID tube at an average velocity of 0.280 ft/sec. The distance between points *O–O* and *B–B* is 2.19 ft.

Time from start of flow, sec	*Dye thickness, in.*
102.0	0.005
83.0	0.006
70.5	0.007
62.6	0.008
55.0	0.009
50.0	0.010
39.2	0.013
28.5	0.018
21.2	0.024
15.3	0.034
10.2	0.055

Compare the velocity profile obtained from this experimental data with that predicted for laminar flow by Eq. (4-44).

4-8. A fluid flows downward in a 2-in. schedule 40 vertical pipeline at an average velocity of 3 ft/sec. The fluid has a density of 1.1 g/cc. A differential manometer is connected to taps in the pipeline located a distance of 30 ft apart. The heavy fluid in the manometer is mercury and the manometer lines and space above the mercury are filled with the fluid which flows in the pipe. The reading on the manometer is 1.2 in. A sample of fluid flowing in the pipe is tested in a viscosimeter made up of two concentric cylinders 10 in. high and having diameters of 4.52 and 4.50 in. A torque of 0.45 ft/lb_f is produced on the inner cylinder when the outer cylinder is rotated at the rate of 128 rpm.

(*a*) Determine if the above data are consistent.

(*b*) What would have been the manometer reading if flow had been upward? Explain.

4-9. From Eqs. (4-87) and (4-93) estimate the maximum point velocity at the center of a rectangular conduit which is 2 by 3 in. (inside measurement) and in which water at 60°F flows with a Reynolds number of 1,000. The Reynolds number is expressed as d_eU/ν where d_e is defined by Eq. (4-92).

4-10. A continuous reactor supplies polystyrene for devolatilization at a rate of 630 lb_m/day. The polystyrene (specific gravity = 0.906, viscosity at given flow rate = 4,000 poises) flows through a 2-in. standard steel pipeline from the reactor to the devolatilizer. The line is 16 ft long. The absolute pressure in the devolatilizer is 2 mm of Hg and the surface of the polystyrene in the reactor is 10 ft above the surface in the devolatilizer.

(*a*) Calculate the friction factor.

(*b*) Determine the pressure loss due to friction.

(*c*) Determine the pressure at the surface of the polystyrene in the reactor.

4-11. A fine steel wire 0.01 in. in diameter is located along the axis of a 1-in.-ID tube. Oil (kinematic viscosity = 200 centistokes, specific gravity = 0.85) flows in this annulus at a Reynolds number of 1,500 based on equivalent diameter.

(*a*) Plot the laminar velocity profile across the cross section of the annulus.

(*b*) Determine the friction factor. (Neglect entrance effects.)

(*c*) If the wire is 10 ft long, determine the tensile force exerted on one end of it by virtue of the shear stresses at its surface.

4-12. A crude-oil pipeline 20 miles long delivers 5,000 bbl/day. The pressure drop over the line is 500 psi. To increase the capacity a parallel pipeline is installed over the last 12 miles. This line is the same size as the original. Both ends of the pipeline are at the same elevation and laminar flow may be assumed. Find the total discharge rate under the new conditions, assuming the same pressure drop of 500 psi.

4-13. Oil with a specific gravity of 1.1 and viscosity of 450 centipoises is flowing through a 4-in.-ID pipeline at the rate of 1,500 bbl/day. The static pressure at station A is 1,000 psig. Determine the static pressure at station B which is 5,000 ft away and at an elevation which is 100 ft lower than that of A.

CHAPTER 5

5-1. By methods similar to those used in Sec. 5-4, modify the energy equation [Eq. (2-55)] to include the fluctuation of both the velocity and the temperature terms. Assume $q' = \Phi = DP/Dt = 0$.

CHAPTER 6

6-1. The local velocity u of an incompressible fluid flowing steadily past a solid surface is a function of U, ρ, μ, τ_w, y, x, where y is the distance from the surface where the velocity u exists and x is the distance measured in the direction of flow from the leading edge or entrance of the system. Group these variables into dimensionless quantities. From the groups derived obtain the groups u/u^* and yu^*/ν where $u^* = \sqrt{\tau_w g_c/\rho}$. Also make the dimensional analysis for the case where the entrance effect is negligible (i.e, x has no effect).

6-2. In studying the disintegration of horizontal liquid jets in a stationary surrounding medium, the variables considered are jet diameter, length of jet from discharge to breakup, jet velocity, surface tension, viscosity and density of liquid comprising the jet, and viscosity and density of the surrounding medium. Arrange these variables in dimensionless groups both considering and neglecting gravitational forces.

6-3. It is proposed to study a fuel-injection nozzle for a furnace by means of a model one-tenth the size of the prototype. The prototype nozzle will discharge 100 lb_m/hr of fuel oil having a viscosity of 4×10^{-5} (lb_f)(sec)/ft^2, a surface tension of 0.002 lb_f/ft, and a density of 1.06 g/ml. Fuel temperature is 60°F. The liquid jet will discharge into air essentially at 60°F. The model will also discharge fuel oil into air at 60°F and 1 atm pressure.

(*a*) What flow rate should be used in the model to simulate actual flow conditions?

(*b*) At the flow rate in (*a*) the liquid jet formed by the model is 8.5 in. long from nozzle outlet to breakup. What length of jet might be expected in the prototype? [NOTE: Experimental results on nozzles indicate that the group $(L/d_w)/\sqrt{\text{We}}$ is a function of the jet Reynolds number, where L is the jet length, d_w is the jet diameter, and We is the Weber number, $\rho U^2 d_w/g_c\sigma$.] Neglect gravity forces.

6-4. Models are used to study the power requirements of mixers. In mixers containing baffles the surface of the fluid remains almost level and the power required is a function of impeller diameter, impeller speed, fluid density, and fluid viscosity. When the mixer is not baffled the liquid surface does not remain level but forms a vortex owing to its rotation. In this case, the power is also a function of the acceleration of gravity. Derive the dimensionless groups for both baffled and unbaffled mixers.

A baffled paddle mixer is to be studied with a model one-tenth the size of the prototype. What speed of paddle rotation in the model will correspond to 100 rpm in the prototype? The fluid in the model is the same density, and is one-quarter the viscosity of the fluid in the prototype. When dynamic similarity exists between model and prototype, what is the ratio between power consumption in the model and that in the prototype?

6-5. In analyzing turbulent flow in tubes Prandtl and von Kármán obtained relationships for the eddy diffusivity of momentum ϵ_M where $\epsilon_M = E_M/\rho$ [see Eq. (7-18)]. Prandtl assumed that ϵ_M is a function of y and du/dy while von Kármán assumed that ϵ_M is a function of du/dy and d^2u/dy^2, where y is the distance from the solid boundary, and u is the point velocity at this location. By means of dimensional analysis derive the dimensionless groups resulting from each of these assumptions. (See L. Prandtl, *NACA TM* 720, 1933, and T. von Kármán, *NACA TM* 611, 1931.)

6-6. The performance of a lubricating oil ring depends on the following variables:

Q = oil delivered in volume per unit time
d_w = the inside diameter of the ring
N = shaft speed, revolutions per minute
μ = oil viscosity
ρ = oil density
σ = surface tension of the oil (force per unit length)

Determine the dimensionless groups in which these variables may be arranged.

6-7. The mass of drops formed by a liquid discharging slowly into air from a vertical thin-walled tube is considered to be a function of the tube diameter, fluid density, acceleration of gravity, the radius of the tube, and the surface tension of the liquid against air. The following data were obtained for water at 15°C.

Tube diameter, in.	*Mass of drop, g*
0.088	0.0375
0.134	0.0526
0.191	0.0712
0.200	0.0755
0.256	0.0923
0.354	0.1151
0.383	0.1362
0.406	0.1461
0.459	0.1703
0.521	0.1969
0.523	0.2023
0.566	0.2210
0.584	0.2339

Estimate the size of drop formed by pure benzene at 20°C when discharged from a tube 0.15 in. in diameter.

6-8. The drag on a torpedo is being studied by using a model in a wind tunnel at a scale of 1:20. What wind velocity should be used in the wind tunnel if the torpedo speed is 10 knots per hour? To what prototype resistance would a model resistance of 10 lb_f correspond? Temperature is 60°F.

CHAPTER 7

7-1. A sucrose solution containing 40 wt % sucrose at 20°C flows in a smooth 6.00-in.-ID level pipe 1,000 ft long. The specific gravity of the sucrose solution is 1.25 and the viscosity is 6.0 centipoises. If the average velocity in the pipe is 10 ft/sec calculate:

(*a*) Friction factor.
(*b*) Velocity gradient at the wall.
(*c*) Shear at the wall.
(*d*) Friction velocity.
(*e*) Thickness of the laminar layer.
(*f*) The complete velocity profile and plot.
(*g*) $U/u_{\max}$ and compare with the value given on Fig. 7-3.

7-2. Do parts (*a*), (*d*), (*f*), and (*g*) of Prob. 7-1 using $d_w/e = 100$. In the case of (*g*), compare the value of $U/u_{\max}$ obtained with that given by Fig. 7-18.

7-3. A 2-in.-ID horizontal pipeline is carrying gasoline (specific gravity = 0.69, viscosity = 0.6 centipoise). The flow rate is 40 gal/min. The difference in pressure between two points 100 ft apart is 1.3 psi.

(*a*) Estimate the relative roughness of the pipe.

(*b*) Draw the velocity profile in the pipe. On the same plot draw the velocity profile for the same flow rate, assuming the pipe to be smooth.

(*c*) For each velocity profile drawn in (*b*) determine the average velocity by integration. Compare this with the true average velocity. Explain any difference.

(*d*) What is the power required for pumping per hundred feet of pipe? Compare this with the power required if the pipe were smooth.

7-4. Air at 1 atm pressure is flowing in a 2-in.-ID smooth tube. At a section in the tube the velocity is 50 ft/sec and the temperature is 150°F. Draw the velocity profile across this cross section of the tube for the following cases:

(*a*) Adiabatic flow. Tube-wall temperature is 150°F.

(*b*) Air being heated. Tube-wall temperature is 250°F and local rate of heat transfer is 1500 Btu/(hr)(ft^2).

(*c*) Air being cooled. Tube-wall temperature is 50°F and local rate of heat transfer is 1500 Btu/(hr)(ft^2).

7-5. A new water main 12.0 in. in diameter carries water at 60°F. The rate of flow is 5 ft/sec. The water main is made of cast iron.

(*a*) Determine the Fanning friction factor.

(*b*) What is the point velocity at a distance of 3 in. from the pipe wall?

(*c*) What is the power required to overcome friction in 10 miles of water main?

7-6. Assuming r_{max} may be obtained from Eq. (4-58), plot the ratio τ_1/τ_2 as a function of r_1/r_2 for flow in concentric annuli.

7-7. What reading in inches of water would be given by a pitot tube inserted at the center of a length of straight pipe conveying air at 70°F psig above normal barometer, and at an average velocity of 10 ft/sec?

7-8. Hydrogen from a catalytic reforming unit is pumped a distance of 175 miles to a petrochemical plant through a 24-in.-ID steel pipeline at a rate of 50,000,000 standard ft^3/day. The gas is delivered to the pump at 25 psig and is to be at 100 psig when it reaches the petrochemical plant. What power is required for pumping? The average pipeline temperature is 70°F. If the flow is adiabatic, estimate the temperature change in the gas as it flows the 175 miles. Consider the pipe wall to be always the same temperature as the gas and the wall thickness to be 1 in.

7-9. A horizontal annulus consists of two concentric tubes. The inner tube has an OD of 1.00 in. and the outer tube has an ID of 2.00 in. The inner tube is movable and

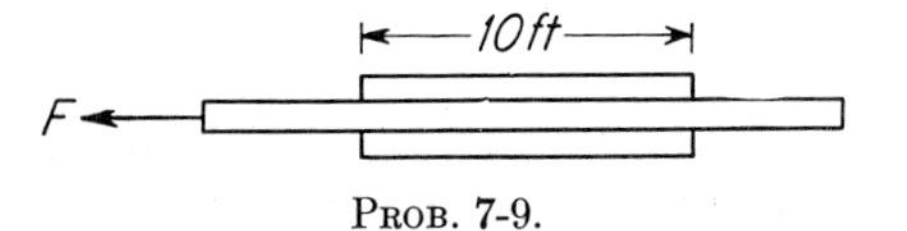

PROB. 7-9.

suspended on frictionless packing. The outer tube is stationary. Calculate the force *F* if

(*a*) Water at 60°F flows in the annulus at the rate of 15 gal/min.

(*b*) Air at 60°F is flowing in the annulus with an average velocity 5 times the average velocity of the water in (*a*).

7-10. Calculate the force *F* exerted on the outer tube for Prob. 7-9 if the inner tube of the annulus is stationary and the outer tube is suspended on frictionless packing.

7-11. Calculate the velocity gradient at each wall of the annulus for parts (*a*) and (*b*) of Prob. 7-9.

7-12. Water at 60°F is flowing in a smooth 6-in.-ID pipe. At what point in the pipe is the velocity equal to the average stream velocity? Calculate using:

(*a*) Re = 100,000

(*b*) Re = 10,000

7-13. Water at 60°F flows through a section of horizontal 6-in.-ID pipe ½ mile long. Static pressure is 120 psig at one end and 75 psig at the other end. Flow is 50,000 gal/hr. Calculate:

(*a*) the friction loss

(*b*) the friction factor

(*c*) the friction velocity

(*d*) the shear stress at the pipe wall

7-14. Water at 60°F flows through a section of 12-in.-ID pipeline 1,000 ft long running from elevation 300 to 270 ft. A pressure gauge at elevation 300 reads 40.5 psig and one at elevation 270 reads 50 psig. Flow is 5 ft^3/sec. Calculate:

(*a*) the friction loss
(*b*) the shear stress at the pipe wall
(*c*) the shear stress 3 in. from the pipe wall
(*d*) the friction factor
(*e*) the friction velocity

7-15. A natural-gas line 23.5 in. ID and 156.56 miles long is transporting 226,178,000 standard ft^3 of gas per 24 hr with a pressure drop of 742.4 to 354.4 psia. The gas is measured at 60°F and 14.735 psia. The gas gravity is 0.600 and the viscosity is 126 micropoises at 750 psia and 113 micropoises at 350 psia at the flowing temperature of 65°F. The critical temperature of the gas is 361°R and the critical pressure is 672 psia. Calculate the gas flow and compare it with the observed gas flow. (NOTE: The compressibility factor of the gas is 0.940 at 360 psia and 0.875 at 760 psia.)

7-16. A city is considering the installation of 10 miles of water main. It is found that if water flows at a velocity of 5 ft/sec in this water main, the pipe diameter must be 8.0 in. for the desired amount of water to be delivered. The water is at a temperature of 60°F.

Cast-iron pipe with 8.0 in. ID costs $1.50 per foot while standard wrought-iron pipe with 8.0 in. ID costs $2.50 per foot. The cast-iron pipe must be replaced after 8 years of service, and the wrought-iron pipe must be replaced after 10 years' service; hence the initial investment must be written off in either 8 or 10 years, depending on whether cast-iron or wrought-iron pipe is used. With both types of pipe the roughness (value of *e*) may be considered to be 1.25 times the roughness when the pipe is new and may be considered constant at this value over the period the pipe is in service.

Determine which pipe would be most economical to install. The cost of power for overcoming friction is 2 cents per hp-hr.

7-17. Gases from a copper-smelting operation contain 7.0 per cent SO_2 (remainder is assumed to be air). They flow to a Cottrel precipitator through a rectangular duct 6 ft by 1 ft and 500 ft long. Gauge pressure at the exit of the duct is 2 in. of water. Gas temperature may be assumed to be 150°F over the whole length of the duct. Volume of gas flowing is 22,500 ft^3/min at 60°F and 1 atm pressure. The barometer is normal.

(*a*) Estimate the pressure of the gases at the entrance to the duct.

(*b*) Assuming the duct to consist of two parallel planes 1 ft apart, draw the velocity profile between the planes.

(*c*) Using the assumption in (*b*) calculate the shear stress at the wall.

7-18. At the Kemano power development in Northern British Columbia water is supplied to the turbines by an 11-ft-diameter steel-lined circular tube. This tube is 10 miles long. The entrance to the tube is at the bottom of an artificial lake. The surface of the lake is 2,600 ft above the exit of the tube. Consider the tube to be smooth. The flow rate is 2,000 ft^3 of water per second. Water temperature is 60°F.

(*a*) Determine the static pressure before a valve placed at the exit of the tube.
(*b*) What is the thickness of the laminar sublayer?
(*c*) What is the shear at the wall?
(*d*) What is the velocity gradient at the wall?

7-19. Field tests on a 15-in. cast-iron water main indicate that the wall roughness has increased to 0.005 ft after many years of service. If the main now carries a flow of 8 ft^3/sec, what increase in flow at the same power input would result from replacing the line with a new cast-iron pipe of the same diameter? The line is horizontal and water temperature is 60°F.

7-20. Two double-pipe heat exchangers contain finned tubes. Exchanger A contains a transverse finned tube which has 5 fins per inch. The root diameter of the fin is 0.75 in. and the fin diameter is 1.75 in. Exchanger B contains a longitudinal finned tube. This tube is 0.75 in. OD and contains 20 longitudinal fins ½ in. high on its surface. The effective heat-transfer length in each exchanger is 10 ft. The outside surface area of each tube is essentially the same. The outer tube of each exchanger is 2.50 in. ID, and 350 lb_m/hr of air at 1 atm pressure and average temperature of 100°F flows in the annular space of each exchanger. What must be the ratio of heat-transfer coefficients in the heat exchangers so that they are equivalent when compared on the basis of heat transferred per unit of pumping power?

CHAPTER 8

8-1. (*a*) What rate of water flow (at 60°F) would be necessary in a 5-ft-diameter pipe such that the laminar sublayer was on the order of 0.5 in. thick?

(*b*) A pitot tube with an opening 0.05 by 0.05 in. is inserted into the laminar sublayer for the flow described in (*a*). Plot the measured impact pressure (psi) when the center of the opening is 0.45, 0.40, 0.35, 0.30, 0.25, 0.20, 0.15, 0.10, and 0.05 in. from the wall.

(*c*) If the measured impact pressure calculated in (*b*) is measured by a differential manometer containing carbon tetrachloride under water, what reading of the manometer would be obtained for the various positions of the pitot tube?

8-2. For a pitot tube with a circular opening determine the ratio of the measured average impact pressure to the impact pressure at the center of the opening. The pitot tube is placed in the laminar sublayer adjacent to the pipe wall such that the center of the opening is located a distance a from the wall where a is the radius of the pitot-tube opening.

8-3. Water at 60°F is flowing at the rate of 2,000 gal/min in a smooth pipe 5 ft in diameter. A pitot tube with a square opening 0.02 by 0.02 in. is inserted in the flowing stream. The impact pressure (static pressure at pitot-tube opening minus static pressure of flowing stream) is measured with the center of the pitot tube located 0.05, 0.04, 0.03, 0.02, and 0.01 in. from the wall. On the same graph, plot the following as a function of the distance of the center of the pitot-tube opening from the tube wall:

(*a*) Measured impact pressure (psi).

(*b*) Velocity corresponding to the measured impact pressure in (*a*).

(*c*) The velocity in the laminar sublayer calculated from Eq. (7-37).

8-4. Solve Prob. 8-3 for air at 60°F and 1 atm pressure flowing in the pipe at the same Reynolds number as the water.

8-5. Solve Prob. 8-3 for mercury at 60°F flowing in the pipe at the same Reynolds number as the water. For mercury at 60°F, $\mu = 1.65$ centipoises and $\rho = 13.56$ g/cc.

8-6. Solve Prob. 8-3 for an oil flowing in the pipe at the same Reynolds number as the water. Viscosity of the oil is 100 centipoises and density is 1.10 g/cc.

CHAPTER 9

9-1. Estimate the amount of oil ($\mu = 150$ centipoises, $\rho = 60$ lb_m/ft^3) which would be lost through a hole ⅛ in. in diameter in the wall of a 6-in. schedule 80 pipe. The static pressure in the pipeline is 400 psig.

9-2. The amount of SAE 10 oil at 60°F coming from a hole ⅛ in. in diameter in a pipe wall has been measured and found to be 1 gal/min. The pipe is 4 in., schedule 40. Determine the static pressure in the pipeline both considering and neglecting entrance effects.

9-3. A capillary viscometer is used to determine the viscosity of glycerin. The capillary is 0.05 in. in diameter and 0.5 in. long. It is connected to the bottom of a vertical cylindrical chamber 2 in. in diameter. At 20°C the viscosity of glycerin is 1,410 centipoises and the specific gravity is 1.263 (20°/20°C). Determine the time required for 10 g of glycerin to be discharged from the capillary when the pressure in the cylinder is 200 psia and the height of glycerin above the entrance of the capillary is 10 in.

9-4. A capillary viscometer is used to measure the viscosity of a certain grade of lubricating oil. The capillary is 0.04 in. in diameter and 0.75 in. long. It is connected to the bottom of a vertical cylindrical chamber 3 in. in diameter. The chamber is filled to a height of 12 in. and is under 100 psia pressure. The specific gravity of the oil is 0.94. If it takes 4.00 sec for 10 cc of oil to be forced through the capillary, what is the viscosity of the oil?

9-5. Two static-pressure taps in a horizontal 2-in. schedule 40 pipe are located 1 in. from the entrance and 25 in. from the entrance, respectively. For water at 60°F flowing at Reynolds numbers of 10,000 and 100,000, determine the difference in static pressure between these points. Compare these with the pressure drop calculated by neglecting entrance effects. Consider the pipe to be smooth, i.e., $e/d_w = 0$.

9-6. An annulus is made up of an outer tube 1.649 in. ID and an inner tube 1.000 in. OD. Carbon tetrachloride at 70°F is flowing in the annulus at a value of Re_2 of 25,000. Plot the shear stress at the outer wall of the annulus as a function of distance from the entrance for both low and high turbulence streams. Also, for both cases, determine the pressure drop between two pressure taps located 10 in. and 100 in. from the entrance to the horizontal annulus.

9-7. A plate fin radiator consists of flat cooling surfaces connected by fins of metal such that the cross section of flow consists of a number of small rectangles. The size

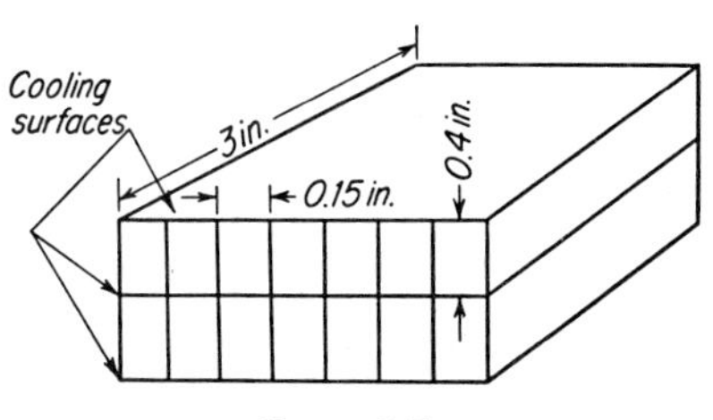

PROB. 9-7.

of the rectangles is 0.15 in. by 0.40 in. The radiator is 3 in. thick. Assuming that the curves in Figs. 9-8 and 9-9 apply for rectangles with d_w replaced by d_e, determine the pressure drop in inches of water for air at 100°F flowing through the radiator at a mass velocity of 15,000 $\mathrm{lb}_m/(\mathrm{hr})(\mathrm{ft}^2)$. The entering air is at atmospheric pressure.

9-8. Kreith and Eisenstadt [*Trans. ASME*, **79**:1070 (1957)] studied the flow of air through short capillaries. They measured the pressure drop across a plate containing 50 holes. The plate was placed in a pipe. The cross-sectional area of the pipe was much larger than the area of the holes. The pressure difference was determined across static pressure taps installed in the pipe wall. A small portion of the data obtained by these workers is shown below.

Hole diameter, in.	Plate thickness, in.	Pressure loss across plate in H_2O†	Air-flow rate in pipe, ft^3/min, at 790 mm Hg and 100°F
0.030	1/16	0.052	0.136
0.0312	1/8	0.70	0.585
0.0138	1/16	0.92	0.114
0.0312	1/4	0.170	0.204
0.0312	1/2	0.485	0.290

† Pressure difference between upstream reservoir and end of capillary.

Determine whether these data are in agreement with pressure losses predicted from Langhaar's theory.

9-9. A small sieve plate column is being used to humidify air. Air flows upward through the column, bubbling through water on the plates and leaving the column at the top at a pressure of 1 atm. Estimate the air pressure at the bottom of the column described below.

Column diameter	1.83 in.
No. of sieve plates	5
Plate thickness	1/16 in.
Height of liquid on plate	1 in.
Hole diameter	1/16 in.
No. of holes	140
Air-flow rate	10 ft^3/min (at 60°F, 1 atm)
Average column temperature	80°F

CHAPTER 10

10-1. Air is flowing in a wind tunnel at a velocity of 23 ft/sec. The temperature is 60°F and the pressure is atmospheric. A flat plate is immersed in the flowing air parallel to the direction of flow.

(*a*) Determine the length of the flat plate if the boundary layer over the complete length of the plate is laminar.

(*b*) Calculate the boundary-layer thickness at a point 4 in. from the leading edge of the plate.

(*c*) Calculate the local drag coefficient at a point 4 in. from the leading edge of the plate.

(*d*) Calculate the total force exerted on the plate using the length of the plate calculated in (*a*). Give this force in terms of pounds per foot of plate width.

10-2. Determine the following quantities at a value of $\text{Re}_x = 100{,}000$:

(*a*) The boundary-layer thickness.

(*b*) The local drag coefficient.

(*c*) The total drag coefficient (for plate length of $2x$).

(*d*) The velocity gradient at the surface.

(*e*) The velocity profile (plot u versus y).

Consider the following cases of flow over a smooth flat plate:

(1) Air at 100°F at a main-stream velocity of 50 ft/sec.

(2) Water at 60°F at a main-stream velocity of 10 ft/sec.

(3) Oil (viscosity = 100 centipoises, density = 58 lb_m/ft^3) at a main-stream velocity of 3 ft/sec.

(4) Mercury (viscosity = 1.21 centipoises, density = 845 lb_m/ft^3) at a main-stream velocity of 1 ft/sec.

10-3. Estimate the drag force exerted on the sides of a building 50 by 100 ft and 50 ft high, by a wind of 25 miles/hr blowing in a direction parallel to the sides of the building. Neglect end effects. The temperature is 70°F. Barometer is normal. What is the total wind load on the building?

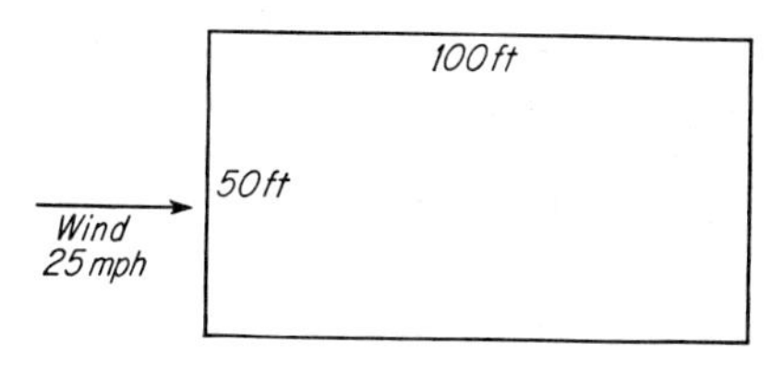

PROB. 10-3.

10-4. A refinery finds that its waste water contains 1 per cent by volume of oil in the form of droplets ranging in size from 0.010 to 0.042 cm in diameter. The oil has a density of 0.92 g/cc. An engineer proposes that the water be dumped into a ditch of rectangular cross section, 8 ft wide and 4 ft deep, and be discharged through a skimming mechanism which removes all the oil that rises to the top. How far down the ditch should the skimmer be placed to remove all the oil droplets? The waste water amounts to 1,500,000 gal/hr.

10-5. A petroleum engineer in the field must estimate the viscosity of the crude oil from a new well. He has no equipment for measuring viscosity, but finds that a spherical pebble falls through the oil at a rate of 2 cm/sec. The pebble is 0.35 in. in diameter. Specific gravity of the oil is assumed to be 1 and the specific gravity of the rock is 2.5. Estimate the viscosity of the oil.

10-6. A rough estimate of the power required for paddle-type agitators may be obtained through the use of the drag coefficient for a fluid impinging on a flat plate at Reynolds numbers greater than 5×10^3. Determine an expression for the power number $pg_c/n^3 d_0^5 \rho$ for a paddle agitator of diameter d_0 with paddles of the size shown.

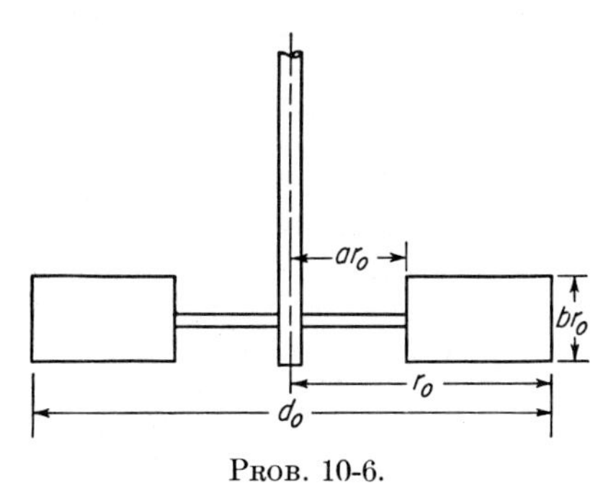

PROB. 10-6.

In the power number, p is the power per paddle, n is the rate of rotation in revolutions per second, and ρ is the density. The quantities a and b are constants giving the size of the paddle.

Estimate the power required for a six-paddle agitator for:

$b = 0.1667$
$a = 0.5$
$r_0 = 2$ ft
$\rho = 1.2$ g/cc
$n = 120$ rpm

10-7. Compare the drag force exerted on a smooth flat plate with that on a plate of the same size covered with sand grains 0.04 in. in diameter. The plates measure 1 by 1 ft and are immersed in air (at atmospheric pressure and 60°F) flowing at 100 ft/sec.

10-8. Air flows upward in a 12-in.-diameter pipe at an average velocity of 5.00 ft/sec. Calculate and plot against diameter, the diameter of a silica particle which will have a settling velocity equal to the air velocity. Make reasonable assumptions for missing data.

10-9. Dust is removed from air in a Cottrell precipitator made from two horizontal charged metal plates which are arranged parallel to each other to form a narrow channel of depth equal to b. A charged dust particle which is carried into the space between the plates by means of an air stream falls toward the lower plate at a constant velocity. The particle is carried forward through the space between the plates at the velocity of the gas stream which is in laminar flow. In order for the precipitator to remove all the dust, the plates must be made long enough for the smallest dust particle to settle from the leading edge of the upper plate to the bottom plate before it is carried out of the channel. If the mean air velocity is 0.5 ft/sec in a channel which is 0.5 in. deep, how long must the channel be for a spherical particle 10 microns in diameter to settle? An electrostatic field exerting a force which is ten times gravity is applied. Assume air at 1 atm pressure and 100°F.

10-10. An impact tube is to be used in an attempt to measure velocity and velocity distributions in a large vessel (30 to 40 ft in diameter) such as the regenerator in a fluid catalytic cracker. It has been decided to build a multiple-opening impact tube for this

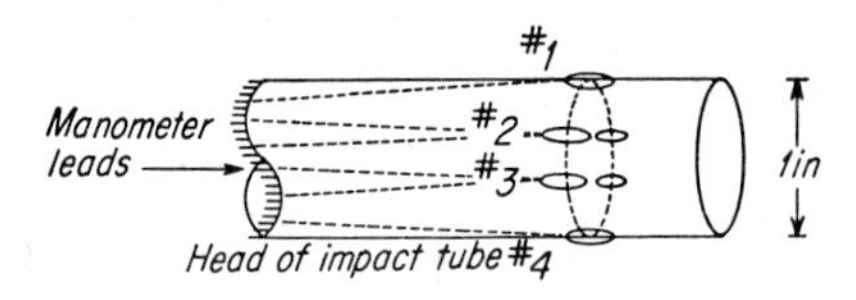

PROB. 10-10.

purpose. The tube is to be constructed from 1-in.-OD bar stock. Six small, equally spaced impact openings are to be located in a plane perpendicular to the axis of the cylinder. A separate static-pressure tap is also to be constructed to permit the measurement of the velocity head. The velocity head is measured by a differential manometer. In order to calibrate the unit it is inserted to the center of a 15-in. line carrying 12,000 lb_m of water/min at 60°F. The ring of impact openings is placed at the exact center of the pipe with one of the openings facing directly upstream. A differential pressure of 2.0 in. of water was read on the gauge when this tap was in use.

(*a*) Calculate the operational coefficient of this impact tube.

(*b*) If opening 2 is placed directly upstream, what is the reading (differential pressure) at taps 1 and 3?

(*c*) If the impact tube is placed in the large vessel with tap 1 facing directly down (vertical) and readings of 0.0 in. and 1.10 in. of water are found on taps 1 and 2 respectively, what is the velocity of the material in the vessel at this point and what is its direction? Assume the density of the material in the vessel is 115 lb_m/ft^3.

10-11. A person is towed across a smooth lake on a 2- by 6-ft horizontal surfboard. Assuming the rider stands vertically, estimate how fast he is moving if the boat pulls with a force of 12 lb_f. Perform the calculation for (*a*) both sides submerged; (*b*) one side submerged. Water temperature is 50°F.

10-12. A counter-current washing column is to be operated with water flowing upward at a velocity of 2 ft/sec. If the packing is to be glass spheres, what is the size of the smallest sphere allowable so that no spheres will be carried out of the bed? Assume glass density is 150 lb_m/ft^3.

10-13. (*a*) Estimate the wind load on a spherical tank having a capacity of 50,000 gal for wind velocities of 20, 40, and 60 miles/hr.

(*b*) For each wind velocity in (*a*) determine the thickness of the boundary layer and plot the velocity profile across the boundary layer at a point 45° from the forward point of stagnation.

(*c*) For each wind velocity in (*a*), determine the pressure at the surface of the tank at a point 45° from the forward point of stagnation. Assume normal barometer and an air temperature of 50°F.

10-14. A weather balloon 6 ft in diameter and containing helium is released at sea level. At all times the helium is at the same temperature as the surrounding air, and the pressure in the balloon is always 2 mm of mercury greater than that of the surroundings. The balloon and its load weigh 2000 g. How long will it take the balloon to reach an altitude of 10,000 ft? Atmospheric pressure and temperature are as follows:

Altitude, ft	Temperature, °F	Pressure, mm
0	59	760
2,500	. .	693
5,000	42	630
7,500	. .	574
10,000	23	523

10-15. A circular cylinder 3 in. in diameter is immersed in air flowing at 60 ft/sec. The air is at 1 atm pressure and 60°F.

(*a*) Plot the pressure at the surface of the cylinder as a function of the angle measured from the forward point of stagnation assuming incompressible nonviscous flow. Compare the curves obtained with those shown in Fig. 10-29.

(*b*) Plot the thickness of the boundary layer on the cylinder as a function of the angle measured from the forward stagnation point.

(*c*) At positions 30° and 60° from the forward stagnation point, calculate the local velocity in the boundary layer at a distance halfway through the boundary layer.

(*d*) What drag force will be exerted on this cylinder per foot of length?

10-16. A water tower consists of a tank supported above the ground. The tank is 15 ft in diameter and 20 ft high. What is the horizontal force exerted on the tank by a wind of 50 miles/hr? The barometer is normal and air temperature is 50°F.

10-17. An agitator is constructed as shown below.

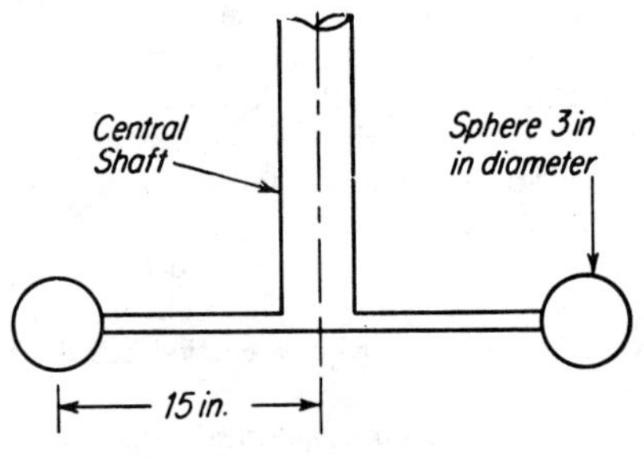

PROB. 10-17.

(*a*) Estimate the power required for rotation when this agitator is rotated in water at 60°F such that the Reynolds number for the spheres is 2×10^5. Neglect the resistance of the arms supporting the spheres.

(*b*) Estimate the power required when the rate of rotation is doubled and the sphere Reynolds number is 4×10^5.

(*c*) Estimate the power required when the agitator is rotated at the same rate as in (*a*) but the diameter of the spheres is doubled.

CHAPTER 11

11-1. Air at a superficial velocity of 40 ft/sec is flowing across a tube bank consisting of 1-in.-OD tubes spaced on 2-in. centers. The temperature is 60°F and the air is at atmospheric pressure. The tube bank contains 15 transverse rows of tubes.

(*a*) Calculate the pressure drop across the bank for equilateral triangular arrangement of tubes. Use Eqs. (11-23), (11-25), and (11-27) and compare.

(*b*) Calculate the pressure drop across the bank for square in-line arrangement of tubes. Use Eqs. (11-24), (11-26), and (11-27) and compare.

(*c*) Calculate the pressure drop considering each tube as a cylinder immersed in a flowing fluid, and obtain drag coefficients as given by Fig. 10-36. In calculating the pressure drop by this method use (1) the superficial velocity as the main-stream velocity and (2) the maximum velocity as the main-stream velocity. Compare the results with those obtained in (*a*) and (*b*).

11-2. Compare the pressure drop across the following two tube banks for air (at 60°F, 1 atm pressure) flowing at a superficial velocity of 40 ft/sec.

(*a*) Circular tubes: 1-in. OD; equilateral triangular arrangement on 2-in. centers; 15 transverse rows.

(*b*) Elliptical tubes: minor axis 1 in., major axis 2 in., triangular arrangement; 1-in. clearance between tubes in transverse rows, ½-in. clearance between tubes in adjacent transverse rows; 10 transverse rows.

11-3. Compare the pressure drop across the following two tube banks for air (at 60°F, 1 atm pressure) flowing at a superficial velocity of 40 ft/sec.

(*a*) Circular tubes: 1-in. OD; equilateral triangular arrangement on 2-in. centers; 15 transverse rows.

(*b*) Transverse-finned tubes: 1-in. root diameter, 2-in. fin diameter, 7 fins per inch; equilateral triangular spacing on 2-in. centers; 15 transverse rows.

CHAPTER 13

13-1. A fluid is flowing at a steady rate with laminar motion through a tube whose walls are kept at constant temperature by means of a steam jacket. Owing to a slight variation of viscosity with temperature, however, the velocity distribution is not parabolic as it is for isothermal flow, but is practically flat and will be considered the same for all points in the pipe. The heat capacity and thermal conductivity may be considered constant. Transfer of heat by conduction parallel to the axis of the pipe may be assumed to be negligible.

(*a*) Develop an expression in terms of the size of the pipe and velocity and physical properties of the fluid for calculating the temperature rise in passing through the tube.

(*b*) Also develop an expression for the local Nusselt number in terms of the variables of the system.

13-2. Solve Prob. 13-1 for flow between infinite parallel planes. Compare the results with the curve labeled "slug flow" in Fig. 13-13.

13-3. Use the Leveque solution to determine the mass-transfer Nusselt number $k_m d_w/D$ for mass transfer from the wall in the entrance to a circular tube.

13-4. Water at 68°F flows into a 1-in.-ID circular tube at the rate of 200 lb_m/hr.

(*a*) A section of the tube 9 in. long is heated so that the wall temperature is 200°F. Estimate the bulk temperature of the water leaving the heated section.

(*b*) A 9-in. section of the tube is replaced by a tube of solid sodium chloride. What will be the bulk concentration of the water stream leaving the section? The solubility of NaCl in water at 68°F is 36 g per 100 cm^3 of solution. The diffusivity of NaCl in water is 1.26×10^{-5} cm^2/sec.

13-5. Dry air at 96°F enters a 2-in.-ID wetted-wall column at the rate of 8 lb_m/hr. The column is 10 in. long and the wall temperature is constant at 55°F at all times. Estimate the temperature and humidity of the air leaving the column. What would the humidity and temperature be if the 2-in.-ID column were replaced by sixteen ½-in.-ID columns?

13-6. Plot the radial temperature distribution at a cross section 5 in. from the entrance of a 1-in.-ID tube, in which water at 100°F is flowing at the rate of 200 lb_m/hr. Tube wall temperature is 200°F.

13-7. Calculate the mean Nusselt number for water flowing in a ⅛-in.-ID tube 30 in. long. The water flows at a rate of 10 lb_m/hr and has an average temperature of 180°F. Compare the results obtained by Eq. (13-24) and Fig. 13-10. Assume that the Grashof number is essentially zero.

13-8. Crude oil is to be heated from 65 to 95°F in a shell-and-tube heat exchanger. The oil flow rate (through the tubes) is 100,000 lb_m/hr. The tubes are 0.5-in.-OD 16 BWG and the tube wall is at 200°F. If the exchanger should be no longer than 10 ft, estimate the number of tube passes if (*a*) there are 20 tubes per pass; (*b*) there are 10 tubes per pass. Use the following properties for the oil.

Specific gravity	1.1
Specific heat	0.5
Viscosity	
At 80°F	100 centipoises
At 200°F	10 centipoises
Thermal conductivity	0.08 Btu/(hr)(ft²)(°F)/ft

Estimate the power required to pump the oil through the tube side of the exchanger.

13-9. The viscosity of an oil as a function of temperature is given below.

Temperature, °F	*Viscosity, centipoises*
50	105
75	75
100	52
125	35
150	24
175	15
200	10

Specific gravity = 1.1
Thermal conductivity = 0.08 Btu/(hr)(ft^2)(°F)/ft
Specific heat = 0.5

This oil at 50°F flows in a 1-in.-ID tube at the rate of 6,000 lb$_m$/hr. After a suitable calming length the tube wall is maintained at a temperature of 200°F. Estimate the length of heating required for the flow to become turbulent.

13-10. Air enters a small circular duct 1.0 in. ID at the rate of 4 lb$_m$/hr. Compare the mean heat-transfer coefficient in a heated length of 10 in. as calculated by two methods.

(*a*) Assume the velocity profile fully developed at beginning of heating.

(*b*) Assume a uniform velocity profile at beginning of heating. Assume constant wall temperature, and air at 70°F, 1 atm pressure.

13-11. An air heater consists of fins placed between two heated plates maintained at 200°F. The ducts for the flow of the air consist of rectangular channels. The fins are made of 16 BWG copper and are 3 by 6 in., and there are eight fins per inch. Assuming

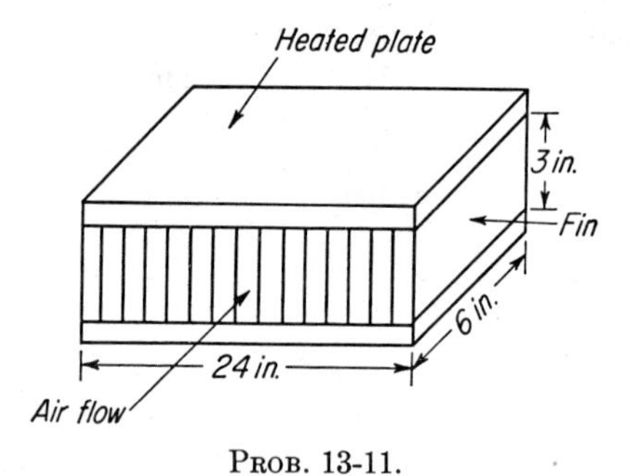

Prob. 13-11.

100 per cent fin efficiency, calculate the temperature of the air leaving the heater. Air is supplied at the rate of 150 lb$_m$/hr at 70°F and 1 atm pressure. Other dimensions are shown in the sketch.

CHAPTER 14

14-1. Water at 100°F is flowing in a 2-in.-ID tube at a Reynolds number of 100,000. Determine the mean heat-transfer coefficient in this tube for each of the entrances shown in Table 14-1. Consider tube lengths of 1, 5, and 10 ft. For each entrance, at what length is the coefficient within 5 per cent of that for an infinite length?

14-2. Determine the heat-transfer coefficients for the finned tubes in the two annuli and for the flow conditions described in Prob. 7-20. From the results of this problem and those of Prob. 7-20, which heat exchanger appears to be most economical assuming that both cost the same to purchase and install?

CHAPTER 15

15-1. Apply Colburn's analogy to heat transfer from the inner tube in an annulus using Eq. (7-97). Compare values of the group $(h_m/C_pG)(C_p\mu/k)^{2/3}$ as predicted by the analogy with those predicted by Eq. (14-18).

15-2. An annulus is made up of an inner tube 1 in. OD and an outer tube 2 in. ID. A 2-ft section of the inner tube is replaced with a tube of solid sodium chloride. Water at 80°F flows through the annulus at an average velocity of 4 ft/sec. Neglecting entrance effects, estimate the concentration of sodium chloride in the water as it leaves the annulus. As in Prob. 15-1 use Colburn's analogy to calculate the mass transfer j factor $(k_m/U)(\mu/\rho D)^{2/3}$.

15-3. In the annulus in Prob. 15-2 a 2-ft section of the inner tube is replaced by a porous material which may be kept wet with water at 55°F. Dry air at 95°F, 1 atm pressure, enters the annulus at the rate of 200 lb_m/hr. Estimate the humidity of the air leaving the annulus.

15-4. Use Prandtl's analogy to predict the amount of evaporation from a wetted-wall column 2 in. ID and 4 ft long. Air is supplied at the rate of 200 lb_m/hr. It enters dry at 100°F and 1 atm pressure. The water on the walls of the column is at the adiabatic saturation temperature (57°F). Using Prandtl's analogy, also estimate the temperature of the air leaving the column. Do these results indicate that the column operates adiabatically?

15-5. Solve Prob. 15-4 using von Kármán's analogy.

15-6. Water at 60°F is flowing in a 2-in.-ID smooth tube at a velocity of 10 ft/sec. Calculate the ratio of the total conductivity of heat to the molecular conductivity of heat at a point in the turbulent core 0.50 in. from the tube wall. Assume that the flow is steady and unaffected by any entrance disturbance. In the turbulent core the velocity distribution may be obtained from Eq. (7-35). State any assumptions made in carrying out the calculations.

15-7. Plot the following quantities as a function of radius for flow in a 2-in.-ID smooth tube:

(*a*) Eddy viscosity
(*b*) Total viscosity
(*c*) Eddy diffusivity of momentum
(*d*) Ratio of total conductivity of heat to molecular conductivity of heat (assume $\epsilon_H = \epsilon_M$).

Consider the following flow conditions:
(1) 30 gal/min of water at 60°F
(2) 700 lb_m/hr of air at 60°F, 1 atm pressure
(3) 70,000 lb_m/hr of crude oil [specific gravity = 1.1, specific heat = 0.5, viscosity = 20 centipoises, thermal conductivity = 0.08 Btu/(hr)(°F)(ft^2)/ft]
(4) 3,000 lb_m/hr of liquid sodium at 1300°F (see Table 16-1)

15-8. In Martinelli's equation (15-61) for predicting the Nusselt number, the three terms in the denominator correspond respectively to the resistance of the laminar layer, buffer layer, and turbulent core. Express these resistances in terms of percentages of the total for the following cases of flow in a 1-in.-ID tube. The Reynolds number is 100,000 in each case. Evaluate fluid properties at the bulk temperature.

Fluid	Bulk temperature, °F	Tube-wall temperature, °F
(*a*) Water	150	250
(*b*) Air	60 (1 atm)	150
(*c*) Liquid sodium	1300	700

15-9. Compute the Stanton number for each of the three cases in Prob. 15-8. Use Martinelli's analogy and compare with results obtained from relationships presented in Chaps. 14 and 16.

15-10. Using Eqs. (15-58), (15-59), and (15-60), draw the temperature profile showing the radial temperature distribution $(T_w - T)/(T_w - T_c)$ as a function of tube radius for each of the three cases given in Prob. 15-8. From the information given in Prob. 15-8 compute the value of T_c in each case. From the temperature profile obtained predict the Nusselt number for each case from Eq. (12-12).

15-11. For high-Prandtl-number fluids the prediction of the heat-transfer coefficient from the velocity distribution depends on knowing the thickness of the laminar sublayer. Show this with Lyon's equation (15-88) for water at 100°F flowing in a 1-in.-ID tube at a Reynolds number of 50,000. Consider values of $y^+ = 4.0$, 5.0, 6.0, and 7.0 for thicknesses of the laminar sublayer.

15-12. For water at 100°F flowing at a Reynolds number of 100,000 in a 2-in.-ID tube plot the total conductivity of heat as a function of radius. Calculate the total conductivity by two methods. (*a*) Assume $\epsilon_H/\epsilon_M = 1$. (*b*) Use values of ϵ_H/ϵ_M from Fig. 15-17.

15-13. Using Lyon's equation (15-88) compute the Nusselt number for both cases considered in Prob. 15-12.

15-14. For cases (*b*) and (*c*) described in Prob. 7-4 estimate the temperature at the axis of the tube.

CHAPTER 16

16-1. Liquid sodium at 400°F flows through a 0.28-in.-ID smooth stainless-steel tube at the rate of 1,400 lb_m/hr.

(*a*) Using the velocity-distribution equations determine the velocity at the center.

(*b*) At a point halfway between the wall and center of the tube determine ϵ_M/ν and K/k where K is the total conductivity of heat. Assume $\epsilon_H = \epsilon_M$.

16-2. In an experimental program to determine the heat-transfer characteristics of molten lead-bismuth eutectic, a steel tube (0.750 in. OD and 0.652 in. ID) is electrically heated by nichrome wire wound uniformly around the tube. Estimate the tube-wall temperatures that might be expected to occur for the following two cases:

	(*a*)	(*b*)
Flow rate, lb_m/hr	2,000	20,000
Temperature of liquid entering heating section, °F	300	600
Net electrical heat flux, Btu/(hr)(ft^2)	2,000	12,000
Length of heating section, ft	4	4

16-3. A steam-heated shell-and-tube heat exchanger is to be designed to heat sodium-potassium alloy (56 per cent Na) from 200 to 400°F. Saturated steam at 500 psia is available to supply the shell side of the exchanger. Use ¾-in.-ID 16 BWG stainless-steel tubes. Flow rate of liquid metal is 10,000 lb_m/hr. Consider a mass velocity in each tube of about 400,000 $lb_m/(hr)(ft^2)$. Determine the number of tube passes, tube length, and the friction loss on the tube side of the exchanger. Assume a steam-condensing coefficient of 2000 $Btu/(hr)(ft^2)(°F)$.

16-4. An insulated double-pipe heat exchanger is made up of an outer stainless-steel tube 1.5 in. OD, 10 BWG wall, and an inner tube 0.75 in. OD, 16 BWG wall. Sodium-potassium alloy (56 per cent sodium) flows countercurrently at the same rate (2,000 lb_m/hr) in both the inner tube and the annulus. The liquid metal flowing in the inner tube enters at 1000°F and leaves at 600°F. That entering the annulus enters at 200°F and leaves at 600°F. What is the length of double-pipe heat exchanger required? What are the power requirements to pump each fluid through the horizontal heat exchanger?

CHAPTER 17

17-1. A plate 6 in. long is immersed in a flowing stream. The total Reynolds number is 100,000. Plot the local heat-transfer coefficient, the thickness of the thermal boundary layer, the thickness of the hydrodynamical boundary layer, and the local Nusselt number as a function of the distance from the leading edge for the following cases:

(*a*) Water at 60°F; plate temperature 150°F.

(*b*) Air at 60°F, 1 atm; plate temperature 212°F.

(*c*) Oil at 60°F, specific gravity 1.1, specific heat 0.5, viscosity 20 centipoises, thermal conductivity, 0.08 $Btu/(hr)(ft^2)(°F)/ft$; plate temperature 200°F.

(*d*) Liquid sodium at 500°F; plate temperature 1000°F.

For each of the above cases determine the total Nusselt number h_mL/k.

17-2. A 10-ft double-pipe heat exchanger contains an outer tube 2.50 in. ID and an inner longitudinal-finned tube with 20 fins. The fins are ½ in. high and are placed on a tube 0.75 in. OD. For air flowing at 500 lb_m/hr, compare the heat-transfer coefficient in this exchanger with that obtained if the fins were discontinuous and were each considered to be a thin plate ½ by 3 in. Consider that a turbulent boundary layer exists on each plate. The air is at atmospheric pressure and at an average temperature of 95°F.

17-3. A tank contains 300 gal of a solution with the following properties:

Specific gravity	1.0
Specific heat	1.0
Viscosity	10 centipoises
Thermal conductivity	0.4 $Btu/(hr)(ft^2)(°F)/ft$

It is proposed to heat this solution by means of an electrically heated agitator which has cylindrical paddles. The paddles are cylinders 1 in. in diameter and 1 ft long mounted on a central shaft 4 in. in diameter. The paddles are maintained at a temperature of 300°F.

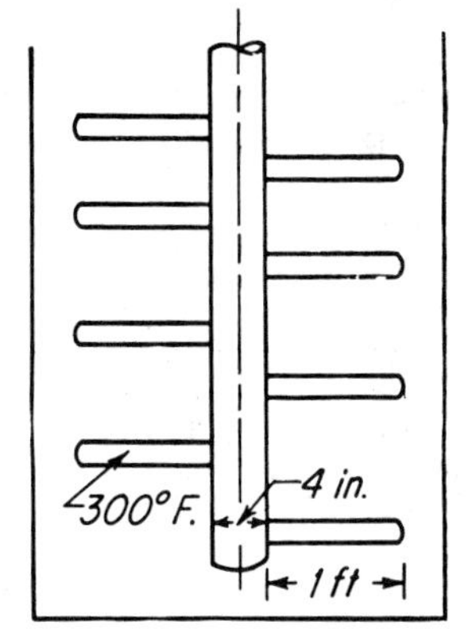

PROB. 17-3.

Estimate the time required to heat the solution from 60 to 250°F. The agitator rotates at 60 rpm and there are a total of eight paddles. Assume complete mixing of the solution at all times. Estimate also the power required to rotate the agitator. What contribution does this power input make in raising the fluid temperature?

17-4. A 1-ft-square thin plate of solid sodium chloride is immersed in a stream of water flowing at 6 ft/sec. The plate is oriented parallel to the direction of flow. Temperature of the liquid and the plate is 100°F. At what rate is the plate losing salt

(*a*) if the boundary layer is completely laminar on the plate?

(*b*) if the boundary layer is completely turbulent on the plate?

17-5. A wet plate 1 ft square is oriented parallel to a stream of dry air flowing at 100 ft/sec. The air is at 1 atm pressure and 100°F. Determine the rate of evaporation from the plate for

(*a*) a completely laminar boundary layer

(*b*) a completely turbulent boundary layer

Consider the cases where the plate is maintained at 100°F and the case where the plate is at a temperature of 57°F (the adiabatic saturation temperature of the air). In the latter case, determine how much heat is transferred to the plate and compare it with the amount of heat required to evaporate the water from the plate.

17-6. From Levy's solution, plot the local Nusselt number hd_0/k as a function of the angle (up to 80°) measured from the forward stagnation point for air (Pr = 0.7) flowing past a cylinder at Reynolds numbers of 23, 244, 597, 39,000, 71,500, 110,000, and 219,000. Assume constant surface temperature. Compare the results with those in Figs. 17-9 and 17-10. Explain any major discrepancies.

17-7. A sharp-edged wedge is placed in an air stream flowing at 50 ft/sec. The angle of the wedge is 20°. The air is at 1 atm pressure and 70°F. Show how the local Nusselt number hx_1/k varies over the first foot of the wedge where

(*a*) the surface temperature is constant

(*b*) the surface temperature is proportional to the distance from the leading edge, i.e., $T_w - T_\infty = Ax_1$ [i.e., $m' = 1$ in Eq. (17-62)]

17-8. Estimate the rate at which a ½-in. sphere of naphthalene will sublime when immersed in an air stream flowing at 40, 60, 100, and 150 ft/sec. Air is at 1 atm pressure and 70°F. The naphthalene may also be considered to be at 70°F.

17-9. Estimate the distance a spherical drop of water originally ⅛ in. in diameter must fall in quiet dry air at 100°F in order to reduce its volume by 50 per cent. The drop always falls at its terminal velocity and its temperature is 57°F.

17-10. A tube bank consists of 1-in.-OD tubes spaced on 2-in. centers in an equilateral triangular arrangement. Air (at 1 atm pressure and 100°F average temperature) flows across the tube bank at a superficial velocity of 30 ft/sec. There are 10 transverse rows of tubes. Determine the heat-transfer coefficient for this tube bank

(*a*) by Eq. (17-86)

(*b*) by Eq. (17-69). In using this equation use both the superficial velocity and the maximum velocity in the determination of the Reynolds number. Compare the results. Tube-wall temperature may be taken as 220°F.

17-11. A rough plate 1 ft by 1 ft is covered with sand grains 0.04 in. in diameter. Air (1 atm pressure, 60°F) flows parallel to the plate. Estimate the mean heat-transfer coefficient for this plate. Compare it with the heat-transfer coefficient for a smooth plate of the same size. Consider an air velocity of 100 ft/sec.

INDEX